Recent Advances in Multifunctional Perovskite Materials

Edited by Poorva Sharma and Ashwini Kumar

Published in London, United Kingdom

Recent Advances in Multifunctional Perovskite Materials
http://dx.doi.org/10.5772/intechopen.100683
Edited by Poorva Sharma and Ashwini Kumar

Contributors
Kawther Laajimi, Mohamed Hichem Gazzah, Jemai Dhahri, Devarajan Uthiran, Arumugam Sonachalam, Sanjeeb Kumar Rout, Rolly Verma, Pankaj Khirade, Anil V. Raut, Megha Unikoth, George Varghese, Karakat Shijina, Hind Neelamkodan, Anshebo Getachew Alemu, Teketel Alemu, Aminta Mendoza, Octavio Guzmán, Izunna Stanislaus Okeke, Priscilla Yahemba Aondona, Amoge Chidinma Ogu, Fabian Ifeanyichukwu Ezema, Eugene Echeweozo, Kurukkal Balakrishnan Subila, Fency Sunny, Linda Maria Varghese, Nandakumar Kalarikkal, Gregory Thien Soon How, Boon Tong Goh, Mohd Arif Mohd Sarjidan, Boon Kar Yap, Eyas Mahmoud, David M. Mulati, Timonah Soita, Franklin Jaramillo, Rafael Betancur, Juan F. Montoya, Esteban Velilla, Daniel Ramírez, Edwin Ramírez, Byunghong Lee, Rira Kang, Tae-Ho Jeong, Parameswar Krishnan Iyer, Mayur Jagdishbhai Patel, Himangshu Baishya, Ritesh Kant Gupta, Rabindranath Garai, Akin Olaleru, Eric Nnditshedzeni Maluta, Joseph Kirui, Olasoji Adekoya, Soon-Gil Yoon, Swathi Ippili, Venkatraju Jella, Hyun You Kim, Hyun-Suk Kim, Chunjoong Kim, Tae-Youl Yang

First published in London, United Kingdom, 2022 by IntechOpen
IntechOpen is the global imprint of INTECHOPEN LIMITED, registered in England and Wales, registration number: 11086078, 5 Princes Gate Court, London, SW7 2QJ, United Kingdom

British Library Cataloguing-in-Publication Data
A catalogue record for this book is available from the British Library

Additional hard and PDF copies can be obtained from orders@intechopen.com

Recent Advances in Multifunctional Perovskite Materials
Edited by Poorva Sharma and Ashwini Kumar
p. cm.
Print ISBN 978-1-80355-318-4
Online ISBN 978-1-80355-319-1
eBook (PDF) ISBN 978-1-80355-320-7

For EU product safety concerns:
IN TECH d.o.o., Prolaz Marije Krucifikse Kozulić 3, 51000 Rijeka, Croatia,
info@intechopen.com or visit our website at intechopen.com.

Meet the editors

Dr. Poorva Sharma is an associate professor at Key Laboratory of Multifunctional Materials and Magnetoelectric Devices, School of Electrical and Electronic Engineering, Luzhou Vocational and Technical College, China. Dr. Sharma received her Ph.D. in Physics from the School of Physics, Devi Ahilya University, India in 2015. Her thesis was devoted to experimental work on perovskite-based multiferroic materials. From 2016 to 2018, she worked as a postdoctoral fellow in the laboratory of Prof. S.X. Cao and Prof. W. Ren at the International Centre of Quantum and Molecular Structures (ICQMS), Department of Physics, Shanghai University. After completing her postdoctoral fellowship, she served as an assistant professor in the Department of Applied Physics, at Nanjing University of Aeronautics and Astronautics, China. Dr. Sharma received a Young Scientist training fellowship from the Madhya Pradesh Council of Science and Technology (MPCST), India; the IUCr Young Scientist Award from the International Union of Crystallography; and Best Paper Awards from the Luzhou Chemistry and Chemical Society and Luzhou Association of Science and Technology, China. Her primary research is devoted to rare-earth orthoferrite, multiferroic, manganite, and metal oxide materials to provide a solution for advanced sustainable, and economical spintronic, memory devices, and magnetic data storage devices. She has been the author/co-author of more than fifty research papers in indexed journals. She has participated in many national and international conferences.

Dr. Ashwini Kumar received a Ph.D. in Physics from Devi Ahilya University Indore, India in 2014. Currently, he is an associate professor at the School of Electrical and Electronic Engineering, Luzhou Vocational and Technical College, China, and a lead member of the Luzhou Key Laboratory of Intelligent Control and Electronic Device Application Technology. He is also the Science and Technology Commissioner of the Luzhou Branch of the National Technology Transfer Southwest Center. He did a postdoctoral at Southeast University, Nanjing (2015-2019) and worked in the field of multiferroic materials. He has published more than forty-five research papers in internationally renowned journals, as well as presented numerous papers at national and international conferences and seminars. Dr. Kumar received a Young Scientist fellowship from the Madhya Pradesh Council of Science and Technology, India in 2014 and a Best Paper Award (first prize) from the Luzhou Chemistry and Chemical Society, China in 2021. He also served as a principal investigator and presided over several national, provincial, and municipal research projects, such as the National Natural Science Foundation (NSFC) of the Foreign Youth Fund, Jiangsu Province Postdoctoral Research Fund, Luzhou Municipal Science and Technology Research Fund, and Central University Basic Innovative Research Fund, among others. His research interests include the development of novel multifunctional materials, perovskite-structured materials, microwave dielectric ceramics, and oxide ceramics.

Contents

Preface

Perovskite materials are attracting increasing attention from condensed matter physicists and the materials science community because of their potential applications in energy, storage, solar cells, and sensing/actuating devices. In perovskite-centered transition metal oxides, the strong interaction between lattice, charge, spin, and/or orbital degrees of freedom provides an excellent playground for adjusting their physical properties. This book introduces the phenomenon and physics of multifunctional perovskite materials.

The book covers many of the exciting areas in multifunctional perovskite materials and applications. It is organized into three sections. Section 1 (Chapters 1–6) discusses various perovskite materials and their synthesis, characterization, and interesting physiochemical properties. Section 2 (Chapters 7–12) presents recent advances in perovskites for solar cells. Finally, Section 3 (Chapters 13–16) focuses on multifunctional materials, including Heusler alloys, metal halides, and hybrid perovskites as well as their applications.

This book is an invaluable resource for researchers, scientists, and academicians working in energy, physics, chemistry, and materials. It is also a source of excellent information for graduate students. I trust that the readers will find this book both enjoyable as well as educationally rewarding. We hope this book contributes in some way to the understanding of the diverse aspects of multifunctional perovskite materials.

We thank Ms. Maja Bozicevic and Mr. Filip Lovricevic from IntechOpen who initiated this book and did a great deal of work to bring it to completion. Finally, we thank all the authors who contributed their informative and in-depth chapters, which made this book a reality.

Poorva Sharma, Ph.D. (Physics) and Ashwini Kumar, Ph.D. (Physics)
Associate Professor,
Luzhou Vocational and Technical College,
Luzhou, China

Section 1

Perovskite Materials and Characterization

Chapter 1

Perovskite Structured Materials: Synthesis, Structure, Physical Properties and Applications

Pankaj P. Khirade and Anil V. Raut

Abstract

There is a constant need for newer exceptional materials with better than ever properties to achieve new prerequisites of the future society and progress inventive industrial improvement. The potential to combine these oxides in composite structures to produce multifunctional materials has rekindled interest in perovskites (ABO_3) compounds over the past 10 years. Because of its intriguing characteristics, such as ferroelectricity, piezoelectricity, superconductivity, multiferroicity, photocatalysis, enormous magnetoresistance, dielectric, ionic conduction characteristics, etc., a huge variety of perovskite types have been thoroughly explored. Current applications for perovskite solids include electronics, geophysics, astronomy, nuclear, optics, medicine, the environment, etc. Perovskite compounds have distinctive features that make them suitable for a variety of commercial and technological applications, including capacitors, non-volatile memories, photo-electrochemical cells, catalysts in contemporary chemistry, actuators and sensors, ultrasonic and underwater devices, drug delivery, spintronics devices, tunable microwave devices, and many others. Potential applications for nano-scale perovskites include energy storage, fuel cells, nanomedicine, molecular computing, nanophotonics adjustable resonant devices, catalysts, and sensors. Nanoscale perovskites have intriguing features that are comparable to or better than those of bulk perovskites. This review includes topics such as perovskite structured materials' chronology, classification, production, crystal structure, special physical properties, and applications.

Keywords: perovskite, ceramics, ferroelectricity, multiferroics, ABO_3

1. Introduction

Perovskites were called for the Russian aristocrat and mineralogist Count Lev Aleksevich Von Perovski (1792–1856), who first discovered the calcium titanium oxide ($CaTiO_3$) structure in the Ural Mountains of Russia in 1839 [1]. The perovskites materials often contain the generic formula ABX3, where X is an anion that bonds to both A and B, two cations of quite different sizes [2]. Although X is frequently oxygen, it is also feasible for it to be other big ions such halides, sulphides, or nitrides. Many oxide compounds from a few homologous perovskite series are known, including $A_{n+1}B_nO_{3n+1}$ Ruddlesden-Popper, $A_nB_nO_{3n+1}$ Dion-Jacobson, $Bi_2A_{n-1}B_nO_{3n+3}$

IntechOpen

Aurivillius series and some others [3–5]. The perovskite structures are exists in different form such as: ABO_3-perovskite (ex: $BaTiO_3$, $CaTiO_3$), A_2BO_4-Layered perovskite (ex: Sr_2RuO_4, K_2NiF_4), $A_2BB'O_6$-Double perovskite (ex: Ba_2TiRuO_6) and $A_2A'B_2B'O_9$-Triple perovskite (ex: $La_2SrCo_2FeO_9$), etc. [6–8]. Because of its intriguing range of features, including as superconductivity, insulator-metal transition, ionic conduction characteristics, dielectric properties, and ferroelectricity, perovskite type oxides have been researched extensively [9–12]. One of the most typical solid-state physics structures, perovskite, contains a considerable variety of anions in addition to the majority of the metal ions from the periodic table. Numerous theoretical and experimental studies have focused on perovskite materials, typically ABO_3, over the past few years. These solids are currently assuming a significant role in fields such as electrical ceramics, refractories, geophysics, material science, astrophysics, particle accelerators, fission-fusion reactors, heterogeneous catalysis, environment, etc. [13–22]. In order to maintain their original crystalline structure, perovskite structured oxides can take significant substitutions in either one or both of their cationic sites (i.e., the A and B sites). Through the partial replacement of the cationic site(s) with foreign metal ions, this property enables the chemical customization of the materials, changing their structural, microstructural, electrical, and magnetic properties [23–25]. Perovskite-like compounds and the oxides of the perovskite type have many uses in physics and chemistry. These materials' physicochemical characteristics are influenced by their pore structure, surface morphology, particle size, exposed lattice plane, lattice defect, and surface morphology [26–29]. Many perovskite-type oxides and perovskite-like oxides have been created and studied so far in order to better understand their physicochemical characteristics. The ideal perovskite is referred to as the cubic perovskite. Due to their straightforward crystal structures and distinctive ferroelectric and dielectric characteristics, this class of materials holds enormous potential for a range of device applications. Perovskites solids, one of the most prevalent and frequently studied minerals, are extensively researched as potential substrate materials [30, 31]. Due to its unique ferroelectric, thermoelectric, pyroelectric, dielectric, and optoelectronic properties, perovskite-structured ceramics (ABO_3, where A and B are two cations) have recently gained popularity on a global scale [32–36]. Perovskite ceramics are used in a variety of exceptional applications, including wireless technology, sensors, actuators, screens, capacitors, random access storage, and adjustable microwave devices [37–44].

2. Perovskite structure

It is known that a perovskite type ABO_3 oxide structure may maintain the stability of almost 90% of the metallic natural elements listed on the periodic table. The crystal perovskite calcium titanate is where the atomic arrangements in this structure were originally discovered ($CaTiO_3$). The naturally occurring $CaTiO_3$ species are depicted in **Figure 1** (accompanied by Lev Aleksevich von Perovski). The majority of ABO_3-type oxides crystallize in the $CaTiO_3$ mineral's (relatively) straight forward form or in a structure quite similar to it. Even though $CaTiO_3$ was later discovered by Megaw in the United Kingdom, this straightforward cubic form has remained the name perovskite [45]. Subsequently, it was quickly verified with Miyake and Ueda's work [46]. Perovskites show brittle toughness, a sub-metallic to metallic sheen, colorless streaks, a cube-like structure, and imprecise cleavage. Brown, gray, black, orange, and yellow are among the colors.

Figure 1.
Lev Aleksevich von Perovski and the perovskite mineral species ($CaTiO_3$).

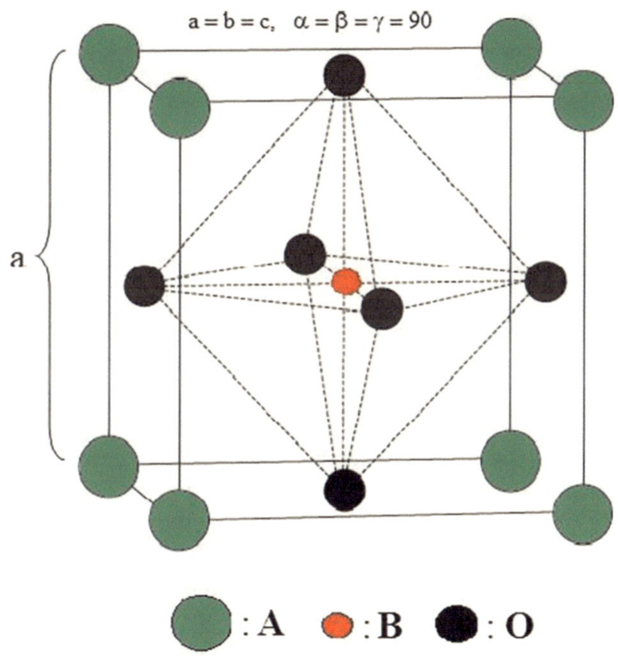

Figure 2.
Typical perovskite material crystal structure.

The structural formula for perovskite is ABO_3, where A and B are cations of varying sizes and O is the anion. Smaller than the B cation is the cation at the A location. According to **Figure 2**, the A atom has a 12 fold co-ordination number while the B atom has a 6 fold co-ordination number.

Divalent A cations typically reside in the corners of a cube at corner location and are 12 fold coordinated by oxygen anions (0, 0, 0). Tetravalent B cations are located in the body's core (½, ½, ½) and are found inside that oxygen octahedron. The position (½, ½, 0) of the oxygen atoms in the cubic lattice's face centre. The structure is typically represented as a three-dimensional network of BO6 octahedra with regular corner links. The A atom's coordination number is 12. A perfect perovskite has a network of shared corner BO6 octahedra with all B-O-B angles at 180 degrees, according to its structure. The proportion of A to B ionic size and the electronic arrangement of the metal ions are two indicators of the structural distortion in perovskite. Perovskite typically exhibits two different structural distortions, one of which is the tilting of the BO6 octahedral and the other of which is the off-centering of the B ion in the BO6 octahedral. The first kind relates to a phase transition that is displacive, and the second type refers to a phase transition that is order-disorder [47, 48]. Tolerance factor (t′), introduced by Goldschmidt, can be used to evaluate the prediction criteria for identifying the formability of perovskite structure [49].

$$t' = \frac{(r_A + r_O)}{\sqrt{2} \cdot (r_B + r_O)} \quad (1)$$

where r_A and r_B are the ionic radii of the A and B cations, respectively, and r_O is the oxygen anion's ionic radius (in units). When t′ is less than one, the BO_6 octahedron tilts; nevertheless, when t′ is more than one, the smaller B cation centres off. Off-centering is mostly caused by larger A and smaller B ions, which causes BO_6 octahedron to compress. The BO_6 octahedron creates a cavity where the B ion tilts more effectively [50]. It has been discovered throughout time that whereas few perovskite-type oxides exhibit the straightforward cubic structure at ambient temperature, many do so at higher temperatures. The ideal perovskite-type structure has a cubic space group Pm3m-Oh [51].

3. Classification of perovskites

Numerous perovskite-based combinations with a variety of physical properties result from the flexibility of the ABO_3 perovskite crystalline structure and its capacity to accommodate a broad range of cations with various oxidation states as well as cation or anion vacancies. The two main categories of oxide phases are the ternary ABO_3 kind and their solid solutions, and the more modern complicated type compounds $(AB'_xB''_y)O_3$, where B′ and B″ are two distinct elements in various oxidation states and x + y = 1. On the basis of oxidation states, the ternary oxides can be divided into oxygen and cation deficient species and $A^{1+}B^{5+}O_3$, $A^{2+}B^{4+}O_3$, $A^{3+}B^{3+}O_3$ [52, 53]. The flowchart below **Figure 3** displays the comprehensive classification.

$A(B_x'B_y'')O_3$ is a complex perovskite type compound that can be separated into A $(B'_{0.67}B''_{0.33})O_3$ compounds that contain twice as much a lower valence state element as a higher valence state element, those that have $A(B'_{0.33}B''_{0.67})O_3$, which has twice as much of the higher valence state element as the lower valence state element. Those with $A(B'_{0.5}B''_{0.5})O_3$, those with equal levels of the two B components $A(B_x'B_y'')O_{3-z}$ and oxygen-deficient phases. The A and B cations' electron orbitals are typically near to 2^+ and 4^+, correspondingly, but in a few unique situations, they may be 3^+ and 3^+ if the B^{3+} cation has a six coordination. The oxygen anion array may be bent or displaced as a result of the valence variation at the A cation site, which will buckle the $(AO3)^{4-}$ layers. The octahedra with B cations at their centres may become distorted as a result of this buckleing. Due to their multiple valencies or unique 3d and 4d electron configurations, transition metal elements are good candidates to fill the B cation position because they have the adaptability needed to withstand this impact. This explains why transition metal oxides typically exhibit exceptional physical properties and feature perovskite-like structures. The oxygen and cation deficient phases will be viewed as

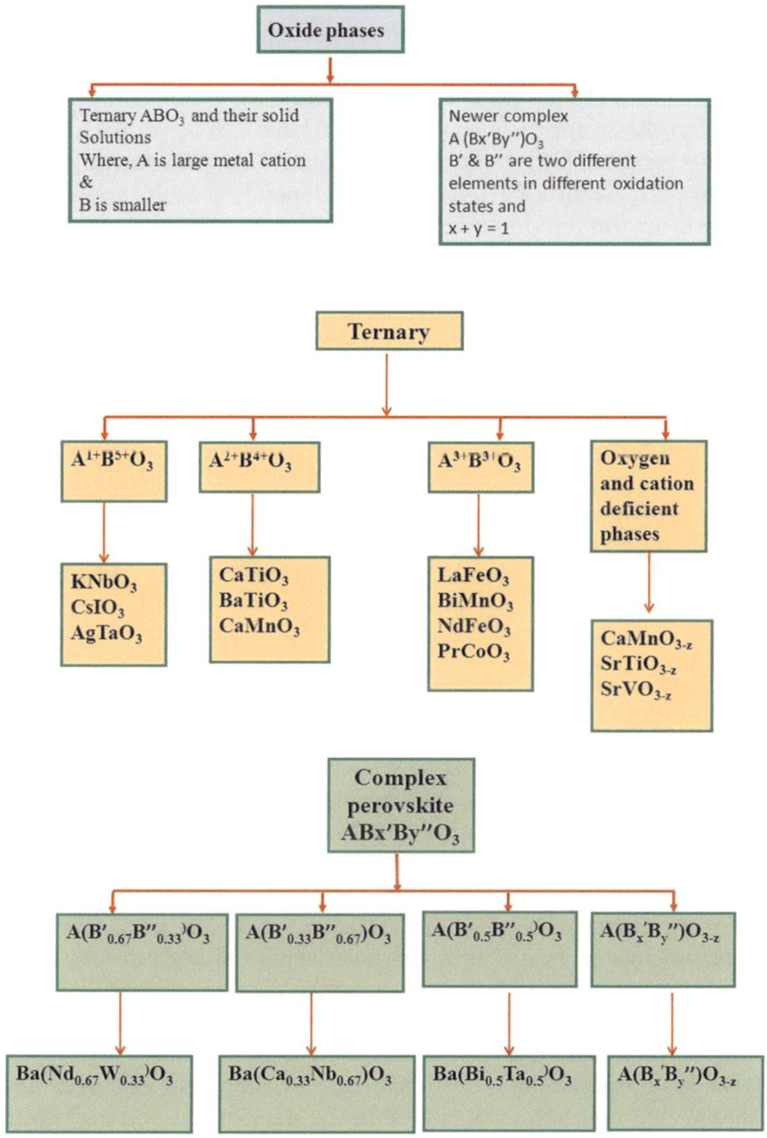

Figure 3.
Sorting out perovskite structures.

having a significant amount of vacancies rather than being just out of stoichiometry. Many of them differ from the complex perovskite compounds, which contain various elements in various valence states, in that they contain B ions of one element in two valence states [54, 55].

4. Synthesis of perovskites

When production of superior ceramic powders for advanced technology the ceramics industries are developing into one of the most important and quickly expanding sectors. Particularly, there will be a lot of interest in the creation of fine ceramic powders with exceptional and uncommon qualities. Regarding the structural and physico-chemical characteristics of perovskite structured materials, the synthesis processes are crucial. A crucial duty is sample synthesis, which can be accomplished via a variety of synthesis approaches [56]. Perovskites can be prepared in various forms like nanocrystalline [57], bulk [58], thin films [32], nanowires [59], nanotubes [60], nanocubes [61], nanorods [62] etc. forms depending on its applications using particular bottom-up and top down approach. There are several methods to synthesize perovskite materials in different forms such as chemical co-precipitation [63], microemulsion [64], hydrothermal [65], solvothermal [66], microwave irradiation [67], spray-pyrolysis [68], chemical vapor deposition [69] etc. The sol-gel auto combustion approach for bulk and nanoparticles of the $BaTiO_3$ solid is highlighted here with a brief explanation of the solid-state reaction.

4.1 Solid state reaction method

The majority of prior research on perovskite materials concentrated on their solid-state reaction-produced structure and physical characteristics. The most popular technique for creating polycrystalline bulk solids from a variety of solid starting materials is the solid-state reaction approach, sometimes referred to as the ceramic method. The solid state reaction pathway offers a wide variety of raw materials, including oxides, carbonates, etc. At room temperature, solids do not react with one another; therefore, it is required to heat them to much higher temperatures, typically between 800 and 2000°C, for the reaction to take place at an acceptable rate. Consequently, in this strategy, both the thermodynamic and kinetic elements are crucial [70]. Precursor materials $BaCO_3$ and TiO_2 were combined to create bulk $BaTiO_3$, and the mixture was then pulverized with a mortar and pestle to reduce the size of the particle sizes and increase the surface area exposed to the reaction. The crushed powder was calcined in furnace over 900°C to form the necessary product. Different wet chemical techniques can be used to overcome the limitations of the traditional solid state reaction approach, including high sintering temperature, secondary phase formation, poor ingredient dispersion, high porosity, and big particle size [71]. Due to its higher surface-to-volume ratio and quantum confinement effects, nanomaterials have different properties than those of bulk materials [72]. It is a widely acknowledged fact that the properties of a bulk material radically alter as it gets closer to the nanoscale. These characteristics are related to the size, shape, and distribution of the particles inside the materials, which in turn are dependent upon the synthesis process.

4.2 Sol-gel auto combustion method

Another name for the sol-gel auto combustion process is low-temperature self-combustion, commonly known as auto-ignition, self-propagation, nitrate-citrate combustion, gel-thermal breakdown technique, etc. [73]. This method uses a sol-gel procedure to create a gel out of an aqueous solution of the necessary metal salts (often nitrates or acetates) and organic fuel. The gel is then ignited to cause combustion, producing a fluffy, volumous end product with a sizable surface area. Fuel is utilized in this procedure as a complexant to create a homogenous precursor called xerogel [74]. In comparison towards other synthesis techniques, a sol-gel auto combustion approach has unquestionable advantages since it can precisely manage the composition, purity, least particle aggregation, homogeneity at the microscopic scale, and sintering temperature [75]. Here, a straightforward procedure for producing nanocrystalline $BaTiO_3$ material is described. Selecting the proper complexant additives is crucial for the production of homogeneous phases. Metal particles of varying sizes can be successfully chelated by complexant agents or fuel, which helps to maintain the specific precipitation needed to maintain compositional homogeneity among the constituents. The complexant substances act as reductants and are oxidized by nitrate ions to produce fuel. According to a literature review, many types of complexant agents are utilized to create perovskite materials. Most commonly used complexants are citric acid ($C_6H_8O_7$), glycine (NH_2CH_2COOH), urea ($CO(NH_2)_2$), ethylene glycol $(CH_2OH)_2$, dextrose ($C_6H_{12}O_6$), acetic acid (CH_3COOH), ascorbic acid ($C_6H_8O_6$), propionic acid (CH_3CH_2COOH) etc. [76]. Here, tetra butyl titanate ($Ti(OC_4H_9)_4$) or titanium isopropoxide $Ti[OCH(CH_3)_2]_4$ can be chosen as a precursor (for Ti^{4+} ions) due to the fact that it is a transition metal alkoxide. It is very reactive because it contains strongly electronegative groups that keep the metal in the highest oxidation state and enable nucleophilic attack on the metal. These alkoxide precursors are highly electrophilic, making them less resistant to condensation, hydrolysis, and other nucleophilic processes. Additionally, the group R from $Ti(OR)_4$ influences the gel's shape (size and surface area of the crystallites) and crystallization behavior. Controlling the condensation path and polymer development requires chemical modification of the transition metal alkoxide with chelating ligands. Tetra butyl titanate was chelated with ethanol (1:2) to produce a highly condensed product and to aid in the gelification process. A unidimensional polymer is produced when too much water is introduced to ($Ti(OC_4H_9)_4$)-ethanol, which causes the O-R ligands to be hydrolyzed preferentially [77]. The metal nitrate to citric acid molar concentration was taken to be 1:5 in accordance with the rules of propellant chemistry, which take into account the oxidizing and reducing valencies of various components. The total valence of the precursors was used for this $Ba(NO_3)_2 = -10$, $C_6H_8O_7 = +18$ and $Ti(OC_4H_9)_4 = +96$. Tetra butyl titanate solution was initially added to a citric acid aqueous solution with a pH of 8, which was then brought to the desired level by adding the necessary amount of ammonia. A yellowish translucent liquid that is designated as solution "A" was produced after being agitated at 80°C for 1 hour. Inorganic compounds (in this case, barium nitrate) were simultaneously dissolved in distilled water while being continuously stirred; the resulting solution, designated as solution "B," is the result. Then, mixtures of solutions "A" and "B" were added. Ammonia was used to bring the pH level to 7 and maintain it there until a translucent liquid was obtained. The solution's viscosity progressively rose after 3 hours of nonstop stirring, after which a stable translucent sol formed. The creation of the gel is started by continuous heating to 110° C. Viscous gel becomes dry gel when heated and constantly stirred. It was discovered that the nitrate-citrate gel's combustion process was autocatalytic, and experimental results revealed that the dried gel made of metal nitrates and citric acid exhibited self-propagating combustion behavior. The entire combustion process was completed in a matter of minutes. To produce the nanocrystalline powders, the resulting powders were dried, crushed, and annealed at 900°C for 5 hours in a muffle furnace. **Figure 4** shows the process for making $BaTiO_3$ nanoceramics using the sol-gel method.

5. Unique properties of perovskites

The ceramic materials with the general chemical formula ABO_3 made of perovskite are utilized as high-value materials in a variety of engineering and technological

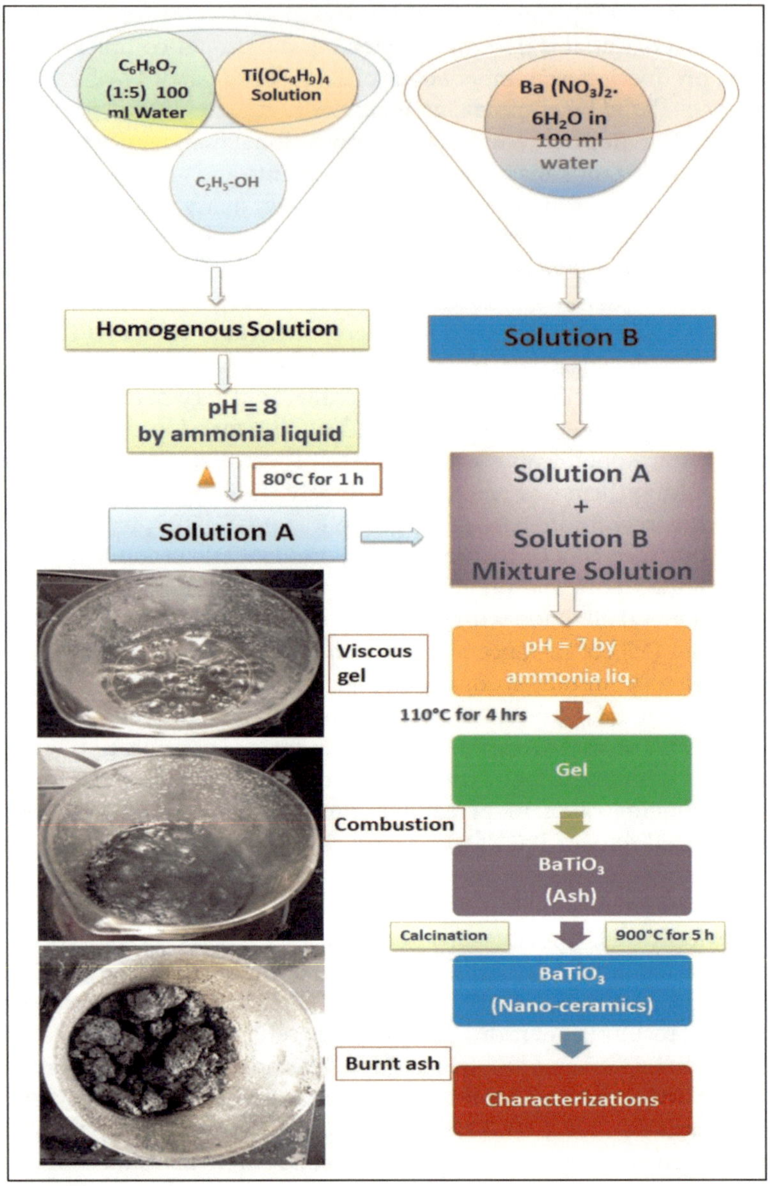

Figure 4.
The scheme of synthesis of $BaTiO_3$ nanoceramics.

applications. Because of the non-stoichiometry of the cations and/or anions, the distortion of the cation configuration, and the mixed valence and valence mixture electronic structure, perovskite-type structures have functional features. Furthermore, it is known that the majority of the naturally occurring metallic elements are stable in perovskite-type oxide structures, and the ability to synthesize multicomponent perovskites by partially substituting cations in positions A and B results in a variety of complicated kinds. These traits are what give perovskites their unique qualities. These include ferroelectric, thermoelectric, multiferroic, superconductive, dielectric, optical, and many more unique properties. A few of them are described in depth here [78–80].

5.1 Dielectric properties

Materials that allow electro-static fields to last a long time are known as dielectric or electrical insulating materials [81]. The basic electrical properties of conductive materials and dielectric materials sharply diverge because the dielectric materials offer a very high resistance to the channel of electric current under the influence of the applied direct-current voltage. Capacitors frequently have layers of these materials added to them to improve performance; the term "dielectric" specifically refers to this use [82]. Ferromagnetic or high dielectric permittivity materials are crucial for the production of electroceramics in engineering and electronics. $BaTiO_3$ and $KNbO_3$ are two examples of perovskites that have underwent substantial research in the past [35, 83]. The ratio of a substance's permittivity to the permittivity of empty space is known as the dielectric constant. A significant dielectric constant (ε), which is a highly nonlinear and anisotropic incident, is based on the collective polar deflections of the metal ions with respect to the oxygen sublattice [84]. A soft-mode model typically describes the phase shift that results in ferroelectricity [85]. Many variety of careers have been taken from the structurally straightforward $BaTiO_3$ by the solid coordination compound $Pb(Zr,Ti)O_3$ to various unique families of materials in order to adjust the dielectric and mechanical properties [86]. These approaches specifically account for the fact the perovskites' adaptability to chemical alteration and docility [87]. The relaxor ferroelectric is one of them. It is undoubtedly based on a multi-element substituted lead titanate ($PbTiO_3$) with the chemical formula $A(B'B'')O_3$ with random metal cation occupants at the A and B sites with various valence states and ionic radii [88, 89]. Large dielectric constants, a clear frequency dispersion, and temperature-dependent dielectric constant variations are all characteristics of relaxor ferroelectrics. For temperatures above the glass transition, these effects result from slow relaxation processes [90]. The effects are based on electrical inhomogeneities and the presence of polar nano-regions, and their length scales for changing composition and spontaneous polarization are 2–5 nm. The lattice component of the response is thought to constitute a local softening of the transverse-optical phonon branch that prevents long-wavelength ($q = 0$) phonon propagation. It is remarkable to notice that for such small length scales, the basic limit, the superpara-electric state, has not yet been attained [91]. PZT and PMN are two general examples of relaxor ferroelectrics [92]. Two common examples of relaxor ferroelectrics are PZT and PMN (eg. $SrTiO_3$) [93]. Perovskites, such as $BaTiO_3$, are one of the potential possibilities for usage in dynamic random access memory (DRAM) and tunable microwave devices due to their high dielectric constant and low dielectric loss [94, 95].

5.2 Superconductivity

When cooled below a specific critical temperature, certain materials exhibit the phenomena of superconductivity, which results in exactly zero electrical resistance and the expulsion of magnetic flux fields [96]. A wide family of materials with a variety of crucial physical characteristics is known as oxide perovskites. Notably, this form of perovskite structure offers a superb structural foundation for the presence of superconductivity. The most well-known examples of superconducting perovskites are high *Tc* copper oxides, although there are other others as well [97]. Many superconducting materials have historically come from intermetallic compounds, however perovskite oxides have recently overshadowed their existence. Sweedler et al. looked at the tungsten bronzes' superconductivity [98]. Bronzes made of cesium, sodium, potassium, rubidium, and tungsten were discovered to be superconductors. Three samples of decreased strontium titanate were examined by Schooley and colleagues to measure superconducting transitions [99]. The reduced crystals were made by heating for a lengthy time in a vacuum between 10^{-5} and 10^{-7} mm Hg. The transitions happened at respective temperatures of 0.25 K and 0.28 K. Additionally, superconductivity has been discovered in the systems' decreased phases $Ba_xSr_{1-x}TiO_3$ and $Ca_ySr_{1-y}TiO_3$ when $x \leq 0.1$ and $y \leq 0.3$ [100]. Superconductors classified as "Type 2" are made of metallic compounds and alloys (except for the elements vanadium, technetium and niobium). This Type 2 group includes the recently found superconducting "perovskites" metal-oxide ceramics, which typically have a ratio of 2 metal atoms to every 3 oxygen atoms. They outperform Type 1 superconductors in terms of transition temperature Tc by a mechanism that is yet not fully understood [101]. The discovery that $Ba(Pb,Bi)O_3$ had a *Tc* of 13 K in 1973 by a DuPont research team led to the development of the first of the oxide superconductors [102].

5.3 Ferroelectricity

> *"You may say anything you like but, we all are made up of ferroelectrics".*
> (B. T. Matthias).

Ferroelectricity is a phenomena in which the introduction of an external electric field induces a spontaneous electric polarization in some materials [103]. Various crystals, including quartz, tourmaline, and Rochelle salt, exhibit piezoelectricity, which was discovered by brothers Pierre and Paul-Jacques Curie in 1880 [104, 105]. This discovery inspired subsequent research in the topic of piezoelectrics, most notably Erwin Schrodinger's work [106, 107]. Erwin Schrodinger originally used "ferroelektrisch" or "ferroelectricity" in 1912 [108]. Joseph Valasek is credited with finding ferroelectricity for his systematic investigation of the magnetic characteristics of ferromagnetic and the dielectric properties of rochelle salt, which he presented at the American Physical Society's annual conference in Washington on April 23, 1920 [106]. The early 1940s saw the discovery of ferroelectricity in materials based on perovskite, such as barium titanate ($BaTiO_3$), which was a significant advance in the field of ferroelectric research [109, 110]. There is a lot of interest in various forms of ferroelectrics as a result of the finding of ferroelectricity in $BaTiO_3$. This discovery opened up new application possibilities for ferroelectric materials [105, 111]. Ven'skev and Zhdanov identified the perovskite family's distribution of ferroelectrics (FE) and antiferroelectrics (AFE) [112]. The FEs cover the entire perovskite range $0.78 \leq t' \leq 1.05$. The AFEs are found to have a restricted distribution ($0.78 \leq t' \leq 1.0$). In addition to classifying ferroic behavior in perovskites using the tolerance factor (t'), Halliyal and Shrout discovered that graphing t against the average electronegativity as indicated by [112];

$$\overline{\chi} = (\chi_{AO} + \chi_{BO})/2 \tag{2}$$

where the electronegativity differences between the A and B cations and oxygen are denoted by χ_{AO} and χ_{BO}, respectively. The endurance of the crystalline phase for a variety of simple and complicated perovskites was discovered. Perovskite compounds with low t′ extended to develop pyrochlore phase(s), and the perovskite phase was stabilized by solid solutions in which t′ was raised [113]. The dielectric constant of ferroelectric materials is about two orders of magnitude more than that of regular dielectric. A well-known ferroelectric substance with a relative dielectric constant of above 2000 is $BaTiO_3$. Ferroelectric materials must include permanent electric dipoles; one of the most common ferroelectrics, barium titanate, is explained for its creation [114]. The spontaneous polarization is a consequence of the positioning of the Ba^{2+}, Ti^{4+}, and O^{2-} ions within the unit cell, as represented in **Figure 5a**.

The O^{2-} anions are at the face centres, the Ti^{4+} ion is in the octahedral void at the body centre, and the Ba2+ ions are in the body corners. Only one of the four octahedral voids in the unit cell is filled, which matches the chemical formula of one titanium for every four species of the other kinds: one barium plus three oxygen. Barium titanate is a cubic crystal with the ion positions mentioned above 120°C [115]. The positive and negative charge centres coincide in this instance, and a spontaneous dipole moment is absent. The Ti^{4+} ion shifts to one side of the body centre if the crystal is cooled below the Curie temperature of 120°C, as indicated by the dotted line in **Figure 5**'s front view (b). A displacement of the nearby oxygen anions also occurs. At normal temperature, the crystal changes from its cubic phase to its tetragonal phase. The tetragonal cell's c/a ratio is almost 1.012. Local dipoles are formed all around the crystal as a result of the positive and negative charge centres no longer lining up. A significant amount of polarization occurs in the solid as a result of the alignment of the dipoles of nearby unit cells. Even when the dipoles of nearby unit cells are aligned, a $BaTiO_3$ crystal typically displays no net polarization at ambient temperature in the absence of an external field. This can be understood by visualizing the behavior of ferroelectric domains in a manner similar to that of ferromagnetic domains. When an electric field is applied, the domains have a tendency to line up in the field's direction, and we notice all the hysteresis loop phenomena, including domain rotation and domain growth. A ferroelectric hysteresis loop is shown in

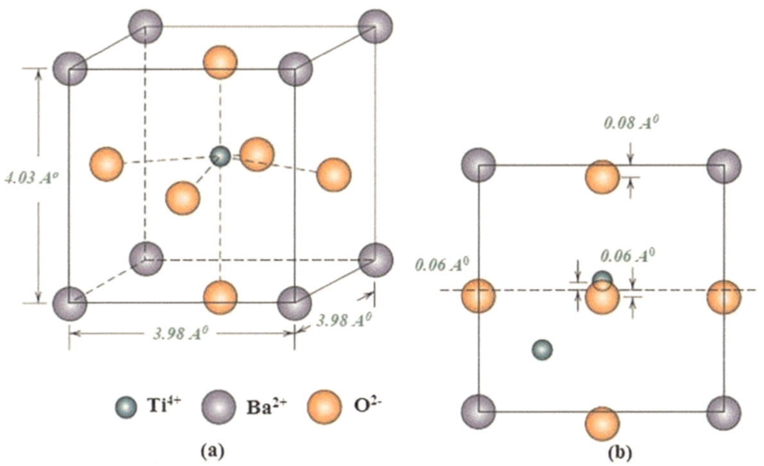

Figure 5.
$BaTiO_3$ crystal unit cell in (a), with the Ti^{4+} ion and O_2 ions migrating out from the centre as indicated by the arrows in (b).

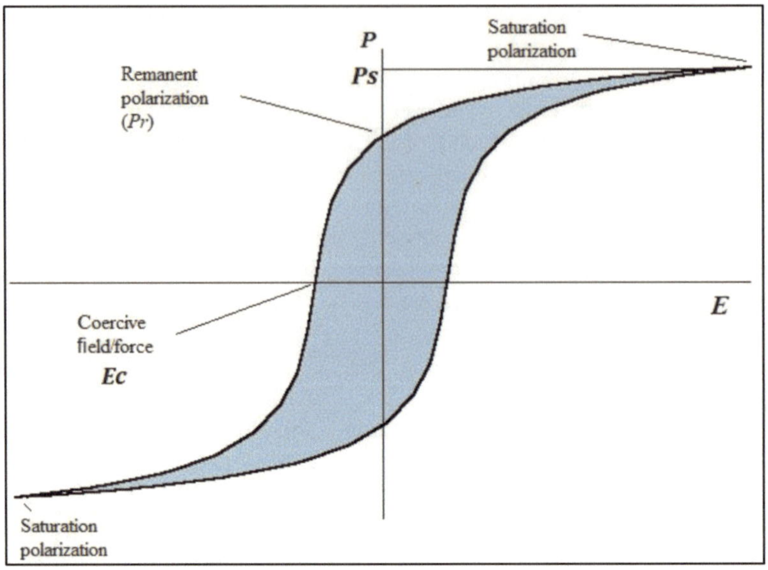

Figure 6.
The ferroelectric material's polarization (P) vs. applied field (E) hysteresis loop.

Figure 6. By extrapolating the linear area of the curve backwards to zero electric field, the spontaneous polarization Ps is obtained [116, 117].

For the creation of sensors, capacitors, memory devices, etc., ferroelectric properties are exploited. Ferroelectric materials have a non-linear feature that tunable capacitors take advantage of to tune the capacitance. Two electrodes and a layer of ferroelectric material sandwich each other in the ferroelectric capacitor. Comparatively speaking to non-tunable dielectric capacitors, ferroelectric capacitors are both tunable and very compact in size. Ferroelectric materials' hysteresis during spontaneous polarization can be used to create ferroelectric RAM and RFID cards. Input devices in ultrasonic imagers, infrared cameras, fire sensors, motion detectors, etc., use ferroelectric material.

5.4 Piezoelectricity

The capacity of some materials to produce an electric charge in response to applied mechanical stress is known as piezoelectricity. The Curie brothers, Pierre and Jacques, discovered the piezoelectric effect in 1880 [118]. They discovered that certain crystals became electrically polarized when subjected to mechanical strain, and the amount of polarization was proportional to the applied strain. The Curies also found that when subjected to an electric field, the same materials distorted. The inverse piezoelectric effect has been coined to describe this [119]. **Figure 7a** and **b** show, respectively, the piezoelectric effect and the inverse piezoelectric effect.

By interacting with one another, electric dipoles spontaneously align in ferroelectric materials, whereas in piezoelectric materials, an additional force is needed. In light of this, all ferroelectrics are also piezoelectrics, but not vice versa [120]. The piezoelectric ceramics, of which PZT is an example, are a significant group of piezoelectric materials in addition to the crystals already described [121]. These are perovskite-based polycrystalline ferroelectric materials, which have tetragonal/rhombohedral

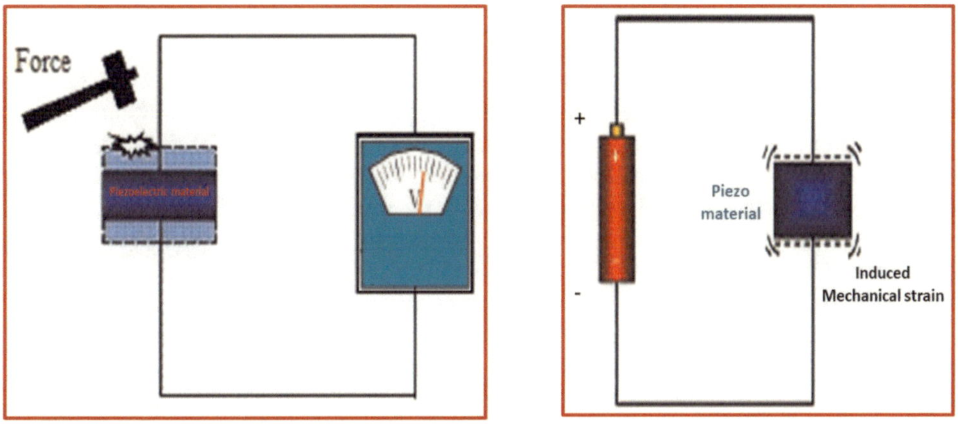

(a) (b)

Figure 7.
(a) Piezoelectric effect and (b) Inverse piezoelectric effect.

crystal structures that resemble cubic shapes. They possess the common formula $A^{2+}B^{4+}O^{2-}{}_3$, where A stands for a big divalent metal ion, like lead or barium, and B stands for a tetravalent metal ion, like titanium or zirconium.

5.4.1 Mathematical modeling

Piezoelectricity is a result of combining Hook's law with the electrical behavior of materials [122]:

$$D = \varepsilon E \tag{3}$$

$$S = sT \tag{4}$$

where, D: electric displacement, ε: permittivity, E: electric field strength, S: strain, s: compliance, and T: stress.

The coupled strain-voltage equation:

$$S = s^E T + d^t E \text{ converse piezoelectric effect} \tag{5}$$

$$D = \varepsilon^T E + dT \text{ direct piezoelectric effect} \tag{6}$$

$$D_{ij,k} = dS_{ij} \tag{7}$$

$$D_{ij,k} = \frac{\partial S_{ij}}{\partial E_k} \text{ piezoelectric coefficient} \tag{8}$$

Natural materials including quartz, cane sugar, collagen, topaz, dna, rochelle salt, wood, and tendon are examples of piezoelectric materials. Man-made crystals and ceramics include langasite, gallium orthophosphate ($GaPO_4$), and quartz analogous crystals ($La_3Ga_5SiO_{14}$), quartz analogic crystal, barium titanate, lead zirconate titanate etc. Acuosto-optic modulators, valves, high voltage and power sources, cigarette lighters, energy harvesting, AC voltage multipliers, piezoelectric motors, actuators, loudspeakers, pressure sensors, force sensors, strain gauges, microphones, pick-ups, actuators, and many other applications use piezoelectric materials.

5.5 Magneto-electronic correlations

In perovskites containing transition metal ions with open *3d* electron shells, magnetism or different orbital (electronic) ordering phenomena are seen. When compared to other electronic states, these *3d* states have a larger ratio of the Coulomb repulsion energy *Ud* to the bandwidth W, which gives them a more localized character and a propensity for insulating states or metal-insulator transitions [123]. Due to the overlap of the different wave functions, these electrons hop and super-exchange via oxygen sites. Thus, non-stoichiometry and the [BO6] octahedra's tilt or distortion have a significant impact on the characteristics and phase diagrams of a perovskite. Additional factors include charge doping, charge/orbital inhomogeneous states that cause enormous response, such as to external magnetic fields, and order/disorder processes of the orbital part of the 3D wave function [23]. But before taking into account such effects, the electronic structure, the number of *3d* electrons, the Hund's Rule coupling, the crystalline electric field or Jahn-Teller splitting of the *3d* electronic configuration a hierarchy of energies that describes the properties of the system.

5.6 Multiferroicity

"Materials should exist, which can be polarized by a magnetic field and magnetized via an electric field." P. Curie. A remarkable class of materials displaying simultaneous ferromagnetic, ferroelectric, and ferroelastic ordering is represented by multiferroics. H. Schmid coined the term "multiferroic" in 1994 [124]. Multiferroic materials with the corresponding properties are shown in **Figure 8**.

These materials are unique in that they have the capacity to simultaneously use both their magnetization and polarization states, a huge potential that would make them excellent candidates for next-generation sensors and memory technology [125, 126]. Numerous multiferroics, such as rare-earth manganites and ferrites, are transition metal oxides with perovskite crystal structure (e.g. $HoMn_2O_5$, $TbMnO_3$, $LuFe_2O_4$) [127, 128]. $Ba_2CoGe_2O_7$, $Ca_2CoSi_2O_7$, $TbFe_3(BO_3)_4$, $CoCr_2O_4$, $FeCr_2O_4$, $MnCr_2O_4$ $NdFe_3(BO_3)_4$ are those substances that exhibit multiferroicity even at ambient temperature [129–132]. Bismuth ferrite ($BiFeO_3$), a rhombohedrally distorted

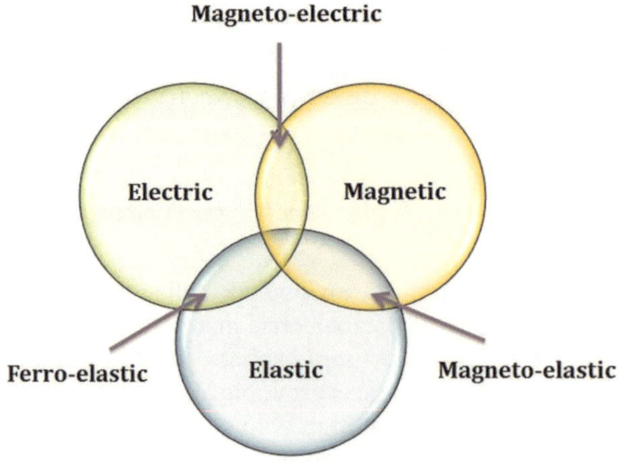

Figure 8.
Multiferroic materials.

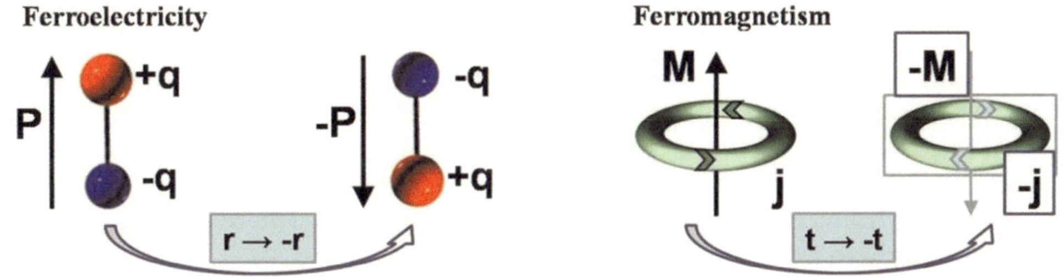

Figure 9.
The prerequisites for ferromagnetism and ferroelectricity (polarization) (unpaired electron spin motion).

perovskite, is one of the many multiferroics that have been studied, and it is receiving constant attention because it exhibits both ferroelectric order and anti-ferromagnetic order over a wide temperature range that is significantly above room temperature [133]. The majority of ferromagnetic materials are typically metals, therefore a ferroelectric material must be an insulator. As a result, the simultaneous occurrence of ferroelectric and ferromagnetic ordering is constrained by the lack of ferromagnetic insulators. Multiferroics are still uncommon even when antiferromagnetic systems are taken into account. A structural deformation from the high symmetry phase, which removes the centre of inversion and enables an electric polarization as seen in **Figure 9**, is the traditional prerequisite for ferroelectricity. Ferroelectricity and ferromagnetism exhibit synchronous time and spatial inversion breaking in **Figure 9**. Unpaired d electrons are necessary for any type of magnetic ordering, however ferroelectric materials like typical perovskite oxides lack this property (ABO_3) have a d^0 configuration on the small B cation. The tendency for the tiny cation to generate a distortion that removes the centre of symmetry is significantly suppressed if the d shell is only partially populated. With some unpaired electrons in the *d* orbitals, magnetoelectric multiferroic materials should therefore exhibit some deformation in their crystal structure. Recently, it was discovered that magnetic spin ordering can create ferroelectricity even in the absence of any structural distortion. As a result, there are now a lot more ferroic materials that could be used [134]. Experimental and theoretical research both point to a new era in the attainability of multiferroicity in transition metals doped $BaTiO_3$ [135]. Extrinsic and intrinsic dopants like excess oxygen vacancies and transition metal cations can be added to these materials to improve their chemical and physical properties [136].

The electrical or magnetic characteristics of perovskites with transition metal ions (TM) doped on the B site are incredibly diverse and intriguing. This variety is more closely linked to the intricate role that transition metal ions play in certain coordinations with oxygen or halides than it is to the chemical flexibility of these compounds [137]. Although unfilled *3d* electron shells of the TM are typically associated with magnetism and electronic correlations, filled *3d* electron shells are associated with strong dielectric characteristics. Due to the limited number of low-symmetry magnetic point groups that permit spontaneous polarization, multiferroicity, the coexistence of spontaneous ferroelectric and ferromagnetic moments, is a rare phenomenon [138]. One of two categories can be used to group all multiferroic materials. Multiferroics of type I and type II. At high temperatures, Type I go through a structural, nonpolar-to-polar phase transition that, in most cases, involves the breaking of inversion symmetry and results in ferroelectricity. At lower temperatures, the emergence of magnetic order takes place below a different phase transition. The main order parameter for type II is the staggered antiferromagnetic) magnetization. Below a certain temperature, magnetic ordering transforms the symmetry group from a non-polar parent phase to a polar magnetic phase. This results in inappropriate ferroelectricity. Additionally, magneto-structural coupling to the crystal structure creates an electrically polar state. However, polar non-centrosymmetric magnetic structures typically originate from the complicated magnetic ordering of geometrically frustrated states or from competing interactions in this situation, where the magnetic and ferroelectric order parameters are intimately related [139].

5.7 Optical properties

Perovskites have emerged as a revolutionary class of materials having excellent optical and photoluminescence properties. W.J. Merz studied the optical properties of single domain crystals of $BaTiO_3$ at various temperatures [140]. The crystal's refractive index was almost constant at ~2.4 between 20 and 90 degrees Celsius and peaked at ~2.46 at 120 degrees. $BaTiO_3$'s index of refraction was also measured by W.N. Lawless and R.C. De Vries at 5893 in the temperature range of 20–105°C; above Curie point, the index increased ~1.3 percent to 2.398 and stayed constant to 160°C [141]. It was discovered that the single BaTiO3 crystal, which is 0.25 mm thick, transmits between 0.5 μ and 6 μ. A weak absorption band was identified near 8, and complete absorption was found for wavelengths longer than 11 μ. Noland reported on the optical characteristics of single crystals of strontium titanate created during flame fusion [142]. The wavelength range where the optical coefficient was obtained was 0.20–17 μ. From 0.55 μ to 5 μ, a transmission of more than 70% was seen. These crystals have an index of refraction of 2.407 at 5893 Å, a dielectric constant of 310, and a loss tangent of 0.00025. Linz and Herrington reported the optical density of $CaTiO_3$ [143]. With the exception of the absorptions being shifted to shorter wavelengths, the absorption properties are remarkably comparable to those of $SrTiO_3$ crystals. High temperature infrared windows have been considered for $BaTiO_3$ and $SrTiO_3$. $SrTiO_3$ is regarded as a superior material for infrared detectors that are optically submerged. The detector-lens combinations are frequently chilled to liquid N_2 and solid CO_2 temperatures to boost sensitivity. Geusic et al. evaluated the electro-optic characteristics of $K(Ta_{0.65}Nb_{0.35})O_3$, $BaTiO_3$ and $SrTiO_3$ in the paraelectric phase [144]. When the distortions of the optical indicatrix are described in terms of the induced polarization, the electro-optic coefficients of these perovskites are almost constant with temperature and from material to material. These experiments also demonstrated the $K(Ta_{0.65}Nb_{0.35})O_3$'s strong electro-optic action at ambient temperature. There has been a lot of interest in materials that can be applied using lasers in recent years. Perovskite laser host materials are widely used. The most often used ion for insertion into somewhat large crystallographic locations seems to be Nd^{3+}. However, compensatory ions are necessary in these substitutions, barring the usage of LaF_3 as a host. Without compensating ions, divalent Tm^{2+} and Dy^{2+} can be substituted in CaF_2, but they are very unstable. Cr^{3+} turned shown to be the best replacement for Al^{3+} at its crystallographic positions. Recent studies have increasingly concentrated on the luminous characteristics of rare earth ion doped perovskite-type oxides. Oxygen phosphors of the perovskite type are exceedingly stable and can consistently function in a variety of conditions. [145–147]. Additionally, Perovskite-type oxide phosphors have been discovered to be a likely contender in field emission display (FED) and plasma display panel (PDP) systems because they are sufficiently conductive to release electric charges on the phosphor particle surfaces [148]. Many perovskite-type oxide phosphors, such as $A^{2+}B^{4+}O_3$ (A = Ca, Sr, Ba; B = Ti, Zr, Si, Hf, etc.) are therefore activated by rare earth ions, such as Sm^{3+}, Tm^{3+}, Pr^{3+}, Eu^{3+}, Tb^{3+} and so forth [149–151] have been made, and their luminous characteristics have also been extensive examination. There aren't many research on photoluminescence (PL) in zirconates, particularly ones with visible emission regions [152]. Phosphors of rare earth ions doped perovskite type oxides, such as $SrHfO_3$:Ce [153] and $CaTiO_3$:Pr [154] X-ray phosphors could be used in many displays, hence current study has concentrated on their luminous qualities. Eu^{3+} is an effective activator ion that emits red or red-orange light in a variety of hosts, including borates [155], niobates [156] and molybdates [157]. Host $BaZrO_3$ is now recognized as a readily available, inexpensive, and environmentally benign photoluminescence (PL) material that produces light in the visible spectrum [158]. This material is promising because of its PL feature, which makes it useful for applications including scitillators, plasma displays, solid state lightning, green photocatalysts, and field emission displays [159–161].

5.8 Colossal magnetoresistance (CMR)

Some materials have a feature called colossal magnetoresistance (CMR) that allows them to drastically alter their electrical resistance when a magnetic field is present (mostly manganese-based perovskite oxides) [162]. CMR was first identified in mixed-valence perovskite manganites in the 1950s by G. H. Jonker and J. H. van Santen [163]. Due to their rich fundamental physics and significant potential for use in spintronics devices, the discovery of the colossal magnetoresistance (CMR) influence in divalent alkaline-earth ion dopant perovskite manganites $Re_{1-x}Ae_xMnO_3$, where Re is a trivalent rare-earth (La, Pr, Sm, etc.), has generated a great deal of interest [164–167].

Different magnetic phases, including insulating antiferromagnetic phases with multiple orbital orders and a ferromagnetic, metallic, orbitally disordered phase, are detected depending on the orbital occupancy of the manganese ions and the related orbital order. Because of the close coupling between spins and orbitals in these compounds, both degrees of freedom have ordering temperatures that are similar in magnitude. In a three-dimensional lattice, the primary spin exchange pathways pass over almost 180 TM-O-TM limits (TM-transition metal). But merely taking into account spins, low dimensionality, magnetic frustration, and quantum processes can also result in some incredibly strange phase diagrams, even ones without magnetic long range order [168]. In fact, in frustrated lattices, quantum fluctuations or a second order energy scale can frequently lift the degeneracy of the magnetic ground state [169, 170]. The nature of CMR manganites, which are strongly correlated electron systems with interactions between the lattice, spin, charge, and orbital degrees of freedom, including the double exchange interaction, Jahn-Taller effect, electronic phase separation, charge ordering, etc., is generally thought to be the root cause of the CMR effect. This theory has been extensively discussed in some review papers [171–175].

6. Applications of perovskites

Calcium titanate ($CaTiO_3$)-like perovskite materials offer fascinating and exceptional physical features that have been thoroughly investigated for use in both theoretical modeling and real-world applications. Due to their extremely stable structure, abundance of compounds, diversity of characteristics, and several useful applications, inorganic perovskite-type oxides are fascinating nanomaterials with a wide range of uses. Current applications for these solids include electronics, geophysics, astronomy, nuclear, optics, medicine, the environment, etc. [176].

Perovskites are useful for many different applications depending on the aforementioned distinctive properties, including thin film capacitors, non-volatile memories, photoelectrochemical cells, recording apps, read heads in hard discs, spintronics devices, laser applications, for windows to block high temperature infrared rays, high temperature heating applications (thermal barrier coatings), frequency filters for wireless technology, non-volatile memorabilia, and spintronics devices. **Table 1** lists additional significant uses for various perovskite-structured materials, along with the corresponding characteristics.

Reference compound	Properties	Applications
$BaTiO_3$	Dielectric ferroelectric	Multilayer ceramic capacitors (MLCCs), sensor, PTCR resistors, embedded capacitance [177–179]
$PbTiO_3$	Pyroelectric piezoelectric	Transducer, pyrodetector, under water devices [180, 181]
$(BaSr)TiO_3$	Non-linear dielectric properties, pyroelectric	Tunable microwave devices, pyrodetector [182]
$Pb(ZrTi)O_3$	Dielectric, pyroelectric, piezoelectric, electro-optic	Nonvolatile memory, ferroelectric memories (FERAMs), surface wave acoustic devices, substrate wave guide devices [183, 184]
$Bi_4Ti_3O_{12}$, high T_c cuprate compounds	Ferroelectric with high Curie temperature superconductivity	High-temperature actuators, FeRAMs [185, 186]
$BaCeO_3$, $BaZrO_3$	Proton conduction	Electrolyte in protonic solid oxide fuel cells (P-SOFCs) [187]
$LaNiO_3$	Chemical	Catalysts [188]
$(La,Sr)MnO_3$	Colossal magnetoresistance	Spintronics devices [189]
$Pb(Mg_{1/3}Nb_{2/3})O_3$	Dielectric	Memorydevice, capacitor [190]
$K(TaMb)O_3$	Pyroelectric, electro-optic	Waveguide device, frequency doubler [191]
$BiFeO_3$	Magnetoelectric coupling, high Curie temperature	Magnetic field detectors, memories [192, 193]
$(La,Sr)BO_3$ (B = Mn, Fe, Co)	Mixed conduction, catalyst	Cathode material in SOFCs, oxygen separation membranes, membrane reactors, controlled oxidation of hydrocarbons [194, 195]
$(K_{0.5}Na_{0.5})NbO_3$, $Na_{0.5}Bi_{0.5}TiO_3$	Ferroelectricity, piezoelectricity	Lead-free piezoceramics [196, 197]
$LaAlO_3$ $YAlO_3$	Host materials for rare-earth luminescent ions,	Lasers Substrates for epitaxial film deposition [198, 199]

Table 1.
Applications of perovskites along with respective properties.

7. Conclusions

This chapter primarily addresses the fundamental concepts of perovskites, their structure, classification of perovskites, their synthesis, their distinctive features, and their applications. Because of its many unique properties, including enormous magnetoresistance, multiferroics, superconductivity, etc., perovskite structured materials have received a great deal of research attention. Perovskites have unique and unusual physical properties that have been extensively studied for both real-world use and theoretical research. Applications for photonic and optoelectronic devices, effective solar cells, fuel cells, spintronics, electrical machines, random - access memory, electrochemical double layer capacitors, and other novel device concepts are just a few of the many novel device concepts that have been made possible by perovskites in the field of materials science.

Acknowledgements

The author thanks Dr. K.M. Jadhav, Senior Professor in the Department of Physics at the Dr. Babasaheb Ambedkar Marathwada University in Aurangabad, India, for a lively discussion and insightful advice.

Conflict of interest

The authors declare no conflict of interest.

Author details

Pankaj P. Khirade[1*] and Anil V. Raut[2]

1 Department of Physics, Shri Shivaji Science College, Amravati, India

2 Department of Physics, Vivekanand Arts, Sardar Dalipsingh Commerce, and Science College, Aurangabad, India

*Address all correspondence to: pankajkhirade@gmail.com; nano9993@gmail.com

References

[1] Cheng Z, Lin J. Layered organic–inorganic hybrid perovskites: Structure, optical properties, film preparation, patterning and templating engineering. CrystEngComm. 2010;**12**: 2646-2662

[2] Warner TE. Synthesis, Properties and Mineralogy of Important Inorganic Materials. Hoboken, New Jersey: John Wiley & Sons; 2012

[3] Schaak RE, Mallouk TE. Prying apart Ruddlesden-Popper phases: Exfoliation into sheets and nanotubes for assembly of perovskite thin films. Chemistry of Materials. 2000;**12**:3427-3434

[4] Uma S, Raju AR, Gopalakrishnan J. Bridging the Ruddlesden–Popper and the Dion–Jacobson series of layered perovskites: Synthesis of layered oxides, A 2–x La 2 Ti 3–x Nb x O 10 (A= K, Rb), exhibiting ion exchange. Journal of Materials Chemistry. 1993;**3**:709-713

[5] Tsunoda Y, Sugimoto W, Sugahara Y. Intercalation behavior of n-alkylamines into a protonated form of a layered perovskite derived from aurivillius phase Bi2SrTa2O9. Chemistry of Materials. 2003;**15**:632-635

[6] Merz WJ. Switching time in ferroelectric BaTiO3 and its dependence on crystal thickness. Journal of Applied Physics. 1956;**27**:938-943

[7] Mackenzie A, Julian S, Diver A, McMullan G, Ray M, Lonzarich G, et al. Quantum oscillations in the layered perovskite superconductor S r 2 Ru O 4. Physical Review Letters. 1996; **76**:3786

[8] Asano H, Hayakawa J, Matsui M. Preparation and properties of triple perovskite La_ {3-3x} Ca_ {1+ 3x} Mn_ 3O_10 ferromagnetic thin films. Applied Physics Letters. 1997;**71**:844-846

[9] Maeno Y, Hashimoto H, Yoshida K, Nishizaki S, Fujita T, Bednorz J, et al. Superconductivity in a layered perovskite without copper. Nature. 1994;**372**:532-534

[10] Torrance J, Lacorre P, Nazzal A, Ansaldo E, Niedermayer C. Systematic study of insulator-metal transitions in perovskites R NiO 3 (R= Pr, Nd, Sm, Eu) due to closing of charge-transfer gap. Physical Review B. 1992;**45**:8209

[11] Bohnke O, Bohnke C, Fourquet J. Mechanism of ionic conduction and electrochemical intercalation of lithium into the perovskite lanthanum lithium titanate. Solid State Ionics. 1996;**91**:21-31

[12] Samara G. Pressure and temperature dependence of the dielectric properties and phase transitions of the ferroelectric perovskites: PbTiO3 and BaTiO3. Ferroelectrics. 1971;**2**:277-289

[13] Waser R, Baiatu T, Härdtl KH. dc Electrical degradation of perovskite-type titanates: I, ceramics. Journal of the American Ceramic Society. 1990;**73**: 1645-1653

[14] Steele B, Powell B, Moody P. Anionic conduction in refractory oxide solid solutions possessing the fluorite, pyrochlore and perovskite structures, in. Proceedings of the British Ceramic Society. 1968;**8**:87-102

[15] Wentzcovitch R, Karki B, Karato S, Da Silva C. High pressure elastic anisotropy of MgSiO 3 perovskite and geophysical implications. Earth and Planetary Science Letters. 1998;**164**: 371-378

[16] Navrotsky A, Weidner DJ. Perovskite: A structure of great interest to geophysics and materials science. Washington DC American Geophysical Union Geophysical Monograph Series. 1989;**45**

[17] Osako M, Ito E. Thermal diffusivity of MgSiO3 perovskite. Geophysical Research Letters. 1991;**18**:239-242

[18] Piel H. High T c superconductors for accelerator cavities. Nuclear Instruments and Methods in Physics Research Section A: Accelerators, Spectrometers, Detectors and Associated Equipment. 1990;**287**:294-305

[19] De AK, Luckscheiter B, Lutze W, Malow G, Schiewer E. Development of glass ceramics for the incorporation of fission products. American Ceramic Society Bulletin;(United States). 1976;**55**: 500-503

[20] Nagata S, Katsui H, Hoshi K, Tsuchiya B, Toh K, Zhao M, et al. Recent research activities on functional ceramics for insulator, breeder and optical sensing systems in fusion reactors. Journal of Nuclear Materials. 2013;**442**:S501-S507

[21] Zhu J, Li H, Zhong L, Xiao P, Xu X, Yang X, et al. Perovskite oxides: Preparation, characterizations, and applications in heterogeneous catalysis. ACS Catalysis. 2014;**4**:2917-2940

[22] Seyfi B, Baghalha M, Kazemian H. Modified LaCoO 3 nano-perovskite catalysts for the environmental application of automotive CO oxidation. Chemical Engineering Journal. 2009;**148**: 306-311

[23] Kreisel J, Glazer A, Jones G, Thomas P, Abello L, Lucazeau G. An x-ray diffraction and Raman spectroscopy investigation of A-site substituted perovskite compounds: The (Na1-xKx) 0.5 Bi0. 5TiO3 (0 x1) solid solution. Journal of Physics: Condensed Matter. 2000;**12**:3267

[24] Xu B, Yin K, Lin J, Xia Y, Wan X, Yin J, et al. Room-temperature ferromagnetism and ferroelectricity in Fe-doped BaTiO 3. Physical Review B. 2009;**79**:134109

[25] Liu Y, Nagra AS, Erker EG, Periaswamy P, Taylor TR, Speck J, et al. BaSrTiO∼ 3 interdigitated capacitors for distributed phase shifter applications. IEEE microwave and guided wave letters. 2000;**10**:448-450

[26] Hur SG, Kim TW, Hwang S-J, Choy J-H. Influences of A-and B-site cations on the physicochemical properties of perovskite-structured A (In 1/3 Nb 1/3 B 1/3) O 3 (A= Sr, Ba; B= Sn, Pb) photocatalysts. Journal of Photochemistry and Photobiology A: Chemistry. 2006;**183**:176-181

[27] Arakawa T, Kurachi H, Shiokawa J. Physicochemical properties of rare earth perovskite oxides used as gas sensor material. Journal of Materials Science. 1985;**20**:1207-1210

[28] Ghazanfari MR, Amini R, Shams SF, Alizadeh M, Ardakani HA. Effect of mechanical alloying synthesis process on the dielectric properties of (Bi 0.5 Na 0.5) 0.94 Ba 0.06 TiO 3 piezoceramics. Materials Research Bulletin. 2015;**68**:260-266

[29] Singh K, Band S, Kinge W. Effect of sintering temperature on dielectric properties of Pb (Fe1/2Nb1/2) O3 perovskite material. Ferroelectrics. 2004;**306**:179-185

[30] Jiang L, Guo J, Liu H, Zhu M, Zhou X, Wu P, et al. Prediction of lattice

constant in cubic perovskites. Journal of Physics and Chemistry of Solids. 2006; **67**:1531-1536

[31] Bokov A, Ye Z-G. Recent progress in relaxor ferroelectrics with perovskite structure. In: Frontiers of Ferroelectricity. Germany: Springer; Journal of Materials Science. 2006;**41**: 31-52

[32] Choi KJ, Biegalski M, Li Y, Sharan A, Schubert J, Uecker R, et al. Enhancement of ferroelectricity in strained BaTiO3 thin films. Science. 2004;**306**:1005-1009

[33] Obara H, Yamamoto A, Lee C-H, Kobayashi K, Matsumoto A, Funahashi R. Thermoelectric properties of Y-doped polycrystalline SrTiO3. Japanese Journal of Applied Physics. 2004;**43**:L540

[34] Chan W-H, Xu Z, Zhai J, Chen H. Uncooled tunable pyroelectric response of antiferroelectric Pb0. 97La0. 02 (Zr0. 65Sn0. 22Ti0. 13) O3 perovskite. Applied Physics Letters. 2005;**87**:192904

[35] Arlt G, Hennings D. Dielectric properties of fine-grained barium titanate ceramics. Journal of Applied Physics. 1985;**58**:1619-1625

[36] Kamalasanan M, Chandra S, Joshi P, Mansingh A. Structural and optical properties of sol-gel-processed BaTiO3 ferroelectric thin films. Applied Physics Letters. 1991;**59**:3547-3549

[37] Uchino K, Nomura S, Cross LE, Newnham RE, Jang SJ. Electrostrictive effect in perovskites and its transducer applications. Journal of Materials Science. 1981;**16**:569-578

[38] Obayashi H, Sakurai Y, Gejo T. Perovskite-type oxides as ethanol sensors. Journal of Solid State Chemistry. 1976;**17**:299-303

[39] Muralt P, Polcawich R, Trolier-McKinstry S. Piezoelectric thin films for sensors, actuators, and energy harvesting. MRS Bulletin. 2009;**34**:658-664

[40] Sebastian MT. Dielectric Materials for Wireless Communication. Amsterdam: Elsevier; 2010

[41] Kingon A, Streiffer S, Basceri C, Summerfelt S. High-permittivity perovskite thin films for dynamic random-access memories. MRS Bulletin. 1996;**21**:46-52

[42] Nenasheva E, Kanareykin A, Kartenko N, Dedyk A, Karmanenko S. Ceramics materials based on (Ba, Sr) TiO3 solid solutions for tunable microwave devices. Journal of Electroceramics. 2004;**13**:235-238

[43] Dimos D, Mueller C. Perovskite thin films for high-frequency capacitor applications 1. Annual Review of Materials Science. 1998;**28**:397-419

[44] Protesescu L, Yakunin S, Bodnarchuk MI, Krieg F, Caputo R, Hendon CH, et al. Nanocrystals of cesium lead halide perovskites (CsPbX3, X= Cl, Br, and I): Novel optoelectronic materials showing bright emission with wide color gamut. Nano Letters. 2015;**15**: 3692-3696

[45] Kay H, Bailey P. Structure and properties of CaTiO3. Acta Crystallographica. 1957;**10**:219-226

[46] Miyake S, Ueda R. Journal of the Physical Society of Japan. 1946;**1**:32

[47] Ramadass N. ABO 3-type oxides—Their structure and properties—A bird's eye view. Materials Science and Engineering. 1978;**36**:231-239

[48] Fatuzzo E, Merz WJ. Ferroelectricity. North-Holland Pub. Co. 1967

[49] Švarcová S, Wiik K, Tolchard J, Bouwmeester HJ, Grande T. Structural instability of cubic perovskite Ba x Sr 1− x Co 1− y Fe y O 3− δ. Solid State Ionics. 2008;**178**:1787-1791

[50] Goncalves-Ferreira L, Redfern SA, Artacho E, Salje EK. Ferrielectric twin walls in CaTiO 3. Physical Review Letters. 2008;**101**:097602

[51] Haines J, Rouquette J, Bornand V, Pintard M, Papet P, Gorelli F. Raman scattering studies at high pressure and low temperature: Technique and application to the piezoelectric material PbZr0. 52Ti0. 48O3. Journal of Raman Spectroscopy. 2003;**34**:519-523

[52] Galasso FS. Structure. Properties and Preparation of Perovskite-Type Compounds: International Series of Monographs in Solid State Physics, Elsevier; 2013

[53] Bhalla A, Guo R, Roy R. The perovskite structure – A review of its role in ceramic science and technology, Material Research. Innovations. 2000;**4**: 3-26

[54] Roth RS. Classification of perovskite and other ABO3-type compounds. Journal of Research of the National Bureau of Standards. 1957;**58**: 75-88

[55] Modeshia DR, Walton RI. Solvothermal synthesis of perovskites and pyrochlores: Crystallisation of functional oxides under mild conditions. Chemical Society Reviews. 2010;**39**: 4303-4325

[56] Ye Z-G. Handbook of Advanced Dielectric, Piezoelectric and Ferroelectric Materials: Synthesis, Properties and Applications. Amsterdam: Elsevier; 2008

[57] Khirade PP, Birajdar SD, Raut A, Jadhav K. Multiferroic iron doped BaTiO3 nanoceramics synthesized by sol-gel auto combustion: influence of iron on physical properties. Ceramics International. 2016;**42**:12441-12451

[58] Iwahara H, Yajima T, Hibino T, Ozaki K, Suzuki H. Protonic conduction in calcium, strontium and barium zirconates. Solid State Ionics. 1993;**61**: 65-69

[59] Xing J, Liu XF, Zhang Q, Ha ST, Yuan YW, Shen C, et al. Vapor phase synthesis of organometal halide perovskite nanowires for tunable room-temperature nanolasers. Nano Letters. 2015;**15**:4571-4577

[60] Alexe M, Hesse D, Schmidt V, Senz S, Fan H, Zacharias M, et al. Ferroelectric nanotubes fabricated using nanowires as positive templates. Applied Physics Letters. 2006;**89**:172907

[61] Parizi SS, Mellinger A, Caruntu G. Ferroelectric barium titanate nanocubes as capacitive building blocks for energy storage applications. ACS Applied Materials & Interfaces. 2014;**6**: 17506-17517

[62] Morozovska AN, Eliseev EA, Glinchuk MD. Ferroelectricity enhancement in confined nanorods: Direct variational method. Physical Review B. 2006;**73**:214106

[63] Mahata MK, Kumar K, Rai VK. Structural and optical properties of Er3+/ Yb3+ doped barium titanate phosphor prepared by co-precipitation method. Spectrochimica Acta Part A: Molecular and Biomolecular Spectroscopy. 2014; **124**:285-291

[64] Sakabe Y, Yamashita Y, Yamamoto H. Dielectric properties of nano-crystalline BaTiO3 synthesized by

micro-emulsion method. Journal of the European Ceramic Society. 2005;**25**: 2739-2742

[65] Zhu X, Zhu J, Zhou S, Liu Z, Ming N, Hesse D. BaTiO3 nanocrystals: Hydrothermal synthesis and structural characterization. Journal of Crystal Growth. 2005;**283**:553-562

[66] Kimura T, Dong Q, Yin S, Hashimoto T, Sasaki A, Sato T. Synthesis and piezoelectric properties of Li-doped BaTiO3 by a solvothermal approach. Journal of the European Ceramic Society. 2013;**33**:1009-1015

[67] Jhung SH, Lee J-H, Yoon JW, Hwang YK, Hwang J-S, Park S-E, et al. Effects of reaction conditions in microwave synthesis of nanocrystalline barium titanate. Materials Letters. 2004; **58**:3161-3165

[68] Lee KK, Kang YC, Jung KY, Kim JH. Preparation of nano-sized BaTiO3 particle by citric acid-assisted spray pyrolysis. Journal of Alloys and Compounds. 2005;**395**:280-285

[69] Kwak B, Zhang K, Boyd E, Erbil A, Wilkens B. Metalorganic chemical vapor deposition of BaTiO3 thin films. Journal of Applied Physics. 1991;**69**:767-772

[70] Lencka MM, Anderko A, Riman RE. Hydrothermal precipitation of lead zirconate titanate solid solutions: Thermodynamic modeling and experimental synthesis. Journal of the American Ceramic Society. 1995;**78**: 2609-2618

[71] Viswanath B, Kundu P, Halder A, Ravishankar N. Mechanistic aspects of shape selection and symmetry breaking during nanostructure growth by wet chemical methods. The Journal of Physical Chemistry C. 2009;**113**: 16866-16883

[72] Takagahara T, Takeda K. Theory of the quantum confinement effect on excitons in quantum dots of indirect-gap materials. Physical Review B. 1992;**46**: 15578

[73] Sutka A, Mezinskis G. Sol-gel auto-combustion synthesis of spinel-type ferrite nanomaterials. Frontiers of Materials Science. 2012;**6**:128-141

[74] Sakka S. Handbook of Sol-gel Science and Technology. 3. Applications of Sol-gel Technology. Germany: Springer Science & Business Media; 2005

[75] Khirade PP, Shinde A, Raut A, Birajdar SD, Jadhav K. Investigations on the synthesis, structural and microstructural characterizations of Ba1-x Sr x ZrO3 nanoceramics. Ferroelectrics. 2016;**504**:216-229

[76] Deganello F, Marcì G, Deganello G. Citrate–nitrate auto-combustion synthesis of perovskite-type nanopowders: A systematic approach. Journal of the European Ceramic Society. 2009;**29**:439-450

[77] Khirade PP, Birajdar SD, Raut A, Jadhav K. Effect of Fe–substitution on phase transformation, optical, electrical and dielectrical properties of BaTiO 3 nanoceramics synthesized by sol-gel auto combustion method. Journal of Electroceramics. 2016;**37**:110-120

[78] Teraoka Y, Nobunaga T, Yamazoe N. Effect of cation substitution on the oxygen semipermeability of perovskite-type oxides. Chemistry Letters. 1988;**17**: 503-506

[79] Radaelli P, Iannone G, Marezio M, Hwang H, Cheong S, Jorgensen J, et al. Structural effects on the magnetic and transport properties of perovskite A 1− x A x′ MnO 3 (x= 0.25, 0.30). Physical Review B. 1997;**56**:8265

[80] Kool TW. Properties of Perovskites and Other Oxides. Singapore: World Scientific; 2010

[81] Zheludev IS. Physics of Crystalline Dielectrics: Volume 1 Crystallography and Spontaneous Polarization. Germany: Springer Science & Business Media; 2012

[82] Von Hippel AR, Morgan SO. Dielectric materials and applications. Journal of The Electrochemical Society. 1955;**102**(3):68Ca

[83] Wiesendanger E. Dielectric, mechanical and optical properties of orthorhombic KNbO3. Ferroelectrics. 1973;**6**:263-281

[84] Anderson JC. Dielectrics. London: Chapman & Hall; 1964

[85] Gouda F, Skarp K, Lagerwall S. Dielectric studies of the soft mode and Goldstone mode in ferroelectric liquid crystals. Ferroelectrics. 1991;**113**: 165-206

[86] Yi G, Wu Z, Sayer M. Preparation of Pb (Zr, Ti) O3 thin films by sol gel processing: Electrical, optical, and electro-optic properties. Journal of Applied Physics. 1988;**64**:2717-2724

[87] Kim JK, Kim SS, Kim W-J. Sol–gel synthesis and properties of multiferroic BiFeO 3. Materials Letters. 2005;**59**: 4006-4009

[88] Ueda I, Ikegami S. Piezoelectric properties of modified PbTiO3 ceramics. Japanese Journal of Applied Physics. 1968;**7**:236

[89] Xiao Z, Ren Z, Liu Z, Wei X, Xu G, Liu Y, et al. Single-crystal nanofibers of Zr-doped new structured PbTiO 3: Hydrothermal synthesis, characterization and phase transformation. Journal of Materials Chemistry. 2011;**21**:3562-3564

[90] Cross LE. Relaxor ferroelectrics. Ferroelectrics. 1987;**76**:241-267

[91] Johnsson M, Lemmens P. Crystallography and Chemistry of Perovskites. Handbook of Magnetism and Advanced Magnetic Materials; New Jersey: John Wiley & Sons; 2007

[92] Cross L, Jang S, Newnham R, Nomura S, Uchino K. Large electrostrictive effects in relaxor ferroelectrics. Ferroelectrics. 1980;**23**: 187-191

[93] Kleemann W, Dec J, Wang Y, Lehnen P, Prosandeev S. Phase transitions and relaxor properties of doped quantum paraelectrics. Journal of Physics and Chemistry of Solids. 2000; **61**:167-176

[94] Kim J-S, Yoon S-G. High dielectric constant (Ba0. 65Sr0. 35)(Ti0. 41Zr0. 59) O3 capacitors for Gbit-scale dynamic random access memory devices. Journal of Vacuum Science & Technology B. 2000;**18**:216-220

[95] Chivukula V, Leung PK. Ferroelectric dielectric for integrated circuit applications at microwave frequencies. Google Patents. 1999

[96] Tinkham M. Introduction to superconductivity. Courier Corporation. 1996

[97] Murphy D, Sunshine S, Van Dover R, Cava R, Batlogg B, Zahurak S, et al. New superconducting cuprate perovskites. Physical Review Letters. 1987;**58**:1888

[98] Sweedler A, Raub CJ, Matthias B. Superconductivity of the alkali tungsten bronzes. Physics Letters. 1965;**15**: 108-109

[99] Schooley J, Hosler W, Cohen ML. Superconductivity in Semiconducting SrTi O 3. Physical Review Letters. 1964; **12**:474

[100] Frederikse H. Superconductivity achieved by Bardeen, Cooper and Schrieffer (BCS)* in 1957 . In the early sixties Gurevich, Larkin and Firsov° dis, in: Electronic Structures in Solids: Lectures presented at the Second Chania Conference. Crete, June 30–July 14. 2013. Held in Chania: Springer; 1968. p. 270

[101] Bardeen J, Cooper LN, Schrieffer JR. Theory of superconductivity. Physical Review. 1957;**108**:1175

[102] Sleight AW. Superconductive barium-lead-bismuth oxides. Google Patents. 1976

[103] Xu Y. Ferroelectric Materials and their Applications. Amsterdam: Elsevier; 2013

[104] Curie J, Curie P. Development by pressure of polar electricity in hemihedral crystals with inclined faces. Bulletin de la Societe Geologique de France. 1880;**3**:90

[105] Haertling GH. Ferroelectric ceramics: History and technology. Journal of the American Ceramic Society. 1999;**82**:797-818

[106] Valasek J. The early history of ferroelectricity. Ferroelectrics. 1971;**2**: 239-244

[107] Scott WT. Erwin Schrodinger. Amherst: Univ of Massachusetts Press; 1967

[108] Busch G. Early history of ferroelectricity. Ferroelectrics. 1987;**74**: 267-284

[109] Kingery WD, Lense E. High technology ceramics: Past, present, and future: The nature of innovation and change in ceramic technology. American Ceramic Society. 1986

[110] Von Hippel A, Breckenridge RG, Chesley F, Tisza L. High dielectric constant ceramics. Industrial & Engineering Chemistry. 1946;**38**: 1097-1109

[111] Lines ME, Glass AM. Principles and Applications of Ferroelectrics and Related Materials. Oxford, England: Oxford University Press; 1977

[112] Venevtsev IN, Zhdanov G. X-ray analysis of ferroelectric soild solutions with perovskite structure, Bulletin of the Academy of Science of the USSR. Physics Series. 1956;**20**:161-167

[113] Randall C, Bhalla A, Shrout T, Cross L. Classification and consequences of complex lead perovskite ferroelectrics with regard to B-site cation order. Journal of Materials Research. 1990;**5**:829-834

[114] Kittel C. Introduction to Solid State Physics. New Jersey: Wiley; 2005

[115] Callister WD, Rethwisch DG. Fundamentals of Materials Science and Engineering: An Integrated Approach. Hoboken, New Jersey: John Wiley & Sons; 2012

[116] Callister WD, Rethwisch DG. Materials Science and Engineering: An Introduction. New York: Wiley; 2007

[117] Mercier JP, Zambelli G, Kurz W. Introduction to Materials Science. Amsterdam: Elsevier; 2012

[118] Katzir S. The discovery of the piezoelectric effect. Archive for History of Exact Sciences. 2003;**57**:61-91

[119] Fukada E, Yasuda I. On the piezoelectric effect of bone. Journal of the Physical Society of Japan. 1957;**12**: 1158-1162

[120] Brockmann T. Piezoelectric materials. In: Theory of Adaptive Fiber Composites. Springer; 2009. pp. 41–67

[121] Sun FP, Chaudhry Z, Liang C, Rogers C. Truss structure integrity identification using PZT sensor-actuator. Journal of Intelligent Material Systems and Structures. 1995;**6**:134-139

[122] Law H, Rossiter P, Simon G, Koss L. Characterization of mechanical vibration damping by piezoelectric materials. Journal of Sound and Vibration. 1996; **197**:489-513

[123] Korneta OB. A systematic study of transport, magnetic and thermal properties of layered iridates. University of Kentucky; 2012

[124] Spaldin NA, Cheong S-W, Ramesh R. Multiferroics: Past, present, and future. Physics Today. 2010;**63**: 38-43

[125] Scott J. Data storage: Multiferroic memories. Nature Materials. 2007;**6**: 256-257

[126] Ramesh R, Spaldin NA. Multiferroics: Progress and prospects in thin films. Nature Materials. 2007;**6**: 21-29

[127] Senff D, Link P, Hradil K, Hiess A, Regnault L, Sidis Y, et al. Magnetic excitations in multiferroic TbMnO 3: Evidence for a hybridized soft mode. Physical Review Letters. 2007;**98**:137206

[128] Wang F, Li C-H, Zou T, Liu Y, Sun Y. Electrically driven magnetic relaxation in multiferroic LuFe2O4. Journal of Physics: Condensed Matter. 2010;**22**:496001

[129] Kézsmárki I, Kida N, Murakawa H, Bordács S, Onose Y, Tokura Y. Enhanced directional dichroism of terahertz light in resonance with magnetic excitations of the multiferroic Ba 2 CoGe 2 O 7 oxide compound. Physical Review Letters. 2011;**106**:057403

[130] Ritter C, Balaev A, Vorotynov A, Petrakovskii G, Velikanov D, Temerov V, et al. Magnetic structure, magnetic interactions and metamagnetism in terbium iron borate TbFe3 (BO3) 4: A neutron diffraction and magnetization study. Journal of Physics: Condensed Matter. 2007;**19**: 196227

[131] Singh K, Maignan A, Simon C, Martin C. FeCr2O4 and CoCr2O4 spinels: Multiferroicity in the collinear magnetic state? Applied Physics Letters. 2011;**99**:172903

[132] Zvezdin AK, Vorob'ev GP, Kadomtseva AM, Popov YF, Pyatakov AP, Bezmaternykh LN, et al. Magnetoelectric and magnetoelastic interactions in NdFe3 (BO3) 4 multiferroics. JETP Letters. 2006;**83**: 509-514

[133] Ederer C, Spaldin NA. Influence of strain and oxygen vacancies on the

magnetoelectric properties of multiferroic bismuth ferrite. Physical Review B. 2005;**71**:224103

[134] Eerenstein W, Mathur N, Scott JF. Multiferroic and magnetoelectric materials. Nature. 2006;**442**:759-765

[135] Liu H, Cao B, O'Connor C. Intrinsic magnetism in BaTiO3 with magnetic transition element dopants (Co, Cr, Fe) synthesized by sol-precipitation method. Journal of Applied Physics. 2011;**109**: 07B516

[136] Chan NH, Sharma R, Smyth DM. Nonstoichiometry in acceptor-doped BaTiO3. Journal of the American Ceramic Society. 1982;**65**:167-170

[137] Lemmens P, Millet P. Spin—orbit—topology, a triptych. In: Quantum Magnetism. Vol. 1. Springer; 2004. pp. 433-477

[138] Schmid H. Multi-ferroic magnetoelectrics. Ferroelectrics. 1994; **162**:317-338

[139] Johnson RD, Radaelli PG. Diffraction studies of multiferroics. Annual Review of Materials Research. 2014;**44**:269-298

[140] Merz WJ. The electric and optical behavior of BaTi O 3 single-domain crystals. Physical Review. 1949; **76**:1221

[141] Lawless W, DeVries R. Accurate determination of the ordinary-ray refractive index in BaTiO3. Journal of Applied Physics. 1964;**35**:2638-2639

[142] Noland JA. Optical absorption of single-crystal strontium titanate. Physical Review. 1954;**94**:724

[143] Linz A Jr, Herrington K. Electrical and optical properties of synthetic calcium titanate crystal. The Journal of Chemical Physics. 1958;**28**:824-825

[144] Geusic J, Marcos H, Van Uitert L. Laser oscillations in Nd-doped yttrium aluminum, yttrium gallium and gadolinium garnets. Applied Physics Letters. 1964;**4**:182-184

[145] Zhang W, Tang J, Ye J. Photoluminescence and photocatalytic properties of SrSnO 3 perovskite. Chemical Physics Letters. 2006;**418**: 174-178

[146] Sun D, Li D, Zhu Z, Xiao J, Tao Z, Liu W. Photoluminescence properties of europium and titanium co-doped BaZrO 3 phosphors powders synthesized by the solid-state reaction method. Optical Materials. 2012;**34**:1890-1896

[147] Zhang H, Fu X, Niu S, Xin Q. Synthesis and photoluminescence properties of Eu 3+−doped AZrO 3 (A= Ca, Sr, Ba) perovskite. Journal of Alloys and Compounds. 2008;**459**: 103-106

[148] Pan Y, Su Q, Xu H, Chen T, Ge W, Yang C, et al. Synthesis and red luminescence of Pr 3+−doped CaTiO 3 nanophosphor from polymer precursor. Journal of Solid State Chemistry. 2003; **174**:69-73

[149] Lu Z, Chen L, Tang Y, Li Y. Preparation and luminescence properties of Eu 3+−doped MSnO 3 (M= Ca, Sr and Ba) perovskite materials. Journal of Alloys and Compounds. 2005;**387**:L1-L4

[150] Cockroft NJ, Lee SH, Wright JC. Site-selective spectroscopy of defect chemistry in SrTiO 3, Sr 2 TiO 4, and Sr 3 Ti 2 O 7. Physical Review B. 1991;**44**:4117

[151] Jia W, Xu W, Rivera I, Pérez A, Fernández F. Effects of compositional phase transitions on luminescence of Sr

1− x Ca x TiO 3: Pr 3+. Solid State Communications. 2003;**126**:153-157

[152] Hwang HY. Perovskites: Oxygen vacancies shine blue. Nature Materials. 2005;**4**:803-804

[153] Retot H, Bessiere A, Kahn-Harari A, Viana B. Synthesis and optical characterization of SrHfO 3: Ce and SrZrO 3: Ce nanoparticles. Optical Materials. 2008;**30**:1109-1114

[154] Pinel E, Boutinaud P, Mahiou R. What makes the luminescence of Pr 3+ different in CaTiO 3 and CaZrO 3? Journal of Alloys and Compounds. 2004; **380**:225-229

[155] Li XX, Wang YH, Chen Z. Photoluminescence properties of GdBa 3 B 9 O 18: Ln 3+(Ln= Eu, Tb) under UV and VUV excitation. Journal of Alloys and Compounds. 2008;**453**:392-394

[156] Wang Y, Sun Y, Zhang J, Ci Z, Zhang Z, Wang L. New red Y 0.85 Bi 0.1 Eu 0.05 V 1− yMyO 4 (M= Nb, P) phosphors for light-emitting diodes. Physica B: Condensed Matter. 2008;**403**: 2071-2075

[157] Neeraj S, Kijima N, Cheetham A. Novel red phosphors for solid-state lighting: The system NaM (WO 4) 2− x (MoO 4) x: Eu 3+(M· Gd, Y, Bi). Chemical Physics Letters. 2004;**387**:2-6

[158] Dhahri K, Bejar M, Dhahri E, Soares M, Graça M, Sousa M, et al. Blue-green photoluminescence in BaZrO 3− δ powders. Chemical Physics Letters. 2014;**610**:341-344

[159] Yuan Y, Zhang X, Liu L, Jiang X, Lv J, Li Z, et al. Synthesis and photocatalytic characterization of a new photocatalyst BaZrO 3. International Journal of Hydrogen Energy. 2008;**33**: 5941-5946

[160] Yamamoto H, Okamoto S. Efficiency enhancement by aluminum addition to some oxide phosphors for field emission displays. Displays. 2000; **21**:93-98

[161] Borja-Urby R, Diaz-Torres L, Salas P, Vega-Gonzalez M, Angeles-Chavez C. Blue and red emission in wide band gap BaZrO 3: Yb 3+, Tm 3+. Materials Science and Engineering: B. 2010;**174**:169-173

[162] Rao CNR, Raveau B. Colossal Magnetoresistance, Charge Ordering and Related Properties of Manganese Oxides. Singapore: World Scientific; 1998

[163] Jonker G, Van Santen J. Ferromagnetic compounds of manganese with perovskite structure. Physica. 1950;**16**:337-349

[164] Kim J-H, Khartsev S, Grishin AM. Epitaxial colossal magnetoresistive La-0.67 (Sr, Ca)(0.33) MnO3 films on Si. Applied Physics Letters. 2003;**82**: 4295-4297

[165] Hsu ST, Li T, Zhang F, Pan W, Zhuang W-W, Evans DR, et al. Buffered-layer memory cell, in. Google Patents. 2006

[166] Lu H, Yang G, Chen Z, Dai S, Zhou Y, Jin K, et al. Positive colossal magnetoresistance in a multilayer p–n heterostructure of Sr-doped LaMnO3 and Nb-doped SrTiO3. Work. 2004;**15**:16

[167] Garg K, Nordblad P, Heinonen M, Panwar N, Sen V, Bondino F, et al. Study of Sb substitution for Pr in the Pr 0.67 Ba 0.33 MnO 3 system. Journal of Magnetism and Magnetic Materials. 2009;**321**:305-311

[168] Ramirez A. In: Buschow KHJ, editor. Handbook of Magnetic Materials.

Vol. 13. Amsterdam: Elsevier; 2001. p. 423

[169] Narlikar AV. Frontiers in Magnetic Materials. Germany: Springer; 2005

[170] Feiner LF, Oleś AM, Zaanen J. Quantum melting of magnetic order due to orbital fluctuations. Physical Review Letters. 1997;**78**:2799

[171] Zener C. Interaction between the d-shells in the transition metals. II. Ferromagnetic compounds of manganese with perovskite structure. Physical Review. 1951;**82**:403

[172] Li X, Zheng R, Li G, Zhou H, Huang R, Xie J, et al. Jahn-Teller effect and stability of the charge-ordered state in La 1− x Ca x MnO3 (0.5⩽ x⩽ 0.9) manganites. EPL (Europhysics Letters). 2002;**60**:670

[173] Yunoki S, Moreo A, Dagotto E. Phase separation induced by orbital degrees of freedom in models for manganites with Jahn-Teller phonons. Physical Review Letters. 1998;**81**:5612

[174] Tokura Y, Tomioka Y. Colossal magnetoresistive manganites. Journal of Magnetism and Magnetic Materials. 1999;**200**:1-23

[175] Tokura Y. Critical features of colossal magnetoresistive manganites. Reports on Progress in Physics. 2006;**69**: 797

[176] Pena M, Fierro J. Chemical structures and performance of perovskite oxides. Chemical Reviews. 2001;**101**:1981-2018

[177] Lee JY, Lee JH, Hong SH, Lee YK, Choi JY. Coating BaTiO3 nanolayers on spherical Ni powders for multilayer ceramic capacitors. Advanced Materials. 2003;**15**:1655-1658

[178] Lewis G, Catlow C, Casselton R. PTCR effect in BaTiO3. Journal of the American Ceramic Society. 1985;**68**: 555-558

[179] Yuk J, Troczynski T. Sol–gel BaTiO3 thin film for humidity sensors. Sensors and Actuators B: Chemical. 2003;**94**: 290-293

[180] Zhang S, Li F, Jiang X, Kim J, Luo J, Geng X. Advantages and challenges of relaxor-PbTiO3 ferroelectric crystals for electroacoustic transducers–A review. Progress in Materials Science. 2015;**68**: 1-66

[181] Chu S-Y, Chen T-Y, Water W. The investigation of preferred orientation growth of ZnO films on the PbTiO3-based ceramics and its application for SAW devices. Journal of Crystal Growth. 2003;**257**:280-285

[182] Feteira A, Sinclair DC, Reaney IM, Somiya Y, Lanagan MT. BaTiO3-based ceramics for tunable microwave applications. Journal of the American Ceramic Society. 2004;**87**:1082-1087

[183] Kingon A. Device physics: Memories are made of.... Nature. 1999;**401**:658

[184] Handoko AD, Goh GK. One-dimensional perovskite nanostructures. Science of Advanced Materials. 2010;**2**: 16-34

[185] Megriche A, Lebrun L, Troccaz M. Materials of Bi4Ti3O12 type for high temperature acoustic piezo-sensors. Sensors and Actuators A: Physical. 1999; **78**:88-91

[186] Funakubo H, Watanabe T, Kojima T, Sakai T, Noguchi Y, Miyayama M, et al. Property design of Bi4Ti3O12-based thin films using a site-engineered concept. Journal of Crystal Growth. 2003;**248**:180-185

[187] Medvedev D, Lyagaeva J, Plaksin S, Demin A, Tsiakaras P. Sulfur and carbon tolerance of BaCeO3–BaZrO3 proton-conducting materials. Journal of Power Sources. 2015;**273**:716-723

[188] Pereñíguez R, González-DelaCruz VM, Holgado JP, Caballero A. Synthesis and characterization of a LaNiO3 perovskite as precursor for methane reforming reactions catalysts. Applied Catalysis B: Environmental. 2010;**93**: 346-353

[189] de Jong MP, Dediu V, Taliani C, Salaneck WR. Electronic structure of La 0.7 Sr 0.3 MnO 3 thin films for hybrid organic/inorganic spintronics applications. Journal of Applied Physics. 2003;**94**:7292-7296

[190] Zhou W, Xiong Y, Zhang Z, Wang D, Tan W, Cao Q, et al. Multilevel resistance switching memory in La2/3Ba1/3MnO3/0.7 Pb (Mg1/3Nb2/3) O3–0.3 PbTiO3 (011) heterostructure by combined straintronics-spintronics. ACS Applied Materials & Interfaces. 2016;**8**: 5424-5431

[191] Wong J, Zhang L, Kakarantzas G, Townsend P, Chandler P, Boatner L. Ion-implanted optical waveguides in KTaO3. Journal of Applied Physics. 1992;**71**:49-52

[192] Rana DS, Kawayama I, Mavani K, Takahashi K, Murakami H, Tonouchi M. Understanding the nature of ultrafast polarization dynamics of ferroelectric memory in the multiferroic BiFeO3. Advanced Materials. 2009;**21**: 2881-2885

[193] Zvezdin A, Logginov A, Meshkov G, Pyatakov A. Multiferroics: Promising materials for microelectronics, spintronics, and sensor technique. Bulletin of the Russian Academy of Sciences: Physics. 2007;**71**: 1561-1562

[194] Shen S-T, Weng H-S. Comparative study of catalytic reduction of nitric oxide with carbon monoxide over the La1-x Sr x BO3 (B= Mn, Fe, Co, Ni) catalysts. Industrial & Engineering Chemistry Research. 1998;**37**:2654-2661

[195] Khine MSS, Chen L, Zhang S, Lin J, Jiang SP. Syngas production by catalytic partial oxidation of methane over (La0. 7A0. 3) BO3 (A= Ba, Ca, Mg, Sr, and B= Cr or Fe) perovskite oxides for portable fuel cell applications. International Journal of Hydrogen Energy. 2013;**38**: 13300-13308

[196] Birol H, Damjanovic D, Setter N. Preparation and characterization of (K0. 5Na0. 5) NbO3 ceramics. Journal of the European Ceramic Society. 2006;**26**: 861-866

[197] Zvirgzds J, Kapostin P, Zvirgzde J, Kruzina T. X-ray study of phase transitions in efrroelectric Na0. 5Bi0. 5TiO3. Ferroelectrics. 1982;**40**:75-77

[198] Karbowiak M, Gnutek P, Rudowicz C. Crystal-field analysis for RE3+ ions in laser materials: III. Energy levels for Nd3+ and Er3+ ions in LaAlO3, YAlO3, and LaGaO3 single crystals–Combined approach to low symmetry crystal field parameters. Chemical Physics. 2012;**400**:29-38

[199] Diehl R, Brandt G. Crystal structure refinement of YAlO3, a promising laser material. Materials Research Bulletin. 1975;**10**:85-90

Chapter 2

Study of the Critical Behavior in $La_{0.67}Ca_{0.18}Sr_{0.15}Mn_{0.98}Ni_{0.02}O_3$ Manganite Oxide

Kawther Laajimi, Mohamed Hichem Gazzah and Jemai Dhahri

Abstract

In order to study the critical behavior of $La_{0.67}Ca_{0.18}Sr_{0.15}Mn_{0.98}Ni_{0.02}O_3$ near room temperature, magnetization measurements were performed. It can be seen from the findings that the specimens show a second order phase transition. Given the relative slope, the 3D-Ising model was deduced as the most suitable model. It was found that the evaluated critical exponents were $\beta = 0.320$, $\gamma = 1.296$ and $\delta = 4.965$ at $T_C = 319$ K. These verify the Broadom's scaling equation $\delta = 1 + \gamma/\beta$, which demonstrates that our values have been proved valid. Near T_C magnetization-strength-temperature, (M-μ_0H-T) results fell in two curves according to critical exponents, obeying the single scaling equation $M(\mu_0H, \varepsilon) = \varepsilon^{\beta} f \pm \left(\mu_0H/\varepsilon^{\beta+\gamma}\right)$ with $\varepsilon = (T - T_c)/T_c$ as the reduced temperature.

Keywords: magnetic materials, critical behavior, spontaneous magnetization, 3D Ising model, Kouvel-Fisher method

1. Introduction

Extensive research on $Ln_{1-x}A_x MnO_3$ (Ln = La; A = Ca, Sr, Ba, Pb, etc.) hole-doped manganite has attracted significant attention, for its specific transport and magnetic properties and its potential for being used over the years for future technological purposes, which are based on the colossal magnetoresistance effect (CMR) [1, 2] and magnetocaloric behavior [3, 4]. For a more detailed understanding of the relationship that exists between the insulator-to-metal transition and the CMR effect, it is necessary to answer two fundamental issues concerning the ferromagnetic (FM)-paramagnetic (PM) phase transition temperature: firstly, the order of the phase transition and, secondly, the common universality class. The detailed study of the critical exponents around the FM-PM transition is essential to address this point [5, 6]. In this way, it can be considered that this transition can be explained through the double exchange phenomenon (DE), and by the percolation process, phase separation [7], electron–phonon pairing [8], the effect of quenched disorder [9] and the observation of the Griffiths phase (GP) [9]. Initially, on the DE model, the description of the critical behavior relied on a long-range mean field theory [10, 11]. Later, the work of Motome-Furulawa [12, 13] which suggested a short-range Heisenberg model for critical behavior comprising only nearest-neighbor exchange, was carried out by taking into account the existence of a short-range interaction at the level of localized spins. In addition, various pertinent experimental studies on critical phenomena confirm that view also the resulting critical exponent data. In accordance with the one obtained by conventional ferromagnetic Heisenberg model. Taking the data from the DC magnetic study, researchers Ghosh et al. [14] have pointed out that critical exponent β is found to be 0.37 at $La_{0.7}Sr_{0.3}MnO_3$ ferromagnetic manganite. (β = 0.365 for the Heisenberg model, in which case the critical exponent β was in close contact to the temperature dependency over the spontaneous magnetization under the Curie temperature T_C) as well as the critical case value of β = 0.374 has been also referred in the ferromagnet DE $Nd_{0.6}Pb_{0.4}MnO_3$ [15]. Nevertheless, a rather elevated value that was β = 0.5 achieved in $La_{0.8}Sr_{0.2}MnO_3$ polycrystals was consistent to the one found in the mean field pattern [16]. In contrast, the small value of the critical exponent in terms of β = 0.14 that was found in $La_{0.7}Ca_{0.3}MnO_3$ monocrystal indicates a first-order rather than second-order PM-FM transition occurring in this system [17]. Between, there was finding of moderate critical value of β = 0.25 in the polycrystalline $La_{0.6}Ca_{0.4}MnO_3$ was in a fair agreement with the values of the tricritical point [18, 19]. Thus, various critical exponents have been seen β ranging between 0.1 and 0.5, now, there are four different types of theoretical patterns, mean-field model (β = 0.5), three-dimensional (3D) Heisenberg (β = 0.365), 3D-Ising (β = 0.325), as well as tricritical mean-field (β = 0.25), which have been employed in order to have some explanation of the critical characteristics provided by manganites. Given the discrepancy between the reported critical values, it is necessary to examine the critical behavior of similar manganite perovskites.

In this chapter, we concentrate on a more detailed evaluation of the critical exponents α, β and γ as well as the Curie temperature T_C for $La_{0.67}Ca_{0.18}Sr_{0.15}Mn_{0.98}Ni_{0.02}O_3$ compound near the FM-PM phase transition temperature by carrying out analyses using different techniques.

2. Experimental details

The polycrystalline sample $La_{0.67}Ca_{0.18}Sr_{0.15}Mn_{0.98}Ni_{0.02}O_3$ was prepared using the sol–gel method. The sol–gel process is a very well known and proven method to generate homogeneous particles having a great quality and thin material. We have produced the polycrystalline $La_{0.67}Ca_{0.18}Sr_{0.15}Mn_{0.98}Ni_{0.02}O_3$ through sol–gel process. During this method, the basic reagents, $C_6H_9La\ O_6$, H_2O, $C_4H_6Ca\ O_{4,2}O$, $C_4H_6MnO_4$, $4H_2O$ and $NiCl_26H_2O$ have been weighed using stoichiometric ratios which were dissolved in distilled water under continued stirring. By adding an excess of citric acid, a sol was obtained by slow evaporation from the addition of ethylene glycol, which rendered the solution complex (with a molar ratio of 1:2:2 with regard to the cations: citric acid:ethylene glycol). Subsequently, every reagents have been diluted, and the mixture has been heated up on a hot plate, which gives a gel form. This gel was then dried and calcined for 5 h at 600°C. After, resultant powders have been annealed at 1000°C during 24 h and compressed to form circular pellets. The resulting pellets have been then sintered in air through 1000°C for approximately 10 h. Using X-ray diffraction, the crystalline structure of the studied powder was verified, confirming further that the specimen actually crystallizes into the rhombohedral structure having $R\bar{3}c$ space group, this was achieved through the use of X-ray diffraction (Siemens D5000 X-ray diffractometer, using monochromatic Cu Kα radiation λ = 1.5406 A°).

We also measured the magnetic isotherms of the sample over a 0–5 T range, approximately the PM-FM, phase transition, through a vibrating sample magnetometer designed in the laboratory of Louis Neel in Grenoble, for precise extraction of ceramic specimen's critical exponent. Isotherm value was adjusted by means of a demagnetization coefficient D that was calculated according to a standardized procedure based on measurement of low field DC magnetization under low temperature ($H = H_{app} - DM$).

3. Scaling analysis

The following power law relation was cited by the scaling hypothesis near the critical region defined by:

$$M_s(T) = M_0(-\varepsilon)^{\beta}, \varepsilon < 0 \tag{1}$$

$$\chi_0^{-1}(T) = \left(\frac{h_0}{M_0}\right)\varepsilon^{\gamma}, \varepsilon > 0 \tag{2}$$

$$M = DH^{\frac{1}{\delta}}, \varepsilon = 0 \tag{3}$$

In which M_0, h_0 and D represent the critical amplitudes while ε stands for the reduced temperature $\varepsilon = (T - T_c)/T_c$. Besides, within the asymptotic critical zone, line with the prediction of the scale formula, the equation of magnetic state is written in the following form:

$$M(\mu_0 H, \varepsilon) = \varepsilon^{\beta} \int \pm \left(\frac{\mu_0 H}{\varepsilon^{\beta+\gamma}}\right) \tag{4}$$

In this case, $\int_-$ and $\int_+$ represent regular analytical functions and above T_c [20, 21]. This last Eq. (4) shows that for the correct choice of the values of β, γ and δ as well as true scaling relations, the scaled $M/|\varepsilon|^{\beta}$ plotted as a function of the scaled $\mu_0 H/|\varepsilon|^{\beta+\gamma}$ will fall in two universal curves for $T > T_c (\varepsilon > 0)$ and the other for $T < T_c$ $(\varepsilon < 0)$ [22]. This makes possible to have a fairly important critical regime criterion.

Composition	Techniques	$T_c(K)$	β	γ	δ	Ref.
$La_{0.67}Ca_{0.33}Mn_{0.98}Ni_{0.02}O_3$	MAP	318.300	0.320	1.269	4.965	This work
	KF	317.900	0.324	1.238	4.820	[23]
Mean field model	Theory		0.5	1	3	[23]
3D-Ising model	Theory		0.325	1.24	4.82	[23]
Tricritical mean field model	Theory		0.25	1	5	[23]
3D-Heisenberg model	Theory		0.365	1.336	4.8	[24]
$La_{0.7}Ca_{0.15}Sr_{0.15}MnO_3$		335	0.32	1.21	4.36	[25]
$Nd_{0.7}Ca_{0.05}Sr_{0.25}Mn_{0.98}Ga_{0.02}O_3$		247	0.308	1.197	4.91	[26]

Table 1.
Critical exponents of the $La_{0.67}Ca_{0.18}Sr_{0.15}Mn_{0.98}Ni_{0.02}O_3$ specimen compared with different theoretical models and previous manganite results.

More generally, it is possible to point to four models that are based on the critical exponent values; Mean-field model (β = 0.5 and γ = 1), 3D-Heisenberg model (β = 0.365 and γ = 1.336), 3D-Ising model (β = 0.325 and γ = 1.24), in addition to Tricritical mean-field model (β = 0.25 and γ = 1). The different models can be seen below in **Table 1**.

4. Critical behavior

More generally, it is necessary to understand the way in which the ferromagnetic-paramagnetic ("FM-PM") phase transition occurs (either first or second order), according to scaling postulate. The magnetic system whose phase transition behavior of second order in the vicinity of the Curie temperature spot is driven by an ensemble of interdependent critical exponents [19] β (linked to the spontaneous magnetization M_S), γ (associated with the starting magnetic susceptibility χ_0), δ (linked to the critical magnetization isotherm to T_c). As already known, it is impossible to find the critical exponents in case of a first-order ferromagnetic phase transition; in fact, the external magnetic field allows this transition to be shifted, leading to a phase depending on the field strength $T_c(H)$ [27]. Magnetization measurement exponents are mathematically defined and given us by the following relationships:

$$M_s(T) = M_0|\varepsilon|^{\beta}\ , \qquad \varepsilon < 0, \qquad T < T_c \tag{5}$$

$$\chi_0^{-1}(T) = \left(\frac{h_0}{M_0}\right)\varepsilon^{\gamma}, \qquad \varepsilon > 0, \qquad T > T_c \tag{6}$$

$$M = DH^{1/\delta}, \qquad \varepsilon = 0, \qquad T = T_c \tag{7}$$

Where ? represents the reduced temperature $(T - T_c)/T_c$, and M_0, h_0/M_0, and D, correspond to the critical amplitudes.

In general, both the critical exponents and the critical temperature may be found easily from the Arott curve. From the Arott-Noakes state equation, $(H/M)^{1/\gamma} = (T - T_c)/T_c + (M/M_1)^{1/\beta}$ [28], where M_1 represents a material constant, the Normal Arrott has demonstrated that the relationship between M^2 as a function of H/M was essentially a function of the average field model in terms of the critical exponent β = 0.5 and γ = 1.0. As a result, the M^2 versus H/M curves must exhibit linear behavior around T_c as well as the line at $T = T_c$ must exactly get through the origin. In addition, one can determine the magnetic transition order using the slope of those lines, following Banerjee's criterion [29]. From these straight lines, we can identify the magnetic transition order. Given that as a result, there is a positive slope associated with the second-order transition, whereas the negative slope is related to the first-order transition.

Figure 1 displays the Arrott plot M^2 versus H/M for the $La_{0.67}Ca_{0.18}Sr_{0.15}Mn_{0.98}Ni_{0.02}O_3$ sample at the temperature range close to T_c. Clearly, currently, the positive slope of curves M^2 versus H/M curves allows to the conclusion of a second order ferromagnetic phase transition. However, the nonlinearity and the appearance of increasing curvature of all Arrott traces, even at high fields, which means that the average field theory is not, satisfied the current phase transition.

Most of the time, charge effect, differential orbital degrees of freedom and lattice, which exists in high field regions, are removed in a ferromagnetic so that the ordering

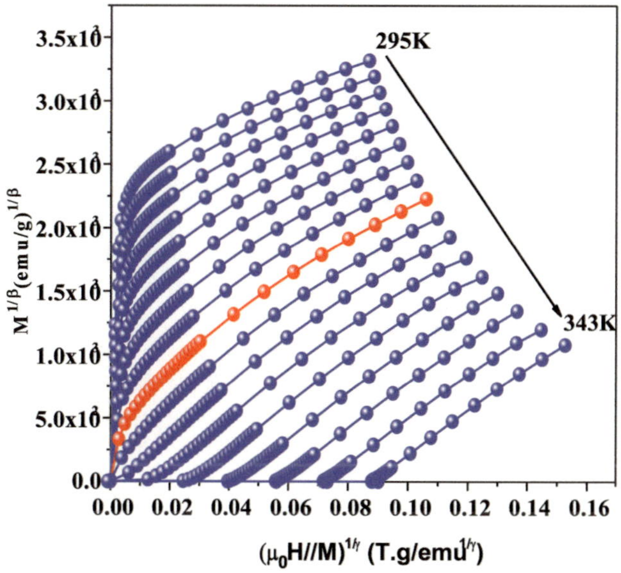

Figure 1.
Typical pattern of M^2 *versus* μ_0H/M *traces, for* $La_{0.67}Ca_{0.18}Sr_{0.15}Mn_{0.98}Ni_{0.02}O_3$ *compound.*

parameter is usually confused with macroscopic magnetization. Thus, in order to achieve the correct values β and γ, a modified Arrott trace requires producing quasi-linear lines of the M^2 versus H/M plots. As illustrated by the **Figure 2(a)–(c)**, it was found that three types of test exponents of tricritical mean field (β = 0.25, γ = 1.0), 3D Ising (β = 0.325, γ = 1.24) as well as 3D Heisenberg model (β = 0.365, γ = 1.336) were employed in order to create a modified Arrott plot.

To match these results, the related slopes (RS) were calculated, given by $RS = S(T)/S(T_c = 319\,K)$. Thus, the relative slopes must be kept at 1 apart from the temperatures, whether the modified Arrott graph displays a set which are absolute parallels.

As already seen in **Figure 3**, the mean field RS, the 3D-Heisenberg and the Tricritical mean filed certainly deviate from the RS = 1 straight line; however the RS of the 3D-Ising model approximated it. As a result, Arrott's third plot shows the best result, amongst these three patterns, reporting that the critical properties of the $La_{0.67}Ca_{0.18}Sr_{0.15}Mn_{0.98}Ni_{0.02}O_3$ specimen can be depicted using the 3D-Ising model.

Subsequently, in the **Figure 2(b)**, the linear extrapolation from the high field region two interceptions with axes $(H/M)^{1/\gamma}$ and $(M)^{1/\beta}$ gives credible values of inverse susceptibility $\chi_0^{-1}(T,0)$ and spontaneous magnetization $M_s(T,0)$, respectively. These temperature dependent values, $\chi_0^{-1}(T,0)$ as a function of T and $M_s(T,0)$ versus T, are displayed in **Figure 4**. As already mentioned by Eqs (5) and (6), based on the experimental results (open line) may be adjusted to two solid graphs (continuous line). It allows to give two novel cases in β = 0.320 with $T_c = 318.300\,K$ and γ = 1.296 with $T_c = 317.819\,K$. Therefore, these findings are very similar to the 3D-Ising model.

Susceptible, these critical exponents and T_c may be more accurately determined by the Kouvel-Fisher (KF) approach [20]:

$$\frac{M_s(T)}{dM_s(T)/dT} = \frac{T - T_c}{\beta} \quad (8)$$

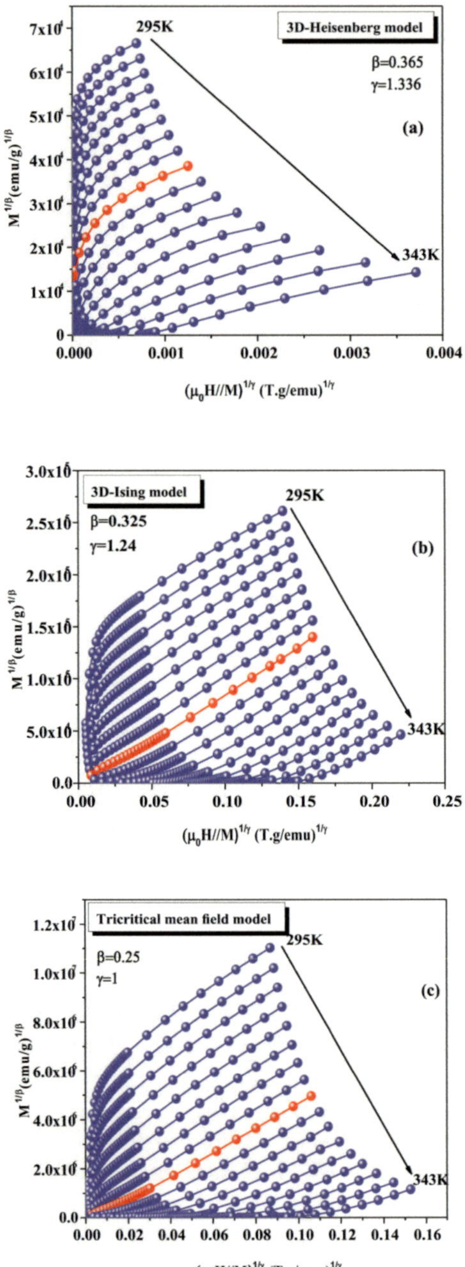

Figure 2.
Modified Arrott graphs: ($M^{1/\beta}$ *versus* $(\mu_0 H/M)^{1/\gamma}$) *over compound* $La_{0.67}Ca_{0.18}Sr_{0.15}Mn_{0.98}Ni_{0.02}O_3$.

$$\frac{\chi_0^{-1}(T)}{\mathrm{d}\chi_0^{-1}(T)/\mathrm{d}T} = \frac{T - T_{\mathrm{c}}}{\gamma} \tag{9}$$

By conforming to this method, $M_s(\mathrm{d}M_s/\mathrm{d}T)^{-1}$ as a function of T with $\chi_0^{-1}(\mathrm{d}\chi_0^{-1}/\mathrm{d}T)^{-1}$ as a function of T is expected to give straight lines of slopes $1/\beta$ and

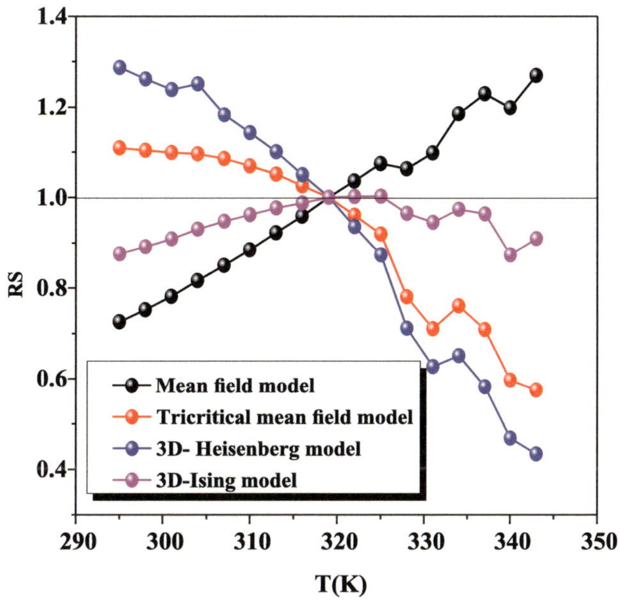

Figure 3.
Variation of RS with temperature of specimen $La_{0.67}Ca_{0.18}Sr_{0.15}Mn_{0.98}Ni_{0.02}O_3$.

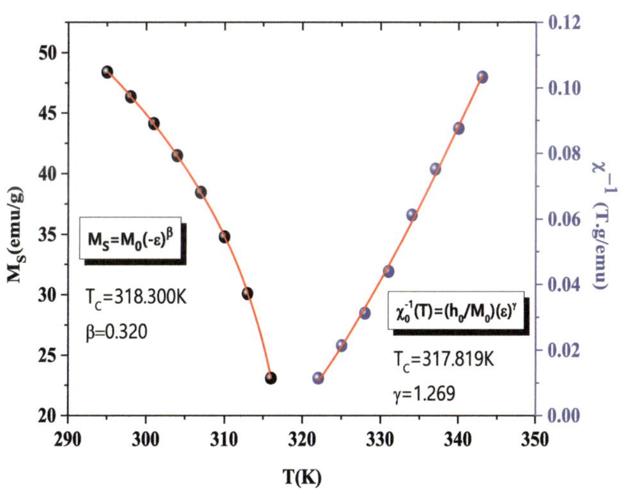

Figure 4.
M_S (left) with χ_0^{-1} (right) versus temperature for $La_{0.67}Ca_{0.18}Sr_{0.15}Mn_{0.98}Ni_{0.02}O_3$ specimen (solid lines are model fits).

$1/\gamma$, respectively. T-axis intercepts fit correctly to T_c if those lines are extrapolated to the zero ordinate. As shown by **Figure 5**, adjustment results using KF method yield the exponents as well as T_c to the deviation: $\beta = 0.324$ with $T_c = 319.900\ K$ and $\gamma = 1.238$ with $T_c = 317.120\ K$.

Certainly, using the KF method, the resulting critical exponent values and T_c and the values obtained by using the modified Arrott of tricritical mean-field model are in agreement. For a better verification on the dependability as regards the above critical exponents, we may consider how the three critical exponents β, γ, and δ relate to each other.

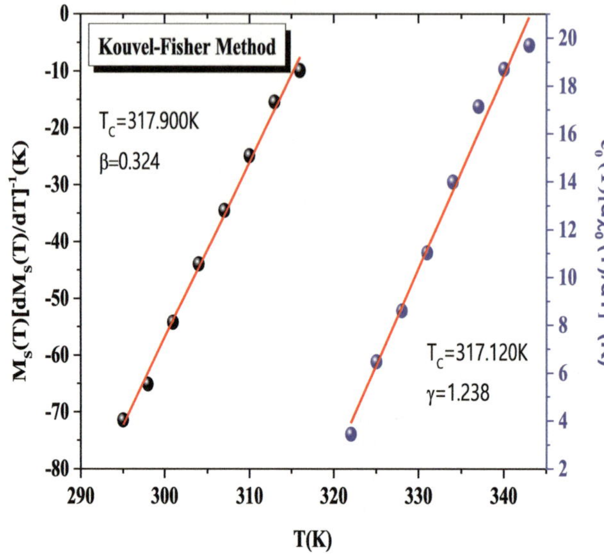

Figure 5.
K–F graphs of $La_{0.67}Ca_{0.18}Sr_{0.15}Mn_{0.98}Ni_{0.02}O_3$ specimen (solid lines represent model fits).

In this place, first we should find out δ value. By the terms of Eq. (7), we can directly get δ value by tracing the critical isotherm at T_c. According to **Figure 6**, we distinguished M versus H plot at 319 K as being critical isothermal from the above discussion. Moreover, it is illustrated through the slide of **Figure 6**, which displays the graph on M versus H in the form of a log–log scale. Using Eq. (7), there is an appropriate result of the full straight edge and a slope of $1/\delta$ is obtained. Based on a straight-line fit, a third critical exponent δ = 4.863 was derived. Following statistical theory, all three of these critical exponents should complete Widom's scale relationship:

$$\delta = 1 + \frac{\gamma}{\beta} \tag{10}$$

Using the above obtained data β and γ As a result, Eq. (10) provides values of δ = 4.965 β and γ as obtained from **Figure 4**, as well as δ = 4.820 β and γ as evaluated using **Figure 5**. Respectfully, the above values are very similar to the ones calculated using the critical isotherm. As a result, the above relationship was proved with a plot of $M(T = T_0)$ as a function of $\mu_0 H^{\beta(\beta+\gamma)} = \mu_0 H^{1/\delta}$ and proving that the curve is linear as illustrated in **Figure 6**.

Given that two critical exponents δ are similar to δ estimated based on critical isotherms near T_c. As a result, the critical exponents obtained in this study absolutely obey the Widom scaling relationship, which entails the implication that both β and γ resultant data agree. In the critical region, the magnetic equation is expressed as follows:

$$M(\mu_0 H, \varepsilon) = \varepsilon^{\beta} f \pm \left(\mu_0 H / \varepsilon^{\beta+\gamma}\right) \tag{11}$$

Where f_+ for $T > T_c$ and f_- for $T < T_c$ are regular functions [30]. As indicated by the Eq. (11), $M\varepsilon^{-\beta}$ as a function of $H\varepsilon^{-(\beta+\gamma)}$ leads to two universal plots: The first plot

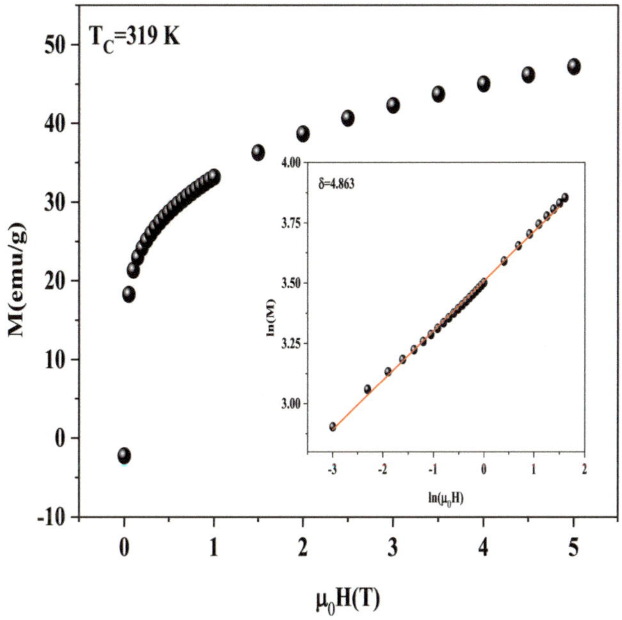

Figure 6.
M versus $\mu_o H$ graph for $La_{0.67}Ca_{0.18}Sr_{0.15}Mn_{0.98}Ni_{0.02}O_3$ specimen that occurs on $T_C = 319$ K. the inset the enclosed shows the similar graph scaled to ln–ln.

for temperature $T > T_c$ and the second plot for temperature $T < T_c$. Therefore, there is a comparison between the obtained results and the scaling theory prediction by Eq. (11). As already seen in **Figure 7**, both experimental data fall on two curves, one over T_c and one under T_c, by correspondence with the scaling theory. This same graph is shown inside the inset in **Figure 7** in the form of a log–log scale. Same, all

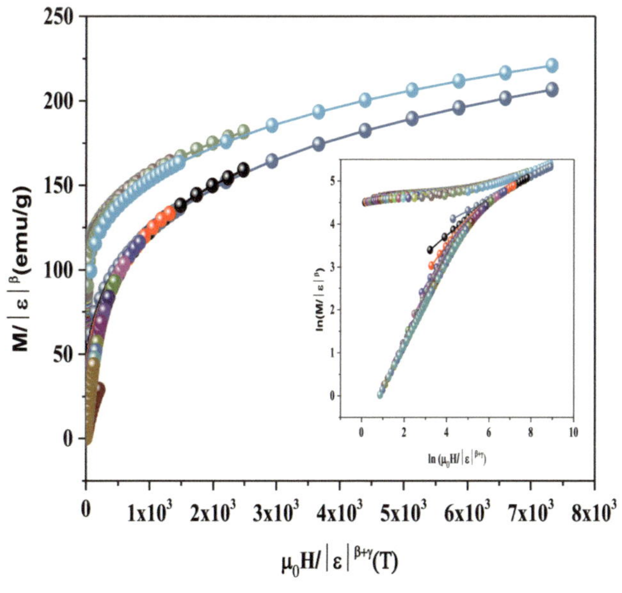

Figure 7.
Renormalized magnetization $M/|\xi|^{\beta}$ versus $\mu_o H/\varepsilon^{\beta+\gamma}$ for the $La_{0.67}Ca_{0.18}Sr_{0.15}Mn_{0.98}Ni_{0.02}O_3$ sample.

points fall in two different curves. We can conclude from these results that the obtained values of the critical exponents and those of T_c are reliable. On the other hand, there is accurate characterization of critical properties in our current system with the 3D-Ising model.

5. Conclusion

In this paper, it was extensively studied by DC magnetization the critical process towards the FM-PM state transition in $La_{0.67}Ca_{0.18}Sr_{0.15}Mn_{0.98}Ni_{0.02}O_3$ polycrystalline manganite. We have extracted the values of critical exponents of our sample by the Modified Arott plot method and the Kouvel-Fisher method. It is interesting to note that the found exponents are quite similar to those expected from the universality class of 3D-Ising model.

Acknowledgements

K. Laajimi would like to express her gratitude and thanks for its funding by the Tunisian Ministry of Higher Education and Scientific Research, which allowed her to carry out her research and provided her with all the necessary assistance.

Author details

Kawther Laajimi[1*], Mohamed Hichem Gazzah[1*] and Jemai Dhahri[2]

1 Quantum and Statistical Physics Laboratory, Faculty of Sciences of Monatsir, Monastir University, Monatsir, Tunisia

2 Laboratory of Condensed Matter Physics and Nanosciences, Faculty of Sciences of Monatsir, Monastir University, Monatsir, Tunisia

*Address all correspondence to: kawther.laajimi@gmail.com and hichem.gazzah@gmail.com

References

[1] Triki M, Dhahri E, Hlil EK. Appearance of Griffiths phase in oxygen deficient $La_{0.4}Ca_{0.6}MnO_{3-\delta}$ oxides. Materials Letters. 2012;**84**:48-51

[2] Laajimi K, Ayadi F, Kchaw M, Fourati I, Khlifi M, Gazzah MH, et al. A new range of specific perovskite-type materials with structural, magnetic and magnetocaloric properties: $La_{0.67}Ca_{0.33-x}Sr_xMn_{0.98}Fe_{0.02}O_3$ ($0.15 \leq x \leq 0.3$). Solid State Sciences. 2021;**119**: 106683

[3] Laajimi K, Khlifi M, Hlil EK, Gazzah MH, Dhahri J. Enhancement of magnetocaloric effect by Nickel substitution in $La_{0.67}Ca_{0.33}Mn_{0.98}Ni_{0.02}O_3$ manganite oxide. The Journal of Magnetism and Magnetic Materials. 2019;**491**:165625

[4] Laajimi K, Khlif M, Hlil EK, Gazzah MH, Ayed MB, Belmabrouk H, et al. Influence of Sr substitution on structural, magnetic and magnetocaloric properties in $La_{0.67}Ca_{0.33-x}Sr_xMn_{0.98}Ni_{0.02}O_3$ manganites. Journal of Materials Science: Materials in Electronics. 2020;**31**: 15322-15335

[5] Laajimi K, Khlif M, Hlil EK, Taibi K, Gazzah MH, Dhahri J. Room temperature magnetocaloric effect and critical behavior in $La_{0.67}Ca_{0.23}Sr_{0.1}Mn_{0.98}Ni_{0.02}O_3$ oxide. Journal of Materials Science: Materials in Electronics;**30**(2019):11868-11877

[6] Khlifi M, Tozri A, Bejar M, Dhahri E, Hlil EK. Effect of calcium deficiency on the critical behavior near the paramagnetic to ferromagnetic phase transition temperature in $La_{0.8}Ca_{0.2}MnO_3$ oxides. Journal of Magnetism and Magnetic Materials. 2012;**324**:2142

[7] Mori S, Chen CH, Cheong SW. Paired and Unpaired Charge Stripes in the Ferromagnetic Phase of $La_{0.5}Ca_{0.5}MnO_3$. Physical Review Letters. 1998;**81**:3972

[8] Gor'kov LP, Kresin VZ. Mixed-valence manganites: Fundamentals and main properties. Physics Reports. 2004; **400**:149

[9] Khelifi J, Tozri A, Issaoui F, Dhahri E, Hlil EK. The influence of disorder on the appearance of Griffiths phase and magnetoresistive properties in $(La_{1-x}Nd_x)_{2/3}(Ca_{1-y}Sr_y)_{1/3}MnO_3$ oxides. Ceramics International. 2014;**40**: 1641-1649

[10] Seeger M, Kaul SN, Kronmüller H, Reisser R. Asymptotic critical behavior of Ni. Physical Review B. 1995;**51**:12585

[11] Kubo K, Ohata N. A quantum theory of double exchange. Journal of the Physical Society of Japan. I. 1972;**33**:21

[12] Motome Y, Furulawa N. Critical temperature of ferromagnetic transition in three-dimensional double-exchange models. Journal of the Physical Society of Japan. 2000;**69**:3785

[13] Motome Y, Furulawa N. Critical temperature of ferromagnetic transition in three-dimensional double-exchange models. Journal of the Physical Society of Japan. 2001;**70**:1487

[14] Ghosh K, Lobb CJ, Greene RL, Karabashev SG, Shulyatev DA, Arsenov AA, et al. Critical Phenomena in the Double-Exchange Ferromagnet $La_{0.7}Sr_{0.3}MnO_3$. Physical Review Letters. 1998;**81**:4740

[15] Hafner J, Hobbs D. Understanding the complex metallic element Mn. II. Geometric frustration in β-Mn, phase

stability, and phase transitions. Physical Review B. 2003;**68**:014408

[16] Mohan CV, Seeger M, Kronmüller H, Murugaraj P, Maier J. Critical behaviour near the **ferromagnetic**-paramagnetic phase transition in $La_{0.8}Sr_{0.2}MnO_3$. Journal of Magnetism and Magnetic Materials. 1998;**183**:348

[17] Shin HS, Lee JE, Nam YS, Ju HL, Park CW. First-order-like magnetic transition in manganite oxide $La_{0.7}Ca_{0.3}MnO_3$. Solid State Communications. 2001;**118**:377

[18] Huang K. Statistical Machanics. 2nd edition. New York: Wiley; 1987

[19] Kim D, Revaz B, Zink BL, Hellman F, Rhyne JJ, Mitchell JF. Tricritical Point and the Doping Dependence of the Order of the Ferromagnetic Phase Transition of $La_{1-x}Ca_xMnO_3$. Physical Review Letters. 2002;**89**:227202

[20] Kaul SN. Static critical phenomena in ferromagnets with quenched disorder. Journal of Magnetism and Magnetic Materials. 1985;**53**:5

[21] Stanley HE. Scaling, universality, and renormalization: Three pillars of modern critical phenomena. Reviews of Modern Physics. 1999;**71**:358

[22] Khadija D, Dhahri N, Dhahri J, Taibi K, Hlil EK, Belmabrouk H, et al. Magnetic, magnetocaloric and critical behavior investigation of $La_{0.7}Ca_{0.1}Pb_{0.2}Mn_{1-x-y}Al_xSn_yO_3$ ($x, y =$ 0.0, 0.05 and 0.075) prepared by a sol-gel method. RSC Advances. 2017;**7**: 43410-43423

[23] Fisher ME. The theory of equilibrium critical phenomena. Reports on Progress in Physics. 1967;**30**:615

[24] Riahi K, Messaoui I, Ezaami A, Cugini F, Solzi M, Cheikhrouhou-Koubaa W, et al. Adiabatic temperature change, magnetic entropy change and critical behavior near the ferromagnetic-paramagnetic phase transition in $La_{0.7}(Ca,Sr)_{0.3}MnO_3$ perovskite. Phase Transitions. 2018;**91**:1474356

[25] Yua B, Suna W, Fana J, Lana X, Zhanga W, Zhua Y, et al. Scaling study of magnetic phase transition and critical behavior in $Nd_{0.55}Sr_{0.45}Mn_{0.98}Ga_{0.02}O_3$ manganite. Materials Research Bulletin. 2018;**99**:393-397

[26] Iqbal M, Khan MN, Khan AA. Structural, magnetic, magnetocaloric and critical behavior studies in the vicinity of the paramagnetic to ferromagnetic phase transition temperature in $LaMnO_{3+\delta}$ compound. Journal of Magnetism and Magnetic Materials. 2018;**465**:670-677

[27] Arrott A, Noakes JE. Approximate equation of state for nickel near its critical temperature. Physical Review Letters. 1967;**19**:786

[28] Banerjee BK. On a generalised approach to first and second order magnetic transitions. Physics Letters. 1964;**12**:16

[29] Kouvel JS, Fisher ME. Detailed magnetic behavior of nickel near its Curie point. Physics Review. 1964;**136**: A1626

[30] Fisher ME, Ma SK, Nickel BG. Critical exponents for long-range interactions. Physical Review Letters. 1972;**29**:917-920

Chapter 3

Low-Doped Regime Experiments in $LaMnO_3$ Perovskites by Simultaneous Substitution on Both La and Mn Sites

Aminta Mendoza and Octavio Guzmán

Abstract

Although extensive substitution studies exist on $LaMnO_3$ perovskites, simultaneous substitution at the La and Mn sites in the low-doping regime is not as common and provides important insights into the subtle balance of various competing effects acting on the crystal structure and the magnetic properties. This chapter presents a study of the evolution of the crystal structure and magnetic properties by simultaneous substitution of magnetic and non-magnetic ions at the perovskite A and B sites, respectively. It will examine some of the ways in which experiments on the evolution of magnetic properties provide a suitable balance scenario between the competitive effects arising from doping at each site. The work discusses the evolution of the Curie-Weiss behavior and the formation of ferromagnetic (FM) clusters above the Curie temperature, whose structure is also dependent on doping.

Keywords: perovskite manganites, electron paramagnetic resonance, magnetic clusters, Jahn-Teller distortion, local magnetism

1. Introduction

Simultaneous substitution in polycrystalline $LaMnO_3$ on the La^{3+} and Mn^{3+} ions by magnetic Dy^{3+} and non-magnetic Zn ions, respectively, offers an adequate scenario for studies on the evolution of structure and magnetic properties due to competing effects. Low-doped regime in $La_{1-x}Dy_xMn_{1-y}Zn_yO_3$ manganites—(La(Dy) Mn(Zn)O_3)—means the amount of substitution (x,y) is kept in the 0.0–0.1 regime for both sites. For both $x = 0, y = 0$, the parent compound $LaMnO_3$ is an insulator composed of ferromagnetic (a,b)-planes of Mn^{3+} ions with orbitals configuration $\{t_{2g}^3, e_g^1\}$, total spin $S = 2$, oriented in basal plane, but antiferromagnetically (AF) coupled along the c axis [1]. This antiferromagnetic ground state structure (A-type) due to the Jahn-Teller (J-T) distortion of the Mn^{3+} ions [2–4] tends to disappear by oxygen excess, and the $LaMnO_{(3+\delta)}$ manganite becomes ferromagnetic. The magnetic A-type structure occurs because of the different orientations of the e_g^1 orbitals within the (a,b) plane lead to FM superexchange, while covalent orientation of these orbitals along the c axis leads to AFM superexchange [5–7], which weakens the ferromagnetism within the (a,b) planes, but the AFM coupling along c axis remains unaltered. For $LaMnO_{(3+\delta)}$, the correlation between crystal structure and magnetism has been discussed in Ref. [8].

For $\boldsymbol{y} = \mathbf{0}$, the magnetic and transport properties of $La_{1-x}Dy_xMnO_3$ compounds with nominal stoichiometry $\boldsymbol{x} = \mathbf{0.05,0.10}$ synthesized by sol-gel method have been studied by the authors in Ref. [9]. Although $LaMnO_3$ and $La_{1-x}Dy_xMnO_3$ compounds were synthesized with the same method, they observed that in $La_{1-x}Dy_xMnO_3$ more La-site vacancies appear. This has been attributed to evaporation of La ions by chemical disorder. The La-site deficiencies, as mentioned, increase the FM double exchange (DE) interaction. Simultaneously, the ionic radius difference between La and Dy ions leads to local distortion, in addition to the J-T distortion caused by Mn^{3+} ions reducing the FM interaction between Mn^{3+} and Mn^{4+} ions. Furthermore, the large magnetic moment of Dy^{3+} ($\boldsymbol{\mu_{eff}}$ $\mathbf{10.83\mu_B}$) randomly polarizes the magnetic Mn sublattice in favor of ferromagnetism. In slightly Dy-doped $LaMnO_3$, the paramagnetic Dy ion can be represented as an impurity between the Mn-Mn spin pairs coupling to Mn ions through 3d and 4f electrons. At large Mn-Dy separation, the interaction between 3d and 4f electrons is smaller than (or just comparable) that between nearby Mn-Mn host ions [10–13], confirmed that through weak coupling between magnetic impurities and their nearest neighbor host ions, the local AFM order in the host spin system will be enhanced. Through small Dy doping, the interactions between Dy impurities are determined by host Mn spins. On the contrary, when the Dy concentration increases, the interaction between the Dy impurities increases, and the lattice becomes much more disordered. Thus, Dy can introduce magnetic disorder that can be important to understand the presence of ferromagnetic clusters above the ferromagnetic ordering.

For $\boldsymbol{x} = \mathbf{0}$, a great deal of work has been conducted concerning the doping of the Mn site in $LaMn_{1-y}Zn_yO_3$ manganites with y in the $\mathbf{0.05}$–$\mathbf{0.4}$ range [14–17]. Partial substitution of Mn by Zn^{2+} $\{3d^{10}\}$ nonmagnetic ions results in the simultaneous presence of Mn^{3+} and Mn^{4+} ions, triggering Zener's DE interaction. Thus, the transformation of Mn^{3+} into Mn^{4+} reduces the antiferromagnetism of the A-type structure, inducing 3-dimensional ferromagnetism [15]. Moreover, the Zn dilution of the Mn sublattice locally weakens magnetism and induces disorder. At the same time, the magnetic La(Dy) sublattice interacts with the local field imposed by the ferromagnetic of the Mn sublattice. Depending on the magnetic nature of the La(Dy) sublattice, the interaction can be FM or AFM; in our case, ferromagnetism is obtained.

2. Crystalline structure

X-ray diffraction (XRD) patterns study of the samples of $La_{1-x}Dy_xMn_{1-y}Zn_yO_3$ family (synthesized by solid reaction method) with $x = 0.0, 0.05, 0.1$ and $y = 0.0, 0.05, 0.1$ for samples $x = 0, y = 0.0$ and $x = 0.0, y = 0.05$ exhibits a trigonal phase, while the other samples correspond to an orthorhombic phase. The effects on the crystal structure through simultaneous substitution on both La and Mn sites are complex. On the one hand, inclusion of Dy [18–20], with smaller ionic radius than La ions, and inclusion of Zn [16, 17], with a greater ionic radius than Mn ions, introduces distortions in the octahedra and, therefore, in the crystallographic structure. On the other hand, introduction of divalent Zn ions induces the change of Mn^{3+} for Mn^{4+} and the J-T distortion of the Mn^{3+} neighboring Zn^{2+} differs from one of the Mn^{3+} no-neighboring Zn^{2+} ions [17]. **Figure 1** shows the XRD pattern for $x = 0$ due to different Zn substitution: $y = 0$ (bottom panel), $y = 0.05$ (middle panel), and $y = 0.1$ (top panel).

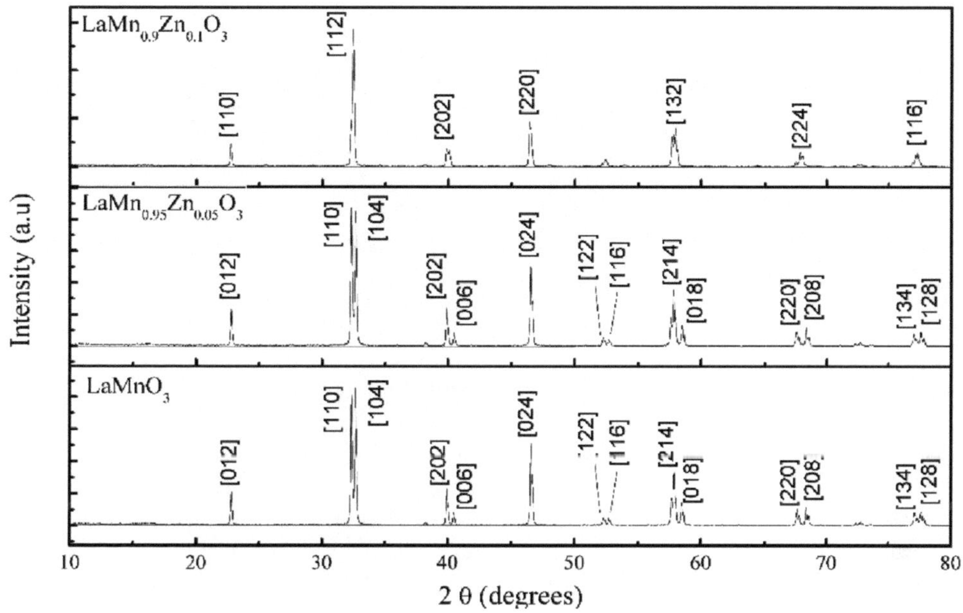

Figure 1.
X-ray diffraction patterns for $La_{1-x}Dy_xMn_{1-y}Zn_yO_3$ (a) $x = 0.0, y = 0.1$; (b) $x = 0.0\ y = 0.05$; and (c) $x = 0\ y = 0.0$. Major Miller indices are indicated.

X-ray diffraction data were refined by the Rietvel method—Fullproff program [21]. Information, such as phase structure, identification of planes, cell volumes, and cell densities were obtained. **Table 1** presents some crystallographic parameters corresponding to $La_{1-x}Dy_xMn_{1-y}Zn_yO_3$ samples.

Figure 2 shows a structural phase transition for $x = 0.0$ samples, when $y > 0.05$ it is from triclinic to orthorhombic. This signals the high structural distortion introduced by Zn ions. Therefore, the sample $x = 0.0, y = 0.1$ relaxes the stress by symmetry breaking. As will be seen later, this behavior is also observed at Dy concentrations for $x = 0.05$, 0.10 series, although for these samples, relaxation is achieved by octahedra tilting increases.

Sample	$x = 0.0$			$x = 0.05$		
	$y = 0.0$	$y = 0.05$	$y = 0.10$	$y = 0.0$	$y = 0.05$	$y = 0.10$
Structure	Trigonal/R–3C			Orthorhombic/Pbnm		
Space group	167			62		
Z	6			4		
a (A)	5.5363	5.5412	5.5459	5.5395	5.5354	5.5354
b (A)	5.5363	5.5412	5.5056	5.4965	5.5017	5.5072
c (A)	13.3670	13.3652	7.7995	7.7765	7.7863	7.7923
Cell volume (Å^3)	354.831	355.400	238.1	236.775	237.124	237.542
Density (g/cm^3)	6.785	6.804	6.807	6.865	6.840	6.828

Table 1.
Structural parameters for $La_{1-x}Dy_xMn_{1-y}Zn_yO_3$ samples. Values in white cells are from Ref. [22].

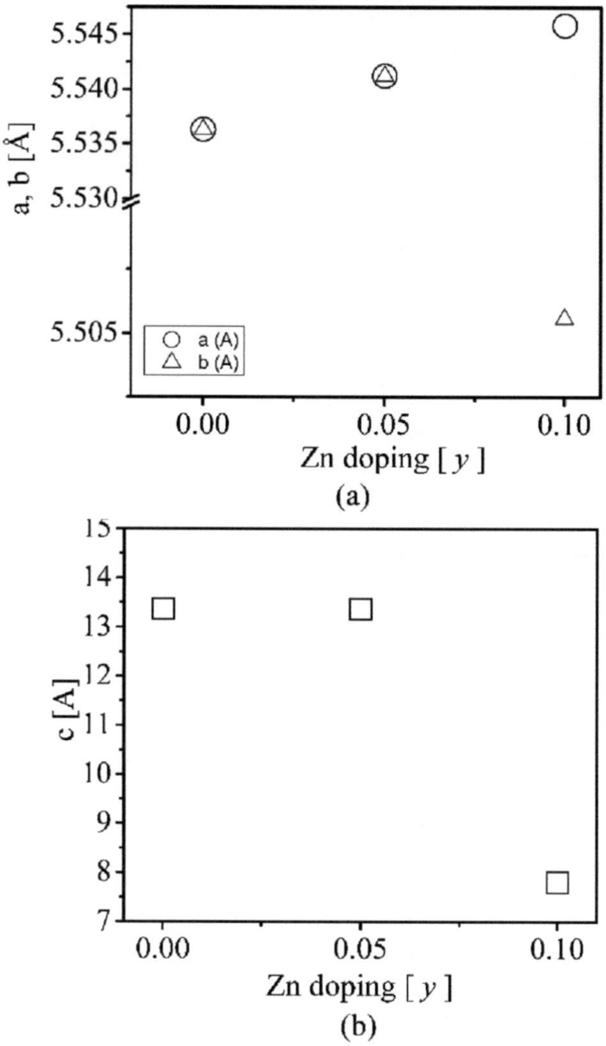

Figure 2.
Variation of unit cell parameter with Zn doping (0.0 < y < 0.1). (a) Cell parameter a *(circle) and* b *(triangle); and (b) Cell parameter* c *(square).* $x = 0.0$ *in all cases.*

Use of 3D reconstruction tools of structures (like *Visualization for Electronic Structural Analysis*) allows obtaining interatomic distances, Mn – O and angles, θ_i in the Mn – O – Mn bonds (**Figure 3**). In the case of manganites, a distortion of structure is directly related to their magnetic properties. This phenomenon, known as the J-T effect, is a structural phase transition driven by the coupling between the orbital state and the vibronic configuration of the crystal lattice. The J-T coupling to the lattice manifests itself in changes in bond lengths Mn – O and θ_i angles, as well as in orbital order [3, 4].

Regarding the measurements of structural distortion in perovskites with orthorhombic structure, two relations have been used to evaluate said distortion from the distances of the manganese with their neighboring oxygen Mn – O – Mn in the plane and with the apical oxygen ions. The octahedra distortion, D_{oct}, can be calculated by [23, 24]:

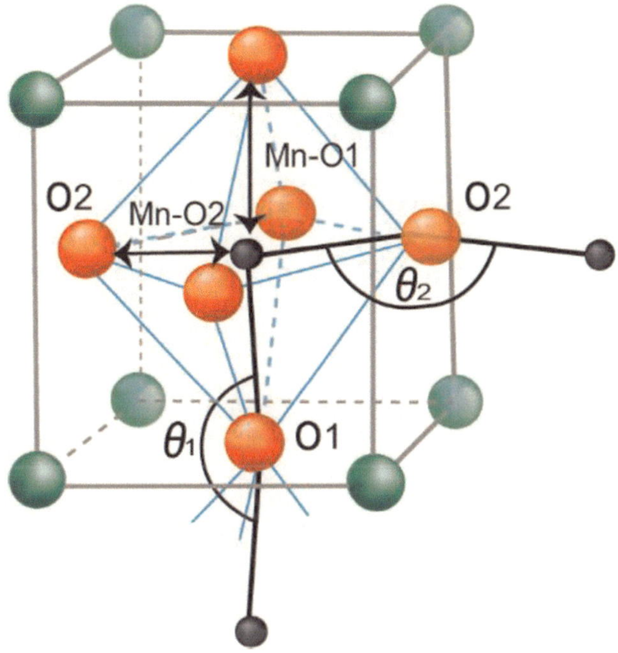

Figure 3.
Unit cell of perovskite structure indicating the bond distances, (Mn/Zn) − O1 *and* $\left(\frac{Mn}{Zn}\right) - O1$ *and bond angles,* θ_1 *and* θ_2*. Atoms: O (red), Mn/Zn (black), and La/Dy (Green).*

$$D_{oct} = 10 \times \frac{\left(\sum_{i=1,\dots 6} \left| (\mathrm{Mn} - \mathrm{O}_i) - \langle \mathrm{Mn} - \mathrm{O} \rangle_{average} \right| \right)}{\langle \mathrm{Mn} - \mathrm{O} \rangle_{average}}, \quad \mathrm{D}_{oct} = 1/6 \times \sum_{i=1,\dots 6} \left[\frac{(\mathrm{Mn} - O_i) - \langle \mathrm{Mn} - \mathrm{O} \rangle_{average}}{\langle \mathrm{Mn} - \mathrm{O} \rangle_{average}} \right]^2 \tag{1}$$

where $\mathrm{Mn} - \mathrm{O}_i$ represents the bond distances between the manganese ion and each of the six oxygens of the octahedra that surround it. These distances will depend, among others, on the ionic radii and mainly on the doping of the ion in site *B*. **Figure 2** shows these distances of the manganese $\mathrm{Mn} - \mathrm{O}_2$ in the plane and $\mathrm{Mn} - \mathrm{O}_1$ with the apical oxygens.

Octahedra distortion, from the energetic point of view known as J-T distortion, is evaluated through the relation [25]:

$$\lambda_{J-T} = \sqrt{\frac{1}{3} \times \left(\sum \left((\mathrm{Mn} - \mathrm{O}_i) - \langle \mathrm{Mn} - \mathrm{O} \rangle_{average} \right)^2 \right)} \tag{2}$$

Although Eqs. (1) and (2) are different ways of evaluating the distortion, both are dimensionless; their range is between [0, 1], and their behavior is similar. In the manganites studied herein, an inverse relationship between distortions and the magnetic moment of individuals or clusters has been observed for $T > Tc$. Where does this structure-magnetism relationship originate? Clearly, the J-T effect results in spontaneous splitting of the energy levels, reducing the total energy of the Mn^{3+} also resulting in a spontaneous distortion of the octahedrons that increase their elastic energy at the cost of a reduction in energy of certain electron orbitals, thus, resulting in a net reduction in energy.

The J-T distortions obtained through Eq. (2) showed anomalous behavior (**Figure 4**): Zn increases in the range of $0 < y < 0.05$ increases the J-T distortion and for $y > 0.05$, a reduction in the J-T distortion is observed (**Figure 4a**). This is due to the difference between the ionic radii of Zn and Mn ions [16]. However, when $x > 0.05$, the accumulated strain of the system is relaxed via octahedral tilting—similar to the introduction of dislocation when the maximum stress tolerated is reached in a crystal [22]. Tilting as a relaxation mechanism is observed in **Figure 4b**, where greater tilting corresponds to samples with lower J-T distortion. The same anomalous behavior is observed by calculating octahedra distortions (Eq. (2)). As will be seen ahead,

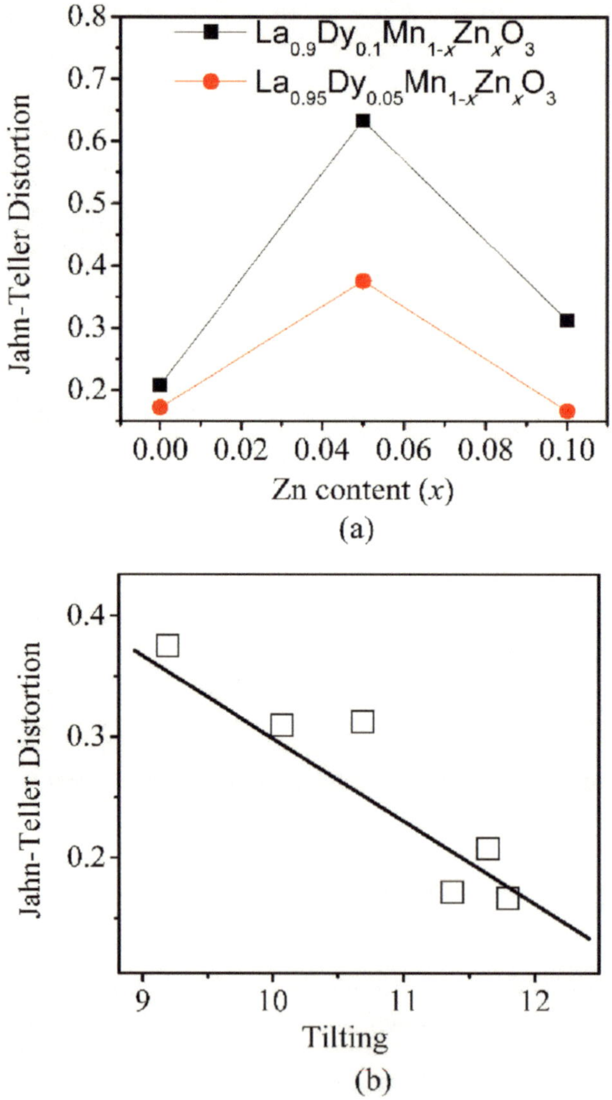

Figure 4.
(a) Jahn-Teller distortion versus Zn composition for x = 0.10 *samples (squares) and* x = 0.05 *samples (circles); and (b) Jahn-Teller distortion vs. Octahedra tilting, ω, by using* $2\omega_i + \theta_i = 180°$ *[22].*

the activation energy that allows the magnetic exchange is accompanied by a charge exchange facilitated by these lattice distortions.

Local distortions affect directly the magnetic properties as a function of T, that is, exchange constants are proportional to the orthorhombic strain [8]. Also, magnetic moments and activation energies depend on distortion [22].

3. Local magnetism-electron paramagnetic resonance (EPR) analysis

Electron paramagnetic resonance is an important technique to study the microscopic nature of local interactions in magnetic materials and, particularly, in manganites [26], which in many cases show short-range interactions for $T > Tc$. This technique contrasts with the magnetization and susceptibility techniques that provide global information about the exchange mechanisms and the possible spontaneous creation of clusters at high temperatures.

The EPR signal corresponds to $d\chi''/dH$, where $\chi = M/H$ is the susceptibility. In the paramagnetic region, where M is linear with H, the double integral of the EPR intensity is proportional to the magnetization. From EPR measurements, we can build the dependence of the inverse of the temperature-dependent susceptibility, $\chi^{-1}(T)$.To carry out the EPR measurements, a Bruker ESP-300 spectrometer was used in the radiation X-band with 9.408 GHz frequency and in a temperature range from 10 to 290 K. **Figure 5a** shows the resonant signals for manganite $La_{0.9}Dy_{0.1}MnO_3$ in the temperature range $220 < T < 290\ K$. The inset of **Figure 5a** is the integral of the EPR signal. It is observed that for high temperatures, the intensity of the EPR signal increases in all cases as the temperature decreases. For $T > T_C$, it is found that the intensity of resonance line fits by the expression [28]:

$$I_{\text{PM}} = I_0 e^{\Delta E/k_B T} \tag{3}$$

where I_{PM} is the intensity extracted from the resonance line, I_0 is a fitting parameter, and ΔE is an activation energy.

Figure 5c shows the variation of the resonant field with temperature. A change in the value of H_r is observed for $T > Tc$. This local signal evidences an FM phase in the PM region [27, 29]. This local magnetism is strongly dependent on doping and oxygen content, as Oseroff et al. [28] show in their work on collective spin dynamics above Tc in manganites ($La_{1-x}Ca_xMnO_3$).

The peak-to-peak EPR linewidth, ΔH_{PP}, can also be used to confirm the presence of short-range interacting magnetic entities. In magnetic resonance, the EPR linewidth is related to the relaxation mechanisms of the magnetic units, whether individual spins or spin-coupled systems. A decrease in ΔH_{PP} as temperature decreases indicates PM behavior of the sample. However, in the case of $La_{1x}D_xMnZnO$ manganites, an increase in ΔH_{PP} is observed as temperature decreases, indicating the presence of short-range interactions. In **Figure 6a**, the arrows indicate the corresponding critical temperatures for each sample. ΔH_{PP} increases as temperature decreases, indicating a greater range of the collective effects, approaching the FM phase. Tovar et al. [8] have found a direct dependence on temperature for the J-T distortion and the EPR linewidth.

The energy transferred in the relaxation process is related to jumps of polarons thermally activated between the Mn^{4+} and Mn^{3+} states. Therefore, the jump rate of the charge carriers will determine the half-life of the spin state and, therefore, determines

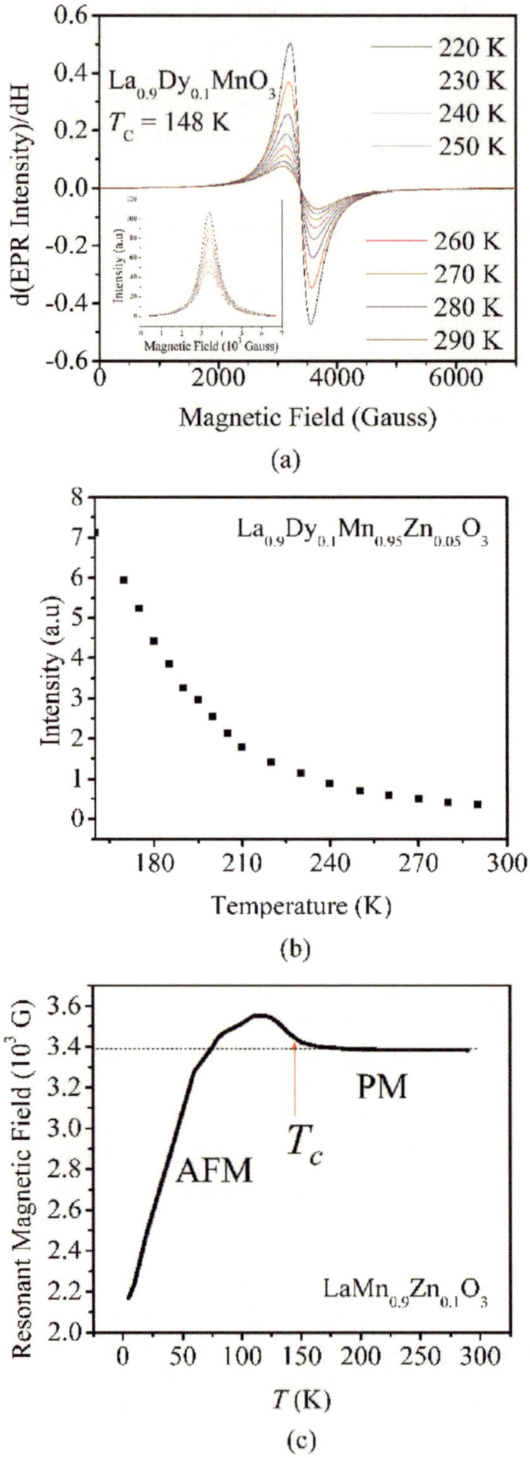

Figure 5.
(a) EPR signal of $La_{0.9}Dy_{0.1}MnO_3$ for different temperatures, Inset. Intensity obtained by integration of the EPR signal; (b) EPR signal intensity as a function of temperature, showing short-range interaction behavior for $T > 181$ K for the sample $La_{0.9}Dy_{0.1}Mn_{0.95}Zn_{0.05}O_3$; and (c) resonant field vs. T for $LaMn_{0.9}Zn_{0.1}O_3$ [27].

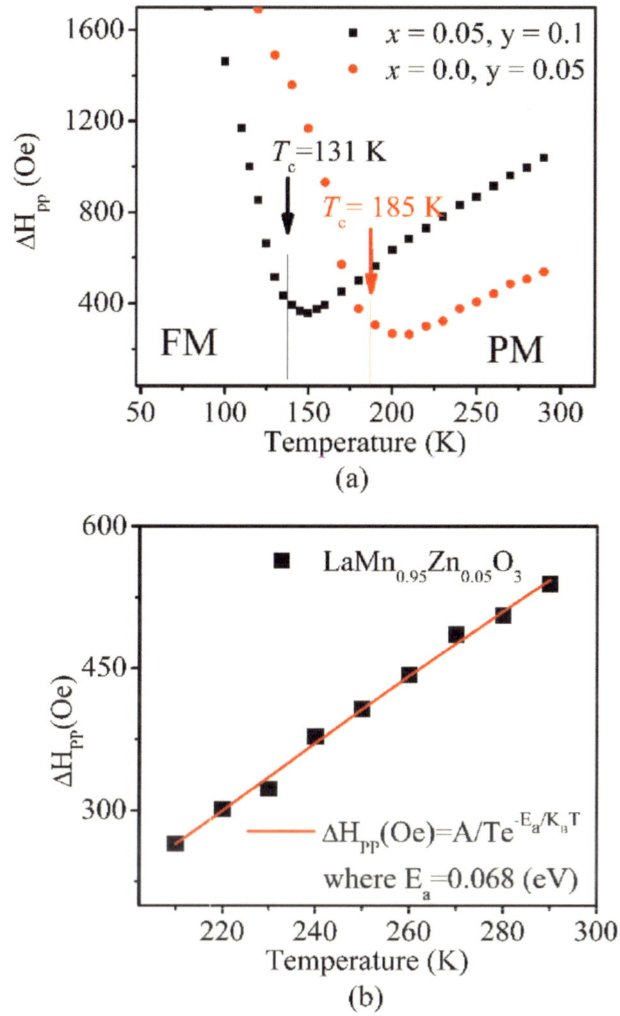

Figure 6.
(a) Temperature dependence on the resonance linewidth; the corresponding T_C values of the $x = 0.0$, $y = 0.05$ (red) and $x = 0.05$, $y = 0.1$ (black) samples are indicated; and (b) linewidth as a function of temperature or the $x = 0.0$, $y = 0.05$ sample; red line corresponds to the fitting line to obtain the activation energy value as $E_a = 0.068$ eV.

the EPR linewidth and the conductivity. The dependence of the EPR linewidth as a function of temperature in the paramagnetic region can be evaluated by [30]:

$$\Delta H_{PP}(T) = \frac{A}{T} e^{-\left(\frac{E_a}{k_B T}\right)} \tag{4}$$

E_a values showed on **Table 2** are obtained by Eq. (4). Previous studies on relaxation modes in mixed-valence manganites have shown that the EPR predominant signal corresponds to $Mn^{3+} - Mn^{4+}$ relaxation and the internal relaxation of the ions through the lattice. Shengelaya et al. [31] showed that the relaxation, R, of the Mn^{4+} (s) with the lattice can be negligible compared with the relaxation Mn^{3+} (σ) with the lattice (L),

Zn content (y)	Dy $x = 0.00$	Dy $x = 0.05$	Dy $x = 0.1$
	Activation Energy E_a (eV)	**Activation Energy E_a (eV)**	**Activation Energy E_a (eV)**
0.00	0.0294(8)	0.0656(7)	0.0596(1)
0.05	0.0683(9)	0.0555(9)	0.0584(5)
0.10	0.0487(7)	0.0477(6)	0.0624(6)

Table 2.
Activation energies as a function of Zn doping (y). White cells from Ref. [27].

while the largest signal corresponds to the relaxation $Mn^{3+} - Mn^{4+}$ and $Mn^{4+} - Mn^{3+}$. Thus, a "Bottleneck" is formed, corresponding to the charge transfer between the two magnetic subsystems (3+ and 4+), while a slower relaxation process, Mn^{3+} with the network occurs. A schematic representation proposed is presented in **Figure 7**.

Furthermore, the intensity of the EPR signal is proportional to the static magnetic susceptibility, χ_e, it is possible to explain the bottleneck regime due to the coupling of the spins (s) and (σ) as:

$$\chi_S = \chi_S^o \frac{1 + \lambda' \chi_\sigma^o}{1 - \lambda'^2 \chi_\sigma^o \chi_S^o} \tag{5}$$

where χ_S^o and χ_σ^o are the ion susceptibilities without exchange of Mn^{4+} and Mn^{3+}, respectively, and λ is the dimensionless coupling constant.

$$\lambda' = \frac{-zJ}{n g_S g_\sigma \mu_B{}^2} \tag{6}$$

with n being the number of spins per cm^3 and $g_S g_\sigma$ being the g-factors of the Mn^{4+} and Mn^{3+} ions, respectively. For $T \gg T_C$, the EPR signal intensity drops faster than predicted by Eq. (5), associated with a transition from the bottleneck to an isothermal regime: $R_{\sigma L} \gg R_{\sigma S}$, where the relaxation, $R_{\sigma S}$, is T independent [31].

For manganites, the contribution of each coexistent phase has been evaluated by means of the EPR technique. In this case, FM clusters also contribute to the magnetization of the material, so that the total magnetization is the result of the PM and FM contributions above T_C by [32]:

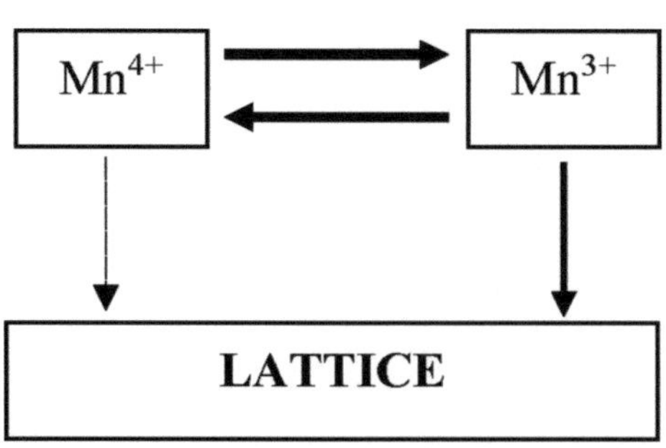

Figure 7.
A block diagram showing the energy flow paths for the Mn^{4+} and Mn^{3+} spin subsystems and the lattice. The relaxation rates $R_{\sigma S}$, $R_{S\sigma}$ represent relaxation between the subsystems. The thickness of the arrows is a measure of the magnitude of the particular relaxation rate [31].

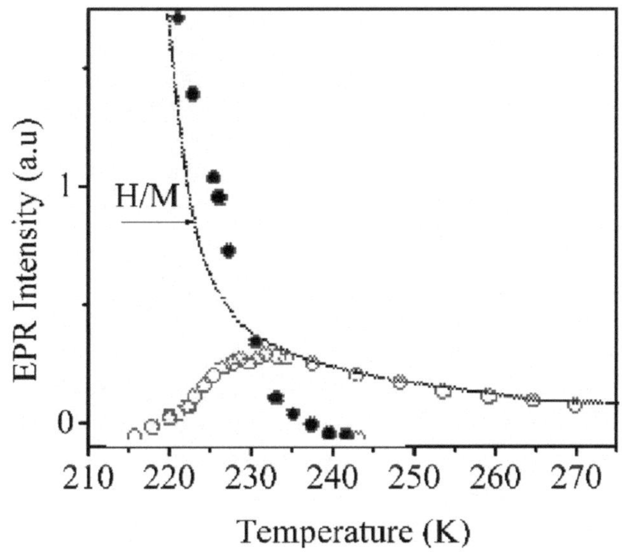

Figure 8.
(◦) $I_{PM}(T)$, (solid line) M/H (T) at 3.5 kG and (•) $I_{FM}(T)$ vs. T for $La_{0.75}Ca_{0.25}MnO_3$. Adapted from Ref. [32].

$$M(H, T) = x(T)\chi_{PM}(T)H + [1 - x(T)]M_{FM}(H, T) \quad (7)$$

where x and $(1 - x)$ are the fractions of PM and FM signals, respectively. The intensity of ESR lines are:

$$I_{PM}(T) \propto x(T)\chi_{PM}(T) \text{ and } I_{FM}(T) \propto [1 - x(T)]M_{FM}(T) \quad (8)$$

For $T \gg Tc$, the EPR signal consists only of a PM line; then $x(T \gg Tc) = 1$, and $M(H, T)/H = \chi_{PM}$, while $x(T < Tc) = 0$ and $M(H, T)/H = \chi_{FM}$.As Eq. (8) shows, it is possible to find the fraction of the FM phase in the samples by subtracting the EPR intensity from the magnetic susceptibility (**Figure 8**). The inset shows the fraction as a function of temperature.

Dormann and Jaccarino [33] proposed the following Huber approximation for a coupled system (clusters) in the PM state:

$$\Delta H_{pp}(T) \propto [\chi_s(T)/\chi_{EPR}(T)]\Delta H_{pp}(\infty) \quad (9)$$

where $\chi_s(T) = C/T$ is the susceptibility of the individual ions M^{3+} and M^{4+}; $\Delta H_{pp}(\infty)$ corresponds to spin-only interactions ($\Delta H_{pp}(\infty) = 2600$ G reported by Causa et al. [34] for perovskites with A = La) and $\chi_{EPR}(T)$ the PM signal of the coupled system.

$$C = N_A\mu_{eff}^2\mu_B^2/3k_B \quad (10)$$

4. Macroscopic magnetism: $M(T, H)$

Thereafter, we will refer to one of the most-used macroscopic techniques to characterize magnetic materials: direct measurement of magnetization as a function of temperature and applied field. This technique offers vast information regarding the type of transition [35–37], transition temperature [38, 39], magnitude of the magnetic moments [22, 38, 40, 41], and the possible presence of clusters in the PM region, which is quite characteristic of manganites. Next, we will present results on the use of this technique in manganites.

4.1 Type of transition

To determine the nature of the FM-PM phase transition (first- or second-order), it is beneficial to use Arrott's plot [42]: $\mu_0 H/M$ vs. M^2. According to the Banerjee criterion curves, showing positive or negative slopes without inflection points are characteristic of second- or first-order transitions, respectively [35, 43]. **Figure 9** shows M vs. T and $\mu_0 H/M$ vs. M^2 for $La_{2/3}Sr_{1/3}MnO_3$ [43]. In **Figure 9b**, in Arrot's plot for $\mu_0 H/M$ vs. M^2, the slope is always positive, indicating the second order of the phase transition for this sample.

4.2 Critical temperature and magnetic moment

Magnetization vs. T is used to find the Curie temperature, T_c, or Neel temperature, T_N, in manganites; also, this technic provides information about the presence of magnetic clusters in PM phase. **Figure 10a** presents the curves M vs. T for the $LaMn_{0.95}Zn_{0.05}O_3$ sample, where T_C= 185 K is obtained from dM/dT. The effect of doping on the Tc value is generally one of the first factors to be evaluated in manganites. For low Dy doping in $La_{1-x}Dy_xMn_{1-y}Zn_yO_3$, two competing effects have been observed: firstly, the large magnetic moment of Dy favors FM by increasing the value of Tc, and secondly, the small ionic radius of Dy (close to 50% of the La ionic radius) distorts the lattice, destroying the FM of the Mn − O − Mn chains, which leads to a reduced Tc value. As seen in **Figure 10b**, Zn determines which of the two effects predominates; at low Zn concentrations, the crystallographic distortion by Dy predominates and for concentrations of Zn > 0.05, breaking of the Mn − O − Mn chains reduces the effect of lattice distortions by Dy and allows the effect due to the magnetic moment of Dy.

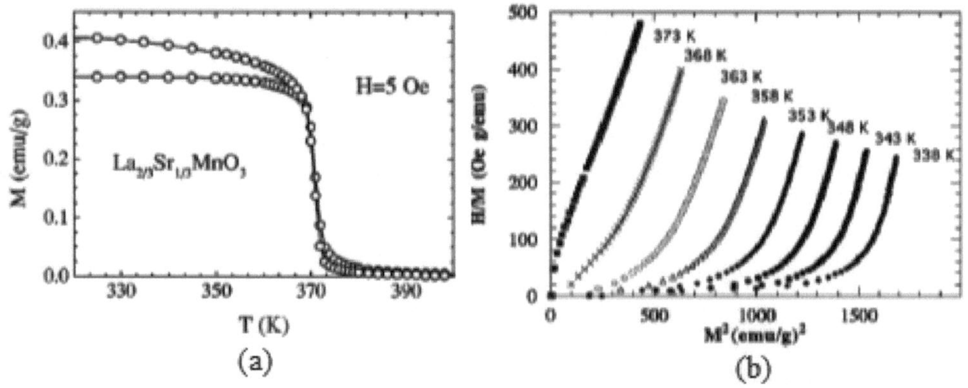

Figure 9.
(a) Temperature dependence on FC magnetization for $La_{2/3}Sr_{1/3}MnO_3$ under an applied magnetic field of $H = 5$ Oe; and (b) Arrott's plot for $\mu_0 H/M$ vs. M^2 for $La_{2/3}Sr_{1/3}MnO_3$ [36].

DOI: http://dx.doi.org/10.5772/intechopen.107309

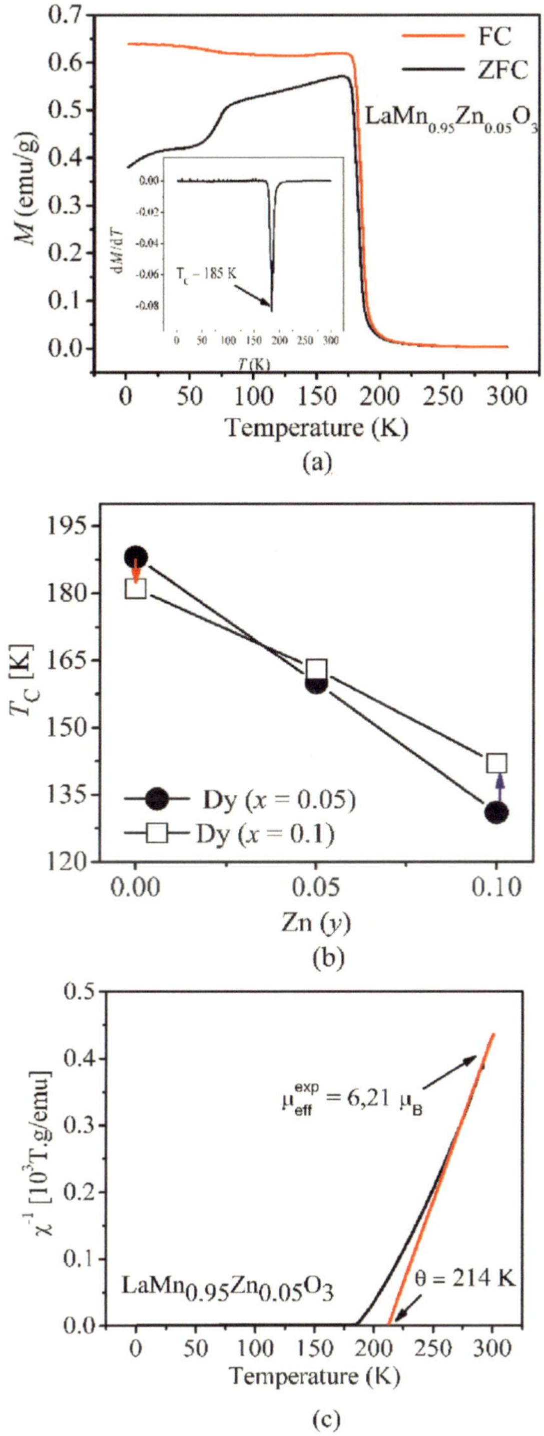

Figure 10.
(a) M *vs.* T *curves for* $x = 0.00, y = 0.05$ *show magnetization measured at field cooling (open) and the zero-field cooling (close), at* $H = 0.001$ T. *Inset. The derivative of magnetization* dM/*d*T *vs.* T *[27]; (b)* T_c *vs.* Zn *composition. Circles for* 0.05 Dy *series and squares for* 0.10 Dy *series [22]; and (c)* χ^{-1}vs.*T for* $x = 0.00, y = 0.05$. *Linear fit (red) at high temperatures. CW temperatures,* θ_{CW}, *is indicated [27].*

In manganites, it has been observed that the inverse of the susceptibility χ^{-1} vs. T does not satisfy the Curie-Weiss law. However, over a wide range of temperatures ($T \gg Tc$), $\chi^{-1}(T)$ is linear and can be described by the Curie-Weiss model (**Figure 10c**); the Curie constant (Eq. (10)) can be used to estimate the effective magnetic moments, as well as the doping effect on the magnetic moment. Overall, for all the samples, it has been found that the experimental effective moments are higher than the theoretical values [38, 39, 41, 44], evidence of the presence of clusters in the material. For the $La_{1-x}Dy_xMn_{1-y}Zn_yO_3$ manganite, the theoretical m values are expressed by [22]:

$$\mu_{eff}^{th} = \left((0.86 - 2y) \left[\mu_{eff}^{th} (Mn^{3+}) \right]^2 + (0.12 + y) \left[\mu_{eff}^{th} (Mn^{4+}) \right]^2 + x \left[\mu_{eff}^{th} (Dy^{3+}) \right]^2 \right)^{\frac{1}{2}} \tag{11}$$

according to the stoichiometric formula.

To visualize the cluster behavior throughout the nonlinear PM region, we calculated the values of $\mu_{eff}^{exp} = 2.83\sqrt{C(T)}\mu_B$ for $T > T_C$ from the difference of $\Delta\chi^{-1}(T)/\Delta T = 1/C(T)$ [38]. **Figure 11b** shows μ_{eff}^{exp} vs. T for $LaMn_{1-y}Zn_yO_3$ that μ_{eff} increases when T decreases in the temperature range corresponding to a $\chi^{-1}(T)$ that has a positive curvature. This suggests an increase in the strength of the exchange coupling with T.

4.3 Coupled moments in a mean-field approximation

The Mean-field theory of coupled moment pairs in an effective molecular field approximation, B_e, has been discussed in the literature [28, 34, 45]. The molecular field constants, as well as the magnetic susceptibilities that depend highly on the three main crystal axes direction, result in a slight deviation from the Curie Weiss (C-W) law.

The effective field, $B_e = 2(z-1)JM/(Ng^2\mu_B{}^2)$ on the coupled moment pair corresponds to a molecular field coefficient, $\lambda = 2(z-1)J/Ng^2\mu_B{}^2$. λ differs from the Weiss model in that z is replaced by $(z-1)$. In the Constant-coupling approximation (CCA) [34, 46, 47], this local field "aligns" the magnetic moments of some Mn^{3+} and Mn^{4+}, resulting in FM clusters $(S_1 + S_2)$- where S_1 and S_2 correspond to the Mn^{3+} and Mn^{4+} spins, respectively, which conform the cluster unit. This can be modeled, even in the paramagnetic region, by a Heisenberg-type isotropic interaction between pairs of Mn^{3+} and Mn^{4+} ions, subjected to the action of the effective field:

$$\mathcal{H} = -2JS_1 \bullet S_2 + g\mu_B(S_1 + S_2) \bullet [H + B_e] \tag{12}$$

where H is the external field. Because the chosen pair is arbitrary, they must all have the same magnetic moment as every other pair. This condition requires that [47]:

$$\tfrac{1}{2}\, Ng\mu_B \langle S_z^{'} \rangle = M \tag{13}$$

with $\langle S_z^{'} \rangle = \langle S_1 + S_2 \rangle$. From Eq. (12) [10], the inverse of susceptibility is obtained as:

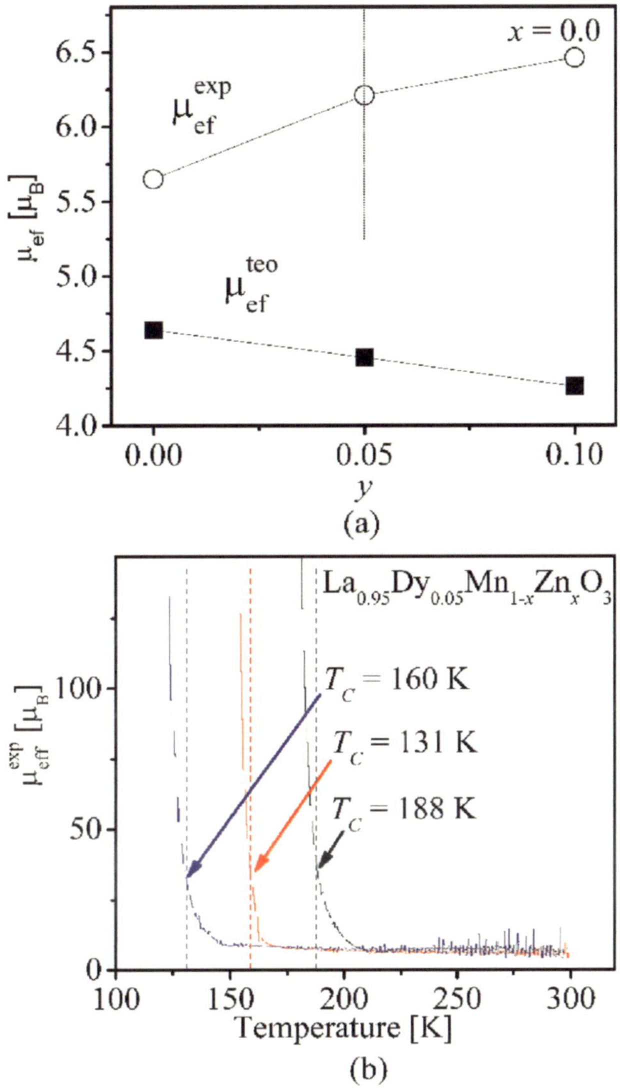

Figure 11.
(a) Theoretical and experimental effective magnetic moments at $T \gg T_C$ for $LaMn_{1-y}Zn_yO_3$; and (b) experimental effective magnetic moments as a function of temperature on $La_{0.95}Dy_{0.05}Mn_{1-y}Zn_yO_3$ for $y = 0.0$ (black), $y = 0.05$ (red), $y = 0.01$ (blue) [22] (b).

$$\chi^{-1} = \left(T - \frac{1}{2}zJ/k_B\right)/C \tag{14}$$

Expanding Eq. (14) around $(j - j_C)$, where $j = J/k_BT$ is the reduced exchange constant, we obtain an expression for the inverse of the susceptibility in terms of T_c:

$$\chi^{-1} = (T - T_c)/\rho C \tag{15}$$

with $\rho = \frac{2}{[J_c(2(z-1)J_c+(z-4)]}$. This expression reproduces a deviation from the linear C-W behavior for FM interactions.

A Heisenberg model over all points of the magnetic lattice (i, j) is insoluble because said magnetic solid has on the order of 10^{22} magnetic moments,

$$H = -\sum_{ij} J_{ij} S_i \bullet S_j \tag{16}$$

If we consider only first neighbors in the FM Heisenberg Hamiltonian, it can be rewritten as [48]:

$$\mathcal{H} = -2J\sum_{n,n'} S_n \bullet S_{n'} - g\mu_B H \sum_n S_n^z \tag{17}$$

From a power series expansion in $j = J/k_B T$ [49], the susceptibility is obtained as:

$$\chi = \frac{Ng^2 {\mu_B}^2 S(S+1)}{3k_B T} \sum_{l=0}^{\infty} a_l j^l \tag{18}$$

with N being the number of sites in the sample, $g\mu_B S$ the magnetic moment associated with the spin, S at each lattice point, z the coordination number, and j being a dimensionless value: $j > 0$ for FM and $j < 0$ for AFM. Finally, Eq. (18) is rewritten as [48]:

$$\chi_0^{-1} = \frac{\mathrm{T}}{\mathrm{C}}(1-\tau)^{4/3} f(\tau) \tag{19}$$

with $\tau = T_C/T$, $f(\tau) = \{(1-b_1\tau)(1-b_2\tau)/(1-a_1\tau)(1-a_2\tau)(1-a_3\tau)\}$ and a_i,b_i parameters obtained by the Padé approximants that only depend on the crystal structure and on the S value. a_i,b_i parameters for $z = 6$ are presented in **Table 3**.

With the coefficients indicated in **Table 3**, Eq. (19) shows a deviation from the straight lines predicted from the C-W theory (**Figure 12a**). The blue line represents experimental data for $La_{0.9}Dy_{0.1}MnO_3$, where clusters are present at $T > T_C$ and J is unknown. For general Heisenberg Hamiltonians, where more than one relevant exchange constant J_i(J_i =1, 2, 3, 4) is required, a high-temperature expansion (HTE) has been developed [50, 51]. **Figure 12b** shows χ^{-1}vs.T experimental and fitting data for $La_{0.95}Dy_{0.05}Mn_{0.9}Zn_{0.1}O_3$. The J value obtained is $J/K_B = 15\ K$. The effect of Dy and Zn doping is evident on J (see **Figure 12c**) [27].

S	a_1	a_2	a_3	b_1	b_2
3/2	−1/21,9677	−1/1.62282	−1/0.38910	−1/1.74300	1/0.38911
2	−1/51.0423	−1/1.73724	*	−1/1.89133	*

Corresponding to conjugate values reported by Gammel et al. [48] ($a_3 = b_2$).*

Table 3.
a_i,b_i coefficients obtained for Z = 6, from Padé approximants.

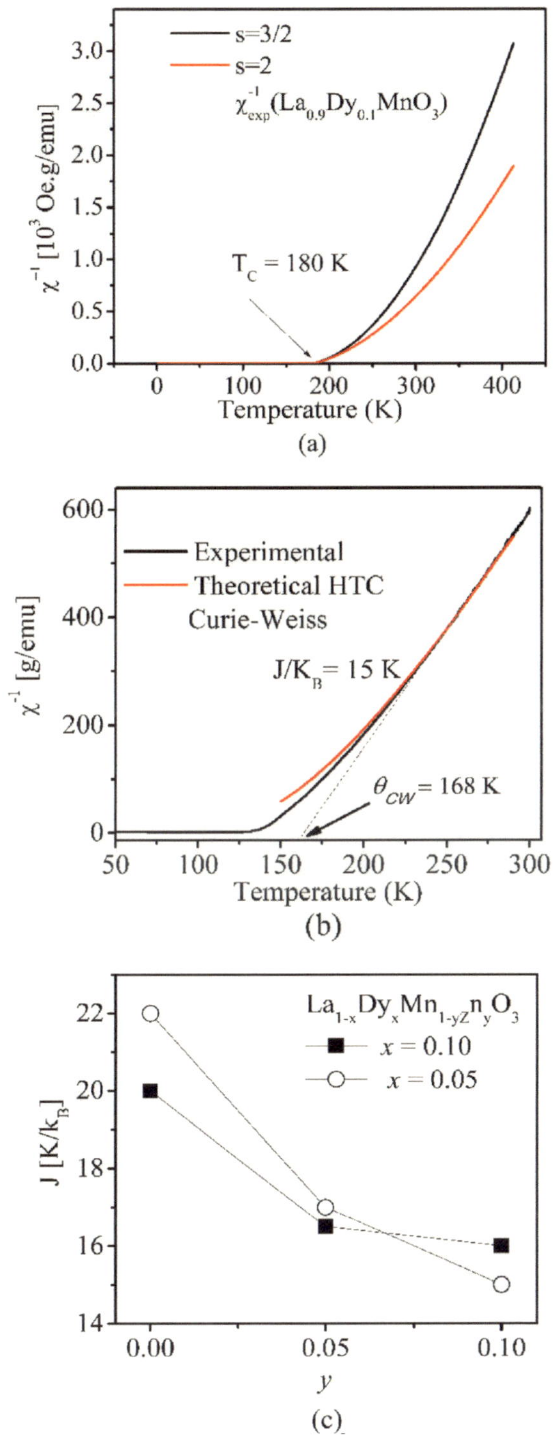

Figure 12.
*χ^{-1} vs.T (a) using Eq. (19) for $z = 6$, $s = 3/2$ (black line), $S = 2$ (red line) and the experimental values [22] for $La_{0.9}Dy_{0.1}MnO_3$ (blue line); (b) $La_{0.95}Dy_{0.05}Mn_{0.9}Zn_{0.1}O_3$ [22]. Black line, experimental data; red line, fit to HTE algorithm; dashed line, C-W model. **J** exchange constant values for both series from HTE algorithms; and (c) **J** values vs. y for $La_{1-x}Dy_xMn_{1-y}Zn_yO_3$ [27].*

Acknowledgements

The authors thank the BC foundation for its support in measuring the magnetic properties of the manganites and thank José Fernando López Toro for authorizing the use of diverse results from his Ph.D. thesis to illustrate the analysis presented herein. This publication was partially supported by the Science Faculty and physics department of the Universidad Nacional de Colombia.

Conflict of interest

The authors declare no conflict of interest.

Author details

Aminta Mendoza[1,2*] and Octavio Guzmán[2]

1 Faculty of Science, Department of Physics, Universidad Nacional de Colombia, Bogotá, Colombia

2 Magnetic Materials and Nanostructures Group, Faculty of Science, Department of Physics, Universidad Nacional de Colombia, Bogotá, Colombia

*Address all correspondence to: gamendozab@unal.edu.co

References

[1] Wollan EO, Koehler WC. Neutron Diffraction Study of the Magnetic Properties of the Series of Perovskite-Type Compounds [(1-x)La, xCa]MnO_3. Physics Review. 1955;**100**:545. DOI: 10.1103/PhysRev.100.545

[2] Bersuker I. The Jah-Teller Effect. Cambridge: Cambridge University Press; 2006. Chapter 8. pp. 479-597. DOI: 10.1017/CBO9780511524769

[3] Van Aken BB, Jurchescu OD, Meetsma A, Tomioka Y, Tokura Y, Palstra TTM. Orbital-Order-Induced Metal-Insulator Transition in $La_{1-x}Ca_xMnO_3$. Physical Review Letters. 2003;**90**:066403. DOI: 10.1103/PhysRevLett.90.066403

[4] Qiu X, Proffen T, Mitchell JF, Billinge SJL. Orbital Correlations in the Pseudocubic O and Rhombohedral R Phases of $LaMnO_3$. Physical Review Letters. 2005;**94**:177203. DOI: 10.1103/PhysRevLett.94.177203

[5] Goodenough JB. Magnetism and the chemical bond. New York, NY: Interscience Publishers; 1963. ASIN: B002TT366A

[6] Booth CH, Bridges F, Snyder GJ, Geballe TH. Evidence of magnetization-dependent polaron distortion in $La_{1-x}A_xMnO_3$, A=Ca, Pb. Physical Review B. 1996;**54**:15606. DOI: 10.1103/PhysRevB.54.R15606

[7] Billinge SJL, DiFrancesco RG, Kwei GH, Neumeier JJ, Thompson JD. Direct Observation of Lattice Polaron Formation in the Local Structure of $La_{1-x}Ca_xMnO_3$. Physical Review Letters. 1996;**77**:715. DOI: 10.1103/PhysRevLett.77.715

[8] Tovar M, Alejandro G, Butera A, Caneiro A, Causa MT, Prado F, et al. ESR and magnetization in Jahn-Teller-distorted $LaMnO_{3-\delta}$: Correlation with crystal structure. Physical Review B. 1999;**69**:10199. DOI: 10.1103/PhysRevB.60.10199

[9] Liu H, Zhang H, Li Y, Chen Y, Chen L, Dong X, et al. Magnetism and Resistances of Slightly Dy Doped $LaMnO_3$ Solid Solutions. Journal of Superconductivity and Novel Magnetism. 2012;**25**:1049-1054. DOI: 10.1007/s10948-011-1348-5

[10] Fabrèges X, Mirebeau I, Bonville P, Petit S, Lebras-Jasmin G, Forget A, et al. Magnetic order in $YbMnO_3$ studied by neutron diffraction and Mössbauer spectroscopy. Physical Review B. 2008;**78**: 214422. DOI: 10.1103/PhysRevB.78:214442

[11] Wehrenfennig C, Meier D, Lottermoser T, Lonkai T, Hoffmann J-U, Aliouane N, et al. Incompatible magnetic order in multiferroic hexagonal $DyMnO_3$. Physical Review B. 2010;**82**: 100414. DOI: 10.1103/PhysRevB.82.100414

[12] Lüscher A, Sushkov OP. Long-range dynamics of magnetic impurities coupled to a two-dimensional Heisenberg antiferromagnet. Physics Review. 2005;**71**:064414. DOI: 10.1103/PhysRevB.71.064414

[13] Engel J, Wesse S. From enhanced to reduced quantum antiferromagnetism by tuning a magnetic impurity. Physical Review B. 2009;**80**:094404. DOI: 10.1103/PhysRevB.80.094404

[14] Ghosh K, Ogale SB, Ramesh R, Greene RL, Venkatesan T, Gapchup KM, et al. Transition-element doping effects in $La_{0.7}Ca_{0.3}MnO_3$. Physical Review B. 1999;**59**:533. DOI: 10.1103/PhysRevB.59.533

[15] Hébert S, Martin C, Maignan A, Retoux R, Hervieu M, Nguyen N, et al. Induced ferromagnetism in $LaMnO_3$ by Mn-site substitution:The major role of Mn mixed valency. Physical Review B. 2002;**65**:104420. DOI: 10.1103/PhysRevB.65.104420

[16] Álvarez-Serrano I, Pico C, Veiga ML. Structural characterization, electric and magnetic behavior of Zn-doped manganites. Solid State Sciences. 2004;**6**: 1321. DOI: 10.1016/j.solidstatesciences.2004.07.022

[17] Tang S, Yue S, Zhang Y. Jahn-Teller distortion induced by Mg/Zn substitution on Mn sites in the perovskite manganites. Physics Letters A. 2003;**319**:530-538. DOI: 10.1016/j.physleta.2003.10.061

[18] Mitra C, Raychaudhuri P, Dhar SK, Nigam AK, Pinto R, Pattalwar SM. Evolution of transport and magnetic properties with dysprosium doping in $La_{0.7-x}Dy_xSr_{0.3}MnO_3$ (x= 0-0.4). . Journal of Magnetism and Magnetic Materials. 1999;**192**:130. DOI: 10.1016/S0304-8853(98)00388-6

[19] Terai T, Kakeshita T, FuKuda T, Saburi T, Takamoto N, Kindo K, et al. Electronic and magnetic properties of $(La\text{-}Dy)_{0.7}Ca_{0.3}MnO_3$. Physical Review B. 1998;**58**:14908. DOI: 10.1103/PhysRevB.58.14908

[20] Xu S, Tong W, Fan J, Gao J, Zha C, Zhang Y. Influence of doped Dy on magnetic and electronic properties in $La_{0.67-x}Dy_xSr_{0.33}MnO_3$. Journal of Magnetism and Magnetic Materials. 2005;**288**:92-105. DOI: 10.1016/j.jmmm.2004.08.022

[21] Rodríguez-Carvajal J. Recent advances in magnetic structure determination neutron powder diffraction. Physica B. 1993;**192**:55-69. DOI: 10.1016/0921-4526(93)90108-I

[22] López-Toro JF, Lezama L, Salazar D, Mendoza A. Influence of Non-magnetic Dilution on the Magnetic Properties of $La_{1-x}Dy_xMn_{1-y}ZnyO_3$ Perovskites at High Temperature. Physica Status Solidi A: Applications and Materials Science. 2021; **219**(1-8):2100513. DOI: 10.1002/pssa.202100513

[23] Alonso JA, Martínez-Lopez MJ, Casais MT, Fernández-Díaz MT. Evolution of the Jahn-Teller Distortion of MnO_6 Octahedra in $RMnO_3$ Perovskites (R = Pr, Nd, Dy, Tb, Ho, Er, Y): A Neutron Diffraction Study. Inorganic Chemistry. 2000;**39**:917-923. DOI: 10.1021/ic990921e

[24] Rodríguez-Carvajal J, Hennion M, Moussa F, Moudden AH, Pinsard L, Revcolevschi A. Neutron-diffraction study of the Jahn-Teller transition in stoichiometric $LaMnO_3$. Physical Review B. 1998;**57**:R3189. DOI: 10.1103/PhysRevB.57.R3189

[25] Blasco J, Ritter C, García J, de Teresa JM, Pérez-Cacho J, Ibarra R. Structural and magnetic study of $Tb_{1-x}Ca_xMnO_3$ perovskites. Physical Review B. 2000;**62**:5609. DOI: 10.1103/PhysRevB.62.5609

[26] Santiago-Teodoro M, Hernández-Cruz L, Montiel-Sánchez H, Álvarez-Lucio G, Flores-González MA, Legorreta-García F. Synthesis, Microstructure and EPR of $CaMnO_3$ and $Eu_xCa_{1-x}MnO_3$ Manganite, Obtained by Coprecipitation. Journal of the Mexican Chemical Society. 2011;**55**(4):204-207. DOI: 2073-4352/11/5/473

[27] López JF. PhD Thesis, Universidad Nacional de Colombia; 2022

[28] Oseroff SB, Torikachvili M, Singley J, Ali S, Cheong S-W, Schultz S.

Evidence for collective spin dynamics above the ordering temperature in $La_{1-x}Ca_xMnO_3$. Physical Review B. 1996; **53**(10):6521. DOI: 10.1103/PhysRevB.53.6521

[29] Eremina RM, Fazlizhanov II, Yatsyk IV, Sharipov KR, Pyataev AV, Zavoisky EK, et al. Phase separation in paramagnetic $Eu_{0.6}La_{0.4-x}Sr_xMnO_3$. Physical Review B. 2011;**84**:064410. DOI: 10.1103/PhysRevB.84.064410

[30] Yang J, Rong X, Suter D, Sun YP. Electron paramagnetic resonance investigation of the electron-doped manganite $La_{1-x}Te_xMnO_3$ ($0.1\leq x\leq 0.2$). Physical Chemistry Chemical Physics. 2011;**13**:16343-16348. DOI: 10.1039/c1cp21807g

[31] Shengelaya A, Zhao G-M y, Keller H, Müller KA. EPR Evidence of Jahn-Teller Polaron Formation in $La_{1-x}Ca_xMnO_{3+y}$. Physical Review Letters. 1996;**77**(26): 5296. DOI: 10.1103/PhysRevLett.77.5296

[32] Ccahuana DL, Winkler E, Prado F, Butera A, Ramos CA, Causa MT, et al. Magnetic phase coexistence in CMR manganites: ESR evidence. Physica B. 2004;**354**:55-58. DOI: 10.1016/j.physb.2004.09.020

[33] Dormann E, Jaccarino V. High temperature EPR line widths in MnO and MnS. Physics Letters. 1974;**48A**:81. DOI: 10.1016/0375-9601(74)90409-5

[34] Causa MT, Tovar M, Caneiro A, Prado F, Ibañez G, Ramos CA, et al. High-temperature spin dynamics in CMR manganites: ESR and magnetization. Physical Review B. 1998; **58**(6):3233. DOI: 10.1103/PhysRevB.58.3233

[35] Banerjee BK. On a generalised approach to first and second order magnetic transitions. Physics Letters. 1964;**12**:16-17. DOI: 10.1016/0031-9163(64)91158-8

[36] Chebaane M, Bellouz R, Oumezzine M, Hlil EK, Fouzri A. Copper-doped lanthanum manganite $La_{0.65}Ce_{0.05}Sr_{0.3}Mn_{1-x}Cu_xO_3$ influence on structural, magnetic and magnetocaloric effects. RSC Advances. 2018;**8**:7186-7195. DOI: 10.1039/C7RA13244A

[37] Manh TV, Shinde KP, Nanto D, Lin H, Pham Y, Razaq DS, et al. Critical behavior and magnetocaloric effect in $La_{0.7}Ba_{0.25}Nd_{0.05}Mn_{1-x}Cu_xO_3$. AIP Advances. 2019;**9**:035345. DOI: 10.1063/1.5079842

[38] Khammassi F, Lopez JF, Chérif W, Mendoza A, Lanceros-Mendez S, Salazar D, et al. Short-range magnetic behavior in manganites $La_{0.93}K_{0.07}Mn_{1-x}Cu_xO_3$ ($0.0\leq x\leq 0.09$) above the Curie temperature. Journal of Physics D. 2021; **54**:175001. DOI: 10.1088/1361-6463/abde6b

[39] Issaoui F, Bejar M, Dhahri E, Bekri M, Lachkar P, Hill EK. Crystal, spin glass, Griffiths phases and magneticaloric properties of the $Sr_{1.5}Nd_{0.5}MnO_4$ compound. Physics B. 2013;**414**:42-49. DOI: 10.1016/j.physb.2012.12.039

[40] Vergara J, Ortega-Hertogs RJ, Madurga V, Sapiña F, El-Fadli Z, Martínez E, et al. Effect of disorder produced by cationic vacancies at the B sites on the electronic properties of mixed valence manganites. Physical Review B. 1999;**60**:1127. DOI: 10.1103/PhysRevB.60.1127

[41] Kamel R, Tozri A, Dhahri E, Hlil EK. Anomalous behavior above the Curie temperature in $(Nd_{1-x}Gd_x)_{0.55}Sr_{0.45}MnO_3$ (x=0, 0.1, 0.3 and 0.5). RSC Advances.

2019;**9**:27541-27548. DOI: 10.1039/c9ra03303c

[42] Standley HE. Introduction to phase transitions and critical phenomena. London: Oxford university Press; 1971

[43] Mira J, Rivas J, Rivadulla F, Vázquez-Vázquez C, A. López-Quintela MA. Change from first- to second-order magnetic phase transition in $La_{2/3}(Ca, Sr)_{1/3}$ MnO_3 perovskites. Physical Review B. 1999;**60**:2998. DOI: 10.1103/PhysRevB.60.2998

[44] Banik S, Banu N, Das I. Evolution from non-Griffiths phase to Griffiths phase: Giant enhancement of magnetoresistance in nanocrystalline $(La_{0.4}Y_{0.6})_{0.7}Ca_{0.3}MnO_3$ compound. Journal of Alloys and Compounds. 2018; **745**:753-760. DOI: 10.1016/j.jallcom.2018.02.243

[45] Oguchi T. A Theory of Antiferromagnetism, II. Progress in Theoretical Physics. 1955;**13**(2):148-160. DOI: 10.1143/PTP.13.148

[46] Allub R, Alascio B. Effect of Disorder on the Magnetic and Transport Properties of $La_{1-x}Sr_xMn0_3$. Solid State Communications. 1996;**99**:9613-9617. DOI: 10.1016/0038-1098(96)00337-7

[47] Smart JS. Effective field theories of magnetism. Philadelphis: Saunders; 1966

[48] Gammel J, Marshall W, Morgan L. An Application of Pade Approximants to Heisenberg Ferromagnetism and Antiferromagnetism. Proceedings of the Royal Society of London A. 1963;**275**: 257-270. DOI: 10.1098/rspa.1963.0169

[49] Rushbrooke GS, Wood PJ. On the Curie points and high temperature susceptibilities of Heisenberg model ferromagnetics. Molecular Physics: An International Journal at the Interface Between Chemistry and Physics. 1958; **1**(3):257-283. DOI: 10.1080/00268975800100321

[50] Schmidt H-J, Lohmann A, Richter J. Eighth-order high-temperature expansion for general Heisenberg Hamiltonians. Physical Review B. 2011; **84**:104443. DOI: 10.1103/PhysRevB.84.104443

[51] Lohmann A, Schmidt H-J, Richter J. Tenth-order high-temperature expansion for the susceptibility and the specific heat of spin-s Heisenberg models with arbitrary exchange patterns: Application to pyrochlore and kagome magnets. Physical Review B. 2014;**89**: 014415. DOI: 10.1103/PhysRevB.89.014415

Chapter 4

Optimal Conditions for Preparation of Perovskite Materials for Optoelectronic Devices

Akin Olaleru, Joseph Kirui, Olasoji Adekoya and Eric Maluta

Abstract

Several data on the preparation of perovskite crystals have been obtained because samples/devices were prepared using films of different qualities. Identifying optimal conditions for perovskite material synthesis and thin film preparation as well as optimizing the properties will go a long way in reducing the disparities in the data obtained. The optimal composition management of various elements of perovskite remains an outstanding research. The chapter will pave the way for the optimum design of the synthesis process of perovskite-based devices for better performance. Further still, the study provides basis for explaining the effective optimizations of synthesis conditions and material properties.

Keywords: optimal conditions, perovskite material, film, properties, optoelectronic devices

1. Introduction

In recent decades, modern technologies in the fields of optical materials and optoelectronics, plus energy storage and up-conversion luminescent applications have received spectacular prominence. They have been instrumental in tackling socio economic needs, including the rising global energy demand, the increasing demand for renewable clean energy, the transition to digitization, the Internet of Things, and the development of multifaceted applications like colorful optoelectronic devices, heat control and agrivoltaics [1–3].

Various materials have been synthesized by many research groups to fabricate optoelectronic and photovoltaic (PV) devices, but, only a few materials have met the basic requirement to manufacture optoelectronic and PV devices. The majority of electronic and optoelectronic devices are made using GaAs, CdTe, Si, $CuInSe_2$ and InP semiconductor materials. Of these materials, the Si semiconductor dominates the market currently [4].

Silicon is globally employed in PV devices [5] and likewise for optoelectronic devices, several silicon materials are employed for manufacturing. During synthesis and fabrication silicon materials require sophisticated equipment which operate at high temperatures. Also, fabrication of crystalline silicon is conducted in an extremely controlled environment to prevent oxidation as the combined oxygen resists the movement/alternation of electrons/charge carriers in PV cell, thereby decreasing the required purity [6, 7]. Consequently, to maintain every basic parameter, the total cost rises and the entire procedure becomes complex.

The demand for better and more sustainable material is increasing either to reduce or replace the dominance of silicon materials. More efficient materials will be needed to meet the growing global need. Hence, for manufacturing of efficient PV cells as well as optoelectronic devices, there is an urgent need to search for the useful combinations of potential materials. The arrival of hybrid perovskite material has created enormous excitement in PV and optoelectronic community.

Hybrid perovskite is a semiconductor which may be described as a class of materials that mix organic, inorganic and halides components. Perovskite has the chemical formula of ABX_3, in which A is an organic or metal cation (such as MA+, FA+, Cs+), B is a metal cation (such as Pb2+, Sn^{2+}), and X is a halogen (such as Cl-, Br-, I-).

Perovskite materials exhibit excellent optoelectronic properties and lower crystallization activation energies, in contrast to silicon [6, 7] Hybrid perovskites, which blend the advantages of excellent optical and electronic properties together with solution-processed manufacturing, have appeared as a new class of revolutionary and innovative optoelectronic materials with the potential for several practical applications.

Furthermore, from an application perspective, this family of materials is employed initially as PV cell material, where they have escalated rapidly to be a rival with silicon in the space of a few years, and have now proven to be useful in nearly all optoelectronic devices as shown in **Figure 1**.

1.1 Motivation behind the optimal control and point of interest

Organic–inorganic halide perovskites have emerged as high performance optoelectronic materials and show still greater potential due to their unique properties

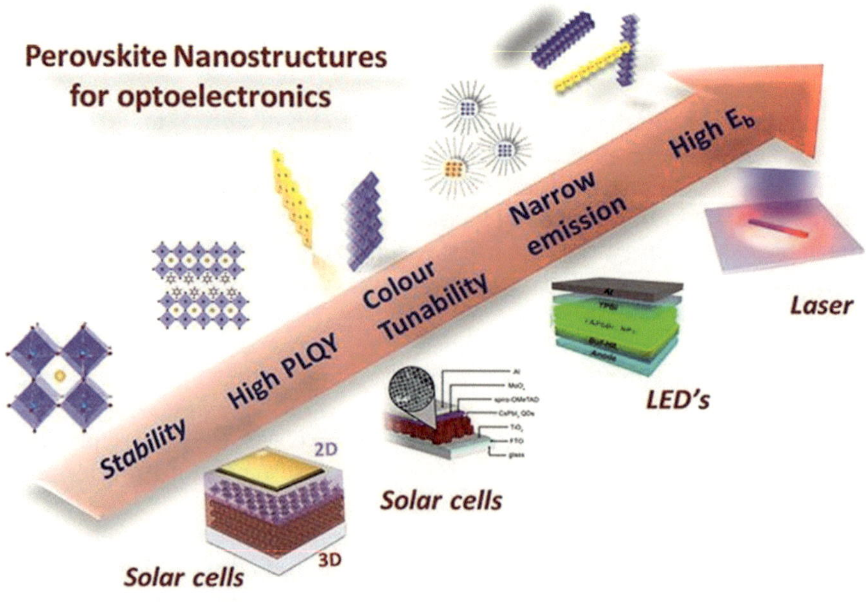

Figure 1.
Potential application in various fields of optoelectronic [8].

such as tunable band gap, long charge carrier diffusion length, high absorption coefficient, large carrier lifetime and ambipolar charge transport. At the same time, this kind of rapid increase in the efficiency of perovskite-based solar cells, for nearly a decade, has not been a walk in the park. The credit goes to the researchers/scientists working around the planet utilizing different device fabrication and design methods [9]. Interestingly, perovskites blend the properties of inorganic materials like high photoluminescence quantum yield, long carrier diffusion lengths, high color purity together with the properties of organic materials such as solution processability at low temperature, and high production yield [10].

Additionally, there are two considerations that increase the market value of this semiconductor material. First of all, crude materials should be cheap. Luckily, the starting materials of perovskites (like C, H, Pb, Cl, I and Br) are earth abundant. Consequently, the cost of raw materials will not hamper the commercial use of perovskite materials.

In the second place, the manufacturing method need to be based on inexpensive equipment and procedures. Once more, formation of perovskites exhibits the affordable characteristic, as a result of uncomplicated preparation of perovskite films (i.e., low formation energy or simple reaction between metal halide and organic halide), low-cost equipment [6].

Thanks to the epic efforts on perovskite materials from perovskite research community, remarkable achievement has been made for both laboratory and large scale fabrication processes especially for PV cells. But, based on the screening from literature [6] there are several issues and parameters that should be considered and optimized as explained in the following paragraphs.

The chemical composition management along with the structure of perovskite with best output need to be broadly investigated, since the crystallization dynamics are highly dependent on the composition of solution, concentration, and solvent type. At the moment, numerous compositions are published [7, 11], and several manufacturing techniques, together with treatment techniques (like types of annealing, and variety of additives) are employed for fabrication of optoelectronic devices. However, there is no one technique that is powerful due to the various compositions and the corresponding numerous physiochemical properties of them. Hence, the role of each technique or its mechanism should be deciphered so as to establish the protocol or the optimal condition for manufacturing optoelectronic devices with greater performance especially in large scale production.

More focus should be on stability, and currently concerted efforts are ongoing to find a lasting solution. But for now, encapsulation is just considered as a complementary technique, and the need for tackling stability concern is to explore perovskite materials with improved intrinsic stability and superb optoelectronic property. Also, the issue of lead toxicity though very minute in quantity can be solved by using effective encapsulation technology to reduce Pb leakage into the surroundings and recycling technique. In conjunction with above issues, the fabrication process of large-scale devices plus fabrication parameters should be broadly studied and meticulously modified to obtain high-quality perovskite films [6, 7, 10].

The main objective in material optimization is to indicate the properties and performance of materials as well as optimal synthesis control prior to fabrication of the devices. This will serve as a reference point/requirement for good repeatability of processing and manufacturing for large-scale production. In addition to optimizing the material composition and fabrication techniques, the sequential development of perovskite film quality such as film coverage, grain size, surface passivation, offer more incremental improvements.

Moreover, the quality of a film for example is controlled by nucleation and crystallization of the material which consequently affects the properties of the film and its stability. Hence, optimizing the perovskite crystals is a noteworthy technique for enhancing the properties of the perovskite film. Controlled formation of perovskite crystal during preparation method is vital in achieving better morphological properties and consequently the material properties. Basically, the morphology and size of the crystals are largely affected by the solvent employed, annealing time and annealing temperature [12, 13].

A lot of data on the preparation of perovskite crystals has been obtained because samples/devices were prepared using films of different qualities. Identifying optimal conditions for perovskite material synthesis and thin film preparation as well as optimization of the properties will go a long way in reducing the disparities in the data obtained. The optimal composition management of various elements of perovskite remains an outstanding research.

The tunability of the hybrid perovskites through their halides and its constituents has many potentials for methodically establishing structure–function relationships to help design novel perovskites. Also, mechanisms of nucleation and growth of these perovskite crystals should be part of these synthetic observation

2. Treatment method: formation of qualitative perovskite film

An extensive various morphologies have already been achieved simply by varying the crystallization parameters thereby attaining uniform nucleation and compact films, especially on smooth substrates. As these films are prepared at low temperature and often by rapid crystallization, it appears that kinetics play a major role in the process.

Treatment technique is the procedure employed to obtain a highly qualitative film with an utmost degree of crystallinity and coverage of a corresponding/associated substrate. The quality of a film in terms of crystallinity, uniformity and coverage is mostly linked to a controlled development of the morphology in the process of film formation.

The morphology of perovskite film is a primal factor in deciding/resolving the charge carrier dynamics such as carrier lifetime, the carrier diffusion lengths, and ultimately the device performance [10, 14].

Morphology control has been improved through careful optimization of processing conditions. Inert processing environments, optimized spin coating conditions and adequate annealing temperature control allowed improvements in film morphology resulting in enhanced efficiency of devices.

For example, poor morphology which is characterized by uneven/irregular grain size, voids/pinholes, low coverage and high roughness leads to low light absorption and ineffective charge transfer. Additionally, it causes easy degradation due to the hydrophilic nature of perovskite. Therefore, from application perspective, it should be stressed that stability is as important as efficiency.

For the energy-efficient PSCs, large grains with interlinked grain boundaries are required, whereas for the high-efficient light-emitting diodes (LEDs), small grains with pinhole-free film morphology are needed. Hence, it is vital to track the crystal growth and control in accordance with the application. Good film morphology implies a smooth, pinhole-free, and compact film as well as favorable interfacial electrical properties [10, 14].

In order to address the issues raised above, the following considerations which are the focus of this chapter must be optimized.

2.1 Heat treatment: nucleation and crystal growth

Perovskite materials show a vast array of film properties such as size of grain, morphology, surface coverage, crystallinity, etc. as a result of different processing techniques. Numerous works have also demonstrated that perovskite films show composition-structure relationships in term of properties [15]. Therefore, in achieving excellent control over the reaction between the inorganic and organic elements, various process parameters have to be included in preparation so as to produce high quality thin film. Chief among these are: solvent engineering, stoichiometry, thermal treatment, and additives.

Irrespective of the manufacturing route, controlling and understanding the preparation parameters, such as precursor's concentration, annealing temperature, and used solvents and additives, are crucial. They play fundamental roles in the film surface quality, coverage, conversion of precursors to perovskite, degree of crystallinity, size of the crystal, hence the performance of the layer and the whole device. In addition, perovskite film coverage is a function of annealing temperature during processing by solution technique. Lower annealing temperatures produce film poor convergence while higher annealing temperatures give rise to decomposition of active layer. In sum, final output of the perovskite PV devices is closely dependent on the perovskite film quality.

2.2 Controlling crystallization process: temperature

Temperature treatment is one of the frequently employed methods to aid this fabrication process. The fundamental difference in the crystal structure and morphology is dependent upon annealing temperature. In the synthesis of perovskite by solution processable method, the temperature plays a key role. The optimum temperature for the growth of perovskite is between 70 and 110°C, and the optimum growth temperature is about 110°C [7, 16].

It has been reported that higher processing temperatures close to or greater than the optimal temperature for crystallization of perovskite may cause simultaneous evaporation of solvent and growth of crystal thereby restraining the processing window along with the repeatability/reproducibility of device fabrication. In light of this, preparation of a controlled crystallization protocol that is in line with low-temperature deposition of precursor films is extremely needed for manufacturing of large-scale perovskite thin film. Thus, it is of paramount importance to investigate novel procedure for fabrication of hybrid perovskites with low-temperature and offer a profound understanding of some basic properties of these materials [13, 17].

Crystallization is a complex process which is designed for preparing a crystalline material from liquid, gaseous, or amorphous solid systems. Nucleation and crystal growth are the two basic stages of great significance which usually occur during the formation of a supersaturated state. Specific to polycrystalline perovskite thin-film growth, the one-step fabrication method has been extensively examined and the exploration of the crystallization mechanism would stimulate the large-scale fabrication procedure. The common crystallization technique entails three steps: (1) the supersaturation of solution; (2) the nuclei formation (3) the crystal growth. By means of the anti-solvent together with the evaporation of the solvent, the solution is supersaturated, then the nucleation process begins, accompanied by the consumption of solute, and the beginning of the crystal growth [18, 19].

As indicated, it is noteworthy that the crystallization process runs under ideal conditions when the precursor deposition is achieved with required properties. Based on the details stated above, it is useful to remember that the variation of two basic steps of the crystallization process – nucleation and crystal growth with the aid of optimization of some parameters like selection of solvent, solution concentration, precursor ratio, annealing temperature – is a vital point for formation of a qualitative and uniform film of perovskite without voids and with a full coverage. Furthermore, it is essential to observe the optimum conditions for high perovskite crystallinity, which enables the separation efficiency of charges, their transport and diffusion length [7].

3. Optimal conditions for hybrid perovskite crystallization

Having understood the importance of high-quality perovskite films as an active layer in a device, we now turn to outline the generality of the protocol for preparation of perovskite material. This will increase reproducibility and enhance the stability.

There are various techniques for regulation and determination of optimal conditions for crystallization as listed in **Figure 2**; they are solvent selection, regulation of solvent evaporation rate, special additions including antisolvents and additives, annealing by solvent and temperature annealing.

3.1 Solvent selection

In the preparation of perovskite, the choice of solvent is a crucial parameter influencing the crystallization and morphology of perovskite. In the selection of preferred solvent, the key perquisite/condition is that the solvent must be polar for easy dissolution of precursors and its physical properties like boiling point and vapor pressure, must to be taken into consideration with respect to the desired crystallization mechanism, that is, rapid or slow. Polar *solvents* such as dimethyl sulfoxide (DMSO), dimethylformamide (DMF), γ-butyrolactone (GBL), *N*-Methyl-2-pyrrolidone (NMP) are the generic solvents used in dissolving precursor and mixing these solvents is a window of opportunity to optimize the crystallization [21].

It is important to carefully control the solvent evaporation as a means to form stepwise crystallization. The dissimilarity in the crystallization process is observed when employing different solvents: (DMF > DMSO > NMP from fast to slow crystallization).

DMF which has high vapor pressure evaporates rapidly during spin coating leading to short drying window for formation of perovskite thin films. While the other solvents like DMSO and NMP have lesser vapor pressure and are tricky to be evaporated, therefore, it highly prolongs the drying window. To that effect, DMF is generally not utilized alone, but it is mixed with other solvents so as to widen/prolong the antisolvent window.

The preferred solvent in these series for formation of qualitative perovskite film is DMSO. Nevertheless, perovskite produced from this solvent has several drawbacks like incomplete conversion (non-reacted PbI_2) and large polydispersity in crystal size. These limitations can be resolved by mixing the solvents [7].

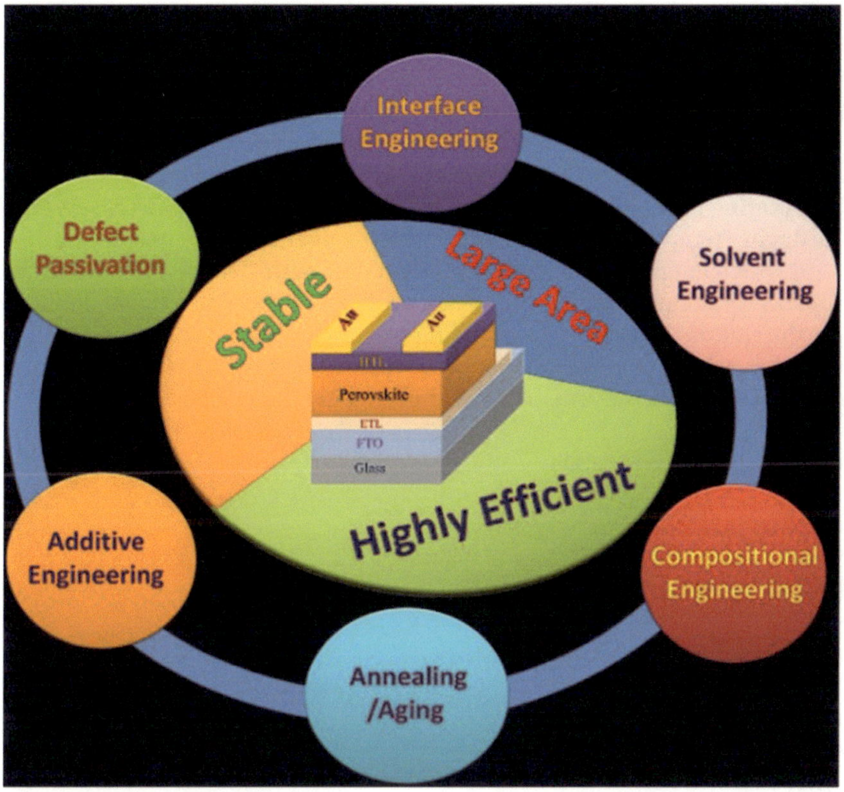

Figure 2.
Parameters to optimize for up-scaling [20].

3.2 Anti-solvent treatment

A solution is a mixture of solute dissolved in a solvent. An anti-solvent is a liquid that does not dissolve the solute but is miscible with the solvent. During preparation of perovskite via solution technique, perovskite precursor is dissolved in solvents.

Antisolvent is commonly added to the perovskite film after its formation or during its growth in order to avoid decomposition and reaction with perovskite. During spin coating procedure, most of the solvent is removed owing to the centrifugal force produced by spinning. But, there is still leftover solvent in the film made by spin coating due to film wetness. For qualitative crystallization, this residual solvent needs to be removed by thermal annealing. But, these solvents evaporate gradually during annealing which may lead to poor film morphology, hence influencing the overall performance of perovskite based devices. To overcome this issue, the use of antisolvent was introduce during spinning operation so as to quickly lessen the solubility of perovskite precursor and facilitate the rapid crystallization, which enhances the performance of devices.

Various anti-solvents were reported in literature [6, 7, 21] such as chlorobenzene (CB), benzene, xylene (XYL), toluene (TL), ethyl acetate etc. The properties of these anti-solvents particularly the boiling points and polarity, occupy an important place in the quality of the films. For instance, if the anti-solvent polarity is as strong as usual solvent, it will dissolve perovskite. The optimal values for suitable antisolvent polarity fall within 2–4.5 [22], above 4.5 is detrimental to the growth of perovskite film. Also, the antisolvent polarity ascertains the miscibility of antisolvent to solvent, which initiates the effect of removal of antisolvent on solvent. The high polarity increases miscibility of the antisolvent with the solvent, which give rise to the high solvent removal rate. However, antisolvent with too low polarity will result in poor solvent removal effect [6]. Moreso, favorable antisolvent depends also on the boiling point. The boiling point of antisolvent checks the rate of perovskite crystal growth [19]. The drying tempo of high boiling point antisolvent is delay in spin coating procedure, which lengthens the period of crystal growth. The presence of antisolvent in the film offers adequate fluidness, which enables the neighboring nuclei bigger and enhances the size of the grain. If the boiling point is small, the antisolvent will melt away too quick, which give rise to poor removal effect of solvent [23]. In sum, the fundamental role of antisolvent are listed below:

- Improving surface coverage
- Increasing grain size and crystallinity
- Minimizing film roughness
- Enhancing photovoltaic performance
- Increasing stability

3.3 Additive treatment

An additive is a substance added to something in small quantities to improve it in some way. It is another parameter affecting the morphology of perovskite. Additives can be added in two ways: one is to add additives in perovskite precursor and the other is to add them in the antisolvent. In additive treatment, the following considerations must be optimized;

First and foremost, additives must be soluble in perovskite dissolvents. If not, they should be added into the antisolvent [24]. The basic roles of additives are as follows:

- Stabilizing the crystal structure
- Improving the uniformity of the film
- Solving the problem of too fast crystallization of perovskite synthesized by antisolvent technique
- Delaying the crystal growth and aid in formation of dense perovskite films with larger grain size.
- Passivating the grain boundaries and prevent non-radiative recombination

Various anti-solvents were reported in literature, such as hydroiodic acid (HI), sulfobetaine zwitterion, Chlorine ion additives, ionic liquid (methyl formate), dimethyl sulfoxide (DMSO), etc.

For instance, the aim of adding additives like HI and sulfobetaine is to stabilize the crystal structure [24], enhance the film uniformity and addressing the issue of too fast crystallization. Also, Terephthalic acid (TPA) [6, 24] and conjugated polymer [6, 24] and the likes are employed to regulate the nucleation and crystal growth of perovskite, increase the crystallinity and stability of perovskite, and minimize the defects in the film. In short, it was observed that the primary role of the additives is to passivate the grain boundaries (GBs) and prevent non-radiative recombination [24].

4. Summary and outlook

The main point of this review is to establish the optimal conditions/control for increasing the repeatability/reproducibility of perovskite thin films fabrication especially for large scale production. The thin film quality synthesized by convectional solution procedure could be poor, which is characterized by irregular grain size, voids, low coverage and high roughness. The application of optimal conditions will greatly solve these problems and increase the reproducibility of preparation. Proper annealing temperature optimizes the nucleation and perovskite crystal growth. Controlling the crystallization of perovskite by the extraction of solvent using antisolvent treatment is very important. The boiling point and polarity of the preferred antisolvent must be moderated that is, not too high or low, otherwise, it will lead to poor morphology of the film. Mixing antisolvents with different polarities can be employed to neutralize the polarity, and hence the optimal crystallization rate will be achieved.

From a practical perspective, actualization of the essential features in the cells and devices directly depends on perovskite film quality (uniformity, absence of voids, high degree of crystallinity and coverage, optimum sizes of crystallites). The considerations and optimal conditions stated in the content of the write up are applicable to all the available techniques of film fabrication with indication of optimization mechanism of perovskite formation and crystallization.

Author details

Akin Olaleru[1*], Joseph Kirui[2,3], Olasoji Adekoya[1] and Eric Maluta[2,3]

1 Physics Department, Yaba College of Technology, Lagos, Nigeria

2 Physics Department, University of Venda, Limpopo, South Africa

3 National Institute for Theoretical and Computational Sciences (NITheCS), Johannesburg, South Africa

*Address all correspondence to: akinolaleru@gmail.com

References

[1] Pecunia V, Occhipinti LG, Hoye RLZ. Emerging indoor photovoltaic technologies for sustainable internet of things. Advanced Energy Materials. 2021;**11**(29):2100698

[2] Forbes, IoT: Number of Connected Devices Worldwide 2012-2025. Available from: https://www.statista.com/statistics/471264/iot-number-of-connected-devices-worldwide/ [Accessed: February 2021]

[3] Available from: https://www.nanowerk.com/spotlight/spotid=60507.php

[4] Hossain A et al. The hybrid halide perovskite: Synthesis strategies, fabrications, and modern applications. Ceramics International. 2021. DOI: 10.1016/j.ceramint.2021.11.313

[5] Cariou R et al. III–V-on-silicon solar cells reaching 33% photoconversion efficiency in two-terminal configuration. Nature Energy. 2018;**3**:326-333

[6] Sun H, Dai P, Li X, et al. Strategies and methods for fabricating high quality metal halide perovskite thin films for solar cells. Journal of Energy Chemistry. 2021;**60**:300-333

[7] Meillaud F et al. Recent advances and remaining challenges in thin-film silicon photovoltaic technology. Materials Today. 2015;**18**:378-384

[8] Kulkarni SA et al. Perovskite nanostructures: Leveraging quantum effects to challenge optoelectronic limits. Materials Today. March 2020;**33**:122-140. DOI: 10.1016/j.mattod.2019.10.021

[9] Srivastava P, Bag M. Elucidating tuneable ambipolar charge transport and field induced bleaching at the CH3NH3PbI3/electrolyte interface. Physical Chemistry Chemical Physics. 2020;**22**:11062-11074. DOI: 10.1039/d0cp00682c

[10] Kumar J, Srivastava P, Bag M. Advanced strategies to tailor the nucleation and crystal growth in hybrid halide perovskite thin films. Frontiers in Chemistry. 2022;**10**:842924. DOI: 10.3389/fchem.2022.842924

[11] Olaleru SA et al. Perovskite solar cells: The new epoch in photovoltaics. Solar Energy. 2020;**196**:295-309

[12] Guo F et al. A generalized crystallization protocol for scalable deposition of high-quality perovskite thin films for photovoltaic applications. Advancement of Science. 2019;**6**:1901067. DOI: 10.1002/advs.201901067

[13] Tombe S et al. The influence of perovskite precursor composition on the morphology and photovoltaic performance of mixed halide MAPbI3-xClx solar cells. Solar Energy. 2018;**163**:215-223

[14] Amrollahi Bioki H et al. Improved morphology, structure and optical properties of $CH_3NH_3PbI_3$ film via HQ additive in PbI2 precursor solution for efficient and stable mesoporous perovskite solar cells. Synthetic Metals. 2022;**283**:116965

[15] Kim H-S, Hagfeldt A, Park NG. Morphological and compositional progress in halide perovskite solar cells. Chemical Communications. 2019;**55**:1192

[16] E3S Web of Conferences ICREN 2020. Vol. 239; 2021. p. 00020

[17] Jeon NJ, Noh JH, Yang WS, Kim YC, Ryu S, Seo J, et al. Compositional

engineering of perovskite materials for high-performance solar cells. Nature. 2015;**517**:476-480

[18] Chen Z, He P, Wu D, Chen C, Mujahid M, Li Y, et al. Processing and preparation method for high-quality opto-electronic perovskite film. Frontiers in Materials. 2021;**8**:723169. DOI: 10.3389/fmats.2021.723169

[19] Hsieh H-C, Yu J, Rwei S-P, Lin K-F, Shih Y-C, Wang L. Ultracompact titanium oxide prepared by ultrasonic spray pyrolysis method for planar heterojunction perovskite hybrid solar cells. Thin Solid Films. 2018;**659**:41-47. DOI: 10.1016/j.tsf.2018.05.002

[20] Tailor NK, Abdi-Jalebi M, Gupta V, Lu H, Dar MI, Li G, et al. Recent progress in morphology optimization in perovskite solar cell. Journal of Materials Chemistry A. 2020;**8**:21356-21386. DOI: 10.1039/D0TA00143K

[21] Cohen B-E et al. Parameters that control and influence the organo-metal halide perovskite crystallization and morphology. Frontiers of Optoelectronics. 2016;**9**(1):44-52. DOI: 10.1007/s12200-016-0630-3

[22] Bu T, Wu L, Liu X, Yang X, Zhou P, Yu X, et al. Synergic interface optimization with green solvent engineering in mixed perovskite solar cells. Advanced Energy Materials. 2017;**7**:1700576

[23] Helian S et al. Strategies and methods for fabricating high quality metal halide perovskite thin films for solar cells. Journal of Energy Chemistry. 2021;**60**:300-333

[24] Ngo TT, Suarez I, Antonicelli G, Cortizo-Lacalle D, Martinez-Pastor JP, Mateo-Alonso A, et al. Enhancement of the performance of perovskite solar cells, LEDs, and optical amplifiers by anti-solvent additive deposition. Advanced Materials. 2017;**29**:1604056

Chapter 5

Role of Surface Defects and Optical Band-gap Energy on Photocatalytic Activities of Titanate-based Perovskite Nanomaterial

Izunna Stanislaus Okeke, Priscilla Yahemba Aondona, Amoge Chidinma Ogu, Eugene Echeweozo and Fabian Ifeanyichukwu Ezema

Abstract

In recent years, water pollution has become one of the major challenges faced by humans because of consistent rise in population and industrial activities. Water pollution due to discharge from cosmetics and pharmaceutical wastes, organic dyes, and heavy metal seen as carcinogens has the potential to disrupt hormonal processes in the body. Different approaches such as chlorination, aerobic treatment, aeration, and filtration have been deployed to treat wastewaters before being discharged into the streams, lakes, and rivers. However, more attention has been accorded to treatment approaches that involve use of nanomaterial due to non-secondary pollution, energy efficiency, and ease of operation. Titanate-based perovskite (TBP) is one of the most frequently studied nanomaterials for photocatalytic applications because of its stability and flexibility in optical band-gap modification. This chapter provided an overview of basic principles and mechanisms of a semiconductor photocatalyst, and current synthesis techniques that have been used in formulating TBP nanomaterial. The effect of reaction conditions and approaches such as doping, codoping, composites, temperature, pH, precursor type, surface area, and morphology on surface defects and optical band-gap energy of TBP nanomaterial was highlighted. Importantly, the impact of surface defects and optical band-gap energy of TBP on its photocatalytic activities was discussed. Finally, how to enhance the degradation efficiency of TBP was proposed.

Keywords: titanate-based perovskites, surface defects, optical bandgap, photocatalysis, organic pollutants

1. Introduction

Currently, water pollution is becoming one of the most serious challenges confronting human beings due to steady increase in population, advancements in industrial activities, and urbanization [1]. Contamination of water bodies through discharge of organic pollutants such as dyes, heavy metals, and petroleum are posing a significant danger to humans as well as the aquatic ecosystem [2]. The effects of these pollutants differ and depend on the source and type; for example, organic dyes and heavy metals have been recognized as carcinogens while cosmetics and pharmaceutical waste products have been identified as endocrine disruptive agents [3]. These agents impede hormonal processes, thereby disturbing regular homeostatic reproduction, advancement, or behavior [4]. Furthermore, presence of dyes in the water bodies blocks sun from penetrating into the water bodies and lessens dissolved oxygen, therefore, causing death to photosynthetic organisms that live in the aquatic system [5].

To mitigate the impact of water pollution, scientists are making efforts to develop approaches to treat wastewaters before being discharged into rivers, streams, and underground water. Approaches such as chlorination, aerobic treatment, aeration, and filtration have been used to treat waste. However, greater attention has been given to treatment processes that involve the use of nanomaterials. Inorganic metal oxides nanoparticles (NPs) have benefits such as no secondary pollution, cost, and energy efficiency and are easily operated [6, 7]. Metal-oxide NPs have been studied for potential degradation of contaminants [8–10], heavy metals [11, 12], and inactivation of bacteria [13–15]. Many other compounds such as metal halides [16], metal nitrides [17], and metal chalcogenides [18, 19] have also been evaluated for photocatalytic applications. Among which, perovskite nanomaterial has drawn so much attention because of its wide variety of properties. Perovskite nanomaterial has indicated a broad range of electro-optical effects, piezo, ferro, and pyro-electrical properties that enable them to exhibit outstanding performances as structural, electronic, and magnetic material [20]. In addition, because of their crystalline structure, perovskites have unique chemical properties that contain a spectrum of cations to generate surface defects, which can balance unstable oxidation states [21].

Titanate-based perovskite material is among the most commonly studied perovskites for photocatalytic applications under visible light. This is because their optical band-gaps energy is easily modified and quite stable for a long period throughout a photocatalytic reaction [22]. A good number of titanium-based perovskites such as $SrTiO_3$, $MnTiO_3$, $BaTiO_3$, $ZnTi_3$, $MgTiO_3$, and $CaTiO_3$ have optical band-gaps energy above 3 eV [23–25], thereby allowing photocatalytic activities only under UV source. On the other hand, TBP, such as $CoTiO_3$ and $NiTiO_3$, have band-gap energy lower than 3 eV but their CB is below the oxidation potential [26, 27], which also limits their chances for photocatalytic applications.

Excellent photocatalyst semiconductor (SC) is expected to have excellent charge carrier mobility and charge separation to prevent recombination of electron (e) and hole (h) generated in the system [28]. Efforts have been made by scientists to enhance photocatalytic activities of SC photocatalyst through various means by modification of physiochemical properties of the nanomaterial. The modification can be done via doping (with metals, nonmetals, and salts), defect engineering, heterostructure, and cocatalysts [29–31]. Research has shown that doping greatly affects the electronic structure of SC nanomaterial [32]. Like TBP, the majority of SC photocatalysts have wide band-gaps energy, which makes it difficult for them to have photocatalytic activation with a visible light source [33, 34]. To improve the efficiency and quantum yield of SC photocatalysts such as TBP, their optical band-gap energy should be altered to respond to visible light sources. Another parameter that affects the photocatalytic activities of this material is presence of defects. Defects seen in a material can be artificial or natural specific structure. Surface defects alter geometric structure, as well as the chemical environment of the host material [35]. It can also serve as charge carrier traps and adsorption sites; the induced electrons can be transferred to the sites and, therefore, prevent recombination of photogenerated $e^{-}h^{+}$ pairs.

This chapter highlighted fundamental principles and mechanisms of SC photocatalysis and recent synthesis techniques that have been deployed in preparing TBP nanomaterial. The influence of reaction conditions and approaches such as doping, codoping, composites, temperature, pH, precursor type, surface area, and morphology on surface defects and optical band-gap energy of TBP nanomaterial was noted. Ultimately, impact of surface defects and optical properties of TBP on its photocatalytic activities against organic pollutants was discussed. Considering progress recorded so far in this area, some perspectives on how to advance and improve degradation efficiency of TBP against organic pollutants were proposed.

1.1 Fundamental principles and mechanisms of photocatalysis

Generally, photocatalysis is initiated when a SC photocatalyst absorbs photons with energy equal to or higher than its optical band-gap energy. Consequently, electrons in the valence band (VB) are excited into the conduction band (CB) leaving holes in the VB. This excitation produces a potential difference midway CB and VB bands, creating reductive and oxidative entities at the CB and VB, respectively. These photo-activated charge carriers can react with H_2O or dissolved oxygen to generate free radicals such as OH and O^{-2} that can degrade pollutants into smaller molecules [36] as shown in Eqs. (1)–(6). It is fundamental that the minimum material CB is located at a higher negative potential compared to the reduction potential for H^+ to H_2, at the same time, it is also essential that the highest VB is located at a higher positive potential compared to the oxidation potential for H_2O to O_2 [22]. **Figure 1** describes the indirect organic pollutant degradation process by SC photocatalyst.

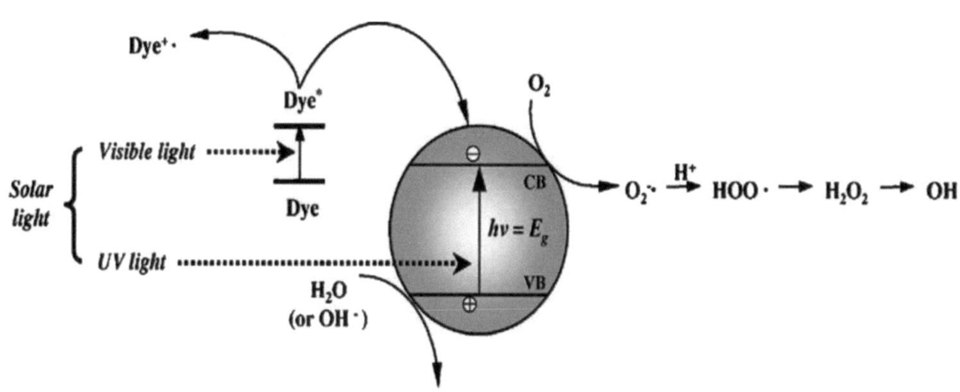

Figure 1.
Pictorial description of indirect organic pollutant degradation process by SC photocatalyst. Adapted with permission [37].

$$SC + hv \rightarrow e^-_{CB} + h^+_{VB} \tag{1}$$

$$h^+_{VB} + H_2O \rightarrow \bullet OH + H^+ \tag{2}$$

$$h^+_{VB} + \bullet OH^- \rightarrow \bullet OH_{ad} \tag{3}$$

$$e^-_{CB} + O_2 \rightarrow \bullet O_2^- \tag{4}$$

$$\bullet OH + pollutants \rightarrow CO_2, H_2O, salts \tag{5}$$

$$\bullet OH_{ad} \rightarrow \bullet OH_{free} + pollutant \rightarrow simple\ oxid.product \tag{6}$$

1.2 Crystal structure of a perovskites

Perovskite material is generally referred to as material whose crystal structure is described by the formula ABO_3, A and B are ions that often have different sizes and O is an ion that is bonded to A and B. It has a cubic structure that contains B cations in a 6-fold orientation encircled by an octahedral of anion while the A cation in a 12-fold cuboctahedral orientation [38]. **Figure 2a and b** describes the idealized cubic perovskite structure. From the crystal structure, B site cations are firmly glued to the oxygen (or other anion) at the same time A site cations interaction with the oxygen is relatively weaker. Based on the nature of the cations residing in the lattice sites, various perovskite crystal geometries can be obtained by modifying these interactions.

2. Synthesis techniques deployed in preparation of titanate-based perovskites nanomaterial

Various synthesis methods have been successfully deployed in the formulation of TBP nanomaterial. Surface defects and optical band-gap energy of TBP can be manipulated by deploying appropriate synthesis methods, which include the temperature, pH of the reaction, types of precursor, and solvents. In this section, we summarized recent methods deployed in synthesizes of TBP in **Table 1** including the precursors used, reaction conditions, and pollutant/application.

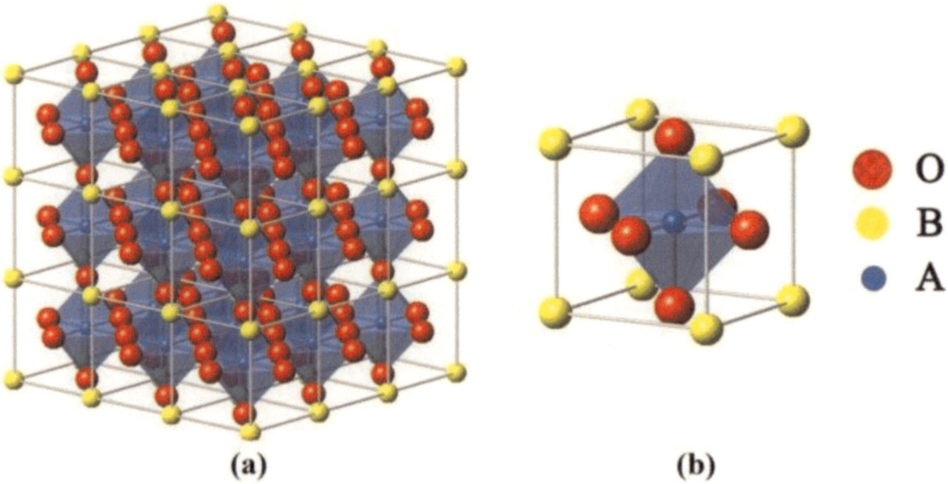

Figure 2.
(a and b) Crystal structure of a perovskite. Adapted from [39], copyright Springer Nature, 2019.

S/no	Method	Sample	Precursors	Light source	Conditions	Pollutant/application	References
1.	Solid state	$SrTiO_3$, $Sr_{1-x}BixTi_{1-x}Fe_xO_3 (0 \leq x \leq 0.4)$	$SrCO_3$ TiO_2 Bi_2O_3 Fe_2O_3	Visible	1200°C for 20 h	H_2 evolution	[40]
2.	One-pot hydrothermal	$SrTiO_3$-rGO composite	GO $Ti(OBu)_4$ $Sr.(NO_3)_2$	UV	180°C for 24 h	H_2 evolution	[41]
3.	Sol–gel	$CaTiO_3$ $CaTi_{1\_x}Cu_xO_3$	$Ca(NO_3)_2$ $Cu(NO_3)_2$ $Ti(OBu)_4$	UV	850°C for 7 h	H_2 evolution	[42]
4.	Hydrothermal	$CaTiO_3$	$Ca(NO_3)_2{\cdot}4H_2O$ $TiCl_4$ (TTC) $Ti(OC_3H_7)_4$ (TIP) $Ti(OC_4H_9)_4$ (TNB)	UV	180°C for 12 h	As(III)	[43]
5.	Modified hydrogenation	$CaTiO_3$	$CaCO_3$ TiO_2	Visible	60°C for 12 h	H_2 evolution	[44]
6.	Solution combustion	Na + co-doped $CaTiO_3$:$Eu_{3}+$	$Ca(NO_3)_2{\cdot}4H_2O$ $Ti(OBu)_4$ Eu_2O_3 Na_2CO_3	UV	900°C for 1 h	methylene blue (MB)	[45]
7	Sol–gel	$CaTiO_3$	(CaC_{l2}) $Ti(OC_4H_9)_4$	Visible	900°C for 2 h	methyl orange and rhodamine degradation	[46]
8	Microwave-assisted hydrothermal	$CaTiO_3$	TiO_2 $(CaCl_2$ $ErCl_3{\cdot}6H_2O$	Visible and near-infrared	180°C for 4 h	MB degradation	[47]
9	solid-state	Fe-doped $CaTiO_3$	$CaCO_3$ TiO_2 $Fe(NO_3)_3$	UV–visible	1400°C for 2 h	MB Degradation	[48]

S/no	Method	Sample	Precursors	Light source	Conditions	Pollutant/application	References
10	One-pot hydrothermal reaction	$CaTiO_3$	$[Ti(C_4H_9O)_4]$ $Ca(NO_3)_2$	Visible	200°C for 24 h	photocatalytic hydrogen generation	[49]
11	Hydrothermal	Rh doped $BaTiO_3$	TiO_2 (anatase) $RhCl_3$	Visible	180°C for 20 h	MB degradation	[50]
12	Hydrothermal	Rh doped $BaTiO_3$	$Ba(OH)_2.8H_2O$ $Rh(NO_3)_3$	Visible	160°C for 42 h	H_2 evolution	[51]
13	Sol–gel	Ag-doped $BaTiO_3$	$Ba(C_2H_3O_2)_2$ $Ti(OBu)_4$ $AgNO_3$	Visible	800°C for 2 h	Rhodamine B degradation	[52]
14	Hydrothermal	$PbTiO_3$, Ag-Fe codoped $PbTiO_3$	$C_{12}H_28O_4Ti$ Pb $(NO_3)_2$ $AgNO_3)$, $(FeH_{18}N_3O_{18}$	UV	400–600°C for 2 h	MB degradation	[53]
15	Hydrothermal microwave-assisted and hydrothermal autoclave	mixed $Bi_4T_{i3}O_{12}$/ $Bi_{12}TiO_{20}$ $Bi_4Ti_3O_{12}/B_{i2}O3$	TiO2 $Bi(NO_3)_3{\cdot}5H_2O$	Visible	700–800°C for 2 h	MB degradation	[54]
16	Microwave-assisted hydrothermal	$Bi_{12}TiO_{20}$	$Ti(OBu)_4$ $Bi(NO_3)_35H_2O$	Visible	180°C for 1 h	Rhodamine B Degradation	[55]

Table 1.
Recent techniques used in synthesize of TBP, samples and precursors used, the reaction conditions, and pollutant/application.

3. Impact of surface defects and optical band-gap energy on photocatalytic activities of titanate-based perovskite nanomaterial

In recent times, substantial work has been done to enhance the physiochemical properties of TBP nanomaterial, more especially surface defects and band-gap energy. This is because of their role in the photocatalytic activities of this group of perovskites. Attempts have been made to modify these two parameters via reaction conditions and synthesis techniques. In this section, the role of reaction conditions on these physiochemical properties was highlighted. The impact of surface defects and band-gap energy on photocatalytic activities of TBP nanomaterial was also discussed.

Zhuang and co [43] reported three different morphologies of $CaTiO_3$ formulated from three different Ti precursors through hydrothermal techniques for removal of As (III). After 40 mins of exposure to UV source, $CaTiO_3$ sample with ($Ti(OC_3H7)_4$) (TIP) as a precursor showed the highest activities (98.4%) in removal of As(III). **Figure 3** indicates the induced changes in the concentration of the As(III) under UV source against different morphologies of $CaTiO_3$. The higher photocatalytic activity of $CaTiO_3$ (TIP) was associated with its fern-like morphology and the higher specific area (108.14^2/g) when compared to other two.

Cai et al. [44] fabricated surface disordered $CaTiO_3$ using modified hydrogenation method. The as-prepared samples were utilized as a model to study roles of surface oxygen vacancies (SOVs) on photocatalytic H_2 evolution under a visible light source. They demonstrated that $CaTiO_3$ hydrogenated at 700°C induced more SOVs with a very high photocatalytic H_2 evolution rate (2.96 mmol g^{-1} h^{-1}). A value that is almost 49 times greater than the value of the pristine $CaTiO_3$ without SOVs. They suggested that the significant increase in H_2 production in the hydrogenated samples was because of the induced SOVs, which lower surface recombination of photogenerated e-h pair and increase charge separation.

Portia et al. [46] studied the influence of annealing temperature on photocatalytic activity of $CaTiO_3$ NPs under the visible light source. The samples were prepared via sol–gel method and were annealed at various temperatures of 500°C, 700°C, and

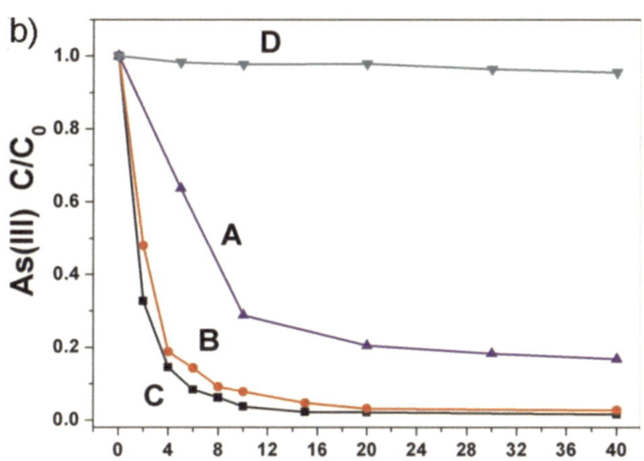

Figure 3.
The induced changes in the concentration of the As(III) under UV-254 nm for different morphologies of CaTiO (A) CaTiO3 TTC (B) CaTiO3 TNB (C) CaTiO3 TIP and (D) No photocatalyst. Reprinted with permission from [43].

900°C. The photocatalytic activities of the samples were assessed against MO and Rhb dyes. The sample annealed at 900°C exhibited the highest degradation efficiency (DE) 88% and 78% for Mo and Rhb, respectively, while sample annealed at 500°C showed DE of 53% and 38% for Mo and Rhb, respectively. The photocatalytic activity results also correlate with the result of UV–visible diffuse reflectance spectroscopy, which indicates the absorption of the sample annealed at 500°C to be 340 nm while the sample annealed at 900°C exhibited outstanding and improved visible—absorption with its peaks at 450 nm. The kinetic fit of the samples against MO and RhB dyes were shown in **Figure 4**. They associated the enhancement in photocatalytic activities of the sample annealed at 900°C to a decrease in the optical band-gap energy of the $CaTiO_3$ sample, specific surface area with porous features as well as a low recombination rate of the $e^{-}h^{+}$ pairs.

Lazono-Sánchez et al. [47] reported influence of Eu^{3+} doping and heat treatment on photocatalytic activities of $CaTiO_3$ under visible and near-infrared light sources. The results showed that samples doped with 1% Er^{3+} exhibited the highest degradation reaction rate against MB ($4.54 \times 10^{-5}\ s^{-1}$), a value, which is about 2.5 times higher than that of undoped sample $1.86 \times 10^{-5}\ s^{-1}$. Furthermore, the as-prepared samples were calcined at 850°C and their photocatalytic activity against MB was also evaluated under UV–vis–NIR irradiation. They reported a significant increase in the photodegradation reaction rate of the calcined sample against MB. The author attributed improvement in the photocatalytic reaction rate of the samples to introducing a new energy state in the optical band-gap of the $CaTiO_3$ and increased crystallinity. In another work, Yang et al. [48] reported photodegradation of MB by Fe-doped $CaTiO_3$ under UV–visible light irradiation. They found that doping $CaTiO_3$ with Fe enhanced its photocatalytic activity against MB. Nevertheless, the photocatalytic activities of Fe doped $CaTiO_3$ sample calcined at a high temperature of 500°C exhibited higher activity of about 100%. Yang et al. associated enhancement in the light absorption of Fe doped $CaTiO_3$ in the visible region with the high calcination temperature.

Recently, Chen et al. [45] reported Na + codoped $CaTiO_3{:}Eu^{3+}$ fabricated via solution combustion method using stoichiometric ratio of $Ca_{1-x}TiO_3{:}Eu_{3+x}$ ($x = 0$, 0.005, 0.01, 0.015, 0.02, and 0.025). The photocatalytic activity of the samples was

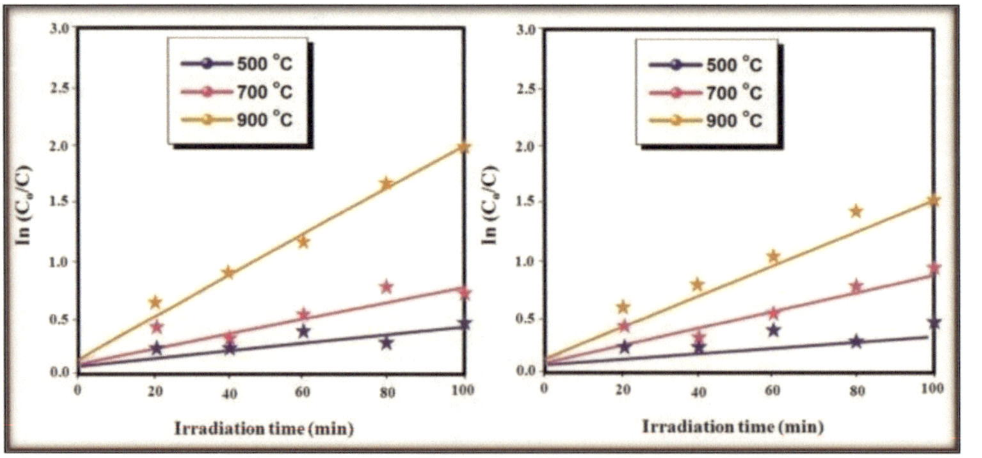

Figure 4.
The kinetic fit of the samples against MO and RhB dyes exposed to a visible light source. Reprinted with permission [46].

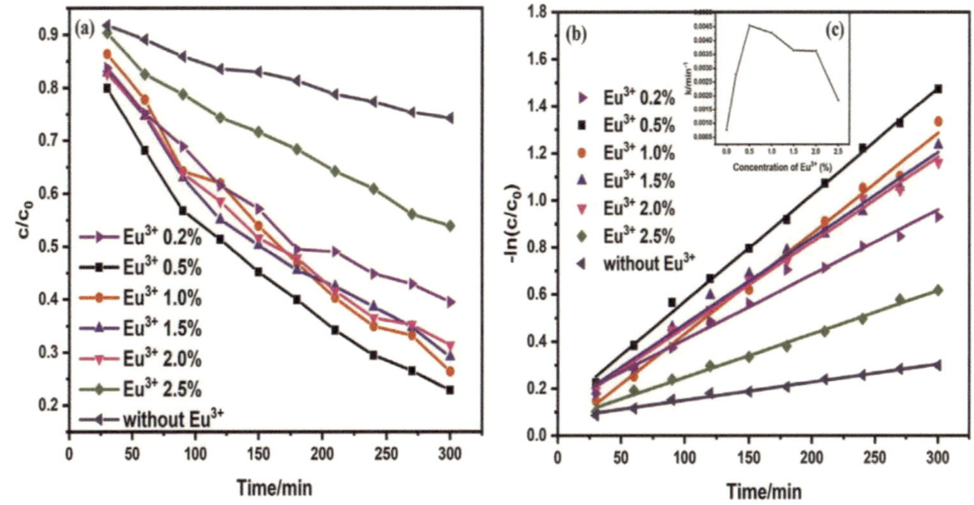

Figure 5.
$Ca_{1-x}TiO_3$:$Eu^{3+}{}_x$ samples degradation of MB under UV light. (b) $Ca_{1-x}TiO_3$:$Eu^{3+}{}_x$ kinetics plots of photocatalytic degradation of MB (c) $Ca_{1-x}TiO_3$:$Eu^{3+}{}_x$ first-order apparent rate constants (k) in photodegradation of MB Reprinted with permission [45].

evaluated against MB under UV light source. **Figure 5**(a, b, and c) indicates the $Ca_{1-x}TiO_3$:$Eu^{3+}{}_x$ degradation of MB, kinetics plots of photocatalytic degradation, and first-order apparent rate constants (k) in photodegradation of MB. They recorded significant improvement in the photocatalytic activity of $CaTiO_3$ after being doped with Eu^{3+}. The $CaTiO_3$ sample doped with 5% Eu^{3+} exhibited the highest DE relative to other samples, which the author attributed to the smaller optical band-gap energy of the sample relative to other samples. They also noted that Eu^{3+} doping led to a decrease in particle size of $CaTiO_3$ hence enhancing its light absorption potential.

Bhat et al. [50] reported the influence of Rh occupying sites in $BaTiO_3$ electronic structure. The samples were prepared by one-pot hydrothermal method and photocatalytic activities of the samples were evaluated against MB under visible light source. In the experiment, the authors found that samples doped with Rh exhibited higher photocatalytic activities than the pristine sample; with 0.5 Rh doped $BaTiO_3$ having the highest photodegradation against MB. They attributed enhancement noted in Rh doped sample to a decrease in band-gap energy. **Figure 6**(a and b) describes the photocatalytic of the sample against MB and the rate constants (K) of pristine $BaTiO_3$ and Rh-doped $BaTiO_3$ against MB. The DFT studies also predicted that an increase in Rh concentration causes a decrease in the optical band-gap energy of the samples, which further increases the absorption light of the sample within the visible region.

Similarly, Khan and co [52] reported effects of 1, 3, and 5% Ag doping on the optical properties and photocatalytic activities of $BaTiO_3$ against Rhb dye under visible light illumination. The decomposition efficiency of $BaTiO_3$, 1, 3, and 5% Ag doped $BaTiO_3$ NPs are 41, 46, 58, and 79%, respectively, as indicated in **Figure 7**. The linear increase in photocatalytic activities of $BaTiO_3$ was credited to the observed decrease in band-gap energy of $BaTiO_3$ NPs as Ag concentration is increased as well as the decrease in the e^-h^+ recombination rate.

Nishioka and Maeda [51] investigated H_2 evolution of Rh doped $BaTiO_3$ nanocrystal synthesized via the hydrothermal method with different precursors. The H_2 evolution was evaluated using aqueous ethanol at wavelength of >420 nm. They claimed that H_2 evolution via photocatalytic reaction of the samples depends strongly on the following; preferred TiO_2 and Rh precursor, precursor ratio, and postheating process. A sample that consists of anatase NPs as its main phase indicated higher photocatalytic

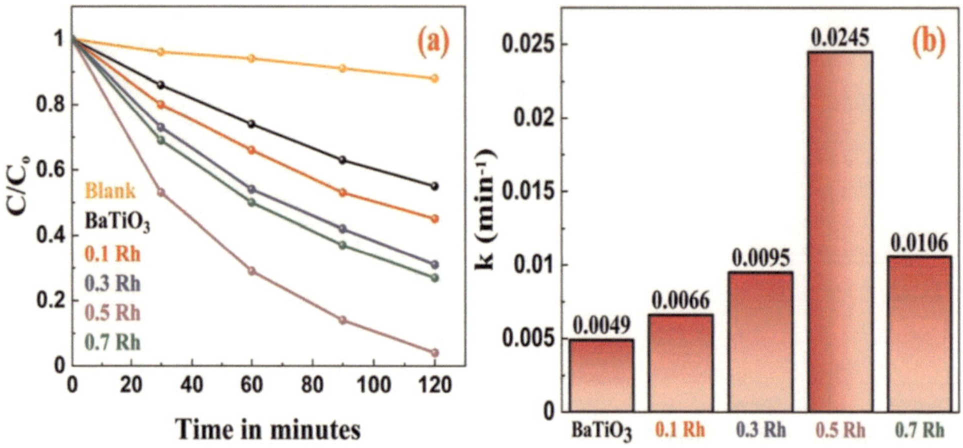

Figure 6.
(a) The photocatalytic degradation of MB and (b) the rate constants (K) of the pristine $BaTiO_3$ and Rh-doped $BaTiO_3$ against MB. Reprinted with permission [50].

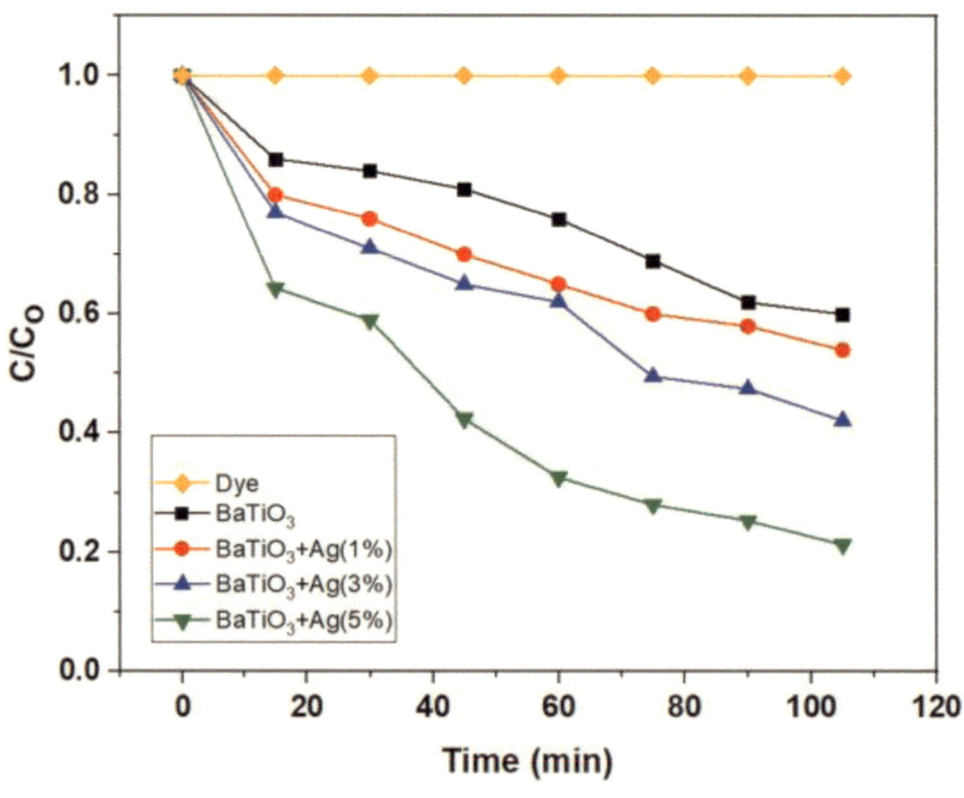

Figure 7.
Decomposition efficiency of the as-prepared samples against Rh B dye under visible light Illumination. Reprinted with permission [52].

activity than sample with rutile precursor. They attributed improved photocatalytic activities of post-treated Rh doped $BaTiO_3$ to higher temperature treatment.

Lu et al. [40] explored the effect of optical and structural properties of $SrTiO_3$-$BiFeO_3$ solid solution on its photocatalytic H_2 production. The as-prepared sample was prepared via solid-state method using stoichiometric ratio as $Sr_{1-x}Bi_xTi_{1-x}FexO_3$ ($0 \leq x \leq 0.4$). The photocatalytic activities of the samples were studied by investigating hydrogen production from Na_2SO_3 (aq) solution under visible light illumination. At wavelength (≥250 nm), an obvious increase in H_2 production was noted in the solid solution relative to pristine $SiTiO_3$. The highest activity was noted in $Sr_{0.9}Bi_{0.1}Ti_{0.9}Fe_{0.1}O3$ sample with mean H_2 production rate of 180 mol/h. This value corresponds to apparent quantum efficiency of 2.28% while pristine $SrTiO_3$ H_2 production activity rate was 17 mol/h. The pristine $SrTiO_3$ value is one order of magnitude lower than that of the solid solution as shown in **Figure 8**. The authors attributed the higher activity rate of a solid solution to the observed larger surface area, which in turn creates more surface reaction sites at the surface. In addition, $SrTiO_3$ indicated a distinct absorption edge in the UV region while the solid solution indicated a large absorption shoulder in the visible light region, which they link to metal-to-metal charge transfer.

He et al. [41] reported photocatalytic activities of $SrTiO_3$/reduced graphene oxide ($SrTiO_3$-RGO) composites, prepared through a one-pot hydrothermal process. The as-prepared sample $SrTiO_3$–0.8%RGO composites exhibited a considerable increase in photocatalytic production of H_2 (363.79 mmol $(g\ h)^{-1}$) than $SrTiO_3$ under UV light exposure. The higher H_2 production rate observed in the composite sample was associated with generation of more reactive sites and low recombination rate of the $e^{-}h^{+}$ caused by a suitable amount of RGO composition in SrTiO3. Moreover, the photoelectrochemical and electrochemical impedance spectroscopy analyses as shown in **Figure 9**(a and b) revealed that $SrTiO_3$–0.8%RGO had higher photocurrent and lower impedance value than $SrTiO_3$. This is an indication of enhanced charge transport, separation of the photo-generated electrons, and holes in the composite sample.

Pei et al. [49] formulated network-like hierarchical nanosized $CaTiO_3$ via one-pot hydrothermal method and polyvinyl alcohol (PVA) as a structure guiding agent. The

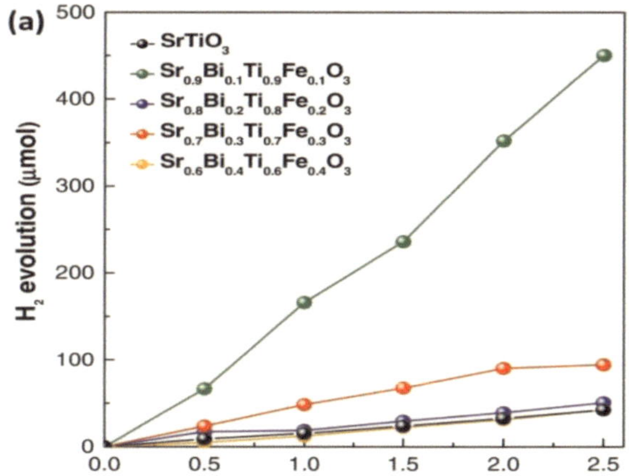

Figure 8.
Photocatalytic H_2 production of the samples when exposed to UV and visible light sources. Reprinted with permission [40] Copyright Elsevier 2017.

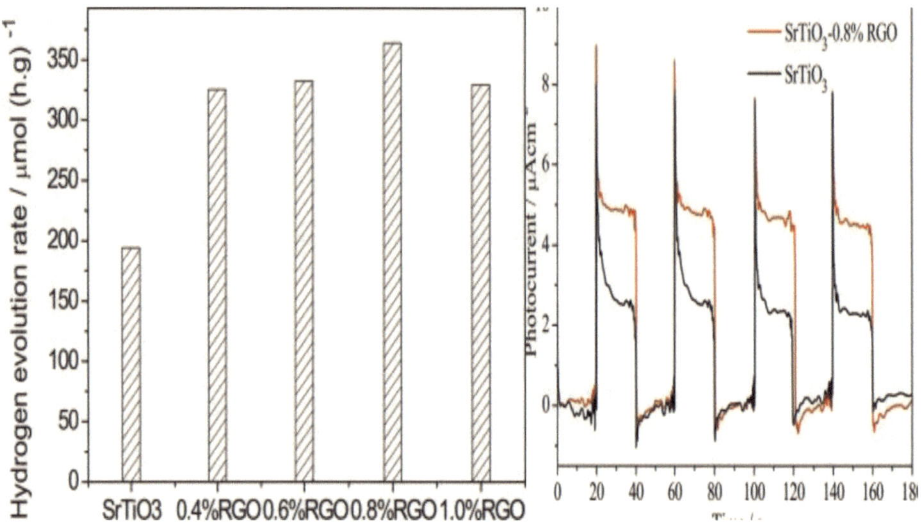

Figure 9.
Photocatalytic H2 production and (b) photoelectrochemical spectroscopy results of the samples. Reprinted with permission [41].

synthesized sample photocatalytic H_2 evolution was evaluated under visible light illumination (>400 nm). It was noted that sample synthesized with PVA had a smaller band-gap value of 2.57 V compared to band-gap of sample synthesized without PVA at 3.55 eV, a value which is typical optical band-gap energy of perovskite $CaTiO_3$. Under visible irradiation PVA-$CaTiO_3$ H_2 generation rate was 14.19 mol h^{-1} g^{-1} while $CaTiO_3$ alone indicated a negligible H_2 generation rate. The H_2 generation rate of the samples is shown in **Figure 10**(i and ii). The authors ascribed the improvement in H_2 generation by PVA-$CaTiO_3$ to presence of defects such as oxygen vacancies (V_o). They found out that V_o creates an energy level at 0.95 eV below the CB, acted as e^- donor, and aided charge transport and separation.

Abirami et al. [53] synthesized undoped $PbTiO_3$ and 0.01, 0.02, and 0.03% Ag–Fe codoped $PbTiO_3$ NPs via hydrothermal method. The photocatalytic activities of the as-prepared samples were evaluated using MB dyes under UV light source. The results showed that 0.02% Ag-Fe codoped $PbTiO_3$ exhibited the highest photocatalytic decoloration of MB (80.4%) at pH 0f 6 within 2 h. From **Figure 11**, it appears that pH

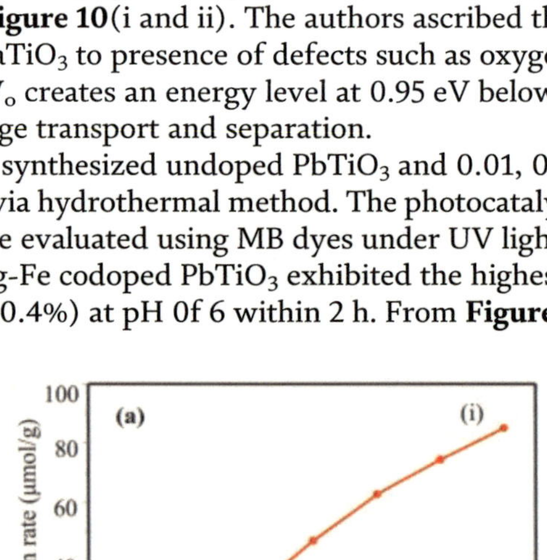

Figure 10.
(i) Depicts the H_2 generation rate of PVA-$CaTiO_3$ and pristine $CaTiO_3$ Reprinted with permission [49].

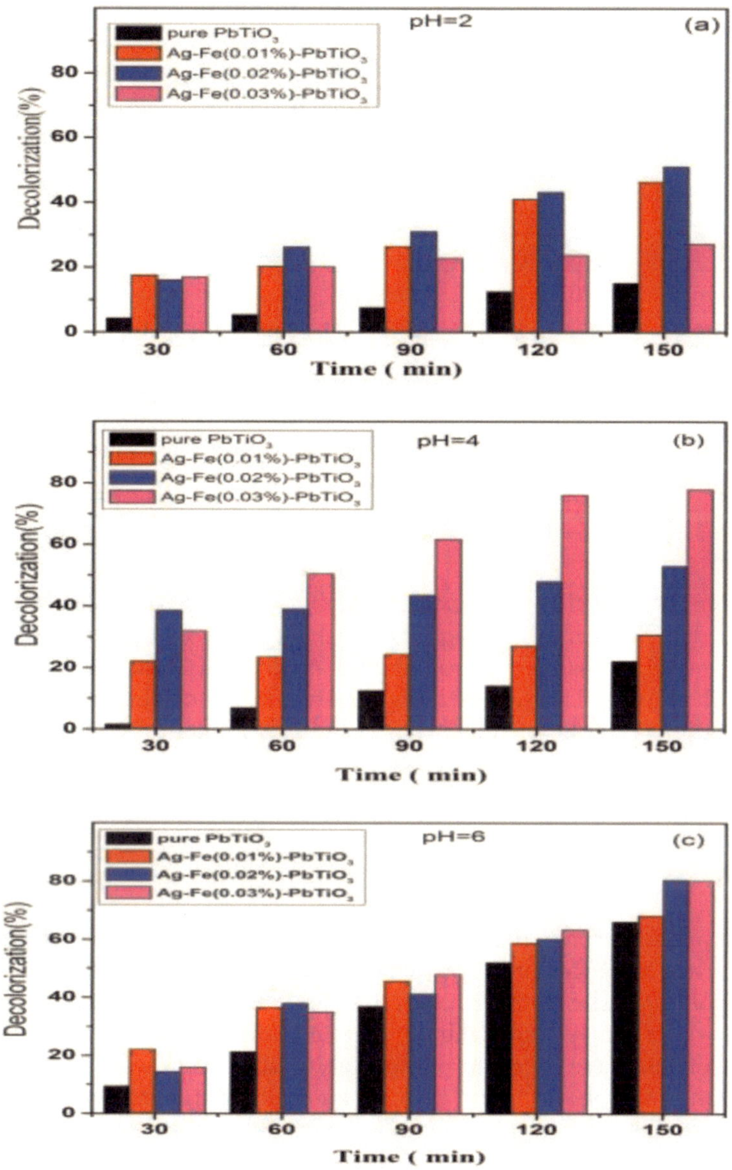

Figure 11.
The decolorization efficiency of the sample at the different pH a) at pH of 2. B) at pH of 4, and c) at pH of 6 Reprinted with permission [53].

also played a critical role in photodegradation of MB; the higher the pH value, the better the discolorization efficiency. According to Devi et al. [56], the adsorption of dyes on SC photocatalysts relies on pH as a result of the changes in the point zero charge of a SC photocatalyst at different pH. The author's associated improvement in the decolorization to reduction in $e^{-}h^{+}$ recombination rate caused by trapping photo-excited electrons by Ag and Fe.

Pirgholi-Givi et al. [54] studied the influence of the synthesis methods on photocatalytic activities of mixed $BiTi_3O_{12}$ and $Bi_{12}TiO_{20}$ NPs. The samples were formulated via microwave-assisted (S1) and hydrothermal synthesis (S2) methods.

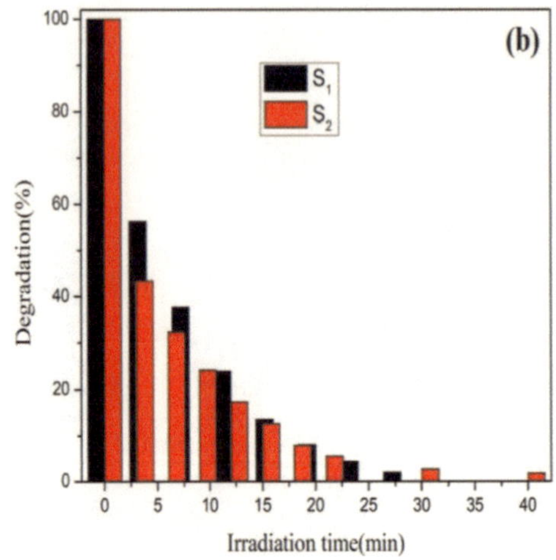

Figure 12.
Photocatalytic activities of S1- sample and S2-sample against MB. Reprinted with permission [54].

Their findings revealed that photocatalytic activity of the S1-sample was about 1.4 times higher than S2-sample at a higher pH value (12.5); S1-sample degraded 98% of MB in 28 min while S2-sample degraded 98% of MB in 40 min. The presence of $Bi_{12}Ti_{12}O_{20}$ phase in $Bi_4Ti_3O_{12}$ substrates elevated charge carrier span, enhanced charge separation, and photodegradation of the MB. **Figure 12** shows a decline in percentage of degradation of the test dye.

4. Conclusions

Use of titanante-based perovskite nanomaterial in photocatalytic application seems to be a promising and effective approach to mitigate the challenges that are associated with water pollution. However, considerable numbers of TBP have large optical bandgap energy, which allows photocatalytic reaction only with a UV source. In addition, the majority of TBP have excellent photocatalytic ability but with a high rate of recombination of the photogenerated e-h^+ pair which significantly affects its efficiency. Some of these challenges made scientists explore ways to optimize and improve photocatalytic activities of TBP material.

In this chapter, we made efforts to avail the overview of basic principles and mechanisms of an SC photocatalyst and recent synthesis techniques that have been deployed in preparing TBP nanomaterial. Special effort was made to highlight the influence of reaction conditions and approaches such as doping, codoping, composites, temperature, and pH on the surface defects and optical properties of TBP nanomaterial. Particularly, how surface effects and optical properties of these materials impact their photocatalytic activities were also discussed. Finally, the future perspective of TBP was proposed in the chapter.

Deploying appropriate synthesis methods and precursors is quite essential as it creates room to control the particle size, crystal structure, shapes, and morphology of a TBP photocatalyst. These physiochemical parameters play a vital role in modification of the electronic bandgap and surface defects of TBP photocatalysts. Much effort should be made to adopt appropriate methods and control over physiochemical properties of these group SC photocatalysts. Attention should also be given to reaction conditions such as temperature and pH of the system during synthesis. Regulating pH and annealing temperature have shown improved crystallinity of TBP, which in turn aids its photocatalytic activities under visible light irradiation. Other strategies that have been efficient in improving photocatalytic activities of TBP material include doping, codoping, and composites with nonmetal, metals, and perovskites material. These strategies have shown to increase charge carrier span, reactive sites, surface oxygen vacancies, decrease in bandgaps, and recombination rate of photogenerated e^{-} - h^{+} pair, improve charge separation, and photocatalytic absorption in the visible light region.

In spite of the progress made in recent times in understanding reaction processes and path involved in degradation of organic wastes and the role of doping TBP with other elements or TBP composite plays in altering its optical band-gap. Nevertheless, the sturdiness of these materials over time and their catalytic active cores require more detailed explanation. Hence there is a need to advance more accurate density-functional models to help optimize the photocatalytic configurations of these materials. In addition, more work should be done to optimize the reaction conditions of TBP nanomaterial, especially pH and temperature selection. Appropriate control to both parameters can enhance its photocatalytic activities as demonstrated by few authors. More work should be done to study photocorrosion of titanate-based perovskite nanomaterial.

Acknowledgements

We thank Engr. Emeka Okwuosa for the generous sponsorship of April 2014, July 2016, July 2018, and July 2021 conferences/workshops on applications of nanotechnology to Energy, Health and Environment, and for providing some research facilities.

Funding information

This research did not receive any specific grant from funding agencies in the public, commercial, or not-for-profit sectors.

Conflict of interest

The authors declare that they have no conflict of interest.

Author details

Izunna Stanislaus Okeke[1,2*], Priscilla Yahemba Aondona[1], Amoge Chidinma Ogu[1], Eugene Echeweozo[3] and Fabian Ifeanyichukwu Ezema[1,4,5]

1 Faculty of Physical Sciences, Department of Physics and Astronomy, University of Nigeria Nsukka, Nigeria

2 National Biotechnology Development Agency, FCT Abuja, Nigeria

3 Department of Physics with Electronics, Evangel University Akaeze, Akaeze, Ebonyi State, Nigeria

4 Nanosciences African Network (NANOAFNET), iThemba LABS-National Research, Western Cape Province, South Africa

5 UNESCO-UNISA Africa Chair in Nanosciences/Nanotechnology, College of Graduate Studies, University of South Africa (UNISA), Pretoria, South Africa

*Address all correspondence to: okekestannie@gmail.com

References

[1] Rana RS, Singh P, Kandari V, Singh R, Dobhal R, Gupta S. A review on characterization and bioremediation of pharmaceutical industries' wastewater: An Indian perspective. Applied Water Science. 2017;7:1-12

[2] Palmate SS, Pandey A, Kumar D, Pandey RP, Mishra SK. Climate change impact on forest cover and vegetation in Betwa Basin, India. Applied Water Science. 2017;7:1-12

[3] Adeogun AO, Ibor OR, Adeduntan SD, Arukwe A. Intersex and alterations in the reproductive development of cichlid, Tilapia Guineensis, from a municipal domestic water supply lake (Eleyele) in south western Nigeria. Science of the Total Environment. 2016;**541**:372-382

[4] Jung C, Son A, Her N, Zoh K, Cho J, Yoon Y. Removal of endocrine disrupting compounds, pharmaceuticals, and personal care products in water using carbon nanotubes: A review. Journal of Industrial and Engineering Chemistry. 2015;**27**:1-11

[5] Inyinbor AA, Adekola FA, Olatunji GA. Liquid phase adsorption of Rhodamine B onto acid treated Raphia hookerie epicarp: Kinetics, isotherm and thermodynamics studies. South African Journal of Chemistry. 2016;**69**:218-226

[6] Kumar A, Kumar S, Krishnan V. Perovskite-Based Materials for Photocatalytic Environmental Remediation. Inamuddin et al. (eds.), Nanophotocatalysis and Environmental Applications, Environmental Chemistry for a Sustainable World. Springer Nature Switzerland AG; 2019 29, DOI: 10.1007/978-3-030-10609-6_5

[7] Nkwachukwu OV, Arotiba OA. Perovskite oxide–based materials for photocatalytic and photoelectrocatalytic treatment of water. Frontiers in Chemistry. 2021;**9**:634630. DOI: 10.3389/fchem.2021.634630

[8] Okeke IS, Agwu KK, Ubachukwu AA, et al. Impact of Cu doping on ZnO nanoparticles phyto-chemically synthesized for improved antibacterial and photocatalytic activities. Journal of Nanoparticle Research. 2020:22. DOI: 10.1007/s11051-020-04996-3

[9] Manvendra P, Rahul K, Kamal K, Todd M, Charles UP Jr, Dinesh M. Pharmaceuticals of emerging concern in aquatic systems: Chemistry, occurrence, effects, and removal methods. Chemical Reviews. 2019;**119**:3510-3673

[10] Marta M, Jolanta K, Iseult L, Marianne M, Jan K, Steve B, et al. Changing environments and biomolecule coronas: Consequences and challenges for the design of environmentally acceptable engineered nanoparticles. Green Chemistry. 2018; **20**:4133-4168

[11] Wenya H, Kelong A, Xiaoyan R, Shengyan W, Lehui L. Inorganic layered ion-exchangers for decontamination of toxic metal ions in aquatic systems. Journal of Materials Chemistry A. 2017;**5**: 19593-19606

[12] Manolis JM, Mercouri GK. Metal sulfide ion exchangers: Superior sorbents for the capture of toxic and nuclear waste-related metal ions. Chemical Science. 2016;7:4804-4824

[13] Okeke IS, Agwu KK, Ubachukwu AA, Madiba IG, Maaza M, Whyte GM, et al. Impact of particle size

and surface defects on antibacterial and photocatalytic activities of undoped and Mg-doped ZnO nanoparticles, biosynthesized using one-step simple process. Vacuum. 2021;**187**:110110. DOI: 10.1016/j.vacuum.2021.110110

[14] Wanjun WYY, Taicheng A, Guiying L, Ho YY, Jimmy CY, Po KW. Visible-light-driven photocatalytic inactivation of *E. coli* K-12 by bismuth vanadate nanotubes: bactericidal performance and mechanism. Environmental Science & Technology. 2012;**46**:4599-4606

[15] Nadine C, Stefanie I, Elisabeth S, Marjan V, Karolin K, David D, et al. Inactivation of antibiotic resistant bacteria and resistance genes by ozone: From laboratory experiments to full-scale wastewater treatment. Environmental Science & Technology. 2016;**50**:11862-11871

[16] Chen J, Dong C, Idriss H, Mohammed OF, Bakr OM. Metal halide Perovskites for solar-to-chemical fuel conversion. Advanced Energy Materials. 2020;**10**:1902433

[17] He F, Wang ZX, Li YX, Peng SQ, Liu B. The nonmetal modulation of composition and morphology of g-C3N4-based photocatalysts. Applied Catalysis B: Environmental. 2020;**269**: 118828

[18] Nie L, zhang q. Recent Progress on crystalline metal chalcogenides as efficient photocatalysts for organic pollutants degradation. Inorganic Chemistry Frontiers. 2017;**4**:1953-1962

[19] Akyüz D, Zunain Ayaz RM, Yılmaz S, et al. Metal chalcogenide based photocatalysts decorated with heteroatom doped reduced graphene oxide for photocatalytic and photoelectrochemical hydrogen production. International Journal of Hydrogen Energy. 2019;**44**:18836-18847

[20] Lacorre P, Goutenoire F, Bohnke O, Retoux R, Laligant Y. Designing fast oxide-ion conductors based on La2Mo2O9. Nature. 2000;**404**:856-858. DOI: 10.1038/35009069

[21] Shimizu T. Partial oxidation of hydrocarbons and oxygenated compounds on perovskite oxides. Catalysis Reviews. 1992;**34**:355-371. DOI: 10.1080/01614949208016317

[22] Asri M, Adnan B, Arifin K, Minggu LJ, Kassim MB. Titanate-based perovskites for photochemical and photoelectrochemical water splitting applications: A review. International Journal of Hydrogen Energy. 2018;**43**: 23209-23220

[23] Xu Y, Schoonen MAA. The absolute energy positions of conduction and valence bands of selected semiconducting minerals. American Mineralogist. 2000;**85**:543e56

[24] Kanhere P, Chen Z. A review on visible light active perovskite-based photocatalysts. Molecules. 2014;**19**: 19995e20022

[25] Wang L, Yang G, Peng S, Wang J, Ji D, Yan W, et al. Fabrication of MgTiO3 nanofibers by electrospinning and their photocatalytic water splitting activity. International Journal of Hydrogen Energy. 2017;**42**:25882e90

[26] Qu Y, Zhou W, Fu H. Porous cobalt titanate nanorod: A new candidate for visible light-driven photocatalytic water oxidation. Chem Cat Chem. 2013;**6**: 265e70

[27] Qu Y, Zhou W, Ren Z, Du S, Meng X, Tian G, et al. Facile preparation of porous NiTiO3 nanorods with

enhanced visible-light-driven photocatalytic performance. Journal of Materials Chemistry. 2012;**22**:16471e6

[28] Sierra GG, Mari'n Alzate N, Arnache O. A novel LaFeO3XNX oxynitride. Synthesis and characterization. Journal of Alloys and Compounds. 2013;**549**:163e9

[29] Ma L, Li N, Wu G, Song G, Li X, Han P, et al. Interfacial enhancement of carbon fiber composites by growing TiO2 nanowires onto amine based functionalized carbon fiber surface in supercritical water. Applied Surface Science. 2018;**433**:560-567. DOI: 10.1016/j.apsusc.2017.10.036

[30] Zhang L, Yu W, Han C, Guo J, Zhang Q, Xie J, et al. Large scaled synthesis of heterostructured electrospun TiO2/SnO2 nanofibers with an enhanced photocatalytic activity. Journal of the Electrochemical Society. 2017;**164**:H651-H656. DOI: 10.1149/2.1531709jes

[31] Zhang L, Qin M, Yu W, Zhang Q, Xie H, et al. Heterostructured TiO2/WO3 nanocomposites for photocatalytic degradation of toluene under visible light. Journal of the Electrochemical Society. 2017;**164**:H1086-H1090. DOI: 10.1149/2.0881714jes

[32] Shenoy S, Bhat DK. Enhanced bulk thermoelectric performance of Pb0.6Sn0.4Te: Effect of magnesium doping. Journal of Physical Chemistry C. 2017;**121**:20696-20703

[33] Parkinson B. An overview of the progress in photoelectrochemical energyconversion. Journal of Chemical Education. 1960;**60**:338-340

[34] Osterloh FE. Inorganic materials as catalysts for photochemical splitting ofwater. Chemistry of Materials. 2008; **20**:35-54

[35] Bai S, Zhang N, Gao C, Y. Xiong defect engineering in photocatalytic materials. Nano Energy. 2018;**52**:296-336

[36] Ajmal A, Majeed I, Malik RN, Idriss H, Nadeem MA. Principles and mechanisms of photocatalytic dye degradation on TiO based photocatalysts: a comparative overview. RSC Advances. 2014;**4**:37003-37026. DOI: 10.1039/C4RA06658H

[37] Rauf MA, Ashraf SS. Fundamental principles and application of heterogeneous photocatalytic degradation of dyes in solution. Chemical Engineering Journal. 2009;**151**: 10-18. DOI: 10.1016/j.cej.2009.02.026

[38] Orlovskaya N, Browning N. Mixed Ionic Electronic Conducting Perovskites for Advanced Energy Systems. Springer Dordrecht; 2004. p. 314. DOI: 10.1007/978-1-4020-2349-1

[39] Gao L, Guan Z, Huang S, Liang K, ChenH ZJ. Enhanced dielectric properties of barium strontium titanate thin films by doping modification. Journal of Materials Science: Materials in Electronics. 2019;**30**:12821-12839

[40] Lu L, Lv M, Liu G, Xu X. Photocatalytic hydrogen production over solid solutions between BiFeO3 and SrTiO3. Applied Surface Science. 2017; **391**:535-541. DOI: 10.1016/j.apsusc.2016.06.160

[41] He G, Zhong Y, Chen M, Li X, Fang Y, Xu Y. One-pot hydrothermal synthesis of SrTiO3-reduced graphene oxide composites with enhanced photocatalytic activity for hydrogen production. Journal of Molecular Catalysis A: Chemical. 2016;**423**:70-76. DOI: 10.1016/j.molcata.2016.05.025

[42] Zhang H, Chen G, Li Y, Teng Y. Electronic structure and photocatalytic properties of copper-doped CaTiO3. International Journal of Hydrogen Energy. 2010;**35**:2713-2716. DOI: 10.1016/j.ijhydene.2009.04.050

[43] Zhuang J, Tian Q, Lin S, Yang W, Chen L, Liu P. Precursor morphology-controlled formation of perovskites CaTiO3 and their photo-activity for As (III) removal. Applied Catalysis B: Environmental. 2014:108-115. DOI: 10.1016/j.apcatb.2014.02.015

[44] Cai J, Cao A, Huang J, Jin W, Zhang J, Jiang Z, et al. Understanding oxygen vacancies in disorder-engineered surface and subsurface of $CaTiO_3$ nanosheets on photocatalytic hydrogen evolution. Applied Catalysis B: Environmental. 2020;**267**:118378. DOI: 10.1016/j.apcatb.2019.118378

[45] Chen M, Xiong Q, Liua Z, Qiu K, Xiao X. Synthesis and photocatalytic activity of Na+ co-doped CaTiO3:Eu3+ photocatalysts for methylene blue degradation. Ceramics International. 2020;**46**:12111-12119

[46] Portia SAU, Rajkumar S, Elanthamilan E, Merlin JP, Ramamoorthy K. Effect of annealing temperature on structural, optical and visible light photocatalytic performance of CaTiO3 catalysts synthesized by simple solgel technique. Inorganic Chemistry Communications. 2020;**119**:108051

[47] Lozano-Sáncheza LM, Obregón S, Díaz-Torres LA, Lee S, Rodríguez-González V. Visible and near-infrared light-driven photocatalytic activity oferbium-doped CaTiO3 system. Journal of Molecular Catalysis A: Chemical. 2015;**410**:19-25

[48] Yang H, Han C, Xue X. Photocatalytic activity of Fe-doped CaTiO3 under UV–visible light.Journal of environmental. Sciences. 2014;**26**: 1489-1495. DOI: 10.1016/j.jes.2014.05.015

[49] Pei J, Meng J, Wu S, Lin Q, Li J, Wei X, et al. Hierarchical CaTiO3 nanowire-network architectures for H2 evolution under visible-light irradiation. Journal of Alloys and Compounds. 2019; **806**:889e896

[50] Bhat DK, Bantawala H, Shenoy US. Rhodium doping augments photocatalytic activity of barium titanate: Effect of electronic structure engineering. Nanoscale Advance. 2020;**2**: 5688. DOI: 10.1039/d0na00702a

[51] Nishioka S, Maeda K. Hydrothermal synthesis of Rhodium-doped barium titanate nanocrystals for enhanced photocatalytic hydrogen evolution under visible light. RSC Advances. 2015;**5**: 100123. DOI: 10.1039/c5ra20044j

[52] Khan MAM, Kumar S, Ahmed JC, Ahamed M, Kumar A. Influence of silver doping on the structure, optical and photocatalytic properties of Ag-doped BaTiO3 ceramics. Materials Chemistry and Physics. 2021;**259**:124058

[53] Abirami R, Senthil TS, Kalaiselvi CR. Preparation of pure PbTiO3 and (Ag–Fe) codoped PbTiO3 perovskite nanoparticles and their enhanced photocatalytic activity. Solid State Communications. 2021;**327**:114232. DOI: 10.1016/j.ssc.2021.114232

[54] Pirgholi-Givia G, Farjami-Shayesteha S, Azizian-Kalandaragh Y. The influence of preparation parameters on the photocatalytic performance of mixed bismuth titanate-based nanostructures. Physica B: Condensed Matter. 2019;**575**:311572. DOI: 10.1016/j.physb.2019.07.007

[55] Yang Z, Fan H, Wang X, Long C. Rapid microwave-assisted hydrothermal synthesis of Bi12TiO20 hierarchica larchitecture with enhanced visible-light photocatalytic activities. Journal of Physics and Chemistry of Solids. 2013; **74**:1739-1744. DOI: 10.1016/j.jpcs.2013.06.020

[56] Devi LG. Reddy KM Enhanced photocatalytic activity of silver metallized TiO2 particles in the degradation of an azo dye methyl orange: Characterization and activity at different pH values. Applied Surface Science. 2010;**256**:3116-3121

Chapter 6

Thermoelectric Nanostructured Perovskite Materials

Megha Unikoth, George Varghese, Karakat Shijina and Hind Neelamkodan

Abstract

The global need for energy production from renewable resources and the effect of greenhouse gas, especially carbon dioxide is increasing day by day. Statistical survey shows that about 60% of the energy lost in vain worldwide, in the form of waste heat. The conversion of this waste into useful energy form will certainly play a major role in alternative energy technologies. Thermoelectric materials (TE) can harvest waste heat and convert this into electrical energy and vice versa. The development of high-efficiency TE materials for waste-heat-recovery systems is necessary to bring vast economic and environmental benefits. The methods of synthesis,that is, control over particle size play an important role in controlling the properties of thermoelectric materials. The nanostructuring of thermoelectric materials can enhance the efficiency by quantum confinement effect and phonon scattering. Perovskites have a long history of being a potential candidate for thermoelectric applications, due to their fascinating electrical, mechanical, and thermal properties. Compared with other thermoelectric materials perovskites have the advantage of eco-friendliness, less toxicity and are highly elemental abundant. Owing to the high thermal conductivity and low electrical conductivity overall performance of perovskites is relatively poor. The hybrid perovskites overcome this difficulty and started to draw the attention to thermoelectric applications.

Keywords: thermoelectric, figure of merit, nanostructuring, power generation, hybrid perovskites

1. Introduction

The imbalance between energy production and demand is increasing day by day. While the conventional resources are being depleted, the challenges of researchers are concentrated on the power generation from renewable energy sources and on the efficient use of available resources. On the other hand, waste heat generation as greenhouse gas especially carbon dioxide is increasing in the environment. In internal combustion engines, only 25% of energy is used for vehicle mobility and accessories, approximately 40% of the fuel energy is wasted as exhaust gas, 30% is dissipated in the engine coolant and 5% is lost as radiation and friction. Here comes the importance of thermoelectric materials! The materials which can harvest heat from combustion of fossil fuels, sunlight, chemical reactions, nuclear decay, vehicles, etc., and convert it

IntechOpen

into electrical energy and vice versa are thermoelectric materials. Thermoelectric power generation technology and the fields are now growing steadily due to their ability to convert heat into electricity and to develop cost-effective and pollution-free forms of energy conversion. A wide variety of thermoelectric materials has been identified and their properties have been explored. Among the oxide-based thermoelectric materials, rare earth-based perovskites are considered to be a potential material due to their fascinating electrical, mechanical, and thermal properties and high value of figure of merit. The thermoelectric materials in nanostructured form can enhance the performance of material by phonon scattering and quantum confinement effects [1–4].

2. Thermoelectric materials

Thermoelectric materials have drawn vast attention due to the direct conversion between thermal and electrical energy. These materials can convert the heat energy to electrical energy and vice versa, thus providing an alternative source for power generation and refrigeration. Statistical survey shows that more than 60% of world's energy loss is in the form of heat. The high-performance thermoelectric materials can easily convert this heat into usable electrical energy. The thermoelectric system is an eco-friendly energy conversion technology with the advantages of high reliability, small size, feasibility in a wide temperature range, and no pollutants. The efficiency of the thermoelectric devices is small compared to Carnot's efficiency. The efficiency of these materials is defined in terms of figure of merit, ZT, which determines the thermoelectric performance.

$$ZT = \frac{S^2\sigma}{\kappa}T \tag{1}$$

where S is the Seebeck coefficient, σ is the electrical conductivity, κ is the thermal conductivity, and T is the absolute temperature. In order to get good performance, the value of ZT should be high. This can be achieved by increasing both the Seebeck coefficient and electrical conductivity and reducing the thermal conductivity.

The physics behind the thermoelectric power generation and the refrigeration is mainly governed by the three fundamental thermodynamic effects-the Seebeck, Thomson, and Peltier effects. When a temperature gradient is applied, an electrical potential gradient is generated, which is the Seebeck effect and is mainly used in the power generation. The Peltier effect is the reverse of the Seebeck effect in which a temperature gradient is established when a current is passed through the material and is used for refrigeration. Thomson heat is absorbed or released internally in the material if the flow of the Peltier heat is balanced by the temperature-dependent Seebeck coefficient. All the three effects are related to the heat transported by the charge carriers in the material as electrons or holes. Seebeck effect illustrating the working of a thermoelectric generator is given in **Figure 1**.

The thermoelectric efficiency (η_P) in the power generation mode as a function of average ZT is given by,

$$\eta_P = \frac{T_H - T_C}{T_H}\left(\frac{\sqrt{1+ZT_M}-1}{\sqrt{1+ZT_M}+\frac{T_C}{T_H}}\right) \tag{2}$$

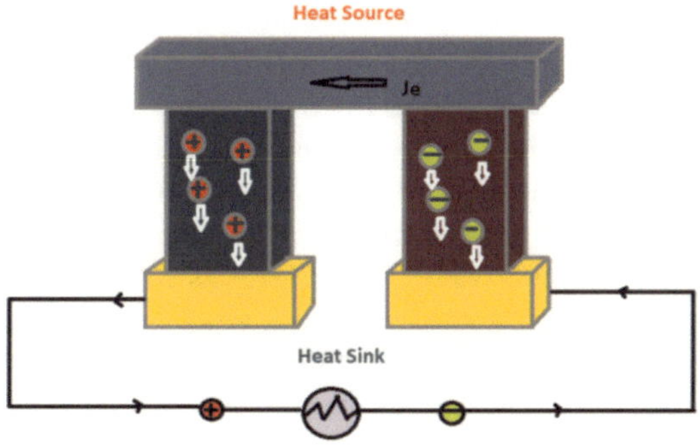

Figure 1.
Illustration of the Seebeck effect, when heat flows through the junction current is generated.

where T_C, T_H, and T_M are the cold side, hot side, and average temperature respectively.

$$ZT_M = \frac{1}{T_H - T_C} \int_{T_C}^{T_H} ZTdT \tag{3}$$

A larger temperature difference can produce higher conversion efficiency, if the value of ZT_M= 3 and ΔT = 400 K, η_P can reach 25%, comparable to that of traditional heat engines. The Seebeck effect is the thermoelectric power generation model and has application in advanced scientific fields. The thermoelectric cooling efficiency (η_C) is given by,

$$\eta_C = \frac{T_H}{T_H - T_C}\left(\frac{\sqrt{1+ZT_M} - \frac{T_H}{T_C}}{\sqrt{1+ZT_M}+1}\right) \tag{4}$$

Similar to thermoelectric power generation, higher ZT_M will produce a large cooling efficiency (η_C). For ZT_M= 3 and ΔT = 20 K, η_C could reach 6%. The Peltier effect is a thermoelectric refrigeration model and is used to cool computer components to keep temperature within the limit or to maintain suitable functioning. The high ZT value is obtained only by increasing the value of S and σ and minimizing the κ. The complex relationship of thermoelectric parameters can be obtained from the Wiedemann-Franz law and Pisarenko relation, which is given by,

$$S = \frac{8\pi^2 K_B^2}{3eh^2} m^* T \left(\frac{\pi}{3n}\right)^{\frac{2}{3}} \tag{5}$$

$$\sigma = \frac{ne^2\tau}{m^*} = ne\mu \tag{6}$$

$$\kappa_{total} = \kappa_{electronic} + \kappa_{lattice} = L\sigma T + \kappa_{lattice} \tag{7}$$

where K_B is the Boltzmann constant, h is the Planck constant, n is the carrier concentration, T is the absolute temperature, e is the electron charge, m* is the effective mass, τ is the relaxation time, μ is the carrier mobility, and L is the Lorenz number. The electronic part of thermal conductivity is proportional to the electrical conductivity. Therefore, simultaneous enlargement of S and σ and the minimization of κ for high ZT values are very difficult. Over the past few decades, there is a lot of progress in the field of thermoelectrics to make an ideal material with ZT value greater than 3. There are many strategies for decoupling the relation between these parameters which includes phonon scattering mechanism, mass fluctuation strategy, rattling strategy, band engineering, 2D superlattice, and Panasonic approach. Energy filtering effects were also used in which an energy barrier was introduced by grain boundaries or nanocomposites [5]. According to the optimal working temperature, TE materials are classified into three – Bi_2Te_3 based low temperature (<400 K) materials, PbTe-based material in the temperature range between 600 K and 900 K, and SiGe-based high temperature(>900 K) materials.

The first generation thermoelectric materials have ZT = 1 and the power generation efficiency is about 4–5%. The second generation materials pushed the ZT value up to 1.7 by nanostructuring and the obtained efficiency is 11–15%. The third generation material is under development and the predicted efficiency will be in the range of 15–20%. The main goal will be to attain ZT ≥3 in future. PbTe is one of the most attractive thermoelectric materials [6].

The Skutterudites, Half-Heuslers, clathrates, and chalcogenides are high-temperature thermoelectric materials. Skutterudites are compounds with general formula MX_3 where M = Co, Rh or Ir and X = P, and As or Sb (e.g., is $CoSb_3$). These materials can influence the phonon transport mechanism, thereby reducing the lattice conductivity to very low level. X. Shi et al. reported that, for $Ba_{0.08}$ $La_{0.05}$ $Yb_{0.04}$ Co_4 Sb_{12} skutterudites has ZT = 1.7 at 850 K [7]. Half-Heuslers are alloys of the form ABX where A-Ti, Zr and Hf, B- CoSb, NiSn, etc. It was reported that, for n-type $Hf_{0.5}Zr_{0.5}NiSn_{0.99}Sb_{0.01}$,$0.8 \leq ZT \leq 1$ was obtained at 600–700° C and for p-type $Hf_{0.5}Zr_{0.5}CoSn_{0.2}Sb_{0.8}$,$0.5 \leq ZT \leq 0.8$ due to the remarkable reduction in the lattice thermal conductivity [8].

The clathrates are low thermal conductivity compounds with Type I having $X_2Y_6E_{46}$ formula and Type II having $X_8Y_{16}E_{136}$ formula, where X and Y are guest atoms, E - Si, Ge, or Sn. $Ba_8Ga_{16}Ge_{30}$ shows a Seebeck coefficient of −45 to -150mVK^{-1} and electrical conductivity of 1500–600Scm^{-1} at 300–900 K. The thermal conductivity of this compound is 1.8 WK^{-1} m^{-1} at 300 K and is reduced to 1.25 W K^{-1} m^{-1} at 900 K which makes ZT = 1.35 [9]. Chalcogenides are compounds with sulfides, selenides, and tellurides present in them (e.g., Bi_2Te_3, PbTe, SnSe, SiGe, etc). Among the oxide materials, $NaCo_2O_4$ has $0.7 \leq ZT \leq 0.8$ at 1000 K [10]. The other new thermoelectric materials include $In_4Se_{3-\delta}$ (ZT =1.48) [33], $In_4Se_{3-x}Cl_{0.03}$ (ZT = 1.53) [34], β-$Cu_{2-x}Se$ (ZT = 1.5) [11] and β-Zn_4Sb_3 (ZT = 1.35) [12]. Low dimensional thermoelectric materials have higher performance than bulk materials because the density of states near the Fermi level is enhanced due to the quantum confinement effects, thereby increasing the thermopower ($S^2\sigma$) and boundary scattering at the interfaces reduces the thermal conductivity more than electrical conductivity. Therefore, by reducing the size of materials to 1D and 2 D, a significant enhancement in the value of ZT is obtained. Hicks and Dresselhaus first improved the value of ZT >1 of 2D Bi_2Te_3 quantum well [13]. He reported that the enhancement of ZT was achieved by the quantum confinement of electrons and holes, which increases the $S^2\sigma$ and the reduction in the thermal conductivity was attributed to various effects, such as scattering of phonons at interfaces, defects, or phonon localization. Venkatasubramanian et al. reported, ZT = 2.4 for Bi_2Te_3-Sb_2Te_3 quantum well superlatttice of 6 nm periodicity [14]. The quantum dot superlattice of PbTe–PbSeTe system developed by Harman and co-workers has ZT = 1.6, which is higher than the bulk (ZT = 0.34) [15].

Hochbaum et.al reported that at room temperature, 50 nm diameter nanowires of Silicon have ZT = 0.6, which is very much higher than the bulk [16]. Boukai et al. noticed that reducing the nanowire's diameter, a significant reduction in thermal conductivity is attained and ZT of 1 at 200 K was reported for nanowires of 20 nm diameter. Nanostructured thermoelectric materials are designed in such a way to introduce nanometer-sized interfaces and polycrystalline into the bulk materials [17]. The lattice thermal conductivity can be reduced by increasing phonon scattering. Nanostructured composites of grain size ~5 nm–10 μm can be fabricated by hot pressing or spark plasma sintering of fine powders [18]. In nanostructured material families (PbTe based nanomaterials, Bi_2Te_3–based and SiGe–based nanocomposites) an enhancement in ZT value is noticed. S. Fan et al. reported that in Bi_2Te_3–based nanocomposites, $Bi_{0.4}$ $Sb_{1.6}Te_3$ has a ZT of 1.8 at 316 K [19]. Biswas et.al suggested that 2% SrTe – containing PbTe nanocomposites have a ZT of 1.7 at 800 K and X.W.Wang et al. studied $Si_{80}Ge_{20}P_2$ nanocomposites and reported the ZT value of 1.3 at 1173 K [20, 21]. Perovskites, as well as their hybrids, started to draw attention as a potential candidate for thermoelectric applications.

3. Perovskites

The first perovskite $CaTiO_3$ was discovered by Gustav Rose in 1839 and named in the honor of an eminent mineralogist Count Lev Alexevich von Perovski. Perovskites are compounds having the structure formula ABC_3, where A - rare earth, alkaline earth, alkali, or large ions, such as Pb^{+2}, Bi^{+3}, B - transition metal ion, and C - O, Fl, Cl, I, etc., commonly seen as in the form of ABO_3. A cation may be monovalent like Li, Na, K, divalent like Ca, Ba, Sr., or trivalent like La, Nd, Pr, which is cubo-octahedrally coordinated with 12 oxygen atoms while B cation as Ti, Ni, Fe, Co, or Mn is octahedrally coordinated with 6 oxygen atoms. The substituted and mixed compounds of the form $A_{1-x}A'_xB_{1-y}B'_yO_3$ also come under this class with distorted non-stoichiometric oxygen deficient configuration. The pseudo-perovskites are a special class of perovskites with empty A-site and BO_3 configurations (e.g., ReO_3 and WO_3). The most abundant materials in the earth's crust are $MgSiO_3$ and $FeSiO_3$ perovskites [22].

The stability of perovskites is determined by a factor called Goldsmith tolerance factor given by, $t = \frac{r_A + r_O}{\sqrt{2}(r_B + r_O)}$, where r_A – ionic radii of A-cation, r_B – ionic radii of B-cation, and r_O-ionic radii of oxygen. The value of t will be unity for an ideal perovskite. For different values of t, these materials have different structures. If $t > 1$, the crystal structure will be hexagonal in which A ions are too big and B ions are too small (e.g., $BaNiO_3$). If $t = 1$ the structure will be cubic with A and B ions having ideal size (e.g., $SrTiO_3$, $BaTiO_3$). The materials with $0.71 < t < 1$ have orthorhombic/rhombohedral crystal structure where A ions are too small to fit into B ion interstices (e.g., $CaTiO_3$/ $GdFeO_3$) and for $t < 0.7$ materials have different structures in which A ions and B ions have similar ionic radii (e.g., $FeTiO_3$) [23].

The coexistence of spin, charge, lattice, and orbital interactions in perovskite materials make them applicable in optoelectronics, spintronics, photocatalysis, sensors, piezoelectric devices, electrode in solid oxide fuel cells, and thermoelectrics. They have a lot of fascinating properties, such as multiferroicity ($BaTiO_3$, $BiFeO_3$), colossal magnetoresistance (manganites), superconductivity (cuprates), ferromagnetism ($SrRuO_3$), metal-insulator transition ($LaMnO_3$), thermoelectricity ($LaCoO_3$), etc.

3.1 Perovskite oxides (ABO_3) as thermoelectric materials

Oxide perovskites have been used as thermoelectric materials due to their low thermal conductivity, high Seebeck coefficient, and electrical conductivity. There are two approaches to enhance ZT value, one is tuning the carrier concentration and another is engineering structure and material properties to decouple the S, σ, and κ. The further modification methods to enhance ZT value are self-doping, nano-engineering, band engineering, and doping shown in **Figure 2**. In perovskite oxides the main candidates which show TE properties were titanates, manganates, and colbatates.

The substituted perovskite compounds of titanates ($Sr_{1-x}A_xTi_{1-y}Nb_yO_3$, where A- Ca, La, Ba, Eu, etc) are promising classes of thermoelectric materials with high figure of merit. It was reported that $SrTi_{0.8}Nb_{0.2}O_3$ thin films have ZT = 0.37 at 1000 K, which was reduced to 0.35 by hot pressing the sample to a temperature of 1073 K. When the material $SrTi_{0.8}Nb_{0.2}O_3$ is grown in nanostructure as superlattice, where a single layer of this material is sandwiched between the several layers of insulating $SrTiO_3$ (STO), a remarkable increase in ZT ~2.4 at 300 K is obtained. In La-doped STO thin films S can be tuned from −120 to -260μVK^{-1}. The La 15% doped STO has a ZT value of 0.28 at 873 K was achieved. Even though the electrical conductivity of STO is increased by La doping, it reduces the lattice thermal conductivity by phonon scattering. It was reported that the substitution of Ce, Ba, Ca, Pr, and Y on A-site and Nb, Ta, Mn, and Co on B-site enhances the electrical conductivity of the sample. The Nb-doped STO($Sr(Nb_xTi_{1-x})O_3$,$0.01 < x < 0.4$), in which substituted Nd^{5+} at Ti^{4+} site will generate carrier electrons and a ZT of ~0.35–0.37 at 1000 K was achieved for 20%Nb. While Mn substitution of $Sr_{1-x}LaxTiO_3$ the S was enhanced from −120 to −180 μV·K^{-1} and the ZT value of 0.07 to 0.15 at 300 K was obtained when the composition changed from $Sr_{0.95}La_{0.05}TiO_3$ to $Sr_{0.95}La_{0.05}Ti_{0.96}Mn_{0.04}O_3$. For $SrTi_{0.9}Ta_{0.1}O_3$ the ZT value obtained was 0.17 at 752 K and for $SrTi_{0.875}Co_{0.125}O_3$ was 0.135 at 300 K. The A-site substitution enhances electrical properties while B-site enhances the Seebeck coefficient value. However, the doping of Y, La, Sm, Gd, and Dy in STO reduces the thermal conductivity. For $(Sr_{0.9}Dy_{0.1})TiO_3$ has a ZT value of 0.22 at 573 K. Mn doped $Sr_{1-x}LaxTiO_3$ can enhance anharmonic lattice vibrations, which result in inelastic phonon-phonon scattering reduces thermal conductivity and offers high electrical conductivity. Compared to ZT of $Sr_{0.95}La_{0.05}TiO_3$ 0.07

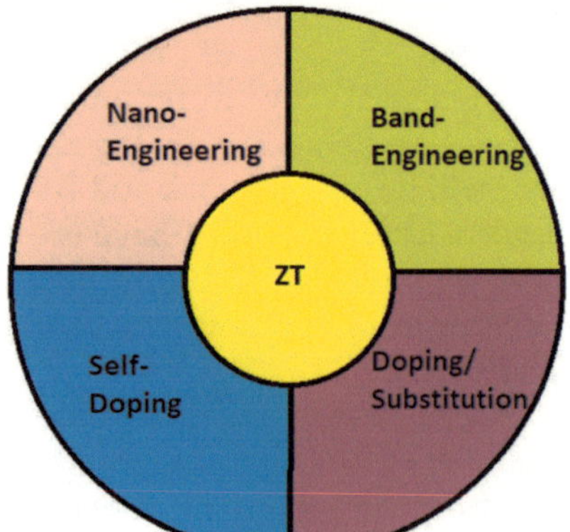

Figure 2.
Methods to enhance the ZT of thermoelectric perovskite materials.

$Sr_{0.95}La_{0.05}Ti_{0.98}Mn_{0.02}O_3$ has 0.15 at 300 K. The effective way to reduce thermal conductivity is rare earth substitution in A-site and Mn substitution in B-site. The data compounds obtained were tabulated in **Table 1** [5].

The thermoelectric properties of Mn substituted perovskites were summarized in **Table 2**. In these perovskite oxides multiple elements are used as A-site dopants in $AMnO_3$ including Yb, Y, La, Ce, Sm, Dy, Tb, Ho, Pr, Ca, Sr, Nd, etc. and for B-site Mo, Ru, Ta, etc. Relatively high ZT values are not achieved in these materials. $CaMn_{0.98}Nb_{0.02}O_3$ shows a significant ZT of 0.32 at 1050 K. It was noticed that the thermal conductivity of Mn-doped samples is low.

Rare earth cobalt oxides are compounds having the stoichiometry $RCoO_3$, R – La, Ce, Pr, Nd, etc. It was also reported that the electrical conductivity of these materials increased with increasing ionic radii of rare earth metals doping ($Pr^{3+} > Nd^{3+} > Tb^{3+} > Dy^{3+}$). The complex spin structure of Co ions in perovskite gives us plenty of opportunities to explore the exotic magnetic phenomenon of these materials. The Co ions in $RCoO_3$ can exist in three different spin states, low spin LS ($t_{2g}^{6}\ e_g^{0}$ for Co^{3+} and $t_{2g}^{5}\ e_g^{0}$ for Co^{4+}), intermediate spin IS ($t_{2g}^{5}\ e_g^{1}$ for Co^{3+} and $t_{2g}^{4}\ e_g^{1}$ for Co^{4+}), and high spin state HS (t_{2g}^{4}, e_g^{2} for Co^{3+} and t_{2g}^{3}, e_g^{2} for Co^{4+}), which can induce spin entropic effect to the perovskite structure and can influence all the magneto-transport properties of the materials [24–26]. It was reported that A- site substituted $LaCoO_3$, has high ZT value. Sr, Na, Pb, and Ba are usually used elements for substitution. For $La_{1-x}Sr_xCoO_3$ the electrical conductivity gets enhanced, and the ZT value of 0.046 to 0.18 was achieved. Pb doped $LaCoO_3$ has Seebeck coefficient of 110 $\mu V{\cdot}K^{-1}$ and ZT of 0.23 to reported. Thermoelectric measurements of doped $ACoO_3$ compounds are tabulated in **Table 3**.

The other B-site cations include iron (Fe), nickel (Ni), tin (Sn), lead (Pb), bismuth (Bi), molybdenum (Mo), ruthenium (Ru), and uranium (U). The thermoelectric measurements are tabulated in **Table 4**. For Fe doped compounds $La_{0.95}Sr_{0.05}FeO_3$ and $Pr_{0.9}Sr_{0.1}FeO_3$, the ZT value obtained are 0.076 and 0.024, while for Ni-doped $LaCo_{0.92}Ni_{0.08}O_{2.9}$ the ZT value of 0.2 was achieved. Double perovskite A_2FeMoO_6 (A-Ca,Sr,K,Ba) was also studied. For ZT ranges from 0.1 to 0.99 was reported. For tin substituted compounds $BaSnO_3$ ZT of 0.65 was theoretically calculated. $Sr_{1-x}Ba_xPbO_3$ ZT of 0.13 was observed. There was no significant high ZT value seen when the B-site is doped with Mo, Ru, and U. The thermoelectric measurement parameters are all tabulated in **Table 4**.

3.2 Hybrid perovskites

Compared to other thermoelectric materials hybrid perovskites have high Seebeck coefficient and low thermal and electrical conductivity. For $CH_3NH_3PbI_3$ at 295 K has $S = 700\ \mu V{\cdot}K^{-1}$, $\kappa = 0.5\ W{\cdot}m^{-1}{\cdot}K^{-1,}$ and ZT value is 10^{-7} due to the low electrical conductivity. The photo-induced or chemical doping strategies were used in these materials to enhance the electrical conductivity. There are many hybrid perovskite materials that show TE applications, such as ABI_3 (A = CH_3NH_3 (MA), NH_2CHNH_2 (FA), and B = Pb, Sn), $CsMI_3$, and $C_6H_4NH_2CuBr_2I$. Theoretical studies on n-type and p-type $CH_3NH_3PbI_3$ hybrid perovskites show ZT value of 0.9 and 1.25. For $(MA)PbI_3$, $(MA)SnI_3$, $(FA)PbI_3$, and $(FA)SnI_3$ n-type materials the reported ZT values are 0.44, 0.45, 0.42, and 0.35 respectively. As the carrier concentration increases ZT also increases. The same $CH_3NH_3PbI_3$ n-type ZT of 2.56 at 800 K can be achieved. The first-principles calculations and semi-classical Boltzmann transport theory showed

Materials	Electrical Conductivity ($S\ cm^{-1}$)	Seeback coefficient (μVK^{-1})	Thermal conductivity ($W\ m^{-1}\ K^{1}$)	Power Factor ($\mu W\ m^{-1}\ K^{-2}$)	ZT	Temperature (K)
$Sr_{0.85}La_{0.15}TiO_3$	400	175	3		0.28	873
reducedgrapheneoxide—$SrTiO_3$	30	−380			0.09	760
$Sr_{0.875}Pr_{0.125}TiO_3$	3700	−80			0.4	323
$Sr_{0.9}\ Dy_{0.1}TiO_3$ R = (La,Sm,Gd, Dy,Y)	500	−160	2.7		0.22	573
$Sr_{0.95}La_{0.05}TiO_3$	150	250	4.2	800	0.15	780
$Sr_{0.9}La_{0.1}TiO_3$	300	−225	3.2		0.21	750
La-doped$SrTiO_3$	80	−300	3.1		0.27	1073
$Sr(Ti_{0.8}Nb_{0.2})O_3$		−200	3.5	1300	0.37	1000
$Ba_{0.3}Sr_{0.6}La_{0.1}TiO_3$		−110	4.4		0.13	420
$Sr_{0.45}Ca_{0.45}La_{0.1}TiO_3$	250	−195	3.7		0.22	850
$Sr_{0.98}La_{0.02}TiO_3$	500	−260	11		0.09	298
$SrTiO_3/SrTi_{0.8}Nb_{0.2}O_3\ /SrTiO_3$	1400	−850	12		2.4	300
$SrTi_{0.9}Ta_{0.1}O_3$	300	−175	4.4		0.17	752
$Sr_{0.95}La_{0.05}Ti_{0.98}Mn_{0.02}O_3$	833	−150	3.9	20	0.15	300
n-type$SrTiO_3$		300			0.7	1400

Table 1.
Thermoelectric studies of doped $ATiO_3$ compounds.

Materials	Electrical Conductivity (S cm^{-1})	Seeback coefficient (μVK^{-1})	Thermal conductivity (W m^{-1} K^{1})	Power Factor (μW m^{-1} K^{-2})	ZT	Temperature (K)
$Tb_{0.1}Ca_{0.9}MnO_3$	−0.18	−140			0.13	950
$Ho_{0.1}Ca_{0.9}MnO_3$	−0.18	−110			0.08	950
$Y_{0.1}Ca_{0.9}MnO_3$	−0.2	−130			0.15	950
$Ca_{0.9}Bi_{0.1}MnO_3$	10	80			0.095	1173
$Ca_{0.85}Pr_{0.15}MnO_{2.98}$	111	−130	1.5		0.17	1100
$Ca_{0.9}Yb_{0.1}MnO_3$	133	−150	1.6		0.16	970
$Pr_{0.3}$ $Sr_{0.7}MnO_3$	250	−75	1.6		0.085	1073
$Ca_{0.96}Bi_{0.04}MnO_3$	66.7	−170	3.6	300	0.086	1000
$CaMn_{0.96}Mo_{0.04}O_3$		−90	3.4		0.012	270
$CaMn_{0.94}Ru_{0.06}O_3$		−140	5.4		0.0085	330
$CaMn_{0.98}Ta_{0.02}O_3$	29	−190			0.05	1000
$CaMn_{0.98}Nb_{0.02}O_3$	31	−255			0.32	1050
$SrMn_{0.7}Ru_{0.3}O_3$	50	−40			0.01	370
$Sr(Mn_{0.975}Mo_{0.025})O_3$	0.13	−120	5		0.013	400
$Ca_{0.8}Nd_{0.2}MnO_3$	280	−62	1.3		0.17	873

Table 2.
Thermoelectric studies of doped $AMnO_3$ compounds.

Materials	Electrical Conductivity (S cm^{-1})	Seeback coefficient (μVK^{-1})	Thermal conductivity (W m^{-1} K^{1})	Power Factor (μW m^{-1} K^{-2})	ZT	Temperature (K)
$(Pr_{0.9}Ca_{0.1})CoO_3$	220	106	1.9		0.047	358
$TbCoO_3$	200	80	1.6		0.05	873
$Ho_{0.9}Ca_{0.1}CoO_3$	20	220	0.75		0.051	573
$La_{0.875}Sr_{0.125}CoO_3$		100	6		0.035	230
$La_{0.95}Sr_{0.05}CoO_3$	20	720	0.037		0.18	300
$La_{0.9}Sr_{0.1}CoO_3$		120	1.5		0.046	300
$La_{0.9}Pb_{0.1}CoO_3$	333	110	0.8		0.23	575
$La_{0.97}Ba_{0.03}CoO_3$	40	80		80	0.08	420

Table 3.
Thermoelectric studies of doped $ACoO_3$ compounds.

that the ZT values of 0.63 and 0.64 for $CsSnI_3$ and CsPb n-type at 1000 K. The thermoelectric properties of hybrid perovskites are tabulated in **Table 5**.

4. Nanostructuring of thermoelectric materials

The materials having at least one of the dimensions in the order of 10^{-9} m are nanostructured materials. If the dimension of material is in nanometer range its surface-to-volume ratio increases and the properties changes drastically. Nanoparticles are highly reactive because they possess large surface energy. Such an increase in surface area in thin films and coating can enhance the sensing property, catalytic activity of surfaces, light trapping in solar cells, surface reactivity, etc. Nanosystems are classified into three two-dimensional (2D), one-dimensional (1D), and Zero dimensional (0D). Nanosheets and superlattices are 2D nanosystems, nanowires, nanorods, and nanotubes are 1D and nanopowders and quantum dots are 0D nanosystems. There are two approaches for the fabrication of nanostructured materials- top-down method and the bottom-up method. Starting from bulk crystalline material and dividing it into small pieces to obtain fine nanosized particles is the top-down method. Ball milling, spin melting, thermal cycling, lithography, etc. [27]. In bottom-up method, nanoparticles are produced from their constituent elements, which are assembled to form dense solids **Figure 3**.

When the particle size is decreased to nano, size-dependent quantum confinement effect arises. Generally, in nanostructures, the energy level spacing increases with decreasing size and is the quantum size confinement effect. This effect influences the optical, electronic, magnetic, thermal, and dynamic properties of the material. Another important size reduction effect is the electron-phonon coupling. With decrease in size, the density of states of both phonons and electrons decreases in size and this decreases the overlap. The combination of density of states and the surface phonon frequencies affect the phonon-electron interaction in the nanostructures. The quantum confinement effect and the phonon scattering have a crucial role in enhancing the efficiency of thermoelectric materials. By nanostructuring, the electrical conductivity of thermoelectric materials can be enhanced by quantum confinement effect

DOI: http://dx.doi.org/10.5772/intechopen.106614

Materials	Electrical Conductivity ($S\ cm^{-1}$)	Seeback coefficient (μVK^{-1})	Thermal conductivity ($W\ m^{-1}\ K^{1}$)	Power Factor ($\mu W\ m^{-1}\ K^{-2}$)	ZT	Temperature (K)
$LaCo_{0.92}Ni_{0.08}O_{2.9}$	33.3	220	0.35		0.2	300
$Pr_{0.9}Sr_{0.1}FeO_3$		140	0.8		0.024	850
$La_{0.95}Sr_{0.05}FeO_3$		230	1.8		0.076	1273
Ca_2FeMoO_6	300	−108	3.2		0.14	1250
$Ca_{1.9}Sr_{0.1}FeMoO_6$	250	−110	3		0.14	1250
$Ca_{1.8}Sr_{0.2}FeMoO_6$	260	−100	2.8		0.14	1250
$Sr_{1.6}K_{0.4}FeMoO_6$		−48	3.1	450	0.24	1250
Ba_2FeMoO_6		−1350			0.995	300
$Ba_{0.998}La_{0.002}SnO_3$	150	−170	4		0.1	1073
$Sr_{0.99}La_{0.01}SnO_3$	1.5	−80	3.6	120	0.05	1073
$BaSnO_3$	300	−130	3.4	1400	0.65	1200
$Ba_{0.4}Sr_{0.6}PbO_3$	250	125	2		0.13	673
$Ba_{0.2}Sr_{0.8}PbO_3$	79	−190	1.8		0.13	680
$BaMoO_3$		−30			0.015	1000
$SrRuO_3$		36	5.3		0.03	1200
$(Sr_{0.95}\ La_{0.05})_2$ $RuErO_6$		−160			0.001	800
$K_{0.991}Ba_{0.009}Ta\ O_3$	333	200			0.03	300
$BaUO_6$	0.1	−170	0.8		0.0002	880

Table 4.
Thermoelectric studies of other doped perovskite oxide.

Materials	Electrical Conductivity ($S\ cm^{-1}$)	Seeback coefficient (μVK^{-1})	Thermal conductivity ($W\ m^{-1}\ K^{1}$)	Power Factor ($\mu W\ m^{-1}\ K^{-2}$)	ZT	Temperature (K)
$CsSnI_3$			0.18		0.63	1000
$CsPbI_3$			0.1		0.64	1000
$(FA)PbI_3$		100			0.43	298
$(FA)SnI_3$		150			0.35	298
$CH_3NH_3PbI_3$-n-type	1.2	80			0.9	330
$CH_3NH_3PbI_3$-p-type	1	50			1.25	330
$C_6H_4NH_2CuBr_2I$	2950	−82	3.25		0.21	363
$CH_3NH_3PbI_3$-n-type	68	−428	0.151	11.3	2.56	800
p-type	25	358	0.04	3.3	1.08	800
$CH_3NH_3SnI_3$	0.001	720	0.08		0.01	295
$CH_3NH_3PbI_3$	10^{-7}	700	0.5		$10–^{7}$	295
$(MA)PbI_3$		200			0.44	298
$(MA)SnI_3$		200			0.44	298

Table 5.
Theoretical ZT value of hybrid perovskites from calculations.

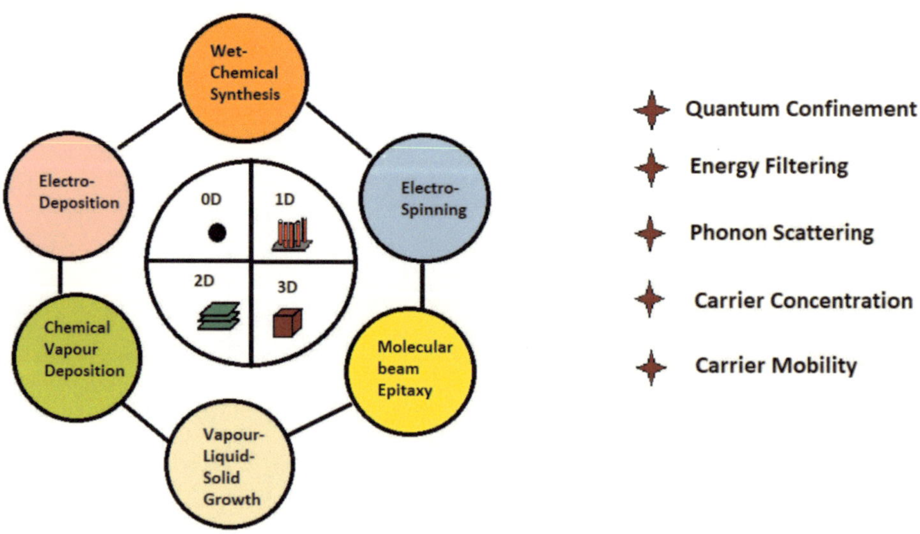

Figure 3.
Bottom-up strategies for nanostructuring of TE materials.

and the thermal conductivity can be reduced by phonon scattering at the interfaces, thereby increasing the figure of merit.

The nanostructured thermoelectric perovskite compounds have been prepared by many techniques which include co-precipitation, mechanical synthesis, solid-state reactions, solution combustion or thermal decomposition, hydrothermal, Pechini, and sol-gel method. Many new methods and improvements in synthesis conditions have been tried by the researchers as the properties of the end product strongly depend on the method of synthesis technique used. The citrate sol-gel auto-combustion method, which is a modified Pechini method based on the polyesterification of ethylene glycol and citric acid for the synthesis of the perovskite nanopowders. The method involves relatively easy synthesis route when compared to the other conventional processes. The control over the end stoichiometry and low operating temperature are the main advantages of this technique.

Popa et al. have synthesized perovskite – $LaMeO_3$ (Me - Co, Mn, Fe) compounds by the polymer complex method and elaborated its advantages. Nonuniformity in particle size, compositional inhomogeneity, and high processing temperature are the main disadvantages when the conventional mixed oxide methods are preferred for the synthesis. Perovskite nanopowders developed through wet-chemical method have relatively high product uniformity and reliable reproducibility. By this method, it is possible to reduce the agglomeration of nanoparticles and can control the particle size. In citrate sol-gel auto-combustion method, at relatively low temperature excellent chemical homogeneity can be achieved. The perovskite nanopowders thus obtained have uniform particle size, which allows sintering to give dense well shaped uniformly grained microstructures [28]. The nanopowders were subjected to characterization techniques including XRD, SEM-EDAX, XPS, particle size analyzer, etc. Finally, dc electrical conductivity, thermal conductivity, and the Seebeck coefficient measurements are carried out using thermoelectric measurement setup.

5. Thermoelectric characterization

The simultaneous measurement of electrical resistivity and Seebeck coefficient was done using ULVAC-ZEM 3. The sample is sandwiched between the electrodes and kept in helium atmosphere at low pressure of 10^{-3} Torr. The resistivity is calculated using four probe method.

$$\rho = \frac{RA}{l} \tag{8}$$

where ρ is the electrical resistivity, R is the resistance, A is the area of cross section and l is the distance between the probes. High impedance current is supplied through the probes connected to upper and lower blocks. Other two probes measure the voltage produced. Seebeck coefficient is determined by measuring the electromotive force generated at the probes. The sample is kept in such a way that a temperature gradient can exist between the two ends. Let T_1 and T_2 be the temperatures at two ends and the electrical potential difference is dV, the Seebeck coefficient can be calculated using the formula,

$$S = \frac{dV}{T_1 - T_2} \tag{9}$$

The measurement is controlled by a computer. The voltage-current measurement is made to check the correct contact of the sample. The thermal conductivity of the sample can be measured by divider bar method. In this method the sample is sandwiched between two metal blocks, heat flows through the sample by measuring the thermal gradient the thermal conductivity can be measured.

6. Conclusion

It is well known from the literature that, the need for thermoelectric material for power generation and refrigeration is increasing day by day. Nowadays, heat generation from automobiles, factories, combustion of fossil fuels, nuclear decay, etc. is increasing a lot and it is necessary to convert these waste heat into useful form. Therefore, development of thermoelectric material which can convert the waste heat into electricity will be a milestone for the modern technology. Even though the ZT value of perovskite was very small, we can tune the properties of these materials by nanostructuring, band gap engineering, and by doping. In these thermoelectric perovskite materials, $SrTiO_3/SrTi_{0.8}Nb_{0.2}O_3$ /$SrTiO_3$ was considered to be the best material with a ZT value of 2.4 at 300 K. Therefore, there are many possibilities in perovskite materials to be a replaceable candidate for TE applications. The hybrid perovskites which are low cost and easily synthesized by energy cost methods can be potential TE material in future at room temperature range. In the case of these materials, $CH_3NH_3PbI_3$-n-type shows a ZT of 2.56 at 800 K, which was only a theoretical calculation. Further in future, we can make all the calculations be true for an alternative energy resource for the world.

Author details

Megha Unikoth[1*], George Varghese[2], Karakat Shijina[3] and Hind Neelamkodan[1]

1 DGMMES Mampad College, Malappuram, Kerala, India

2 Mar Ivanios Autonomous College, Thiruvananthapuram, Kerala, India

3 Department of Physics, NSS College Cherthala, Kerala, India

*Address all correspondence to: meghaunikoth@gmail.com

References

[1] Megha U. Synthesis and Studies of Nanostructured Thermoelectric Perovskite Materials. Kerala, India: University of Calicut; 2016

[2] Sun Y, Liu Y, Li R, Li Y, Bai S. Strategies to improve the thermoelectric figure of merit in thermoelectric functional materials. Frontiers in Chemistry. 2022;**10**:865281

[3] Julian Goldsmid H. Improving the thermoelectric figure of merit. Science and Technology of Advanced Materials. 2021;**22**:280

[4] Zhu Q, Wang S, Wang X, Suwardi A, Chua MH, Soo XYD, et al. Bottom-Up engineering strategies for high-performance thermoelectric materials. Nano-Microletters. 2021;**13**:119

[5] Wu T, Gao P. Development of perovskite-type materials for thermoelectric application. Materials. 2018;**11**:999

[6] Zhang X, Zhao L-D. Thermoelectric materials: Energy conversion between heat and electricity. Journal of Materiomics. 2015;**1**:92

[7] Shi X, Yang J, Salvadoor JR, Chi M, Cho JY, Wang H. Multiple-filled skutterudites: high thermoelectric figure of merit through separately optimizing electrical and thermal transports. Journal of AmericanChemical Society. 2011;**133**: 7837

[8] Liu W, Yan X, Chen G, Ren Z. Recent advances in thermoelectric nanocomposites. Nanoenergy. 2012;**1**: 42-56

[9] Saramat A, Svensson G, Palmqvist AEC, Stiewe C, Mueller E, Platzek D. Large thermoelectric figure of merit at high temperature in Czochralski-grown clathrate Ba8Ga16Ge30. Journal of Applied Physics. 2006;**99**:023708

[10] Lee M, Viciu L, Li L, Wang Y, Foo ML, Watauchi S, et al. Large enhancement of the thermopower in Na_xCoO_2 at high Na doping. Nature Materials. 2006;**5**:537-540

[11] Liu H, Shi X, Xu F, Zhang L, Zhang W, Chen L, et al. Copper ion liquid-like thermoelectrics. Nature Materials. 2012;**11**:422-425

[12] Snyder GJ, Christensen M, Nishibori E, Caillat T. Disordered zinc in Zn_4Sb_3 with phonon-glass and electron-crystal thermoelectric properties. Nature Materials. 2004;**3**:458-463

[13] Hicks LD, Dresselhaus MS. Effect of quantum-well structures on the thermoelectric figure of merit. Physical Review B. 1993;**47**:1993-12727

[14] Venkatasubramanian R, Siivola E, Colpitts T, O'Quinn B. Thin-film thermoelectric devices with high room-temperature figures of merit. Nature. 2001;**413**:597-602

[15] Harman TC, Taylor PJ, Walsh MP, LaForge BE. Quantum dot superlattice thermoelectric materials and devices. Science. 2002;**297**:2229-2232

[16] Hochbaum AI, Chen R, Delgado RD, Liang W, Garnett EC, Najarian M. Enhanced thermoelectric performance of rough silicon nanowires. Nature. 2008;**451**:163-167

[17] Boukai AI, Bunimovich Y, Tahir-Kheli J, Yu J-K, Goddard WA III, Heath JR. Silicon nanowires as efficient

thermoelectric materials. Nature. 2008; **451**:168-171

[18] Vineis CJ, Shakouri A, Majumdar A, Kanatzidis MG. Nanostructured thermoelectrics: Big efficiency gains from small features. Advanced Materials. 2010;**22**:3970-3980

[19] Fan S, Zhao J, Guo J, Yan Q, Ma J, Hng HH. p-type Bi0.4Sb1.6Te3 nanocomposites with enhanced figure of merit. Advanced Physics Letters. 2010; **96**:182104

[20] Biswas K, He J, Zhang Q, Wang G, Uher C, Dravid VP, et al. Strained endotaxial nanostructures with high thermoelectric figure of merit. Nature Chemistry. 2011;**3**:160-166

[21] Wang XW, Lee H, Lan YC, Zhu GH, Joshi G, Wang DZ. Enhanced thermoelectric figure of merit in nanostructured ?n-type silicon germanium bulk alloy. Applied Physics Letters. 2008;**93**:193121

[22] Wolfram T. Electronic and Optical Properties of d-band Perovskites. Cambridge: Cambridge University Press; 2012

[23] Goldschmidt-CCP14, Distortion of perovskites. www.ccp14.ac.uk/ccp/web-mirrors/pki/uni/pki/members/schinzer

[24] Megha U, Shijina K, Varghese G. Nanosized LaCo 0.6 Fe 0.4 O 3 perovskites synthesized by citrate sol gel auto combustion method. Processing and Application of Ceramics. 2014;**8**(2): 87-92

[25] Megha U. Effect of Bi and Sr doping on morphological and magnetic properties of LaCo0.6Fe0.4O3 nanosized perovskites. Bulletin of Materials Science. 2016;**39**(1):125-313

[26] Megha U, Varghese G, Shijina K. Room temperature AC impedance and dielectric studies of Bi and Sr doped PrCo0.6Fe0.4O3 perovskites. Processing and Application of Ceramics. 2017;**11**(1): 52-59

[27] Pradeep T. Nano: The Essentials. United States: McGraw-Hill Education; 2007

[28] Popa M, Kakihana M. Synthesis of lanthanum cobaltite (LaCoO3) by the polymerizable complex route. Solid State Ionics. 2002;**151**:251-257

Section 2

Perovskites in Solar Cells

Chapter 7

Recent Development of Lead-Free Perovskite Solar Cells

Anshebo Getachew Alemu and Teketel Alemu

Abstract

Recently, the world energy demand has been raised up dramatically. Numerous energy sources have been developed to satisfy the urgent energy desires and to overcome the world energy crisis. Among them, solar energy has been considered an efficient energy source for current energy requirements. Nowadays, the lead-based perovskite solar cells achieved excellent power conversion efficiency exceeding 29.1%. However, to address major problems such as toxicity and underprivileged stability, several hardworks were made toward the replacement of lead-free perovskite material in perspective of device's performance and stability. In this book chapter, we summarize material, dimensions, stability, and the current achievement of lead-free solar cells. Finally, we review the remaining challenges and future perspective for development of lead-free perovskite solar cells.

Keywords: material, dimensions, stability, lead-free photovoltaics

1. Introduction

Nowadays, among renewable energy alternatives, solar energy is the most abundant and has minimum impact on the environment compared with nonrenewable sources such as natural gas, fossil fuels, and nuclear energy. The development of photovoltaic has made possible the change of sunlight into electrical energy with high power conversion efficiencies with low cost [1, 2]. The demand of energy from the photon energy is primary significance because it is clean, renewable, abundant, and natural [3, 4]. Among the innovative photovoltaic, perovskite solar cells have hastily enhanced to the frontline for electricity production [5]. Solar-cells-based tandem perovskite achieved world high efficiency of 29.15% in the photovoltaic research field [6] and low production cost [7–9]. However, the increased concern of lead toxicity for extensive use in addition to the distress of disposal, widespread research effort has been dedicated to the path of lead-free PSCs [10, 11]. Due to encounters such as instability in ambient conditions [12], lack of accuracy in thickness [13], and device-incompatible solution growth processes [14, 15], PSCs have not yet gained sufficient trust for commercial applications.

Therefore, novel groups of lead-free halide PSCs have been discovered for substituting lead (Pb) with other elements such as antimony (Sb) [16], bismuth (Bi) [17], germanium (Ge), [18, 19], indium (In) [20, 21], tin (Sn), [22, 23], and double halide perovskite (**Figure 1**) possession the inherent perovskite properties

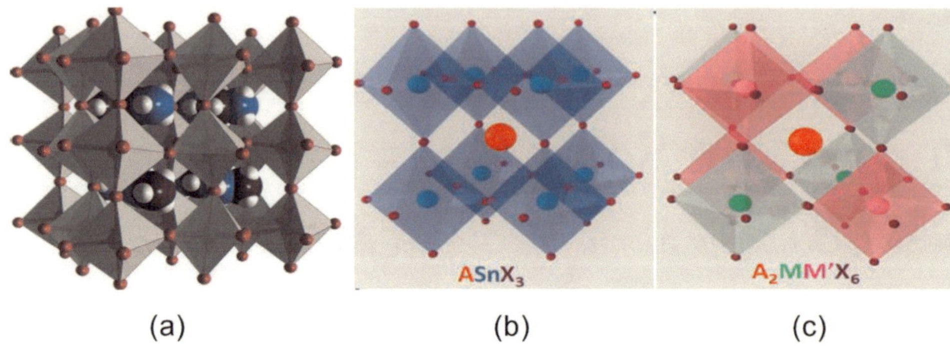

Figure 1.
Structure of (a) Pb-perovskite (b) tin-halide perovskites (c) double-halide perovskites. X is a halide; M and M′ stand for monovalent and trivalent metals, respectively [24, 25].

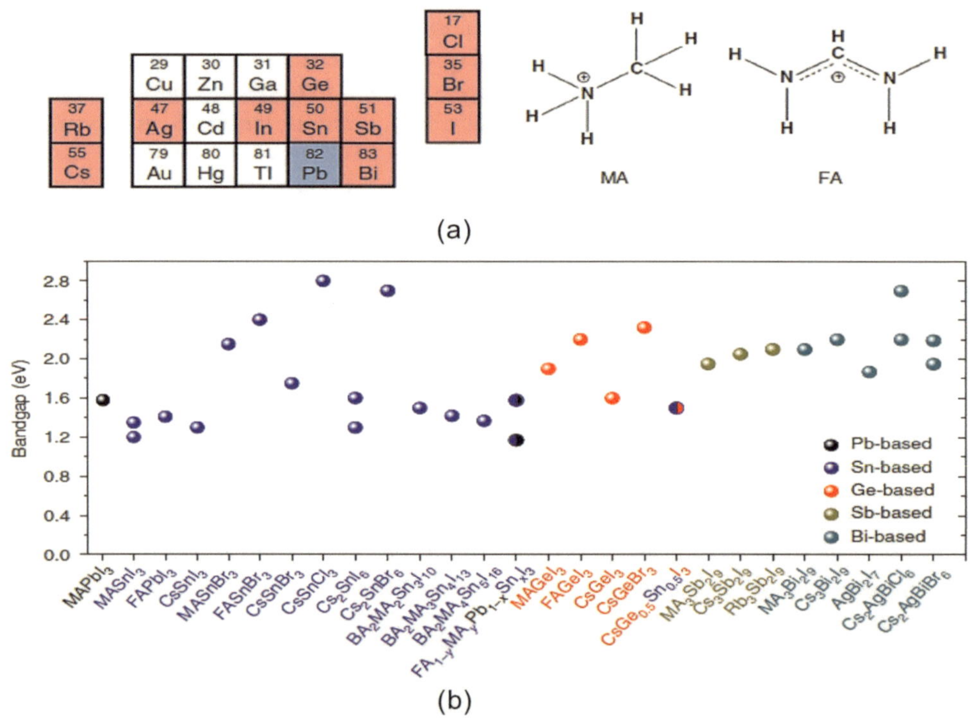

Figure 2.
Solar cell absorbers materials .A-site cations (organic MA and FA or inorganic Cs and Rb), metals, and halides (I, Br, Cl) for perovskite structure. b/ band gaps of different materials solar cells should have band gaps from 1.1 to 2.0 eV [26–31].

unchanged. Furthermore, these alternative Pb-free materials show significant advantages such as highlight absorption coefficients, higher charge carrier mobilities, and narrow optical band gap compared with lead-based perovskites as shown in **Figure 2** [32].

This book chapter contains of the following sections: (1) introduction of metal halide PSCs, (2) origin of lead-free perovskite solar cells, (3) lead-free Pb-free materials, (4) dimensions of Pb-free materials, (5) limitations of Pb-lead materials, and (6) future prospective have also been discussed.

2. Origin of lead-free perovskite solar cells

Later a revolutionary report by Kojima et al. [33], and a successive breakthrough by Kim et al. and Lee et al. [34, 35] showed that lead-based perovskite solar cells have significantly transmuted the field of photovoltaics. Different generations exhibit perovskite solar cells' extraordinarily high power conversion efficiency because of high absorption coefficient [36] and defect tolerance [37] and to the long exciton diffusion length [38] of perovskite materials.

This noteworthy improvement in PSC solar cells restricted by critical intrinsic device instability of PSCs [39, 40] and toxic to the environment restricted for breakthrough outdoor application [41–46]. According to the World Health Organization (WHO), report Pb through metabolism, and children are at a particularly high risk of Pb poisoning the human body cannot purge [47, 48]. Furthermore, Pb can simply spread into the air, water, and soil [49, 50].

More recently, the scientific community has been pointed for Pb-material replacements that have been reported promising efficiency progress [51–53]. **Figure 3a** bar graph shows the recent the highest certified power conversion efficiency (PCE) of the different types of photovoltaics, perovskite solar cells (PSC, 29.1%), organic solar cells (OSC, 17.4%), dye-sanitized solar cells (DSC, 13.8%), CIGS (23.4%), CdTe (22.1%), polycrystalline siliconcells (PCSSC, 22.3%), and monocrystalline silicon solar cells (MCSSC, 26.7%). **Figure 3b** shows that reported PCEs of Pb-based, Bi-based, and Sn-based PSCs from the preliminary stage of development to date. Sn-based PSCs (Sn-PSCs) have thus far shown the greatest prospects though there are fewer reports on Sn-PSCs compared with those on Pb-PSCs, the PCE and stability of Sn-based PSCs have been enhanced quickly. For example, Sn-based perovskites such as cesium tin iodide ($CsSnI_3$), formamidinium tin iodide ($FASnI_3$), and methylammonium tin iodide ($MASnI_3$), have direct band gaps of approximately 1.3 eV, 1.41, and 1.20 eV, respectively, which are narrower than Pb-based perovskites [31, 54].

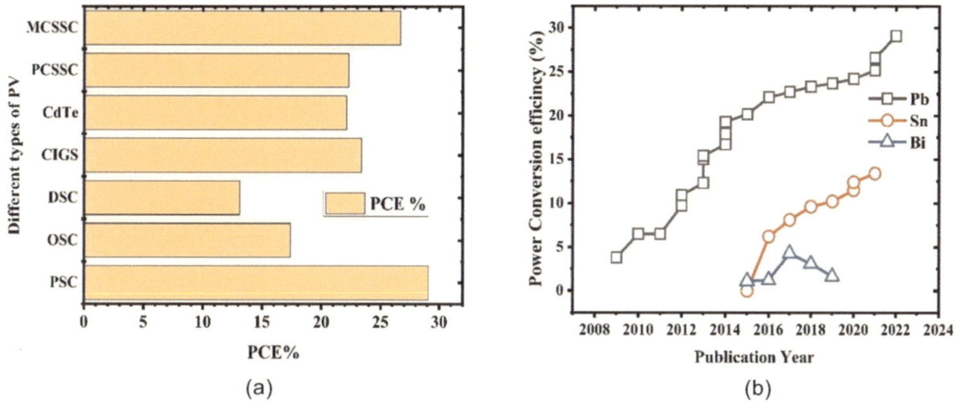

Figure 3.
(a) Highest certified PCEs of the different types of photovoltaics. (b) Reported PCEs of Pb-based, Bi-based, and Sn-based PSCs.

3. Materials

The existence of Pb is an urgent problem in contradiction of the final application of PSCs and the toxicity of lead disturbs the operational of the blood kidneys, liver, testes, brain, and nervous system [55]. In order to address toxicity and poor stability lead (Pb)-based perovskite, several hard works were made toward the replacement of lead-free perovskite material. In this section, we give comprehensive review of the Pb-substitutes, such as Ge-based perovskite, bismuth-based perovskite, Sn-based perovskite, alkaline-earth metals perovskite, transition-metal-based perosivikate, and heterovalent perovskite.

3.1 Ge-based perovskite

Ge, is finest substitute for lead in the same group as Pb. For example, $AGeI_3$ perovskite group, $CsGeI_3$ has the narrowest band gap of about 1.6 eV (**Figure 2**), but $MAGeI_3$ and $FAGeI_3$ have direct band gaps of 1.9–2. 2 eV. The band gap (Eg) of the $MAGeX_3$ perovskites estimated, and the Eg of $MAGeI_3$ was 1.61 eV, in contrast to 2.81 eV and 3.76 eV for the Br and Cl anions, correspondingly. And high-quality $CsSn_{0.5}Ge_{0.5}I_3$ perovskite films with a band gap of 1.5 eV solar cells reported an incredible PCE of 7.11% [56]. The germanium-based-perovskite instability due to from the tendency of Ge^{2+} to oxidize into Ge^{4+} [57]. The new result was achieved $CsSn_{0.5}Ge_{0.5}$ I_3, which provided a PCE of 7.11% with improved stability concerning the $CsSnI_3$ [58]. Furthermore, germanium in 0.75 MA0.25$Sn_{1-x}Ge_xI_3$ was newly provided PCE of 7.9% [59] as shown in **Table 1**.

Germanium halide perovskites	Dimensionality	Band gap (eV)	PCEs (%)	References
$RbGeCl_3$x H_2O	3D	3.84	—	[60]
$RbGeBr_3$	3D	2.74	—	[60]
(RbxCs1-x)$GeBr_3$	3D	2.4	—	[61]
$CsGeCl_3$	3D	3.4–3.67	—	[61]
$CsGeBr_3$	3D	2.32–2.4	—	[61]
CsGe(BrxCl1-x)$_3$	3D	2.65	—	[62]
$CsGeI_3$	3D	1.53–1.63	0.11	[62]
CH3NH_3GeCl_3	3D	3.74–3.76	—	[61]
CH3NH_3GeBr_3	3D	2.76–2.81	—	[60]
CH3NH_3GeI_3	3D	1.9–2.0	0.2	[61]
CH(NH_2)2GeI_3	3D	2.2–2.35	—	[61]
$MFOGeI_3$	3D	2.5	—	[60]
$GUAGeI_3$	3D	2.7	—	[61]
$TMAGeI_3$	3D	2.8	—	[60]
$IPAGeI_3$	3D	2.7	—	[61]

Table 1.
Germanium halide perovskites PCEs in photovoltaic devices.

3.2 Bismuth-based perovskite

Bi is a nontoxic and has the comparable properties to Pb and has satisfactory tolerance factor rule and enhanced the stability. It has unique properties such as 0D dimensionality, indirect band gaps, and mobilities (**Figure 3a**) [60, 61]. According to Park report [63], Bi-based perovskite $A_3Bi_2I_9$ (A to be Cs and MA) as a photovoltaic absorber has estimated band of $A_3Bi_2I_9$ to be ca. 2.1 eV for MA and 2.2 eV for Cs, and the exciton binding energy as 70 meV (while Pb perovskite is at 2550 meV). The reported photovoltaic parameters as (for $CS_3Bi_2I_9$ PCE =1.09%, FF = 0.6, Voc = 0.85 V, J_{sc} = 2.15mAcm^{-2}, 0.12%, for $MA_3Bi_2I_9$ FF = 0.33, Voc = 0.68 V, Jsc = 0.52mAcm^{-2}). This low efficiency because of the number of reasons, such as excess reactant residue, extra band-gap states, poor morphology, and interface contact. Therefore, the Bi-based perovskite is ideal as solar cell absorbers due to low mobilities and good stability [62, 63] as shown in **Table 2**.

3.3 Sn-based perovskite

Tin (Sn) is group 14 element less-toxic metal having comparable properties [71]. As the primary report on Sn-based perovskite solar cells, power conversion efficiency (PCE) of 6% [72, 73]. Later the tin-based perovskites have been dominant and providing the highest efficiency of 13.24% [74]. The representative cesium tin iodide ($CsSnI_3$), formamidinium tin iodide ($FASnI_3$), and methylammonium tin iodide ($MASnI_3$) have direct band gaps of 1.3, 1.2, and 1.41 eV [75, 76], respectively. Recently, Sn-based perovskite highest ($MAPbI_3$) 0.4 ($FASnI_3$) 0.6 15.08% [70] efficiency reported (**Table 3**) [69, 77–92].

Bismuth halide perovskites	Dimensionality	Band gap (eV)	PCEs (%)	References
$(CH_3NH_3)3Bi2I_9$	0D	1.94–2.11	0.42	[64]
$(CH_3NH_3)3Bi2I_{9-x}Clx$	—	2.4	0.003	[65]
Cs_3Bi2I_9	0D	1.8–2.2	1.09	[66]
$HDABiI_5$	1D	2.05	0.027	[67]
$CsBi3I_{10}$	2D	1.77	0.40	[68]
$LiBiI_4$ 5 H_2O	1D	1.7–1.76	—	[69]
$MgBi_2I_8$ H_2O	1D	1.7–1.76	—	[69]
$MnBi_2I_8$ H_2O	1D	1.7–1.76	—	[69]
$KBiI_4H_2O$	1D	1.7–1.76	—	[69]
$Cs_2AgBiCl_6$	3D	2.2–2.77	—	[70]
$Cs_2AgBiBr_6$	3D	1.95–2.19	—	[70]
Cs_2AgBiI_6	3D	1.6	—	[70]
Cs_2AuBiX_6	3D	1.6	—	[70]
$K_3Bi_2I_9$	2D	2.9	—	[70]
$Rb_3Bi_2I_9$	2D	1.89–2.1	—	[70]
$Cs_3Bi_2Br_9$	2D	2.50	—	[70]

Table 2.
Bismuth halide perovskites and the highest obtained PCEs.

Perovskite	Dimensionality	Band gap (eV)	PCE (%)	References
$CH_3NH_3SnBr_3$	3D	2.15–2.2	4.27	[77]
$CH_3NH_3SnIBr_2$	3D	1.75	5.73	[78]
$CH_3NH_3SnI_2Br$	3D	1.56	5.48	[79]
$CH_3NH_3SnI_3$	3D	1.27–1.35	5.23	[80]
$CH(NH_2)2SnI_2Br$	3D	1.68	1.72	[81]
$CH(NH_2)2SnI_3$	3D	1.4–1.41	6.22	[82]
$CsSnBr_3$	3D	1.75–1.8	2.1	[83]
$CsSnIBr_2$	3D	1.63–1.65	3.2	[84]
$CsSnI_2Br$	3D	1.37–1.41	1.67	[85]
$CsSnI_3$	3D	1.27–1.31	3.31	[86]
$CsSnI_2.95F0.05$	3D	1.3	8.51	[87]
Cs_2SnCl_6	3D	3.9	0.07	[69]
Cs_2SnBr_6	3D	2.7	0.04	[69]
Cs_2SnI_6	3D	1.26–1.62	0.86,	[69]
$Cs_2SnI_3Br_3$	3D	1.43	3.63	[69]
$MAPb_{0.85}Sn0.15I_316$	3D	N/A	10.10	[70]
$MA_{0.5}FA_{0.5}Pb_{0.75}Sn_{0.25}I_3$	3D	1.33	14.35	[70]
$MAPb_{0.5}Sn_{0.5}I_3$	3D	1.18	13.60	[70]
$MAPb_{0.5}Sn_{0.5}I_3$	3D	1.12	10.0	[70]
$(MAPbI_3)_{0.4}(FASnI_3)_{0.6}$	3D	1.20	15.08	[70]

Table 3.
Tin halide perovskites and the highest obtained PCEs.

For example, more recently in 2016, Li et al. fabricated an inverted structured device with $MAPb_{0.5}Sn_{0.5}I_3$, achieved PCE of 13.6% [93]. And Liao et al. employed $(MAPbI_3)_{0.4}(FASnI_3)_{0.6}$ in inverted device and achieved a PCE of 15.08% [94]. Finally, Sn-based perovskite has become a hopeful alternative material for replacement the Pb-based perovskite. Though, instability, low PCE and certain degree of toxicity problems hasty more research to find other substitute materials, which can be more stable with less toxicity.

3.4 Alkaline-earth metals perovskite

Another friendly unleaded perovskite such as magnesium (Mg), calcium (Ca), strontium (Sr), and barium Ba) are also interesting candidate for Pb. The alkaline-earth metals and their compounds are usually low cost and have advantage to industrial applications [68, 95].

3.4.1 Magnesium halide perovskite

Magnesium halide perovskite is low effective masse, reasonable absorption coefficients, and direct band gaps. $AMgI_3$ perovskites, the band gap was predicted to be

Perovskite	Dimensionality	Band gap (eV)	PCE (%)	Crystal system (space group)	References
$CH_3NH_3MgI_3$	—	1.5	—	Tetragonal	[70, 96]
CH(NH2)2MgI_3	—	0.9	—	Trigonal(P3m1)	[70, 96]
$CsMgI_3$		1.7		Orthorhombic	
$CH_3NH_3CaI_3$	—	2.95,		Tetragonal	[70, 96]
$CH_3NH_3CaI_3$-xClx	—	—		—	[70, 97]
$CH_3NH_3SrI_3$		3.6		Tetragonal	[70, 97]
$CH_3NH_3BaI_3$		3.3		Tetragonal	[70, 97]

Table 4.
Alkaline-earth metal halide perovskites: NB. Dimensionality and PCE values have not been reported.

tunable using different A-site cations with band gaps of 0.9 eV ($CH(NH_2)_2MgI_3$), 1.5 eV ($CH_3NH_3MgI_3$), and 1.7 eV($CsMgI_3$) (**Table 4**). Until now magnesium halide perovskites have not been applied as materials in solar cells, which might be because of the sensitivity toward moisture [69, 95].

3.4.2 Calcium halide perovskite

Calcium halide perovskite is low-cost, nontoxic, abundant in the Earth's crust. The divalent Ca^{2+} ion has suitable ionic radius (100 pm) similar to Pb^{2+} (119 pm) to exchange lead in the perovskite structure. It is the high band gap, the low mobility, and the instability. This material is not appropriate for photovoltaic applications due to environmental instability but might be probable candidates for charge-selective contacts [69, 97].

3.4.3 Strontium halide perovskite

Strontium halide perovskite is an impartially less toxic, inexpensive, highly abundant alkaline-earth metal with an ionic radius (Sr^{2+}:118 pm) very similar to lead (Pb^{2+}:119 pm), which makes strontium an appropriate candidate for homovalent substitution of lead in the perovskite without affecting the crystal structure. It exhibits an underprivileged stability under ambient conditions because of its hygroscopic nature. It recommended a potential application as charge-selective contact material [70, 97].

3.4.4 Barium halide perovskite

Barium halide perovskite is the stable Ba^{2+} metal cation shows a slightly larger ionic radius (135 pm) compared with Pb^{2+} (119 pm). It is expected to have a similar crystal structure as $CH_3NH_3PbI_3$. According DFT calculations predicted $CH_3NH_3BaI_3$ to form stable perovskite materials with an estimated band gap of 3.3 eV. It is sensitivity to moisture; it hampers the synthesis characterization and applicability in photovoltaics [97].

3.5 Transition-metal-based perovskites

There is significant interest in the field of transition metal halide perovskites rises from the rich chemistry and high abundance of metals [95]. Divalent transition metals Cu^{2+}(73 pm), Fe^{2+} (78 pm) Zn^{2+}, and Pd^{2+}(86 pm)) have been as the replacement of Pb perovskites photovoltaic devices [77]. Their small ionic radii and good tolerance factor of 1, 3D structures. This material has potential for photovoltaic applications in bulky crystal [98].

3.5.1 Copper halide perovskite

Copper halide perovskite is less-toxic, low-cost earth abundant. The divalent Cu^{2+}gets particular attention for replacement for Pb^{2+}due to ambient stability and the high absorption coefficient in visible region. According to Cortecchia et al. report, a noticeable photoluminescence with higher bromine contents resulting from the in-situ formation of Cu^{+} ions and the consistent charge carrier recombination at the charge traps [99].

3.5.2 Iron halide perovskite

Iron halide perovskite is smaller ionic radius of the Fe^{2+} (78 pm) compared to Pb^{2+} (119 pm) hampers the development of 3Dstructures [95]. The limitations of iron halide perovskites are the multiple oxidation states of iron that hinder the constancy reaction, i.e., oxidation of $Fe^{2+}$to Fe^{3+} comparable to tin and germanium perovskite [99, 100]. Therefore, iron halide perovskite has not been suitable for solar applications (**Table 5**).

Perovskite	Dimensionality	Band gap (eV)	PCE (%)	References
$(p\text{-}F\text{-}C_6H_5C_2H_4NH_3)_2CuBr_4$	2D	1.74	0.51	[99, 101]
$(CH_3(CH_2)_3NH_3)_2CuBr_4$	2D	1.76	0.63	[99]
$(CH_3NH_3)_2CuCl4$	2D	2.48	—	[102, 103]
$(CH_3NH_3)_2CuCl2Br_2$	2D	2.12	0.017	[102]
$(CH_3NH3)_2CuClBr_3$	2D	1.90	—	[102]
$(CH_3NH3)_2CuCl0.5Br_3.5$	2D	1.80	0.0017	[104, 105]
$(CH_3NH3)_2FeCl4$	2D	—	—	[104]
$(C_2H_5NH_3)_2FeCl_4$	2D	—	—	[104]
$(C_3H_7NH_3)_2FeCl_4$	2D	—	—	[104]
$(C_6H_5CH_2NH_3)_2FeCl_4$	2D	—	—	[104]
$(CH_3NH_3)_2FeCl_2Br_2$	2D	—	—	[104]
$(CH_3NH_3)_2FeCl3Br$	2D	—	—	[104]
$(CH_3NH_3)_2PdCl_4$	2D	—	—	[104]
$(C_8H_{17}NH_3)2PdCl_4$	2D	—	—	[104]

Table 5.
Optical data of transition metal halide perovskites.

3.5.3 Palladium halide perovskites

Palladium halide perovskites is only a few studies on palladium-based perovskite have been reported so far [106]. In additional the investigation of palladium halide perovskite confirm that the general formula A_2PdX_4, Where A is an organic aliphatic cation (RNH_3^+) such as $CH_3NH_3^+$ [107] and n-octyl ammonium [106] and X is a halide. It characterized 2D layered structures contain an alternating organic and inorganic layers [107]. Thus, palladium halide perovskites solar cell has not been reported. (**Table 5**) [108].

3.6 Heterovalent substitution

Among the ideal substitute of lead heterovalent substitution is a second viable approach toward lead-free perovskite. It is standby of the divalent lead cation with a cation in a diverse valence such as mono-, tri-, or tetravalent cation. Then, two different procedures such as the mixed-valence approach and heterovalent substitution accompanied with a significant change in the structure from ABX_3-type to $A_3B_2X_9$-type to maintain charge neutrality [109–111].

3.6.1 Thallium halide perovskites

It is a p-block metal with a Tl^+ cation isoelectronic to Pb^{2+} ($6s^26p^0$ electronic configuration). The monovalent Tl^+ cation, though, cannot substitute the divalent Pb^{2+} metal cation directly due to the violation of the charge neutrality. According to Giorgi et al. report, thallium halide perovskites ($CH_3NH_3Tl_{0.5}Bi_{0.5}I_3$) is projected to be a potential alternative solar cell material (**Table 6**). The thallium-based compounds are presumably no substitute to lead-based perovskites in terms of photovoltaic applications due to the toxicity of thallium [110, 111].

3.6.2 Gold halide perovskite

It is similar to thallium-based via the mixed-valence approach. Subsequently, gold has to be existing in grouping of mono valent $Au^+(5d_{10}, t_{2g}^6e_g^4)$ and trivalent $Au^{3+}(5d^8, t_{2g}^6e_g^2)$ to form ABX_3-type perovskite structures, like in the case of $Cs_2AuIAu^{III}X_6$(X = Cl, Br, I) compounds. Moreover, hybrid gold halide perovskites

Perovskite	Dimensionality	Band gap (eV)	PCE (%)	References
$Cs_2AuIAuIIICl_6$	3D	2.04	—	[69]
$Cs_2AuIAuIIIBr_6$	3D	1.60	—	[69, 104]
$Cs_2AuIAuIIII_6$	3D	1.31	—	[112]
$[NH_3(CH_2)7NH_3]2[(AuII_2)(AuIII I4)(I_3)2]$	2D	0.95	—	[112]
$[NH_3(CH_2)8NH_3]2[(AuII_2)(AuIII I4)(I_3)2]$	2D	1.14	—	[69]
CsTlF3	3D	—	0.0017	[69]
$(CsTl^+{}_{0:5}Tl^{3+}{}_{0:5}F_3)$ $CsTlCl_3$	3D	2.5	—	[69]
$(CsTl^+{}_{0.5}Tl^{3+}{}_{0.5}F_3Cl_3)$ $CH_3NH_3Tl_{0.5}I_3$	3D	1.6	—	[69]

Table 6.
Optical data of gold and thallium halide perovskites.

have been investigated such as $[NH_3(CH_2)_7NH_3]_2[(AuII_2)(Au^{III}I_4)$ $(I_3)2]$, and $[NH_3(CH_2)_8NH_3]_2[(AuII_2)$ (AuIII I4) $(I_3)2]$ as **Table 6** [69, 112].

3.6.3 Antimony halide perovskite

Antimony halide perovskite is potential alternative to lead perovskite for photovoltaic applications to address problems of chemical stability and toxicity. The valance three Sb^{3+}metal cation, isoelectronic to $Sn^{2+}(4d^{10}\ 5s^2)$ and has a comparable s^2 valence electronic arrangement as $Pb^{2+}(5s^2$ lone pair), and equivalent electronegativity (Sb:2.05, Sn:1.96, Pb:2.33) but considerable lesser ionic radius (76 pm) compare to the valance two Sn^{2+}(110 pm) and Pb^{2+} (119 pm) metal cations [69, 113].

3.7 Lanthanide and actinide halide perovskites

It is another substituent for Pb2+ giving rise toward lanthanide and actinide halide perovskites. According to Liang and Mitzi report, europium halide perovskites: $CH_3NH_3EuI_3$ is a 3D ABX3-type perovskite with a tetragonal distorted structure of BX6 corner-connected octahedral, which can be synthesized via a diffusion based solid-state synthesis route from CH_3NH_3I and EuI_2.The applicability of materials, in optoelectronic devices, limited by the environment sensitivity. However, another group such as lanthanide perovskite exhibits remarkable optical characteristics; it is possible candidates as new light absorbing materials for solar application. In addition, lanthanides such as (Ce^{3+}, Dy^{3+}, Er^{3+}, Eu^{3+}, Gd^{3+}, La^{3+}, Lu^{3+}, Pr^{3+}, Nd^{3+}, Sm^{3+}, Tm^{3+}), and actinides (Am^{3+}, Bk^{3+}, Pu^{3+}) have been functioning in halide double perovskites, but until now no reported study on their solar application [69, 111, 113, 114].

3.8 Ferroelectronic perovskite

It is mainly as magnetic material but investigation of ferroelectronic perovskites have the potential to be employed as light absorbers. Moreover, they can be treated as a type of Pb-free perovskites. According to Nechache et al. (2009), Jiang et al. (2020) report, Bi_2CrFeO_6 as solar material, attaining the PCE of 8% [115].

4. Dimension

The structural chemistry and dimensionality of perovskites are significantly influenced by the performance of solar devices. Depending on the dimensionality, the crystal structures of perovskites can be divided into three categories [116, 117]. As the nonstop novelty of preparation methods has developed, (PCE) of solar cells with three-dimensional (3D) lead (Pb) perovskites has rapidly rushed from 3.9% to over 29.1% within nearly one decade, which incomparable to monocrystalline silicon solar cells [6]. In this section, we summarize applicable dimensions such as 0D, 1D, 2D, and 3D.

4.1 Zero-dimensional Pb-free PSC

It is emerging class of material condensed dark current, and improved environmental stability compared with different dimension perovskites. To create a stable 0D compound is based on the assumption that a quantum-well structure would result in

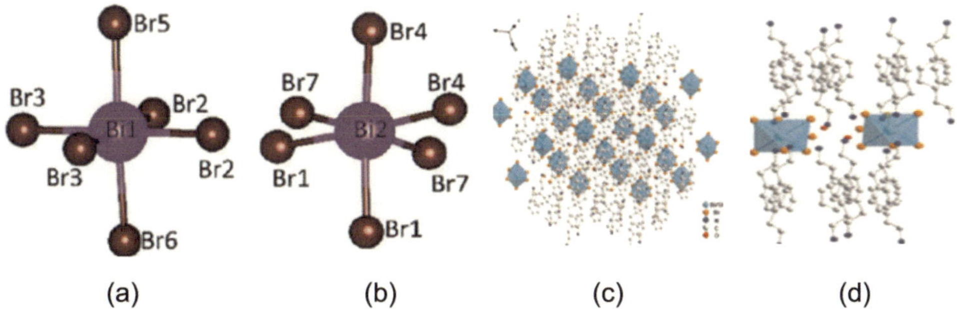

Figure 4.
Chemical structure of (a) Bi$_1$, (b) Bi$_2$, (c) [(Bi/Sb) Br$_6$]$_3$– and (d) Bi/Sb ((Bi/Sb azure, Br orange, N blue, C gray [120].

stronger quantum confinement. The low-dimensional perovskites are proven to inhibit ion migration, sensitivity for humidity and chemical stability. It is required to develop, Pb-free, quality samples for optoelectronic function [118, 119]. For example, Cs_3BiBr_6crystal has orthorhombic space group P_{bcm} to form a 0D perovskite structure (**Figure 4**). An otherair-stable, mixed antimony—bismuth perovskite,(C_8NH_{12}) $4Bi_{0.57}Sb_{0.43}Br_7{\cdot}H_2O$, was synthesized, which reported 0D structure with isolated [(Bi/Sb) Br_6]$_3$–, Bi/Sb metal-halide octahedrons(**Figure 4**) [120, 121].

4.2 One-dimensional Pb-free PSCs

Recently 2D and 3D perovskites in various dimensions have investigated in different perspective. But current research attention been diverted to the lower dimensions (1D and 2D). According to Zhou et al. report, bulk assemblies of 1D and 0D core—shell quantum confined materials, reproducible low-dimensional tin bromide PSCs.

For example, novel set of one-dimensional (TMHD)$BiBr_5$ (TMHD=N,N,N,N-tetramethyl-1,6-hexanediammonium), comprising infinite 1D chains of $BiBr_6$ octahedra and organic TMHD cations as shown **Figure 5** [118]. Another 1D structured Rb_2CuBr_3 was reported with the orthorhombic space group P*nma* where the Cu atom was shown to be harmonized by four Br atoms as shown in **Figure 5c** and **d** [122, 123].

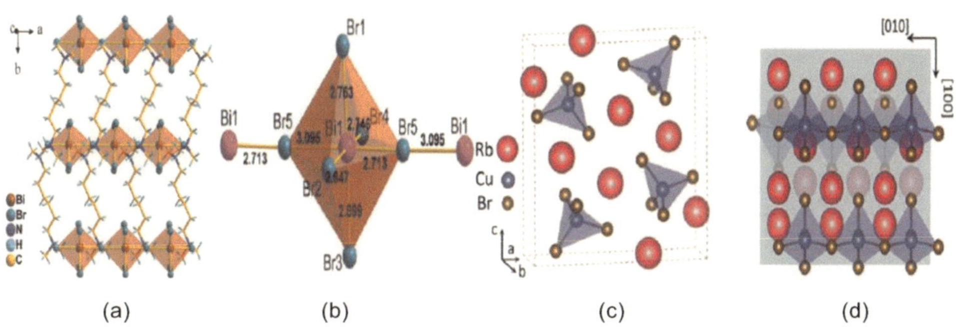

Figure 5.
Structural (a) packing. (b) unit of inorganic BiBr$_6$ octahedra [118, 122]. (c) Crystal structure of Rb$_2$CuBr$_3$, (d) Rb$_2$CuBr$_3$ structure as viewed down the a-axis (red, blue, and brown indicate Rb, Cu, and Br atoms, respectively). (e, f) Isosurface plots of the wave function |Ψ|2 of CBM and VBM [123].

4.3 Two-dimensional Pb-free PSCs

Counter 2D layered structure in the monoclinic framework with the P21 space group is that of (BA) $2CsAgBiBr_7$ which behaves like a quantum-constrained structure where the perovskite layers look like to the quantum wells and the massive cations as the expected obstructions (**Figure 6a** and **b**) [124]. The Crystal show high symmetry in the orthorhombic space group P_{nma} required properties. Another Zhang et al. explained bidoped two-layered tin-based halide perovskite series $PEA2Sn1 - xBixBr_4 + x$. For undoped $PEA2SnBr_4$, the $[SnBr_6]4-$ octahedra show up in the form of sheets lying between the huge natural moieties of PEA+, exhibits as brand of 2D layered morphology as shown **Figure 6c** and **d** [125].

4.4 Three-dimensional Pb-free PSCs

Up to this point; most research work has been conducted to PSCs with a 3D structure and ABX3 stoichiometry. Nonetheless, since rearrangement into various phases is energetically good in the ABX3 has been a blast in the synthesis of a several perovskite crystals shape in various designs and morphology as displayed in **Figure 7**. There has been a boom in the synthesis of several that crystallize in different structures and morphology into lower dimensions for attaining stability [126, 127].

5. Stability

The most common attention grabbed limits of Pb-based perovskites poor stability for photovoltaic application. The stability challenge of perovskites is being addressed through the use of low-dimensional perovskites as well as improved device

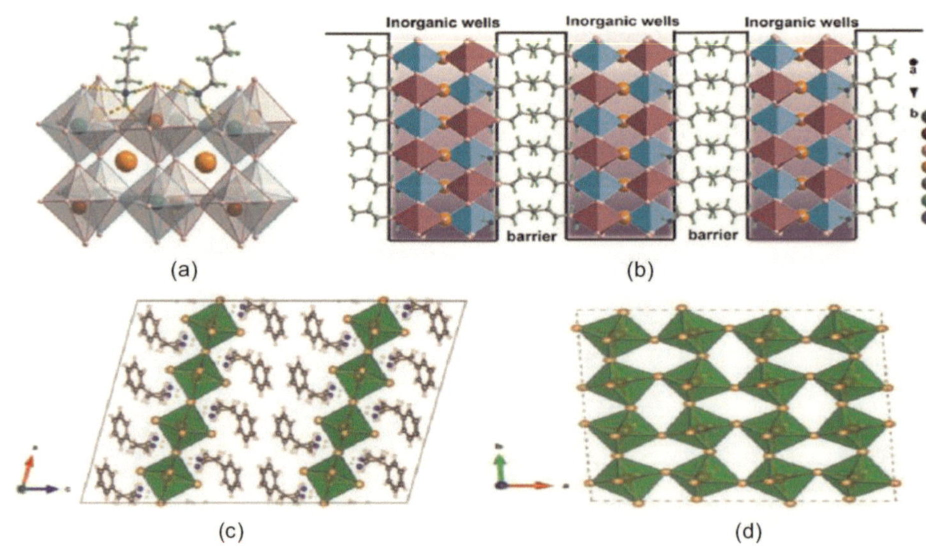

Figure 6.
2D perovskite (a) BA cations are bound to two kinds of octahedra through N − H···Br hydrogen bonds, as shown by the dashed lines. (b) (BA)2CsAgBiBr$_7$ describing the 2D perovskite quantum-confined motif (c) crystal-structural diagram of PEA2SnBr$_4$ (d) distorted [SnBr$_6$]4− octahedron [124, 125].

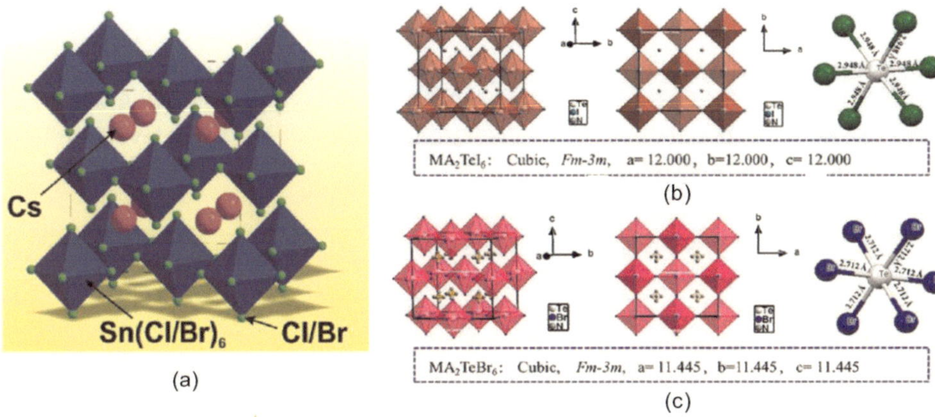

Figure 7.
3D crystal structure perovskites $Cs_2SnCl_6 - xBrx$, (a) MA_2TeI_6; and (b) MA_2TeBr_6; and (c) discrete $[Sn(Cl, Br)_6]2$ [126, 127].

engineering. The development of low-toxicity ideal Pb-free materials should have low toxicity, narrow direct band gaps, high optical absorption coefficients, high mobilities, low exciton-binding energies, long charge-carrier lifetimes, and better stability. The intrinsic tolerance of perovskite to humidity, light, and temperature has made an embarrassment of device applications that rendering for commercialization [128].

5.1 Moisture stability

Upon exposure to moisture, lead-based perovskite damaged by forming coordinate bonds with the H_2O molecule, thus subsequent in the chemical decomposition and the structure of perovskite structure. The perovskite layer with insufficient time for device operation with the moisture, oxygen, air, and high energy photon. The organic halide would remain the hydrolysis and release HI. The HI would be constantly disbursed with the oxygen and the photon. The decomposition reaction described in Eqs. (1)–(4) as bellow with acceptable amount of moisture.

$$CH_3NH_3PbI_3(s) \overset{H_2O}{\Leftrightarrow} PbI_2(s) + CH_3NH_3I(aq) \tag{1}$$

$$CH_3NH_3I(aq) \Leftrightarrow CH_3NH_2(aq) + HI(aq.) \tag{2}$$

$$4HI(aq) + O_2 \Leftrightarrow 2I_2(s) + 2H2O(l) \tag{3}$$

$$2HI(aq) \Leftrightarrow H_2(g) + I_2(s) \tag{4}$$

There is new lead-free Cs_2PdBr_6 was reported to moisture stable after no indication of chemical decomposition was detected even after immersion in water for 10 min [129]. Another type leadless perovskiteCs_2NaBiI_6 was found to hold all its properties after contact to humid air for 5 months, and no deprivation peak was observed [130]. Alternatively, $FA_4Ge^{II}Sb^{III}Cl1_2$ showed no change when visible to 60% humidity for up to 3 months. This natural stability occurring in Pb-free SCs thus demonstrates to be one of the foremost reasons for them to be pitched as excellent candidates dignified to bring about the next big wave in Pb-free perovskite optoelectronics application.

5.2 Photostability

In PSCs mesoscopic device structure was used to photo-generated electrons transportation. Though, it is sensitive to ultraviolet (UV) light. According to Snaith *et al.* report, photoinduced instability of PSCs degraded quicker unshaded device under sun light described as Eqs. (5)–(8).

$$CH_3NH_3PbI_3(s) \overset{hv}{\Leftrightarrow} PbI_2(s) + CH_3NH_2\uparrow + HI\uparrow \tag{5}$$

$$2I^- \Leftrightarrow I_2 + 2e^- \tag{6}$$

$$3CH_3NH_3^+ \overset{hv}{\Leftrightarrow} 3CH_3NH_2\uparrow + 3H^+ \tag{7}$$

$$I^- + I_2 + 3H^+ + 2e^- \Leftrightarrow 3HI\uparrow \tag{8}$$

In recent times, innovative group of lead-free material, for example, $Cs_2AgBiBr_6$ indicators do not give the impression to harmed even subsequently nonstop experience to X-ray radiation in neighboring conditions [131]. The essential driver for this is a consolidation of higher actuation energy (a few times that of organic–inorganic half breeds) and high dispersion hindrances for constituent particles suggesting lesser opportunity of underlying unwinding and thus, greater dependability [132]. $(BA)_2CsAgBiBr_7$ additionally shows an extraordinarily steady reaction even on nonstop openness to X-beams [133].

5.3 Thermal stability

In normal state, direct lighting of PSCs will increase operation temperature of panel. The temperature as high as 85°C that the ecological temperature is 40°C. According to Conings et al. reports, the intrinsic thermal stability of $MAPbI_3 - xClx$ was found that the degradation happened at 85°C condition. This means that PSCs may not be widely used in actual daytime if the device temperature exceeds 85°C. The perovskite phase would transit from lower symmetry to higher symmetry (orthorhombic-tetragonal-cubic). Recently Weber et al. also reported $MAPbBr_3$ and MAPbCl3 could maintain better symmetry than $MAPbI_3$ from −40–85°C.Recently new group of lead-free perovskites, for instance, $(MA)_2AgBiBr_6$ [134] is steady up to ~550 K which is not exactly that of $Cs_2AgBiBr_6$ (stable up until~700 K) [135]. The short fall of Pb appear as attractive variable also since the lead-containing counterpart ($MAPbBr_3$) is just steady until ~490 K. On practically equivalent to lines, tellurium-based A2TeX6 SCs were additionally observed to be entirely steady up to ~270°C alongside being phase tolerant to moisture and air [127].

6. Recent advances

The stability and harmfulness challenge in organic– inorganic hybrid lead halide perovskites cells being addressed through low-poisonousness without lead materials. Similar as the Pb-containing perovskites, the helpful properties for such lead-free solar devices would be appropriate direct band gaps, high assimilation coefficients, high mobilities, low exciton restricting energies, long charge transporter lifetimes.

lead-free perovskite	Voc(V)	Jsc (mA cm2)	FF (%)	η (%)	References
$FASnI_3$	0.64	21.95	0.73	10.16	[116, 117, 136]
$CsSnI_3$	0.86	23.2	65	12.96	[137]
PEAxFA1 − $xSnI_3$ + NH_4SCN	0.94	17.4	75	12.4	[115]
$(FA)_{0.75}(MA)_{0.25}$ SnI_3 + 10% SnF_2	0.61	21.2	62.7	8.12	[138]
PEA_2SnI4	0.61	22.0	70.1	9.41	[139]
$(BA_{0.5}$ PEA0.5$)_2FA_3Sn_4I_{13}$	0.60	21.82	66.73	8.82	[140]
AVA_2FAn − 1SnnI_3n + 1	0.61	21.0	68.8	8.71	[141]
$CsSnBr_3$	0.85	21.23	58	10.46	[137]
$FASnI_3$ + 1% $EDAI_2$	0.58	21.3	0.72	8.9	[136]
$FASnI_3$ + 5% PHCl	0.76	23.5	64	11.4	[120]
0.92$FASnI_3$ + 0.08PEAI +10% SnF_2	0.53	24.1	71	9.0	[142]
$MAPb_{0.5}$Sn0.5I_3	0.75	36.30	75	13.60	[143]
MA0.5$FA_{0.5}$ $Pb_{0.75}$ $Sn_{0.25}I_3$	0.78	22.44	82	14.35	[144]
$(MAPbI_3)_{0.4}(FASnI_3)_{0.6}$	0.706	26.86	79.5	15.08	[145]
$CsSnCl_3$	0.87	19.82	56	9.66	[146]

Table 7.
Lead-free perovskite solar cell parameters.

The Sn-based perovskite have got much interest along with all pb-free solar cells, because of their fundamentally the same as properties to lead-based ones, and the most encouraging exhibition accomplished by gadgets utilizing this class of materials. In **Table** 7, recorded a solar cell parameters based on pb-free perovskites. The controlled crystallization of lead-free perovskite material shows improved performance in solar oriented cells, which is roused toward the manufacture of without pb perovskite film-based sun-powered cells [101, 103, 105].

7. Conclusions and future prospective

It is known until now Pb-based solar cells could not be open commercial market because of the poor stability and Pb toxic nature. In this concern the device showed PCE of 8.12%–15.08% as shown in **Table** 7. These obtained results showed the excellent optical properties of the lead-free perovskite materials and suggested their potential as light absorbers in the construction of PSCs.Finally, we have concluded the recent development of lead-free perovskite materials in perspective of solar cell application. Lead-free perovskites empower to circumvent the problems of instability and toxicity to improve commercial production. Simultaneously, this study overview understanding of the fundamental challenges behind the efficiency, stability, and environmental of lead-free PSCs. Here, we are looking forward the materials, dimension, and the further development of lead-free perovskite materials and PSCs. Finally, we summarize the latest highest performance of lead-free perovskites. In spite of faster development of lead-free perovskites, we believe that the efficiency based on lead-free perovskite materials can breakthrough over 15% after further comprehensive study, and we should keep forward research until the achievement of commercial leadless perovskite solar cell.

Acknowledgements

The authors acknowledge Intech open access Dr. Sara Tikel for invitation and constructive comments.

Conflict of interest

The authors declare no conflict of interest.

Author details

Anshebo Getachew Alemu[1*] and Teketel Alemu[2]

1 Department of Physics, Samara University, Samara, Ethiopia

2 Department of Chemical Engineering, Wachamo University, Hosaena, Ethiopia

*Address all correspondence to: agetachew2013alemu@yahoo.com

References

[1] Adjogri SJ, Meyer EL. A review on Lead-free hybrid halide perovskites as light absorbers for photovoltaic applications based on their structural, optical, and morphological properties. Molecules. 2020;**25**:5039

[2] Mutalikdesai A, Ramasesha SK. Emerging solar technologies: Perovskite solar cell. Resonance. 2017;**22**:1061-1083

[3] Heo JH, Han HJ, Kim D, Ahn TK, Im SH. Hysteresis-less inverted $CH_3NH_3PbI_3$ planar perovskite hybrid solar cells with 18.1% power conversion efficiency. Energy & Environmental Science. 2015;**8**:1602-1608

[4] Snaith HJ, Abate A, Ball JM, Eperon GE, Leijtens T, Noel NK, et al. Anomalous hysteresis in perovskite solar cells. Journal of Physical Chemistry Letters. 2014;**5**:1511-1515

[5] Park N-G. Perovskite solar cells: An emerging photovoltaic technology. Materials Today. 2015;**18**:65-72

[6] NREL Solar Cell Efficiency Chart. 2020. Available from: https://www.nrel.gov/pv/cell-e_ciency.html

[7] Wang L, Wang K, Zou B. Pressure-induced structural and optical properties of organometal halideperovskite-based formamidinium lead bromide. Journal of Physical Chemistry Letters. 2016;**7**: 2556-2562

[8] Wang F, Bai S, Tress W, Hagfeldt A, Gao F. Defects engineering for high-performance perovskitesolar cells. NPJ Flex. Electron. 2018;**2**:1-14

[9] Chen Y, Zhang L, Zhang Y, Gao H, Yan H. Large-area perovskite solar cells—A review of recent progressand issues. RSC Advances. 2018;**8**:10489-10508

[10] Wenderott JK, Raghav A, Shtein M, Green PF, Satapathi S. Local optoelectronic characterization of solvent-annealed, Lead-free, Bismuth-based perovskite films. Langmuir. 2018; **34**:7647-7654

[11] Ke W, Kanatzidis MG. Prospects for low-toxicity lead-free perovskite solar cells. Nature Communications. 2019;**10**: 965

[12] Ghosh S, Pradhan B. Lead-free metal halide perovskite nanocrystals: Challenges, applications, and future aspects. Chem. Nano. Mater. 2019;**5**: 300-312

[13] Chen L, Tan YY, Chen ZX, Wang T, Hu S, Nan ZA, et al. Toward long-term stability: Single-crystal alloys of cesium-containing mixed cation and mixed halide perovskite. Journal of the American Chemical Society. 2019;**141**: 1665-1671

[14] Lv Q, Lian Z, He W, Sun J-L, Li Q, Yan Q. A universal top-down approach toward thickness- controllable perovskite singlecrystalline thin films. Journal of Materials Chemistry C. 2018; **6**:4464-4470

[15] Lei Y, Chen Y, Zhang R, Li Y, Yan Q, Lee S, et al. A fabrication process for flexible single-crystal perovskite devices. Nature. 2020;**583**:790-795

[16] Cai T, Shi W, Hwang S, Kobbekaduwa K, Nagaoka Y, Yang H, et al. Lead-free $Cs_4CuSb_2Cl_{12}$ layered double perovskite nanocrystals. Journal of the American Chemical Society. 2020; **142**:11927-11936

[17] Kumar Chini M, Goverapet Srinivasan S, Tailor N, Yukta K, Salahub

D, Satapathi S. Lead-free, stable mixed halide double perovskites $Cs_2AgBiBr_6$ and $Cs_2AgBiBr_{6-x}Clx$A detailed theoretical and experimental study. Chemical Physics. 2020;**529**:110547

[18] Morad V, Shynkarenko Y, Yakunin S, Brumberg A, Schaller RD, Kovalenko MV. DisphenoidalZero-dimensional Lead, tin, and germanium halides: Highly emissive singlet andTriplet self-trapped excitons and X-ray scintillation. Journal of the American Chemical Society. 2019;**141**:9764-9768

[19] Umadevi D, Watson GW. Quasiparticle GW calculations on Lead-free hybrid germanium iodide Perovskite$CH_3NH_3GeI_3$ for photovoltaic applications. ACS Omega. 2019;**4**:5661-5669

[20] Kung P-K, Li M-H, Lin P-Y, Jhang JY, Pantaler M, Lupascu DC, et al. Lead-free double perovskites for perovskite solar cells. Solar RRL. 2020;**4**:1900306

[21] Ning W, Gao F. Structural and functional diversity in Lead- free halide perovskite materials. Advanced Materials. 2019;**31**:e1900326

[22] Meng X, Wang Y, Lin J, Liu X, He X, Barbaud J, et al. Surface-controlled oriented growth of $FASnI_3$ crystals for efficient Lead-free perovskite solar cells. Joule. 2020;**4**:902-912

[23] Qian WH, Cheng XF, Zhou J, He JH, Li H, Xu QF, et al. Lead-free perovskite $MASnBr_3$-based memristor for quaternary information storage. InfoMat. 2020;**2**:743-751

[24] Vargas B, Torres-Cadena R, Reyes-Castillo DT, Rodríguez-Hernández J, Gembicky M, Menendez-Proupin E, Solis-Ibarra D. Chemical diversity in lead-free, layered double perovskites: a combined experimental and computational approach. Chemistry of Materials. 2020;**32**:424-429

[25] Filippo DA. The Prospect of Lead-free perovskite photovoltaics. ACS Energy Letters. 2021;**6**:1586-1587

[26] Stoumpos CC, Kanatzidis MG. Halide perovskites: Poor man's high-performancesemiconductors. Advanced Materials. 2016;**28**:5778-5793

[27] Ke W, Stoumpos CC, Kanatzidis MG. "Unleaded" perovskites: Status quo and future prospects of tin-based perovskite solar cells. Advanced Materials. 2018;**19**:1803230

[28] Tsai H et al. High-efficiency two-dimensional Ruddlesden-popper perovskite solar cells. Nature. 2016;**536**: 312-316

[29] Mao L, Stoumpos CC, Kanatzidis MG. Two-dimensional hybrid halide perovskites: Principles and promises. Journal of the American Chemical Society. 2019;**141**:1171-1190

[30] Giustino F, Snaith HJ. Toward lead-free perovskite solar cells. ACS Energy Letters. 2016;**1**:1233-1240

[31] Liu Y. The development of low toxic and high efficient solar cells. Journal of Physics: Conference Series. 2020;**1653**: 012002

[32] Gomez A, Wang Q, Goni AR, Campoy-Quiles M, Abate A. Ferroelectricity-free lead halide perovskites. Energy & Environmental Science. 2019;**12**:2537-2547

[33] Kojima A, Teshima K, Shirai Y, Miyasaka T. Organometal halide perovskites as visible-light sensitizers for photovoltaic cells. Journal of the American Chemical Society. 2009;**131**: 6050-6051

[34] Kim HS, Lee CR, Im J-H, Lee K-B, Moehl T, Marchioro A, et al. Lead iodide perovskite sensitized all-solid-state submicron thin film mesoscopic solar cell with efficiency exceeding 9%. Scientific Reports. 2012;**2**:591

[35] Lee MM, Teuscher J, Miyasaka T, Murakami TN, Snaith HJ. Efficient hybrid solar cells based on Meso-super structured Organometal halide perovskites. Science. 2012;**338**:643-647

[36] Lee JW, Seol DJ, Cho AN, Park NG. High-efficiency perovskite solar cells based on the black polymorph of HC $(NH2)_2$. PbI_3. Advanced Materials. 2014; **26**:4991-4998

[37] Yin W, Shi J, Yan T. Unusual defect physics in $CH_3NH_3PbI_3$ perovskite solar cell absorber. Applied Physics Letter. 2014;**104**:063903

[38] Xing G, Mathews N, Sun S, Lim SS, Lam YM, Gratzel M, et al. Long-range balanced electron, hole- transport lengths in organic-inorganic $CH_3NH_3PbI_3$. Science. 2013;**342**:344-347

[39] Jeong M, Choi IW, Go EM, Cho Y, Kim M, Lee B, et al. Stable perovskite solar cells with efficiency exceeding 24.8% and 0.3-V voltage loss. Science. 2020;**369**:1615-1620

[40] Jeong J, Kim M, Seo J, Lu H, Ahlawat P, Mishra A, et al. Pseudo-halide anion engineering for $FAPbI_3$ perovskite solar cells. Nature. 2021;**592**:381-385

[41] Min H, Kim M, Lee S-U, Kim H, Kim G, Choi K, et al. Efficient, stable solar cells by using inherent bandgap of phase formamidinium lead iodide. Science. 2019;**366**:749-753

[42] Kim M, Kim GH, Lee TK, Choi IW, Choi HW, Jo Y, et al. Methylammonium chloride induces intermediate phase stabilization for efficient perovskite solar cells. Joule. 2019;**3**:2179-2192

[43] Jeon I, Shawky A, Lin H-S, Seo S, Okada H, Lee JW, et al. Controlled redox of lithium-ion endohedral fullerene for efficient and stable metal electrode-free perovskite solar cells. Journal of the American Chemical Society. 2019;**141**: 16553-16558

[44] Jeon I, Ueno H, Seo S, Aitola K, Nishi Kubo R, Saeki A, et al. Lithium-ion endohedral fullerene (Li + @C_{60}) dopants in stable perovskite solar cells induce instant doping and anti-oxidation. Angewandte Chemie. 2018;**57**: 4607-4611

[45] Lyu M, Yun J-H, Chen P, Hao M, Wang L. Addressing toxicity of Lead: Progress and applications of low-toxic metal halide perovskites and their derivatives. Advanced Energy Materials. 2017;**7**:1602512

[46] Abate A. Perovskite solar cells go Lead free. Joule. 2017;**1**:659-664

[47] Hu H, Dong B, Zhang W. Low-toxic metal halide perovskites: Opportunities and future challenges. Journal of Materials Chemistry A. 2017;**5**:11436-11449

[48] Mahaffey KR. Environmental lead toxicity. Environmental Health Perspectives. 1990;**89**:75-78

[49] Wani AL, Ara A, Usmani JA. Lead toxicity: A review. Interdisciplinary Toxicology. 2015;**8**:55-64

[50] Li X, Zhang F, He H, Berry JJ, Zhu K, Xu T. On-device lead sequestration for perovskite solar cells. Nature. 2020;**578**: 555-558

[51] Il J, Kyusun K, Efat J, Minjoon P, Hyung-W L, Eric WGD.

Environmentally compatible Lead-free perovskite solar cells and their potential as light harvesters in energy storage systems. Nanomaterials. 2021;**11**:2066

[52] Ke W, Stoumpos CC, Kanatzidis MG. "Unleaded" perovskites: Status quo and future prospects of tin-based perovskite solar cells. Advanced Materials. 2019;**31**:1803230

[53] Jiewei L, Masashi O, Shinya Y, Taketo H, Ryosuke N, Yoshihiko K, et al. Lead-Free Solar Cells based on Tin Halide Perovskite Films with High Coverage and Improved Aggregation. Angewandte Chemie. 2018;**130**(40): 13405-13409. DOI: 10.1002/anie.201808385. Available from: https://onlinelibrary.wiley.com/doi/am-pdf/10.1002/anie. 201808385

[54] Stoumpos CC, Malliakas CD, Kanatzidis MG. Semiconducting tin and lead iodide perovskites with organic cations: Phase transitions, high mobilities, and near-infrared photo luminescent properties. Inorganic Chemistry. 2013;**52**:9019

[55] Nordberg GF, Fowler BA, Nordberg M. Handbook on the Toxicology of Metals. Elsevier. 2015;**4**:75-112

[56] Hussein S. Numerical Simulation of 30% Efficient Lead-Free Perovskite $CsSnGeI_3$-Based Solar Cells. Materials. 2022;**15**:3229. DOI: 10.3390/ma15093229

[57] Krishnamoorthy T, Ding H, Yan C, Leong WL, Baikie T, Zhang Z, et al. Lead-free germanium iodide perovskite materials for photovoltaic applications. Journal of Materials Chemistry A. 2015;**3**: 23829

[58] Montiel KA, Yang C, Andreasen CH, Gottlieb MS, Pfefferkorn MR, Wilson LG, et al. Lead-free perovskite thin film solar cells from binary sources. IEEE Photovoltaic Specialists Conference. 2019:1183

[59] Chen M, Ju MG, Garces HF, Carl AD, Ono LK, Hawash Z, et al. Highly stable and efficient all-inorganic lead-free perovskite solar cells with native-oxide passivation. Nature Communications. 2019;**10**:16

[60] Sean TT, Ghada A, Rashad FK, Ghada IK. Progress towards lead-free, efficient, and stable perovskite solar cells. DOI: 10.1002/ jctb.6830

[61] Min C, Qingshun D, Chuanxiao X, Xiaopeng Z, Xiaopeng Z, Zhenghong D, et al. Lead-free flexible perovskite solar cells with interfacial native oxide have >10% efficiency and simultaneously enhanced stability and reliability. ACS Energy Letter. 2022;**7**(XXX):2256-2264

[62] McCall KM, Stoumpos CC, Kostina SS, Kanatzidis MG, Wessels BW. Strong electron-phonon coupling and self-trapped excitons in the defect halide perovskites A3M2I9 (a=Cs, Rb; M=Bi, Sb). Chemistry of Materials. 2017;**29**:4129-4145

[63] Park B-W, Philippe B, Zhang X, Rensmo H, Boschloo G, Johansson EMJ. Bismuth Based Hybrid Perovskites $A_3Bi_2I_9$ (A: Methyl ammonium or Cesium) for Solar Cell Application. Advanced Materials. 2015;**27**:6806

[64] Lyu M, Yun JH, Cai M, Jiao Y, Bernhardt PV, Zhang M, et al. Organic–inorganic bismuth (III)-based material: A lead-free, air-stable and solution-processable light-absorber beyond organolead perovskites. Nano Research. 2016;**9**:692

[65] Kamminga ME, Stroppa A, Picozzi S, Chislov M, Zvereva IA, Baas J, et al. Polar nature of $(CH_3NH_3)_3Bi_2I_9$ perovskite-like hybrids. Inorganic Chemistry. 2017;**56**:33

[66] Zhang YD, Feng J. $CH_3NH_3Pb_{1-x}Mg_xI_3$ perovskites as environmentally friendly photovoltaic materials. AIP Advances. 2018;**8**:015218. DOI: 10.1063/ 1.5006241

[67] Peresh EY, Sidei VI, Zubaka OV, Stercho IP. $K_2(Rb_2,Cs_2,Tl_2)TeBr_6(I_6)$ and $Rb_3(Cs_3)Sb_2(Bi_2)Br_9(I_9)$ perovskite compounds. Inorganic Materials. 2011; **47**:208

[68] Johansson MB, Zhu H, Johansson EMJ. Journal of Physical Chemistry Letters. 2016;**57**:3467

[69] Sebastian FH, Gregor T, Thomas R. Progress on lead-free metal halide perovskites for photovoltaicapplications: A review. Monatshefte fuer Chemie. 2017;**148**:795-826

[70] Chu Z, Liguo G, Shuzi H, Tingli M. Current advancements in material research and techniques focusing on Lead- free perovskite solar cells. Chemistry Letters. 2017;**46**:1276-1284

[71] Hao F, Stoumpos CC, Cao DH, Chang RPH, Kanatzidis MG. Controllable Perovskite Crystallization at a Gas–Solid Interface for Hole Conductor-Free Solar Cells with Steady Power Conversion Efficiency over 10%. Nature Photonics. 2014;**8**:489

[72] Pathak SK, Abate A, Leijtens T, Hollman DJ, Teuscher J, Pazos L, et al. Towards long-term photo stability of solid-state dye sensitized solar cells. Advanced Energy Materials. 2014;**4**: 1301667

[73] Ma L, Peijun G, Mikael K, Ioannis S, Yihui H, Claudine K, et al. Organic cation alloying on intralayer A and interlayer A' sites in 2D hybrid Dion-Jacobson lead bromide perovskites. JACS. 2020;**142**(18):8342-8351

[74] Jokar E, Chien C-H, Tsai C-M, Fathi A, Diau EW. Robust tin-based perovskite solar cells with hybrid organic cations to attain efficiency approaching 10%. Advanced Materials. 2019;**31**: 1804835

[75] Nishimura K, Kamarudin MA, Hirotani D, Hamada K, Shen Q, Iikubo S, et al. Nano Energy. 2020;**74**:104858

[76] Noel NK, Stranks SD, Abate A, Wehrenfennig C, Guarnera S, Haghighirad A-A, et al. Lead-free organic-inorganic tin halide perovskites for photovoltaic applications. Energy & Environmental Science. 2014;**7**: 3061-3068

[77] Jung M-C, Raga SR, Qi Y. Properties and solar cell applications of Pb-free perovskite films formed by vapor deposition. RSC Advances. 2016;**6**:2819

[78] Serafettin De A, Oz N, Mustafa C, Cebrail O, Ka M. Recent Progresses in Perovskite Solar Cells. Vol. 10. London: INTECH; 2017. pp. 278-304. Available from: https://www.intechopen.com/chapters/52374

[79] Feng J, Xiao B. Effective masses and electronic and optical properties of nontoxic MASnX3 (X = Cl, Br, and I) Perovskite structures as solar cell absorber: A theoretical study using HSE06. Journal of Physical Chemistry C. 2014;**118**:19655

[80] Yu Y, Zhao D, Grice CR, Meng W, Wang C, Liao W, et al. Thermally evaporated methyl ammonium tin triiodide thin films for lead-free perovskite solar cell fabrication. RSC Advances. 2016;**6**:90248

[81] Yokoyama T, Cao DH, Stoumpos CC, Song T-B, Sato Y, Aramaki S, et al. Overcoming short-circuit in lead-free $CH_3NH_3SnI_3$ perovskite solar cells via

kinetically controlled gas–solid reaction film fabrication process. Journal of Physical Chemistry Letters. 2016;7:776

[82] Luis KO, Shenghao W, Ka Y, Sonia RR, Yabing Q. Fabrication of semi-transparent perovskite films with centimeter-scale superior uniformity by the hybrid deposition method. Energy and Environmental Science. 2014;7:3989

[83] Liao W, Zhao D, Yu Y, Grice CR, Wang C, Cimaroli AJ, et al. Lead-free inverted planar formamidinium tin triiodide perovskite solar cells achieving power conversion efficiencies up to 6.22%. Advances Materials. 2016;**28**:9333

[84] Maughan AE, Kurzman JA, Neilson JR. Hybrid inorganic–organic materials with an optoelectronically active aromatic cation: $(C_7H_7)_2SnI_6$ and $C_7H_7PbI_3$. Inorganic Chemistry. 2015;**54**: 370

[85] Peedika KL, Bhargava P. Composition dependent optical, structural and photoluminescence characteristics of cesium tin halide perovskites. RSC Advances. 2016;**6**: 19857

[86] Gupta S, Bendikov T, Hodes G, Cahen D. $CsSnBr_3$, a lead-free halide perovskite for long-term solar cell application: insights on SnF_2 addition. ACS Energy Letters. 2016;**1**:1028

[87] Sabba D, Mulmudi HK, Prabhakar RR, Krishnamoorthy T, Baikie T, Boix PP, et al. Impact of anionic Br^- substitution on open circuit voltage in lead free perovskite ($CsSnI_{3-x}Br_x$) solar cells. Journal of Physical Chemistry C. 2015;**119**:1763

[88] Huiyun W, Peng Q, Ye L, Yingfeng H, Mingzeng P, Xinhe Z. XiaohuLiu Challenges and strategies of all-inorganic lead-free halide perovskite solar cells. Ceramics International. 2022;**48**(5): 5876-5589. DOI: 10.1016/j.ceramint.2021.11.184

[89] Yihui H, Liviu M, Hee JJ, Kyle MM, Michelle C, Constantinos CS, et al. High spectral resolution of gamma-rays at room temperature by perovskite $CsPbBr_3$ single crystals. Nature Communications. 2018;**9**:1609

[90] Zhang M, Lyu M, Yun J-H, Noori M, Zhou X, Cooling NA, et al. Low-temperature processed solar cells with Formamidinium tin halide perovskite/fullerene heterojunctions. Nano Research. 2016;**9**:1570

[91] Koh TM, Krishnamoorthy T, Yantara N, Shi C, Leong WL, Boix PP, et al. Formamidinium tin-based perovskite with low E_g for photovoltaic applications. Journal of Materials Chemistry A. 2015;**3**:14996

[92] Lee SJ, Shin SS, Kim YC, Kim D, Ahn TK, Noh JH, et al. Fabrication of Efficient Formamidinium Tin Iodide Perovskite Solar Cells through SnF_2–Pyrazine Complex. Journal of the American Chemical Society. 2016;**138**: 3974

[93] Li Y, Sun W, Yan W, Ye S, Rao H, Peng H, et al. 50% Sn-based planar perovskite solar cell w ith power conversion efficiency up to 13.6%. Advances Energy Materials. 2016;**6**: 1601353

[94] Liao W, Zhao D, Yu Y, Shrestha N, Ghimire K, Grice CR, et al. Fabrication of efficient low-bandgap perovskite solar cells by combining formamidinium tin iodide with methyl ammonium lead iodide. Journal of the American Chemical Society. 2016;**138**:12360

[95] Filip MR, Giustino F. Computational screening of homovalent lead

substitution in organic–inorganic halide perovskites. Journal of Physical Chemistry C. 2016;**120**:166

[96] Uribe JI, Ramirez D, Osorio-Guille'n JM, Osorio J, Jaramillo F. $CH_3NH_3CaI_3$ perovskite: synthesis, characterization, and first-principles studies. Journal of Physical Chemistry C. 2016;**120**:16393

[97] Pazoki M, Jacobsson TJ, Hagfeldt A, Boschloo G, Edvinsson T. Effect of metal cation replacement on the electronic structure of metalorganic halide perovskites: Replacement of lead with alkaline-earth metals. Physical Review B. Condensed Matter and Materials Physics. 2016;**93**:144105

[98] Kumar MH, Dharani S, Leong WL, Boix PP, Prabhakar RR, et al. Lead-free halide perovskite solar cells with high photocurrents realized through vacancy modulation. Advanced Materials. 2014; **41**:7122

[99] Boix PP, Agarwala S, Koh TM, Mathews N, Mhaisalkar SG. Perovskite solar cells: Beyond methylammonium Lead Iodide. Journal of Physical Chemistry Letters. 2015;**6**:898

[100] Ye H-Y, Zhang Y, Fu D-W, Xiong R-G. A molecular perovskite solid solution with piezoelectricity stronger than lead zirconatetitanate. Angewandte Chemie, International Edition. 2014;**53**: 11242

[101] Cui X-P, Jiang K-J, Huang J-H, Zhang Q-Q, Su M-J, Yang L-M, et al. Cupric bromide hybrid perovskite heterojunction solar cells. Synthetic Metals. 2015;**209**:247

[102] Volonakis G, Filip MR, Haghighirad AA, Sakai N, Wenger B, Snaith HJ, et al. Lead-free halide double perovskites via heterovalent substitution of noble metals. Journal of Physical Chemistry Letters. 2016;**7**:1254

[103] Cortecchia D, Dewi HA, Yin J, Bruno A, Chen S, Baikie T, et al. Lead-Free $MA_2CuCl_xBr_{4-x}$ Hybrid Perovskites. Inorganic Chemistry. 2016;**55**:1044

[104] Riggs SC, Shapiro MC, Corredor F, Geballe TH, Fisher IR, McCandless GT, et al. Single crystal growth by self-flux method of the mixed valence gold halides $Cs_2[Au^{I}X_2][Au^{III}X_4]$ (X=Br,I). Journal of Crystal Growth. 2012;**355**:13

[105] Han J, Nishihara S, Inoue K, Kurmoo M. On the nature of the structural and magnetic phase transitions in the layered perovskite-like $(CH_3NH_3)_2[Fe^{II}Cl_4]$. Inorganic Chemistry. 2014;**53**:2068

[106] Kataoka S, Banerjee S, Kawai A, Kamimura Y, Choi J-C, Kodaira T, et al. Layered Hybrid Perovskites with Micropores Created by Alkylammonium Functional Silsesquioxane Interlayers. Journal of the American Chemical Society. 2015;**137**:4158

[107] Huang TJ, Thiang ZX, Yin X, Tang C, Qi G, Gong H. $(CH_3NH_3)_2PdCl_4$: A compound with two-dimensional organic–inorganic layered perovskite structure. Chemistry—A European Journal. 2016;**22**:2146

[108] Wang M, Wang W, Ma B, Shen W, Liu L, Cao K, et al. Lead-free perovskite materials for SolarCells. Nano-Micro Letters. 2021;**13**:62

[109] Retuerto M, Emge T, Hadermann J, Stephens PW, Li MR, Yin ZP, et al. Synthesis and properties of charge-ordered thallium halide perovskites, $CsTl^{+}{}_{0.5}Tl^{3+}{}_{0.5}X_3$ (X = F or Cl): Theoretical precursors for superconductivity. Chemistry of Materials. 2013;**25**:4071

[110] Suta M, Wickleder C. Photoluminescence of $CsMI_3:Eu^{2+}$ (M = Mg, Ca, and Sr) – a spectroscopic probe on structural distortions. Journal of Materials Chemistry C. 2015;**3**:5233

[111] Giorgi G, Yamashita K. Alternative, lead-free, hybrid organic–inorganic perovskites for solar applications: A DFT analysis. Journal of Chemistry Letters. 2015;**44**:826

[112] Castro-Castro LM, Guloy AM. Organic-based layered perovskites of mixed-valent gold(I)/Gold(III) iodides. Angewandte Chemie, International Edition. 2003;**42**:2771

[113] Yang RX, Butler KT, Walsh A. Assessment of hybrid organic–inorganic antimony sulfides for earth-abundant photovoltaic applications. Journal of Physical Chemistry Letters. 2015;**6**:5009

[114] Li T, Luo S, Wang X, Zhang L. Alternative lone-pair ns_2-cation-based semiconductor beyond Lead halide perovskites for optoelectronic applications. Advanced Materials. 2021; **33**:2008574

[115] Jiang X, Wang F, Wei Q, Li H, Shang Y, Zhou W, et al. Ultra-high open circuit voltage of tin perovskite solar cells via an electron transporting layer design. Nature Communications. 2020; **11**:124

[116] Wu T, Liu X, He X, Wang Y, Meng X, Noda T, et al. Efficient and stable tin-based perovskite solar cells byintroducing π-conjugated Lewis's base. Science China: Chemistry. 2020;**63**: 107-115

[117] Meng X, Wu T, Liu X, He X, Noda T, Wang Y, et al. Highly reproducible and efficient $FASnI_3$ perovskite solar cells fabricated with Volatilizable reducing solvent. Journal of Physical Chemistry Letters. 2020;**11**:2965-2971

[118] Naveen KT, Shaoni K, Pranjal M, Albert T, Christian K, Hanlin H, et al. Advances in Lead-free perovskite SingleCrystals: Fundamentals and applications. ACS Materials Letters. 2021;**3**:1025-1080

[119] Liu Y, Xu Z, Yang Z, Zhang Y, Cui J, He Y, et al. Inch-size 0D-structured Lead-free perovskite single crystals for highly sensitive stable X-ray imaging. Matter. 2020;**3**:180-181

[120] Wang C, Gu F, Zhao Z, Rao H, Qiu Y, Cai Z, et al. Self-repairing tin-based perovskite solar cells with a BreakthroughEfficiency over 11% Adv. Maternité. 2020;**32**:e1907623

[121] Tang YY, Liang ML, Chang BD, Sun HY, Zheng KB, Pullerits T, et al. Lead-free double halide perovskite Cs_3BiBr_6 with well-defined crystal structure and high thermal stability for optoelectronics. Journal of Materials Chemistry C. 2019;7:3369

[122] Ji CM, Wang P, Wu ZY, Sun ZH, Li LN, Zhang J, et al. Inch-size single crystal of a Lead- free organic-inorganic hybrid perovskite for high-performance photodetector. Advanced Functional Materials. 2018;**28**:1705467

[123] Yang B, Yin L, Niu G, Yuan JH, Xue KH, Tan Z, et al. Lead-free halide Rb_2CuBr_3 as sensitive X-Ray scintillator. Advanced Materials. 2019;**31**:e1904711

[124] Xu Z, Liu X, Li Y, Liu X, Yang T, Ji C, et al. Exploring Lead-Free Hybrid Double Perovskite Crystals of $(BA)_2CsAgBiBr_7$ with Large Mobility-Lifetime Product toward X-Ray Detection. Angewandte Chemie, International Edition. 2019;**58**:15757-15761

[125] Zhang RL, Mao X, Cheng PF, Yang Y, Yang SQ, Wumaier T, et al. Bismuth doped lead-free two-dimensionaltin-based halide perovskite single crystals. Journal of Energy Chemistry. 2019;**36**:1-6

[126] Zhou J, Luo J, Rong X, Wei P, Molokeev MS, Huang Y, et al. Lead-free perovskite derivative $Cs_2SnCl_6{-}xBrx$ single crystals for narrowband photodetectors. Optical Materials. 2019; **7**:1900139

[127] Ju Zheng X, Yin J, Qiu Z, Türedi B, Liu X, Dang Y, et al. Tellurium-based double perovskites A2TeX6 with tunable band gap and long carrier diffusion length for optoelectronic applications. ACS Energy Letters. 2019;**4**:228-234

[128] Dai WB, Xu S, Zhou J, Hu J, Huang K, Xu M. Leadfree, stable, and effective double $FA_4GeIISbIIICl_{12}$ perovskite for photovoltaic applications. Solar Energy Materials & Solar Cells. 2019;**192**: 140-146

[129] Zhou L, Liao JF, Huang ZG, Wang XD, Xu YF, Chen HY, et al. All-Inorganic Lead Free Cs_2PdX_6 (X=Br,I) Perovskite Nanocrystals with Single Unit Cell Thickness and High Stability. ACS Energy Letters. 2018;**3**:2613-2619

[130] Zhang C, Gao LG, Teo S, Guo ZL, Xu ZH, Zhao S, et al. Design of a novel and highly stable lead-free Cs_2NaBiI_6 double perovskite for photovoltaic application. Sustainable Energy and Fuel. 2018;**2**:2419-2428

[131] Xu Z et al. Exploring Lead-Free Hybrid Double Perovskite Crystals of $(BA)_2$ $CsAgBiBr_7$ with Large Mobility-Lifetime Product toward X-Ray Detection. Angewandte Chemie. 2019; **58**:15757-15761

[132] Yin L, Wu H, Pan W, Yang B, Li P, Luo J, et al. Controlled cooling for synthesis of $Cs_2AgBiBr_6$ single crystals and its application for X-ray detection. Advances Optical Materials. 2019;**7**: 1900491

[133] Dang YY, Tong GQ, Song WT, Liu ZH, Qiu LB, Ono LK, et al. Interface engineering strategies towards $Cs_2AgBiBr_6$ single-crystalline photodetectors with good Ohmic contact behaviours. Journal of Materials Chemistry C. 2020;**8**:276-284

[134] Wei F, Deng Z, Sun S, Zhang F, Evans DM, Kieslich G, et al. Synthesis and Properties of a Lead-Free Hybrid Double Perovskite: $(CH_3NH_3)_2AgBiBr_6$. Chemistry of Materials. 2017;**29**: 1089-1094

[135] He Q, Zhou C, Xu L, Lee S, Lin X, Neu J, et al. Highly stable organic antimony halide crystals for X-ray scintillation. ACS Materials Letters. 2020;**2**:633-638

[136] Meng X, Wang Y, Lin J, Liu X, He X, Barbaud J, et al. Surface-controlled oriented growth of $FASnI_3$ crystals for efficient Lead-free perovskite solar cells. Joule. 2020;**4**:902-912

[137] Zhao B, Abdi-Jalebi M, Tabachnyk M, Glass H, Kamboj VS, Nie W, et al. High open-circuit voltages in tin-rich low- bandgap perovskite-based planar heterojunction photovoltaics. Advanced Materials. 2017;**29**:1604744

[138] Sabba D, Mulmudi HK, Prabhakar RR, Krishnamoorthy T, Baikie T, Boix PP, et al. Impact of anionic Br− substitution on open circuit voltage in lead freeperovskite ($CsSnI_{3-x}Br_x$) solar cells. Journal of Physical Chemistry C. 2015;**119**:1763-1767

[139] Wang F, Jiang X, Chen H, Shang Y, Liu H, Wei J, et al. 2D-quasi-2D-3DHierarchy structure for tin perovskite

solar cells with enhanced efficiency andStability. Joule. 2018;**2**:2732-2743

[140] Qiu J, Xia Y, Zheng Y, Hui W, Gu H, Yuan W, et al. 2DIntermediate suppression for efficient Ruddlesden−popper (RP) phase Lead-free perovskite solar cells. ACS Energy Letters. 2019;**4**:1513

[141] Xu H, Jiang Y, He T, Li S, Wang H, Chen Y, et al. Orientation regulation of tin-based reduced-DimnsionalPerovskites for highly efficient and stable photovoltaics. Advanced Functional Materials. 2019;**29**: 1807696

[142] Song TB, Yokoyama T, Stoumpos CC, Logsdon J, Cao DH, Wasielewski MR, et al. Importanceof reducing vapor atmosphere in the fabrication of tin-BasedPerovskite solar cells. Journal of the American Chemical Society. 2017; **139**:836-842

[143] Yang Z, Rajagopal A, Chueh C-C, Jo SB, Liu B, Zhao T, et al. Stable Low-Bandgap Pb–Sn Binary Perovskites for Tandem Solar Cells. Advanced Materials. 2016;**28**:8990

[144] Li Y, Sun W, Yan W, Ye S, Rao H, Peng H, et al. 50% Sn-based planar perovskite solar cell with power conversion efficiency up to 13.6%. Advanced Energy Materials. 2016;**6**: 1601353

[145] Changlei W, Zhaoning S, Yue Y, Dewei Z, Awni RA, Grice CR, et al. Synergistic effects of thiocyanate additive and cesium cations on improving the performance and initial illumination stability of efficient perovskite solar cells. Sustainable Energy and Fuels. 2018;**29**:7

[146] Jokar E, Chien CH, Fathi A, Rameez M, Chang YH, Diau EWG. Slow surface passivation and crystal relaxation withadditives to improve device performance and durability for tin-basedperovskite solar cells. Energy & Environmental Science. 2018;**11**: 2353-2362

Chapter 8

Thin Film Solution Processable Perovskite Solar Cell

Mayur Jagdishbhai Patel, Himangshu Baishya, Ritesh Kant Gupta, Rabindranath Garai and Parameswar Krishnan Iyer

Abstract

Perovskite has emerged as a promising light-harvesting material for solar cells due to its higher absorption coefficient, bandgap tunability, low-exciton binding energy, and long carrier diffusion length. These lead to high power conversion efficiency >25% for thin film-based perovskite solar cells (PSCs). Additionally, PSCs can be fabricated through simple and cost-effective solution processable techniques, which make this technology more advantageous over the current photovoltaic technologies. Several solution-processable methods have been developed for fabrication of PSCs. In this chapter, the advantages and disadvantages of various solution processable techniques and their scope for large-scale commercialization will be discussed.

Keywords: perovskite, solar cells, thin film, solution processable, commercialization

1. Introduction

Solar cell technologies have grown in the past few decades across four different generations. The first generation consisted of wafer-based photo active layer which was dominated by silicon wafer and is continuing to conquer the photovoltaic market. However, its high energy and cost of production has allowed the thin film technologies to gain attention among the research community. In the second-generation thin film based inorganic materials are being utilised to develop solar cells, but the efficiency has not reached the first-generation solar technology and the material production cost is also on the higher side. To reduce the cost of production further, organic and hybrid materials are being used in the third generation. The main advantage of this generation is that it allows photoactive layers to be deposited using low-cost solution processable techniques including spin coating, dip coating, etc. [1]. Further, large area fabrication is also facilitated by techniques such as doctor blading, inkjet printing, etc. [2, 3].

This advantage has also been strategically utilised in fabrication of perovskite solar cell which is the most promising and growing solar cell technology [4–6]. These techniques facilitate quick deposition of perovskite on any substrate at low temperature processing. Further, the solution processing techniques offer added advantage over the well-known thermal evaporation methods which requires complex vacuum systems. It has been also observed that the high-performance perovskite solar cells

(PSCs) are generally fabricated using one of the solution processing techniques. This chapter compiles all solution-processing techniques that are being utilised for the fabrication of perovskite solar cell. Each technique has been explained in details covering the advantages and disadvantages as well as its way in controlling the crystallinity of the deposited perovskite film for photovoltaic application.

2. Solution processable techniques

2.1 Anti-solvent dripping: one step deposition technique

Anti-solvent dripping is sub part of spin coating process. Spin coating is a batch process that spreads a liquid film onto a rotating substrate using centrifugal force. This spin coating technique is categorised into two types: (a) one-step process and (b) two-step process. This method is successfully employed to develop a small area and large area PSCs of 0.1 cm^2 and 1 cm^2, respectively. From the above-mentioned techniques, the solution processable one step deposition (OSD) technique is widely accepted to develop PSCs from the laboratory to the industrial level. The formation of perovskite in the one-step process involves two stages: (a) evaporation of excess solvent in the active layer and (b) crystallisation of the active layer [7]. Its popularity stems from its ease of use and low cost of equipment. However, like any other method, this technique suffers a significant drawback. This is primarily due to a lower substrate coverage area and the formation of a rough and porous surface at interface layer of the perovskite solar cell. This method relies on uniformity of film thickness and morphology control to achieve a desirable film quality. This method also faces the challenge of reducing pinholes in the perovskite film. This has a drastic impact on the optoelectronic properties of the PSCs resulting in poor performance [8].

2.1.1 Antisolvent in chemistry

In chemistry, antisolvent precipitation is a well-known method of crystallising a substance. In **Figure 1a** the antisolvent precipitation is illustrated. The unique part of the antisolvent method is its applicability for the manufacture of PSCs. In this method, a non-dissolving liquid, or anti-solvent, is dropped onto a spinning substrate containing a perovskite solution to quickly remove specific solvents like DMF and GBL (**Figure 1b**) [9]. The treatment causes rapid nucleation in the film and converts it to a homogeneous intermediate film. Annealing the substrate then results in smooth perovskite film.

2.1.2 Antisolvent in perovskite

To address all of the aforementioned issues in OSD, the Antisolvent dripping (ASD) method is used to control crystal growth kinetics and film quality. A dense layer of larger grain size (100–500 nm) $CH_3NH_3PbI_3$ crystals is observed after successful ASD treatment. The significant formation of $CH_3NH_3PbI_3$ crystals by ASD method has catapulted perovskite research into a whole new realm. For the preparation of high-quality Pb-based perovskite crystalline films, anti-solvents such as benzene, toluene, ethanol, methanol, acetonitrile, benzonitrile chloroform, isopropyl alcohol, ethylene glycol, and chlorobenzene have been utilised [10, 11]. This method was first reported by Jeon and his team, who discovered that using an antisolvent in the

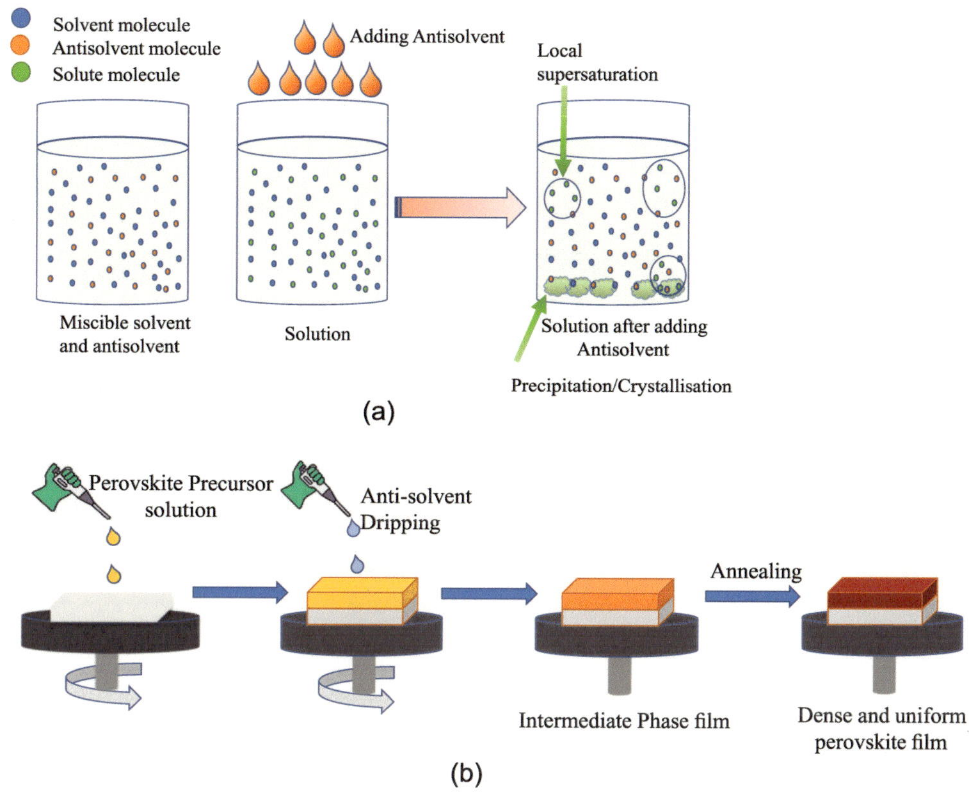

Figure 1.
(a) Schematic representation of antisolvent assisted precipitation/or crystallisation. (b) Simple representation of anti-solvent treatment for the OSD technique.

fabrication of perovskite films resulted in better-quality, dense films with big grain size. The comparison of the various perovskite materials performance with different antisolvents have been shown in the **Table 1**.

The ASD method can significantly alter the morphology of the $MAPbI_3$ film's surface. During the deposition of the perovskite layer by the OSD method, many voids and pinholes were observed. On the other hand, the ASD method reveals lesser pinholes with large grains, densely packed $MAPbI_3$ crystals due to its smooth and homogeneous surface morphology [12]. However, the anti-solvent preparation process must be done at the correct time and in the correct quantity as well as with a high level of proficiency. Uncontrolled crystallisation can also result in pinholes and higher defect density, which can reduce device efficiency and stability.

The performance of PSCs and their reproducibility is significantly improved when the $MAPbI_3$ film is of higher quality and covers the entire surface area by using ASD method. The addition of favourable additives has facilitated perovskite crystal growth to improve the morphology, stability, excitonic, and optoelectronic properties of hybrid inorganic-organic perovskite films. Although it has been discovered that the solution-processable technique is capable of producing the ideal perovskite film, the quality of the film may be compromised due to factors such as temperature, precursor solubility, atmosphere, and annealing time [20].

Zhou et al. used an antisolvent-solvent extraction process to study the crystallisation behaviour of mix halide perovskites at room temperature. A small amount of

Perovskite material	Anti-solvent	Antisolvent volume (μL)	Post annealing condition	Device structure	PCE, (%)	Ref.
$MAPbI_3$	Toluene	130	80°C, 10 min	FTO/NiOx/ perovskite/ PCBM/ Rhodamine/ Au	21.18	[4]
$MAPbI_3$	Di-isopropyl ether	1000	100°C, 10 min	FTO/c-TiO_2/ mp-TiO_2/ perovskite/ Spiro-MeOTAD/ Ag	19.07	[12]
$MAPbBr_3$	Toluene	60	100°C, 10 min	FTO/c-TiO_2/ mp-TiO_2/ perovskite/ Spiro-MeOTAD/ Au	7.54	[13]
$FASnI_3$	Diethyl ether	—	70°C, 20 min	ITO/ PEDOT:PSS/ perovskite/ C60/BCP/Ag	5.41 ± 0.46	[14]
$(CH_3NH_3)_3Sb_2I_9$	Chlorobenzene	300	70°C, 20 min	ITO/ PEDOT:PSS/ perovskite/ PCBM/C60/ BCP/Al	2.7	[15]
$Cs_5(MA_{0.17}FA_{0.83})_{95}$ $Pb(I_{0.83}Br_{0.17})$	Diethyl ether	800	Post annealing free	FTO/bl-TiO_2/ mp-TiO_2/ perovskite/ Spiro-MeOTAD/ Au	19.5	[16]
$(FAPbI_3)_{0.85}(MAPbBr_3)_{0.15}$	Trifluorotoluene	110	100°C, 90 min	FTO/bl-TiO_2/ mp-TiO_2/ perovskite/ Spiro-MeOTAD/ Au	20.3	[17]
$FA_{0.75}MA_{0.25}SnI_3$	Toluene	—	100°C, 10 min	ITO/ PEDOT:PSS/ perovskite/ C60/BCP/Al	6.36 ± 0.64	[18]
$FA_{0.75}MA_{0.25}SnI_3$	Chlorobenzene	350	45°C, 10 min; then 65°C,20 min; and 100°C, 10 min	ITO/ PEDOT:PSS/ perovskite/ C60/BCP/Ag	7	[19]

Table 1.
PCE comparison of various PSCs using different antisolvents.

solvent diffuses within a large amount of antisolvent in this strategy. This antisolvent-solvent extraction method achieves supersaturation state and nucleation for the crystallisation procedure, and nucleation rate can be improved by magnetic stirring of antisolvent. The antisolvent DEE (anhydrous diethyl ether), and magnetic stirrers were used to introduce advection in the antisolvent bath. The antisolvent-solvent extraction process is a straightforward method for producing high-quality perovskite films with improved morphology [21].

Xiao et al. demonstrated a fast, single-step, solution-based deposition crystallisation technique that allows control over the dynamics of nucleation and grain growth of $CH_3NH_3PbI_3$, resulting in rapid and repeatable fabrication of high-quality perovskite thin films. In this method, a DMF solution of $CH_3NH_3PbI_3$ perovskite is spin-coated on a substrate, followed by a second solvent, such as chlorobenzene (CBZ), applied on top of the wet film during the spin coating process to induce fast crystallisation. The second solvent is important for lowering the solubility of $CH_3NH_3PbI_3$ and promoting crystal nucleation and growth within the thin film [22].

For inverted planar perovskite solar cells, Liu et al. reported effective and stable green mixed anti-solvent engineering. This green mixed anti-solvent technique can improve the surface morphology of perovskite films and passivate the grain boundary of perovskite thin films [23].

2.1.3 Modified antisolvent treatment

Modification, particularly during the anti-solvent treatment, is absolutely necessary to manage perovskite crystal growth in a humid environment and obtain a highly efficient device. Currently, the anti-solvent is dropped 15 s after the fast-spinning programme begins, with chlorobenzene, toluene, and diethyl ether being the most common anti-solvents used in PSC fabrication [24]. Determining an appropriate time delay for quenching anti-solvents is the most important part of anti-solvent treatment. The turbid point during the spin deposition stage must be identified in order to achieve this goal. During spin coating, the turbid point is the point at which the precursor film turns from transparent to turbid. The time delay increases as the RH level rises. This result is linked to the previously discussed solvent evaporation dynamics. According to Wang et al., the turbid point appears at approximately 9, 15, 19, and 20 s for RH0, RH50, RH70, and RH90%, respectively. In order to develop PSCs under high-humidity conditions, it appears that determining an accurate dripping time is an inconvenient procedure [25].

Thus, there are three possible strategies to avoid the antisolvent dripping time window: (a) creating an anti-solvent mixture by combining traditional non-polar solvents, such as diethyl ether and chlorobenzene, with other polar solvents (e.g., R-OH); (b) finding another anti-solvent that is suitable for high-humidity processing and completely substituting the commonly used anti-solvent; and (c) applying preheating treatment to the solution [26]. Anti-solvent mixes can be divided into two types: those with (1) a small amount of polar solvent and those with (2) a large amount of polar solvent. Because the anti-solvent contains a polar solvent, the adsorbed H_2O molecules on the perovskite intermediate layer can be dissolved and removed concurrently with DMF and excess DMSO. As a result of the direct contact between the electron transfer layer and the hole transfer layer, a homogenous perovskite intermediate is generated, which gradually converts into a smooth and pinhole-free film, limiting charge recombination [27].

Another viable strategy for forming a high-quality perovskite film in high humidity is to replace the standard anti-solvent, as previously described. For high-humidity PSC production (up to RH75%), Troughton et al. utilised ethyl acetate. This was due to the fact that ethyl acetate, unlike other anti-solvents like toluene, chlorobenzene, and diethyl ether can absorb a significant amount of moisture in the air. Because of these properties, ethyl acetate can absorb moisture from the air and prevent it from interacting with the intermediate phase of perovskite. As a result, regardless of the RH level, uniform and smooth films can be created. Another effective strategy for making the anti-solvent treatment humidity insensitive is to speed up the rate of solvent evaporation by pre-annealing the substrate, causing the turbid point to appear earlier. According to Wang et al., the anti-solvent can be applied to the perovskite film 2 s after the spinning protocol starts, regardless of the humidity level, to facilitate the fabrication of highly efficient PSCs when the substrate is pre-heated to 70°C [28].

2.2 Hot-casting

Hot-casting technology has emerged as an excellent tool for the deposition of high-quality perovskite thin films with notable benefits including rapid crystallisation, a quick film formation process, increased grain size, preferred crystal orientation, and low defect density [29]. Within the framework of nucleation growth theory, a direct formation mechanism is proposed that constitute a driving force for the phase change provided by a high substrate temperature leading to an ultrashort crystallisation process. Meanwhile, a sufficient thermal energy allows atoms to diffuse in a liquid without forming an intermediate phase. Nie et al. were the first to report this method, and they went on to improve the film quality by tweaking deposition parameters like substrate temperature, annealing temperature, and precursor composition. This technology has recently been expanded to include the deposition of organic–inorganic hybrid, all-inorganic, lead-free, and low dimensional perovskite films [30].

2.2.1 Fundamentals of nucleation and crystal growth

The goal of the hot-casting technology is to spin coat a hot precursor solution over a substrate at higher-temperature. The crystal growth is influenced by factors such as substrate temperature, solution concentration, solvent, and supersaturated environment. In **Figure 2a**, the LaMer diagram represents these deposition parameters correlated with the nucleation and growth processes [31].

2.2.2 Classical nucleation and classical growth

The Volmer-Weber model, the Frank-van der Merwe model, and the Stranski Krastanov model are three classic thin film nucleation and growth models [32, 33]. The Volmer-Weber model is valid for the nucleation and growth of most polycrystalline thin films if the substrate temperature is sufficiently high and the deposited atoms have a certain diffusion capability.

According to classical nucleation theory, to initiate the crystallisation, the solution must be supersaturated. Nucleation can be either heterogenous or homogenous. Heterogeneous nucleation is defined as nucleation with preferential nucleation sites, which means that new phases form preferentially within certain regions of a liquid phase. The system must overcome an energy barrier known as the maximum free energy of critical nucleation (G) during the thermodynamic nucleation process.

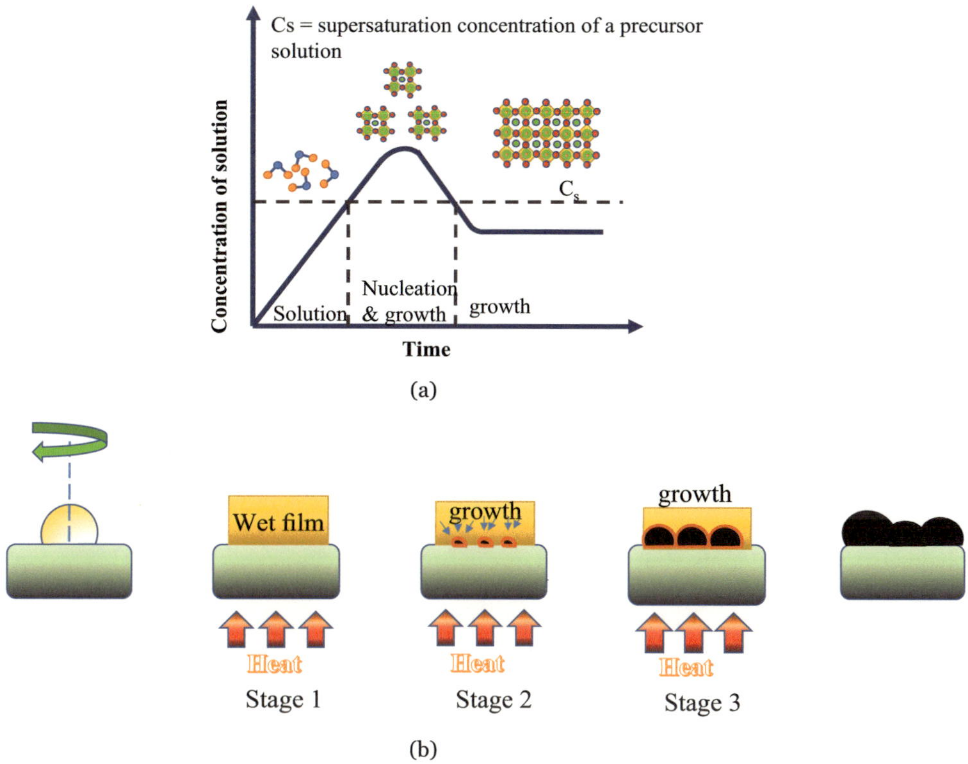

Figure 2.
(a) Lar Mer model for nucleation and growth of perovskite thin films, (b) a schematic illustration of nucleation and growth of perovskite films at each stage.

Heterogeneous nucleation is defined as nucleation with preferential nucleation sites, which indicates that new phases form preferentially within some parts of a liquid phase. The system must surpass an energy barrier known as the maximal free energy of critical nucleation (G) during the thermodynamic nucleation process.

2.3 Temperature and thermal annealing

2.3.1 Direct formation mechanism

Nucleation growth is influenced by the substrate temperature and thermal annealing. When the substrate temperature is low (less than 100°C), the perovskite film formation process includes three steps- the initial solution stage, the transition-to-solid film stage, and the transformation stage from intermediates to a perovskite film [34]. However, when the substrate temperature is raised to 100–180°C, enough thermal energy is provided to speed reactant diffusion and contact, hence no intermediate phase development occurs during the hot-casting process leading to the direct formation of perovskite film.

2.3.2 Substrate temperature

The supersaturation of the solution is affected by the substrate temperature, which changes the nucleation rate and film shape. The high boiling solvent supports a stable development of the perovskite crystal with big crystal grains when the substrate temperature is higher than the crystallisation temperature of the perovskite phase. A large-grain perovskite layer improves device performance by lowering the defect density and boosting the mobility which suppresses charge trapping [35]. The effect of Substrate temperature on various perovskite material has been provided in the **Table 2.**

2.3.3 Thermal annealing

In hot-casting technique, Yang et al. described the Volmer-Weber growth mechanism for the creation of island-like grains and the transition to a compact perovskite film. Thermodynamic energy is important in accelerating perovskite crystallisation and precursor diffusion, which directly affects the film shape (**Figure 2b**) [46].

Thermal energy can lower the surface tension between the precursor solution and the substrate, increasing the probability of large grain domain with fewer surface defects. As the temperature rises, isolated like grains get larger and begin to form bonds with one another, finally generating a high-quality perovskite film with no pinholes. Thermal annealing not only increases material transport inside the film and facilitates solvent evaporation, it also improves charge extraction in the active layer [47]. Through

Perovskite material	Device structure	Temperature	Voc, (V)	Jsc, (mA/cm^2)	FF, (%)	PCE, %	Ref.
$MAPbI_3$	FTO/c-TiO_2/m-TiO_2/Perovskite/spiro-OMeTAD/Ag	175	1.07	21.32	70	16.01	[36]
$MAPbI_{3-x}Cl_x$	FTO/PEDOT:PSS/perovskite/PCBM/Al	190	0.954	24.21	61	14.11	[37]
$FA_{0.25}MA_{0.75}PbI_3$	FTO/c-TiO_2/perovskite/spiro- OMeTAD/Au	240	0.97	22.6	66	14.6	[38]
$FAPbI_{3-x}Cl_x$	FTO/c-TiO_2/perovskite/spiro-OMeTAD/Ag	175	0.96	18.93	66	12.07	[39]
$CsPbI_2Br$	ITO/SnO_2/perovskite/PTAA/MoO_3/Al	340	1.19	14.54	74	13.8	[40]
Ag_2BiI_5	FTO/c-TiO_2/m-TiO_2/perovskite/PTAA/Au	100	0.69	6.04	62	2.60	[41]
$BA_2MA_3Pb_4I_{13}$	FTO/PEDOT:PSS/perovskite/PCBM/Al	150	1.01	16.76	74	12.52	[42]
$(CPEA)_2MA_4Pb_5I_{16}$	FTO/c-TiO_2/perovskite/spiro-OMeTAD/Au	120	0.99	19.92	60	11.86	[43]
MA_2PbI_4	FTO/c-TiO_2/perovskite/spiro-OMeTAD/Au	150	1.06	21.00	76	16.92	[44]
$(GA)(MA)_3Pb_3I_{10}$	FTO/c-TiO_2/perovskite/spiro-OMeTAD/Au	100	1.00	20.70	66	13.87	[45]

Table 2.
Photovoltaic parameters of various hot-casted PSCs.

quick solvent removal, flash infrared annealing (FIRA) can enhance the development and crystallisation of perovskite films. In comparison to the traditional thermal annealing, short heating pulses, compared to traditional thermal annealing, can greatly minimise the deterioration of organic components, even at extremely high temperatures [48]. The grain size distribution along the periphery of the perovskite film made by hot casting is greater than the core size distribution, and the particle size distribution is annular, as proposed by Ren et al. During the evaporation of the solvent, a compensatory flow from the centre of the solution to the edge is thought to occur spontaneously, resulting in a greater concentration and larger particle size near the edges.

2.4 Precursor chemistry

2.4.1 Additives

The wettability of the substrate and consequently the film formation is affected by the viscosity and solvation ability of the solvent. The rate of solution evaporation is determined by the solvent's boiling point and vapour pressure. Therefore, a selection of extremely polar aprotic organic solvents is required for a weakly soluble inorganic lead salt. Lewis-base character of some highly polar aprotic organic solvents can cause a solvent-solute coordination, which can alter the perovskite crystallisation process. The colloidal skeleton of the perovskite precursor solution is made up of numerous coordination complexes, and it is viewed as a colloidal cluster with a soft colloidal skeleton [49, 50]. Additives (Cl^- and Br^-) influence the size of the colloidal clusters. According to Liao et al., adding chlorine to perovskite films improves their optoelectronic capabilities and environmental resilience. The addition of 10% Cl^- to a $MAPbI_3$ precursor solution increases film uniformity and coverage greatly [51].

2.4.2 Ageing time and solvent

The size and shape of colloidal clusters, as well as the nucleation and growth process, can be affected by the ageing duration of the precursor solution. Mohite et al. found a substantial link between film crystallinity and ageing time. The crystallinity and grain size of perovskite films were greatly improved when the precursor solution was aged for more than 24 h, according to the findings. The precursor solution gradually develops big seeds (or crystals) as it matures. Grain development, phase purity, surface uniformity, and trap state density of the perovskite film have all been shown to be considerably affected by precursor ageing. Meanwhile, the phase development and crystalline characteristics of perovskite films are influenced by the solvents and content of the precursor. To deposit perovskite films, Wang et al. employed gamma-butyrolactone (GBL) and DMF as co-solvents [36]. The grain size and photovoltaic characteristics of the device produced by the GBL solvent were found to be much lower than those produced by the DMF solvent. Iyer et al. also used DMSO as a Lewis base adduct to control perovskite crystallisation and grain development [37].

2.4.3 Composition and other factors

Processing parameters such as substrate temperature, rotation speed, and thermal annealing can be used to fine-tune the perovskite film morphology. Through heat energy and centrifugal force, these processing factors influence solute diffusion and perovskite crystallisation behaviour. The residual organic residue is evaporated during thermal annealing. To create a high-quality perovskite layer, Moon et al. employed MAI and lead acetate ($PbAc_2$) as precursors. By using by-product gas ($3MAI + PbAc_2$ $MAPbI_3 + 2MAAc$) to remove the $PbAc_2$ residue, the crystal development can be accelerated to generate a fully covered, pinhole-free, and highly crystalline perovskite film [52]. Janssen et al. used a mixture of $PbAc_2$, PbI, and MAI to make a high-quality perovskite layer. In ambient condition, Huang et al. used methylammonium acetate (MAAc) as a general solvent to produce high-quality perovskite films. To facilitate solvent evaporation, a constant substrate temperature (100°C) was used, resulting in supersaturation and fast nucleation and crystal formation [53].

2.4.4 Atmosphere

The crystallinity and surface morphology of as-cast perovskite films are affected by various deposition circumstances. One benefit of hot-casting method is that the deposition is not affected by the processing environment. Therefore, the perovskite films may be produced in ambient air, and the device revealed great stability under high humidity [54]. Mori et al. combined a gas flow with hot-casting method to make $MAPbI_3$ films in ambient circumstances (relative humidity = 42–48%) [55]. The flowing gas can greatly expedite mass transfer and eliminate thickness non-uniformity due to the difference in centrifugal force between the centre and edge of the substrate. Yang et al. also used a combination of hot-casting and methylamine (MA) gas treatment to create dense and homogeneous perovskite films at high relative humidity. With MA gas treatment, porous and rough $MAPbI_3$ perovskite films made by hot casting can be turned into dense and high-quality films. In addition, Hao et al. used non-destructive ethanol/chlorobenzene to treat $MAPbI_3$ perovskite films, which resulted in coarsening of the perovskite grains and lateral grain expansion of the $MAPbI_3$ perovskite films. To overcome the challenges of temperature gradient and moisture intrusion during the deposition process, Cheng et al. developed a thermal radiation hot-casting method [56].

2.5 Advantages of hot-casting technique

2.5.1 Grain size, orientation, and film thickness

To create reasonably thick and preferentially oriented large-grain perovskite films, hot-casting process has been frequently used. The grain boundaries of the perovskite films are reduced due to the higher grain size. At the same time, the absorption, charge transport, and crystallinity of perovskite films are all positively affected. Increases in grain size, on the other hand, may increase the density of unwanted pinholes, resulting in direct contact between the HTL and ETL and leakage current [57]. The device's performance will be severely harmed by the voids in the perovskite coating. A thick perovskite layer helps to gather enough light absorption across the visible light spectrum [58]. To tune the thickness of the perovskite layer, the concentration of the precursor solution and the rotation speed can be varied. The inorganic halide octahedrons are joined at a shared apex and stretched in a 2D direction to form 2D perovskites when a long-chain organic cation layer is placed into the inorganic framework to deviate the tolerance factor from a value of 1. The introduction of an organic chain will make charge extraction and collecting more difficult. Controlling the growth direction of a 2D perovskite film is critical for carrier transport. A hot-casting technique can produce a selectively oriented development of 2D perovskites [59, 60].

2.5.2 Defects and recombination

The performance of the PSCs is harmed by nonradiative recombination in the following ways. Radiative recombination occurs when electrons return to the valence band. Minority recombination occurs at the interface as holes (electrons) are transported back to the perovskite layer. The porous perovskite layer will then come into direct contact with the functional layers, resulting in carrier recombination. A nonradiative recombination process is caused by a number of defect states in the device forming recombination centres to trap the carriers. The major avenue for a carrier loss is through some deep-level traps. Furthermore, the nonradiative recombination of carriers within the device has a direct impact on the PSCs' V_{oc}. As a result, forming low-defective perovskite thin films is critical for preventing nonradiative recombination. Large grain sizes help limit the number of grain boundaries in hot-casted perovskite films, which reduces charge trapping [30].

2.6 Efficiency and stability

2.6.1 Device performance

Three key limiting considerations for the actual deployment of PSCs are PCE, stability, and cost. **Table 1** shows how PCEs have progressed in previous years while employing a hot-casting approach. The two most common light harvesters in PSCs are $MAPbI_3$ and $FAPbI_3$, which have energy bandgaps of 1.55 and 1.47 eV and theoretical maximum PCEs of 31.3 and 32.5%, respectively [61]. Nie et al. were the first to describe the manufacture of perovskite films with a millimetre grain size, resulting a PCE of 17.48% with no hysteresis. Marks et al. used a hot casting technique to manage Cl^- incorporation and reported a PCE of 18.2% for a small area (0.09 cm^2) and 15.4% for a big area (0.09 cm^2) (1 cm^2). We believe that in the future, the PCE of PSCs produced using hot-casting technology will catch up to that produced using a solution approach. Meanwhile, the perovskite composition, device structure, and encapsulation all play a role in improving PSC long-term stability [62].

2.6.2 Stability

H_2O and O_2 have a direct impact on PSC device performance and stability throughout film deposition, device testing, characterisation, and storage. In the case of $MAPbI_3$, water vapour can dissolve the perovskite material, and MAI is dissolved to generate a combination of MA and HI; however, HI will either react with O_2 to form H_2O and I_2, or self-decompose. After being exposed to moisture, $MAPbI_3$ continues to degrade. The inability hinders the commercialization of PSC devices. Improving long-term stability requires adjusting the ABX_3 perovskite content and boosting crystalline quality [42, 63]. Long-chain organic cations have been routinely used as a site cation substitution in this regard. The organic and inorganic layers alternatively form a layered structure in these 2D perovskites, which have great long-term stability. As long-chain organic cations are hydrophobic, the bigger organic cations in the 2D perovskite crystal structure improve humidity stability. Pure 2D PSCs are stable, but have a low PCE. Combining 3D perovskite and 2D perovskite yields exceptional optoelectronic characteristics and stability. The intrinsic performance of the 2D perovskite deposited using a hot-casting approach can be maintained for a long time, showing greater stability in humid and other environmental conditions [64, 65].

2.7 Two-step coating

Mitzi et al. first introduces the two-step coating method in order to improve the morphology and quality of the active perovskite layer [66]. Coating is done in the first phase by using conventional spin coating process, and the second step uses other coating methods such as immersion and spin coating depending on the materials' requirements.

2.7.1 Immersion method

The organic and inorganic components of the perovskite material are treated separately in the immersion method. The perovskite material's inorganic salt (PbI_2) is first spin coated on the substrate at a particular RPM. The spin-coated PbI_2 film substrate is then immersed in an organic salt (MAI) precursor solution for a period of time. After taking the substrates from the precursor solution, rinse them using the same solvent and concentration that was used to prepare the organic precursor solution. This step is used to remove any surplus organic material from the substrate's surface. Finally, the ultimate perovskite film is obtained by annealing the substrate for a few minutes at a specific temperature. The procedure is depicted schematically in **Figure 3**. Grätzel et al. used this strategy for the first time in 2013 to obtain $MAPbI_3$ film in order to optimise the morphology of active perovskite material for photovoltaic devices [67].

The concentration of inorganic compound in the spin coated film, the concentration of organic salt in the precursor solution, and the immersion duration all seem to have a significant impact on the growth of the perovskite film, which defines the morphology and quality of the ultimate perovskite film.

After immersion in the precursor solution, two distinct reaction pathways convert the spin coated film to the ultimate perovskite film. The first is the solid-liquid interface conversion reaction, which happens when the precursor solution concentration is low. Due to the low concentration of the MAI precursor solution, MAI tends to diffuse into the PbI_2 structure during immersion of the PbI_2 coated film. Finally, the reaction of MAI with the PbI_2 film at the interface produces the ultimate $MAPbI_3$ perovskite film. The conversion reaction at the solid-liquid interface is show by Eq. (1).

$$PbI_2\,(s) + CH_3NH_3^+\,(sol) + I^-\,(sol) \rightarrow CH_3NH_3PbI_3\,(s) \tag{1}$$

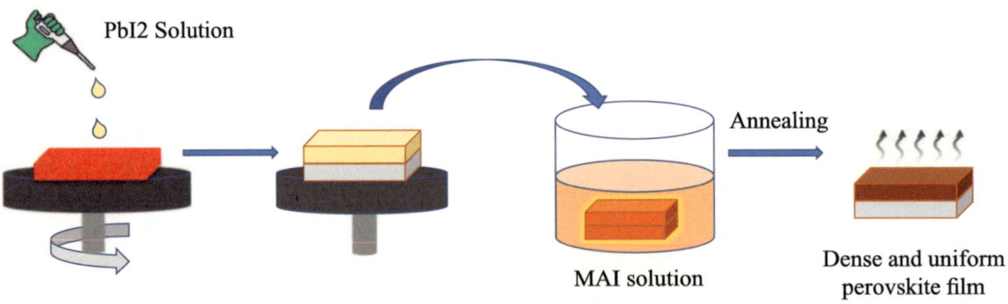

Figure 3.
Schematic diagram of immersion coating process of perovskite ($MAPbI_3$) film.

In the above reaction initially, MAI diffuses into the PbI_2 structure as well as the reaction start occurs at the interface of PbI_2. Once the perovskite crystal is formed at the surface of PbI_2 the diffusion of the organic cation MAI into the PbI_2 structure is at a standstill and the conversion reaction is completed [68].

When the concentration of organic cation is high enough (>10 mg/mL), another reaction called dissolution-recrystallization evolution mechanism occurs. As the concentration of organic cation in this process is high, it causes rapid crystallisation of perovskite on the surface of PbI_2. As a result, organic cation diffusion into the PbI_2 structure is totally stopped after a few times, as stated by Eq. (2). According to Yang et al., a high concentration of the organic cation MAI leads to production of the lead iodide complex PbI_4^{2-}, as shown in the Eq. (3). The excess iodine in the system began to dissolve previously created $MAPbI_3$ crystals and PbI_2 components that had not been covered during the crystallisation process, and the reaction continued until the system could not reach thermal equilibrium [69].

$$CH_3NH_3PbI_3\,(s) + I^-\,(sol) \rightarrow CH_3NH_3^+\,(sol) + PbI_4^{2-}\,(sol) \tag{2}$$

$$PbI_2\,(s) + 2I^-\,(sol) \rightarrow PbI_4^{2-}\,(sol) \tag{3}$$

Following these two reactions, when the concentration of excess PbI_4^{2-} approaches the supersaturation state, the gradual re-crystallisation of $MAPbI_3$ via the reaction mechanism shown in Eq. (4) begins anew.

$$PbI_4^{2-}\,(sol) + CH_3NH_3^+\,(sol) \rightarrow CH_3NH_3PbI_3\,(s) + I^-\,(sol) \tag{4}$$

2.7.2 Sequential spin coating method

The two-step spin coating method, also known as sequential spin coating, is a well-studied laboratory technique for perovskite thin film deposition. Im et al. presented this technology in 2014 to fabricate high efficiency (>16%) perovskite solar cells by growing cuboid perovskite grains of controlled size [70]. The organic precursor is spin coated above the inorganic layer at a certain spin-rpm in this technique, which starts with the inorganic part of the perovskite material being spin coated onto the substrate. The final perovskite thin film is obtained by periodically annealing the substrate, which exhibits the perovskite film's growth by changing colour during the

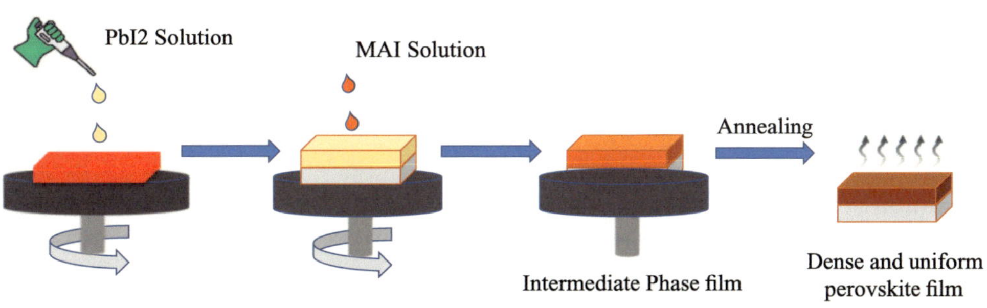

Figure 4.
Schematic diagram of Sequential spin coating process of perovskite ($MAPbI_3$) film.

process. As long as the concentration of organic precursor is high enough, the dissolution crystallisation mechanism produces the resulting perovskite film. **Figure 4** illustrates the perovskite film fabrication process using this technique.

Huang et al. devised a two-step spin coating process to produce a perovskite film with fewer pinholes at low temperatures. The inorganic precursor is spin coated onto the substrate and dried on the hot plate in the first stage. After that, the inorganic film substrate is spin coated with the organic precursor solution. The solvent used to dissolve the organic component of the perovskite material should have a low solubility of the inorganic part in this approach, such that no inorganic part washed out during the organic precursor coating. Due to the inter-diffusion of spin coated organic and inorganic assembly layers, the substrate ultimate compact, pin-hole free perovskite film is achieved after thermal annealing [71].

Panzer et al. revealed the reaction mechanism for obtaining the ultimate perovskite layer using a two-step spin coating method. According to their findings, the reaction mechanism can be broken down into five steps: perovskite capping layer formation, change in organic layer concentration during solvent evaporation, initial caping layer dissolution, rapid dissolution recrystallization, and complete conversion of residual PbI_2 to perovskite crystal. The perovskite capping layer begins to form immediately in the second step of the spin coating process, i.e., on the surface of PbI_2 after the introduction of MAI, which initiates the formation of the $MAPbI_3$ crystal. The size of perovskite grains and the time it takes to produce them both decrease as the MAI concentration rises [72]. Furthermore, the reaction temperature has a major impact on the production of perovskite crystals and their grain sizes. Solvent evaporation occurs in the organic precursor during thermal annealing. During this method, a dense perovskite crystal layer is first generated on top of the PbI_2 layer to prevent MAI from penetrating the PbI_2 structure and preventing further perovskite crystal formation in the system [72]. The concentration of MAI solution continues to rise due to the suppression of MAI solution penetration onto PbI_2 and the volatile nature of its precursor solvent. As the concentration of MAI rises, so does the concentration of iodine ions, which react with the perovskite layer on the surface of PbI_2 to generate complex lead iodine PbI_4^{2-} as a by-product. The degradation of the previously produced perovskite layer occurs primarily at grain boundaries and smaller grains in this process [72]. Finally, when the concentration of volatile PbI_4^{2-} approaches the supersaturation point it start recrystallisation of perovskite grain with all the uncovered inorganic component PbI_2 converted into $MAPbI_3$ crystal.

Two-step coating, also known as sequential deposition, has a number of advantages, including increased processing window flexibility. Solvent engineering, concentration variation, annealing time, annealing temperature adjustment, spin rpm variation etc. can be used to optimise each deposition stage separately, resulting in a superior morphological perovskite film. Furthermore, depending on the solubility of the additive, different additives can be doped directly into the precursor solution to form a defect-free perovskite layer. These approaches also produce a high-quality perovskite film with great reproducibility. When comparing the performance of solar cell devices, sequential deposition delivers a comparable efficiency to the single-step deposition procedure.

Aside from these benefits, the two-step deposition process has a few disadvantages. The formation of a perovskite layer in this approach relies heavily on molecular exchange. The organic component is frequently overreacted in the second step of the deposition technique, making it difficult to precisely control the chemical composition of the film [73]. Another issue with this approach is incomplete inorganic

Methods	Device architecture	J_{sc}, (mA/cm^2)	V_{oc}, (V)	FF, (%)	PCE, (%)	Ref.
Immersion	FTO/c-TiO_2/$MAPbI_3$/Spiro-OMeTAD/Ag	22.49	1.08	67	16.21	[67]
	FTO/c-TiO_2/m-TiO_2/$MAPbI_3$/Spiro-OMeTAD/Au	19.4	1.01	65.8	12.9	[74]
	FTO/TiO_2/$CsPbBr_3$/Carbon	5.99	1.33	65.7	5.25	[75]
	TCO/c-TiO_2/m-TiO_2/$MAPbI_3$/Spiro-OMeTAD/Au	20	0.993	73	15	[76]
	FTO/c-TiO_2/m-TiO_2/$MAPbI_3$/Spiro-OMeTAD/Au	19.42	1.03	66	13.2	[77]
Sequential spin coating	ITO/SnO_2/$CsFAMAPbI_2Br$/PCBM/Ag	22.11	1.13	81.53	20.35	[78]
	ITO/c-TiO_2/$FAPbI_3$/PCBM/Au	24.03	1.08	72	18.77	[79]
	ITO/PTAA/$MAPbI_3$/PCBM/Al	23.41	1.05	78	19.5	[80]
	ITO/TiO_2/$FAPbI_3$/Spiro-OMeTAD/Au	24.3	1.12	81.1	22.1	[81]
	FTO/c-TiO_2/m-TiO_2/$FA_xMA_{1-x}PbI_{2.55}Br_{0.45}$/Spiro-OMeTAD/Au	23.72	1.075	78.8	20.11	[82]

Table 3.
Photovoltaic parameters of different perovskite material using two step deposition method.

compound conversion to perovskite crystal, which affects overall molecular exchange with the organic component. **Table 3** shows some of the best results obtained using these two-step coating methods.

The methods described above are only suitable for making solar cells on a laboratory scale because a large amount of precursor solution is wasted during the spinning process, and the deposited film does not provide uniformity across the entire area of the film, resulting in device performance degradation.

3. Techniques for large area device fabrication

The performance of PSCs in large area modules must be maintained for practical implementation in the industrial environment. The methods described above are exclusively used to create small-scale laboratory devices. The performance of a perovskite solar cell declines as the active area grows. With the scaling up of device area, the PCE value reduces by 0.8% in most large-scale devices [83]. The PCE is lost due to an increase in series resistance caused by the huge transparent substrate's resistance. Furthermore, when the active area of the device grows, it becomes more difficult to maintain homogeneity across all layers, which has a significant impact on its PCE. The scaling up of perovskite solar cells with morphology manipulation for high quality pin hole free perovskite layer is the most significant and challenging of all the deposition layers.

Several attempts have been made to scale up PSCs towards commercialization. For the fabrication of large area devices, numerous solution-based scalable deposition processes have been developed that can retain overall good device performance. Blade coating, slot die coating, bar coating, spray coating, and inkjet printing, screen printing are examples of these processes.

3.1 Blade coating

For the printing of large area devices, the blade coating process is a relatively simple and inexpensive technique. The blade is utilised to spread the precursor solution throughout the substrate. A micrometre screw is used in the system to adjust the distance between the blade and the substrate by rotating it, allowing the thickness, homogeneity, and crystallinity of the deposited layer to be determined. This approach can be utilised to deposit not just the active perovskite layer, but also other charge transport layers [84]. **Figure 5(a)** illustrates a schematic illustration of the deposition procedure. This approach allows for a slower drying of the wet film, allowing for a broader coverage and higher-quality perovskite film. The creation of a bigger grain perovskite coating as a result of the slow solvent drying process enhances the carrier diffusion length and thus the device's PCE [85]. In the film creation process of these technologies, two significant regimes exist: the evaporation regime and the Landau-Levich regime, which are depicted in **Figure 5(b)** and **(c)**, respectively.

The thickness of the film tends to decrease in the evaporation regime as the substrate's movement speed increases. This occurs as a result of the solute's shorter residence duration, which results in a lower accumulation amount. In the Landau-Levich regime, the thickness of the deposited layer grows as the coating speed increases, owing to the viscous force that pulls the liquid film out and then dries [86]. The Landau-Levich regime is advantageous in terms of practical use. However, if the solvent present in the precursor solution has a high surface-tension, the perovskite material can generate islands with a high roughness surface [87].

3.2 Slot die coating

The slot die coating process is similar to blade coating in that it uses a different coater to apply a thin and homogeneous film. The coater is made up of a head that includes a downstream and upstream die. The precursor solution is initially pushed to the die head using a syringe pump in this procedure. The solution forms a liquid bridge between the head and the substrate during the procedure. As the substrate begins to move, a moist layer of solution forms. The flow rate of the solution, coating

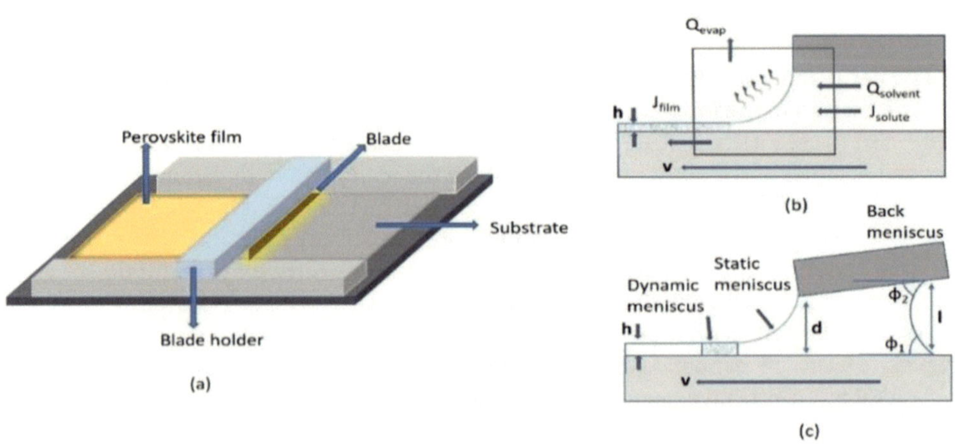

Figure 5.
(a) Schematic diagram of blade coating process, (b) schematic diagram of evaporation regime, (c) schematic diagram of Landau-Levich regime.

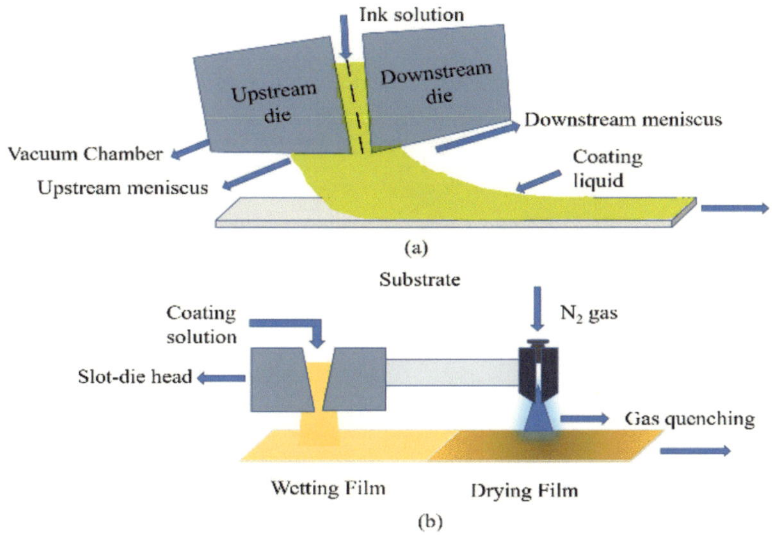

Figure 6.
(a) Schematic diagram of slot-die coating process, (b) schematic diagram of gas quenching annealing of slot-die coated film.

speed, spacing between the head and the substrate, viscosity of the solution, and other factors all influence the quality and shape of the deposited film in this process [88]. **Figure 6(a)** depicts a schematic diagram of the coating process. This method can be used to deposit the inorganic part of the perovskite material in the first step of sequential deposition. Again, annealing can be done by N_2 gas quenching in this process for solvent evaporation to fabricate a pin hole free perovskite layer which is shown by **Figure 6(b)**. It is the only approach that has shown perovskite solar cell role-to-role manufacturing.

Apart from the many advantages of using the slot die coating method to fabricate high-quality large module perovskite films, it also has some failure mechanisms that prevent the development of good morphological films. The failure mechanism includes a low flow limit of the solution, which causes a breakup of the downstream meniscus, resulting in a discontinuity in the wet film [89]. The creation of an air entrainment defect in the wet film is caused by the breaking of upstream meniscus bubbles inside the film. Flooding or dripping occurs when the ink flow to the head is greater than the coating speed, leading in a progressive build-up of ink at the coating head, preventing the desired film thickness from being attained.

3.3 Bar coating

A well-known method for producing high-efficiency PSCs is bar coating, often known as D-bar coating. The procedure is similar to blade coating, but it spreads the solution across the substrate with a cylindrical bar. In this case, the precursor solution is put onto a bar with a small cylindrical tube. As the substrate moves below the cylindrical bar, a wet film is formed due to the solution passing through the cylinder's wire gap [90]. **Figure 7** depicts a schematic diagram of the coating procedure. After annealing, the deposited solution is converted into the ultimate film. The deposited film thickness and quality is directly proportional to the amount of solution passing through the wire gap of the cylinder and the type of solvent used to prepare the

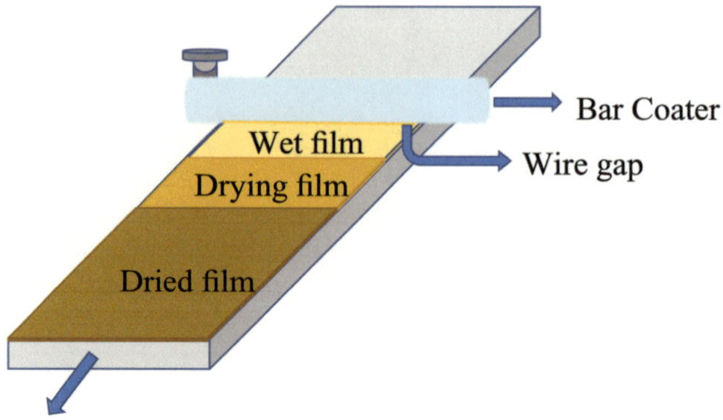

Figure 7.
Schematic diagram of Bar coating process.

solution in this technique, providing for high repeatability and minimal precursor solution loss [91].

3.4 Spray coating

Spray coating is a low-temperature coating technology that is advanced. A nozzle sprays the tiny solution droplets over the pre-heated substrate at a high speed in this process. Pneumatic spraying, which provides the solution droplet as a rapid gas flow, is the most commonly used spray coater. The spray coating process consists of four steps: creation of the droplet at the nozzle, transportation of the droplet to the substrate, coalescence of the droplet on the substrate, and drying [92]. The droplets are formed through the nozzle in the first step, which is known as atomization. The substrate is ready for the annealing procedure to obtain the ultimate film after the droplet coalescence on the wet film. **Figure 8** depicts a schematic diagram of the overall coating process.

To achieve thorough wetting of the substrate, the solvent utilised in this approach must have a low surface tension and contact angle. Another way to produce a

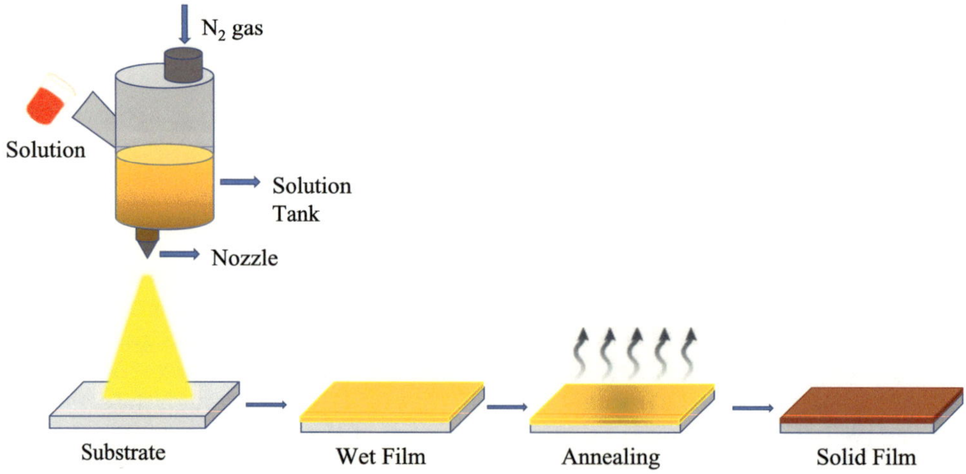

Figure 8.
Schematic diagram of spray coating process.

completely wet substrate is to use a pre-heated substrate, which lowers the surface tension and reduces the contact angle between the solution and the substrate [93]. Furthermore, the type of nozzle utilised, the pressure of the gas jet propeller, the distance between the nozzle and the substrate, and the temperature of the pre-heated substrate all have a significance in obtaining a uniform deposited film. An ultrasonic spray coater was utilised to create a high-quality smooth perovskite coating with homogeneous droplets.

3.5 Ink-jet printing

Ink-jet printing is a common method for fabricating electrical devices, particularly optoelectronics. This approach has a cost advantage over others due to its maskless on demand printing and, more crucially, contactless high-resolution printing. The printing procedure operates in the same way as spray coating. A piezo electronic transducer and a detector are also used to control the droplet size and trajectory, enabling this system to deposit material with precise patterning [94]. This technique is divided into two groups based on how ink droplets are emitted. Continuous ink-jet printing (CIP) and drop-on-demand ink-jet printing is two of them (DOD).

As the name implies, in continuous ink-jet printing, the solution droplet flows continuously towards the substrate under the influence of gravity. When the droplets fall from the nozzle, they acquire up an electric charge. The charged droplets are subsequently sent through a deflection coil, which directs them. A tiny voltage is applied between the nozzle and the ground to achieve this. **Figure 9(a)** depicts a schematic of the complete printing process. In this method, a piezoelectric transducer (PZT) provides an appropriate frequency for regular separation of the droplets, and the separation force involved is surface tension. When no printing is necessary, this approach also recycles the depositing material by collecting it in a reservoir [95]. Aside from that, because it is a non-contact printing technique, it enables excellent film deposition on both rough and curved substrates.

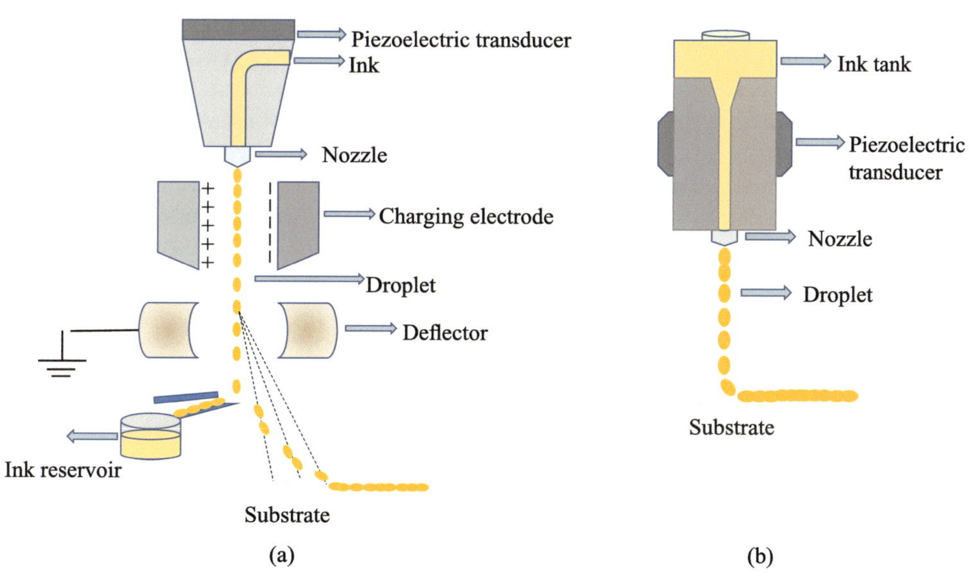

Figure 9.
(a) Schematic diagram of CIP process, (b) schematic diagram of DOD printing process.

Again, DOD printing is a modern, high-precision printing technology that can print with a single drop precession, allowing it to save a lot of material that would otherwise be wasted throughout the CIP process which is depicted in the **Figure 9(b)**. In this technology, a computer programme controls the movement of either the printing head or the substrate. Due to the contraction of the ink contain volume, the printing material is ejected out the nozzle at a definite pressure pulse. The pressure pulse production method is split into two parts: piezoelectric DOD and thermal DOD.

In piezoelectric DOD, an ink droplet is created by passing an impulse current over the transducer, causing PZT to deform mechanically. The majority of the printing industry employs this printing process because it allows for variable actuation pulses to adjust the velocity and size of ink droplets released from the nozzle. The thermal DOD, on the other hand, uses thermal evaporation to create the ink droplet. This is accomplished by passing electricity through the small resistive heater. When the temperature rises above the boiling point of the printing ink, the vapour entrapment causes bubbles to form. When the heater's power is turned off, the bubble begins to collapse because of heat transfer to the surrounding tank depending on the temperature difference [96]. Due to the difficulty in generating ink bubbles for high vapour pressure solution ink, this approach is not appropriate for large-area printing.

3.6 Screen printing

The screen-printing process is used to print a pattern that has previously been created on a thread or steel mesh. When the ink has a high viscosity, this approach provides the best printing pattern. The printing ink is spread over the patterned mask with a squeegee, which prints the pattern on the substrate. Due to the high viscosity of the printing fluid, this method produces a somewhat thick film [97]. **Figure 10** depicts a schematic diagram of the printing process. Depending on the printing technique, this printing technology is divided into two categories: flatbed and rotary screen printing.

The printing is done in a stepwise manner in the flatbed method, with the screen held extremely close to the top of the substrate and the paste transferred over the screen by the squeegee. The screen is then lifted or transferred to continue the printing process over the entire substrate after the printing is completed. For roll to roll and large-area printing, the recurrence of this procedure is not suitable. However, rotary screen printing is a low-cost, high-efficiency large-area printing process for generating rapid, precise patterns. The squeegee and paste are stored in a folded tube in this way. The stationary squeegee constantly spreads the paste across the substrate

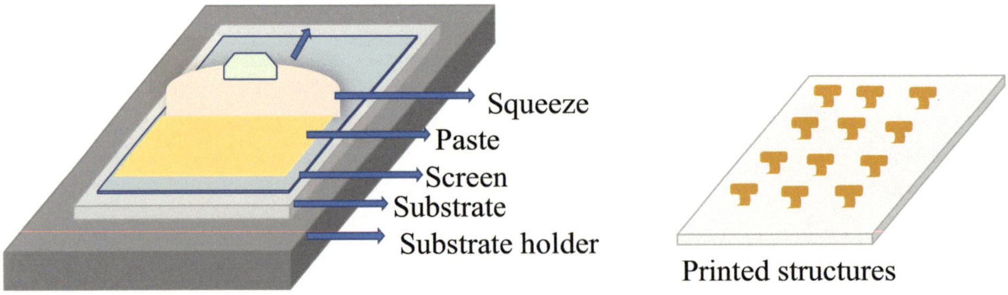

Figure 10.
Schematic diagram of screen-printing process and printed patterns.

Methods	Device architecture	Device area (cm^2)	PCE, (%)	Ref.
Blade coating	ITO/PTAA/$MAPbI_3$/C60/BCP/Cu	57.2	14.6	[99]
	ITO/PTAA/$MAPbI_3$/C60/BCP/Cu	63.7	16.40	[100]
	FTO/TiO_2/$MAPbI_3$/Spiro-OMeTAD/Au	47	14.7	[101]
Slot-die coating	FTO/NiMgLiO/$FA_{0.83}Cs_{0.17}PbI_{2.83}Br_{0.17}$/LiF/C60/BCP/Bi/Ag	20.77	16.63	[102]
	FTO/SnO_2/ $Cs_{0.15}FA_{0.85}Pb(I_{0.83}\,Br_{0.17})_3$/Spiro-OMeTAD/Au	57.5	16.22	[103]
	ITO/c-TiO_2/$CH_3NH_3PbI_{3-x}Cl_x$/ Spiro-OMeTAD/Au	168.75	10	[104]
Bar coating	FTO/SnO_2/$GA_xMA_{1-x}PbI_3$/spiro-OMeTAD/Au	16	13.85	[105]
	FTO/SnO_2/$(FAPbI_3)_{0.875}(CsPbBr_3)_{0.125}$/spiro-OMeTAD/Au	25	17.01	[106]
	FTO/SnO_2/$(FAPbI_3)_{0.95}(CsPbBr_3)_{0.05}$/Spiro-OMeTAD/Au	19.69	17.94	[107]
Spray coating	ITO/PEDOT:PSS/$CH_3NH_3PbI_3$/C60/BCP/Cu	56.25	16.4	[108]
	FTO/SnO_2/C60/$Cs_xFA_{1-x}PbI_yBr_{3-y}$/Spiro-OMeTAD/Au	25	14.7	[109]
	FTO/TiO_2/$MAPbI_{3-X}Cl_X$/PTAA/Au	40	15.5	[110]
Ink-jet printing	PEN/Ag(NWs)/PEDOT:PSS/$CH_3NH_3PbI_3$/$PC_{71}BM$/PEI/Ag(NWs)	180	10.68	[111]
	FTO/TiO_2/$MAPbI_3$/Spiro-OMeTAD/Au	4	13.27	[112]
Screen printing	FTO/TiO_2/ZrO_2/$CH_3NH_3PbI_3$/carbon	70	10.74	[113]
	FTO/TiO_2/ZrO_2/$5PVA_XMA_{1-X}PbI_3$/carbon	49	10.4	[114]

Table 4.
A comparison of the performance of several perovskite large area modules prepared utilising various large scale techniques.

through the mesh as the tubular screen rotates with the substrate, allowing for complete printing [98].

Table 4 shows some of the best results obtained by all of the large-scale deposition processes.

Not only these methods are utilised to fabricate active perovskite layers, but they also provide high-quality deposition of all transport layers, including metal electrodes. However, the type of solvent employed in the precursor solution, the solute concentration, printing speed, surface tension, and viscosity of the solution ink, among other factors, all play key role in achieving high-quality pin hole-free morphological film deposition.

4. Conclusion and prospects

PSCs have been a front runner in competition towards the commercially available photovoltaic technologies. Despite the extensive effort and significant progress in the large area fabrication of perovskite based photovoltaic, the large-area efficiency still lags behind those of small area devices. One of the main reasons for this gap between large and small area PSCs is the difficulty in large area processing of uniform and high-quality perovskite thin-films. This chapter covers all the solution processing techniques used in the fabrication of perovskite solar cells including the large-area coating techniques. The role of each technique is achieving homogeneous, pinhole-free and large grain-sized perovskite thin-films have been explained. The understanding the crystallisation kinetics of perovskite in one or more of these techniques will help in developing commercially viable large-area perovskite based solar modules.

Conflict of interests

There are no conflicts to declare.

Author details

Mayur Jagdishbhai Patel[1], Himangshu Baishya[2], Ritesh Kant Gupta[2], Rabindranath Garai[1] and Parameswar Krishnan Iyer[1,2]*

1 Department of Chemistry, Indian Institute of Technology Guwahati, Guwahati, Assam, India

2 Centre for Nanotechnology, Indian Institute of Technology Guwahati, Guwahati, Assam, India

*Address all correspondence to: pki@iitg.ac.in

References

[1] Yan J, Saunders BR. Third-generation solar cells: A review and comparison of polymer: Fullerene, hybrid polymer and perovskite solar cells. RSC Advances. 2014;**4**(82):43286-43314. DOI: 10.1039/c4ra07064j

[2] Krebs FC. Fabrication and processing of polymer solar cells: A review of printing and coating techniques. Solar Energy Materials and Solar Cells. 2009;**93**(4):394-412. DOI: 10.1016/j.solmat.2008.10.004

[3] Gupta RK, Garai R, Afroz MA, Iyer PK. Regulating active layer thickness and morphology for high performance hot-casted polymer solar cells. Journal of Materials Chemistry C. 2020;**8**(24):8191-8198. DOI: 10.1039/d0tc00822b

[4] Garai R, Gupta RK, Hossain M, Iyer PK. Surface recrystallized stable 2D–3D graded perovskite solar cells for efficiency beyond 21%. Journal of Materials Chemistry A. 2021;**9**(46):26069-26076. DOI: 10.1039/d1ta06901b

[5] Hailin Z, Ji X, Yao H, Fan Q, Yu B, Li J. Review on efficiency improvement effort of perovskite solar cell. Solar Energy (Phoenix, Ariz.). 2022;**233**:421-434. DOI: 10.1016/j.solener.2022.01.060

[6] Teixeira CO, Castro D, Andrade L, Mendes A. Selection of the ultimate perovskite solar cell materials and fabrication processes towards its industrialization: A review. Energy Science & Engineering. 2022;**10**(4): 1478-1525. DOI: 10.1002/ese3.1084

[7] Murugan P, Hu T, Hu X, Chen Y. Current development toward commercialization of metal-Halide perovskite photovoltaics. Advanced Optical Materials. 2021;**9**(17):2100390. DOI: 10.1002/adom.202100390

[8] Jeon NJ, Noh JH, Kim YC, Yang WS, Ryu S, Seok SI. Solvent engineering for high-performance inorganic-organic hybrid perovskite solar cells. Nature Materials. 2014;**13**(9):897-903. DOI: 10.1038/nmat4014

[9] Park N-G. Methodologies for high efficiency perovskite solar cells. Nano Convergence. 2016;**3**(1):15. DOI: 10.1186/s40580-016-0074-x

[10] Liu M, Johnston MB, Snaith HJ. Efficient planar heterojunction perovskite solar cells by vapour deposition. Nature. 2013;**501**(7467):395-398. DOI: 10.1038/nature12509

[11] Jeon NJ, Noh JH, Yang WS, Kim YC, Ryu S, Seo J, et al. Compositional engineering of perovskite materials for high-performance solar cells. Nature. 2015;**517**(7535):476-480. DOI: 10.1038/nature14133

[12] Wang L-Y, Deng L-L, Wang X, Wang T, Liu H-R, Dai S-M, et al. Di-isopropyl ether assisted crystallization of organic-inorganic perovskites for efficient and reproducible perovskite solar cells. Nanoscale. 2017;**9**(45):17893-17901. DOI: 10.1039/c7nr06410a

[13] Zheng X, Chen B, Wu C, Priya S. Room temperature fabrication of CH3NH3PbBr3 by anti-solvent assisted crystallization approach for perovskite solar cells with fast response and small J–V hysteresis. Nano Energy. 2015;**17**:269-278. DOI: 10.1016/j.nanoen.2015.08.023

[14] Liao W, Zhao D, Yu Y, Grice CR, Wang C, Cimaroli AJ, et al. Lead-free

inverted planar formamidinium tin triiodide perovskite solar cells achieving power conversion efficiencies up to 6.22. Advanced Materials. 2016;**28**(42):9333-9340. DOI: 10.1002/adma.201602992

[15] McDonald C, Ni C, Maguire P, Connor P, Irvine JTS, Mariotti D, et al. Nanostructured perovskite solar cells. Nanomaterials. 2019;**9**(10):1481. DOI: 10.3390/nano9101481

[16] Karuppuswamy P, Boopathi KM, Mohapatra A, Chen HC, Wong KT, Wang PC, et al. Role of a hydrophobic scaffold in controlling the crystallization of methylammonium antimony iodide for efficient lead-free perovskite solar cells. Nano Energy. 2018;**45**:330-336. DOI: 10.1016/j.nanoen.2017.12.051

[17] Paek S, Schouwink P, Athanasopoulou EN, Cho KT, Grancini G, Lee Y, et al. From nano- to micrometer scale: The role of antisolvent treatment on high performance perovskite solar cells. Chemistry of Materials. 2017;**29**(8):3490-3498. DOI: 10.1021/acs.chemmater.6b05353

[18] Liu X, Yan K, Tan D, Liang X, Zhang H, Huang W. Solvent engineering improves efficiency of lead-free tin-based hybrid perovskite solar cells beyond 9%. ACS Energy Letters. 2018;**3**(11):2701-2707. DOI: 10.1021/acsenergylett.8b01588

[19] Liu J, Ozaki M, Yakumaru S, Handa T, Nishikubo R, Kanemitsu Y, et al. Lead-free solar cells based on tin Halide perovskite films with high coverage and improved aggregation. Angewandte Chemie (International Ed. in English). 2018;**57**(40):13221-13225. DOI: 10.1002/anie.201808385

[20] Lee K-M, Lin C-J, Liou B-Y, Yu S-M, Hsu C-C, Suryanarayanan V. Effect of anti-solvent mixture on the performance of perovskite solar cells and suppression hysteresis behavior. Organic Electronics. 2019;**65**:266-274. DOI: 10.1016/j.orgel.2018.08.048

[21] Nie Z, Yin J, Zhou H, Chai N, Chen B, Zhang Y, et al. Layered and Pb-free organic-inorganic perovskite materials for ultraviolet photoresponse: (010)-oriented $(CH_3NH_3)_2MnCl_4$ thin film. ACS Applied Materials & Interfaces. 2016;**8**(41):28187-28193. DOI: 10.1021/acsami.6b08962

[22] Xiao M, Huang F, Huang W, Dkhissi Y, Zhu Y, Etheridge J, et al. A fast deposition-crystallization procedure for highly efficient lead iodide perovskite thin-film solar cells. Angewandte Chemie. 2014;**53**(37):9898-9903. DOI: 10.1002/anie.201405334

[23] Jiang J, Yang X, Huang Y, Li M, Tao Q, Fei M, et al. Improvement in solar cell efficiency based on the MAPbI3 films extracted by a mixed anti-solvent. Applied Physics Letters. 2020;**117**(20):203901. DOI: 10.1063/5.0019360

[24] Cao X, Zhi L, Li Y, Fang F, Cui X, Ci L, et al. Fabrication of perovskite films with large columnar grains via solvent-mediated Ostwald ripening for efficient inverted perovskite solar cells. ACS Applied Energy Materials. 2018;**1**(2):868-875. DOI: 10.1021/acsaem.7b00300

[25] Ciro J, Betancur R, Mesa S, Jaramillo F. High performance perovskite solar cells fabricated under high relative humidity conditions. Solar Energy Materials and Solar Cells. 2017;**163**:38-42. DOI: 10.1016/j.solmat.2017.01.004

[26] Yang F, Kamarudin MA, Zhang P, Kapil G, Ma T, Hayase S. Enhanced crystallization by methanol additive in antisolvent for achieving high-quality $MAPbI_3$ perovskite films

in humid atmosphere. ChemSusChem. 2018;**11**(14):2348-2357. DOI: 10.1002/cssc.201800625

[27] Gedamu D, Asuo IM, Benetti D, Bast M, Ka I, Cloutier SG, et al. Solvent-antisolvent ambient processed large grain size perovskite thin films for high-performance solar cells. Scientific Reports. 2018;**8**(1):12885. DOI: 10.1038/s41598-018-31184-0

[28] Ahn N, Son D-Y, Jang I-H, Kang SM, Choi M, Park N-G. Highly reproducible perovskite solar cells with average efficiency of 18.3% and best efficiency of 19.7% fabricated via Lewis base adduct of lead(II) iodide. Journal of the American Chemical Society. 2015;**137**(27):8696-8699. DOI: 10.1021/jacs.5b04930

[29] Soe CMM, Nie W, Stoumpos CC, Tsai H, Blancon J-C, Liu F, et al. Understanding film formation morphology and orientation in high member 2D ruddlesden–popper perovskites for high-efficiency solar cells. Advanced Energy Materials. 2018;**8**(1):1700979. DOI: 10.1002/aenm.201700979

[30] Liao K, Li C, Xie L, Yuan Y, Wang S, Cao Z, et al. Hot-casting large-grain perovskite film for efficient solar cells: Film formation and device performance. Nano-Micro Letters. 2020;**12**(1):156. DOI: 10.1007/s40820-020-00494-2

[31] Peng ZA, Peng X. Nearly monodisperse and shape-controlled CdSe nanocrystals via alternative routes: Nucleation and growth. Journal of the American Chemical Society. 2002;**124**(13):3343-3353. DOI: 10.1021/ja0173167

[32] Mulheran PA, Blackman JA. The origins of island size scaling in heterogeneous film growth. Philosophical Magazine Letters. 1995;**72**(1):55-60. DOI: 10.1080/09500839508241614

[33] Yu HZ, Thompson CV. Grain growth and complex stress evolution during Volmer–Weber growth of polycrystalline thin films. Acta Materialia. 2014;**67**:189-198. DOI: 10.1016/j.actamat.2013.12.031

[34] AnayaM GLJF, Calvo ME, López C, Míguez H. Photophysical analysis of the formation of organic-inorganic trihalide perovskite films: Identification and characterization of crystal nucleation and growth. The Journal of Physical Chemistry. C, Nanomaterials and Interfaces. 2016;**120**(5):3071-3076. DOI: 10.1021/acs.jpcc.6b00398

[35] Nie W, Tsai H, Asadpour R, Blancon JC, Neukirch AJ, Gupta G, et al. Solar cells. High-efficiency solution-processed perovskite solar cells with millimeter-scale grains. Science (New York, N.Y.). 2015;**347**(6221):522-525. DOI: 10.1126/science.aaa0472

[36] Wang S, Wang L, Zhang L, Chang L, Wang L, Wang J. Construction of high performance CH3NH3PbI3-based solar cells by hot-casting technique. Solar Energy Materials and Solar Cells. 2017;**163**:120-124. DOI: 10.1016/j.solmat.2017.01.009

[37] Afroz MA, Gupta RK, Garai R, Hossain M, Tripathi SP, Iyer PK. Crystallization and grain growth regulation through Lewis acid-base adduct formation in hot cast perovskite-based solar cells. Organic Electronics. 2019;**74**:172-178. DOI: 10.1016/j.orgel.2019.07.007

[38] Liu T, Zhou Y, Hu Q, Chen K, Zhang Y, Yang W, et al. Fabrication of compact and stable perovskite films with optimized precursor composition in the fast-growing procedure. Science China Materials. 2017;**60**(7):608-616. DOI: 10.1007/s40843-017-9044-y

[39] Aguiar JA, Alkurd NR, Wozny S, Patel MK, Yang M, Zhou W, et al. In situ

investigation of halide incorporation into perovskite solar cells. MRS Communications. 2017;**7**(3):575-582. DOI: 10.1557/mrc.2017.52

[40] Wang Z, Liu X, Lin Y, Liao Y, Wei Q, Chen H, et al. Hot-substrate deposition of all-inorganic perovskite films for low-temperature processed high-efficiency solar cells. Journal of Materials Chemistry A. 2019;**7**(6):2773-2779. DOI: doi.org/10.1039/c8ta09855g

[41] Ghosh B, Wu B, Guo X, Harikesh PC, John RA, Baikie T, et al. Superior performance of silver bismuth iodide photovoltaics fabricated via dynamic hot-casting method under ambient conditions. Advanced Energy Materials. 2018;**8**(33):1802051. DOI: 10.1002/aenm.201802051

[42] Tsai H, Nie W, Blancon JC, Stoumpos CC, Asadpour R, Harutyunyan B, et al. High-efficiency two-dimensional Ruddlesden-Popper perovskite solar cells. Nature. 2016;**536**(7616):312-316. DOI: 10.1038/nature18306

[43] Zhao W, Gan X, Ke L, Guo L, Liu H. 2D multilayered perovskites based on 4-chlorophenylethylamine for solar cell application. Solar Energy. 2020;**196**:1-9. DOI: 10.1016/j.solener.2019.12.021

[44] Zhu X, Xu Z, Zuo S, Feng J, Wang Z, Zhang X, et al. Vapor-fumigation for record efficiency two-dimensional perovskite solar cells with superior stability. Energy & Environmental Science. 2018;**11**(12):3349-3357. DOI: 10.1039/c8ee02284d

[45] Zhang Y, Wang P, Tang MC, Barrit D, Ke W, Liu J, et al. Dynamical transformation of two-dimensional perovskites with alternating cations in the interlayer space for high-performance photovoltaics. Journal of the American Chemical Society. 2019;**141**(6):2684-2694. DOI: 10.1021/jacs.8b13104

[46] Zheng YC, Yang S, Chen X, Chen Y, Hou Y, Yang HG. Thermal-induced Volmer–Weber growth behavior for planar heterojunction perovskites solar cells. Chemistry of Materials: A Publication of the American Chemical Society. 2015;**27**(14):5116-5121. DOI: 10.1021/acs.chemmater.5b01924

[47] Zadeh NJ, Zarandi MB, Nateghi MR. Effect of crystallization strategies on CH3NH3PbI3 perovskite layer deposited by spin coating method: Dependence of photovoltaic performance on morphology evolution. Thin Solid Films. 2018;**660**:65-74. DOI: 10.1016/j.tsf.2018.03.038

[48] Sanchez S, Christoph N, Grobety B, Phung N, Steiner U, Saliba M, et al. Efficient and stable inorganic perovskite solar cells manufactured by pulsed flash infrared annealing. Advanced Energy Materials. 2018;**8**(30):1802060. DOI: 10.1002/aenm.201802060

[49] Kulkarni SA, Baikie T, Boix PP, Yantara N, Mathews N, Mhaisalkar S. Band-gap tuning of lead halide perovskites using a sequential deposition process. Journal of Materials Chemistry. A, Materials for Energy and Sustainability. 2014;**2**(24):9221-9225. DOI: 10.1039/c4ta00435c

[50] Tsai H, Nie W, Lin YH, Blancon JC, Tretiak S, Even J, et al. Effect of precursor solution aging on the crystallinity and photovoltaic performance of perovskite solar cells. Advanced Energy Materials. 2017;**7**(11):1602159. DOI: 10.1002/aenm.201602159

[51] Liu Z, Hu J, Jiao H, Li L, Zheng G, Chen Y, et al. Chemical reduction of intrinsic defects in thicker heterojunction

planar perovskite solar cells. Advanced Materials (Deerfield Beach, Fla.). 2017;**29**(23):1606774. DOI: 10.1002/adma.201606774

[52] Shin GS, Choi WG, Na S, Gökdemir FP, Moon T. Lead acetate based hybrid perovskite through hot casting for planar heterojunction solar cells. Electronic Materials Letters. 2018;**14**(2):155-160. DOI: 10.1007/s13391-018-0042-1

[53] Chao L, Xia Y, Li B, Xing G, Chen Y, Huang W. Room-temperature molten salt for facile fabrication of efficient and stable perovskite solar cells in ambient air. Chem. 2019;**5**(4):995-1006. DOI: 10.1016/j.chempr.2019.02.025

[54] Christians JA, Miranda Herrera PA, Kamat PV. Transformation of the excited state and photovoltaic efficiency of CH3NH3PbI3 perovskite upon controlled exposure to humidified air. Journal of the American Chemical Society. 2015;**137**(4):1530-1538. DOI: 10.1021/ja511132a

[55] Lei B, Eze VO, Mori T. High-performance CH3NH3PbI3 perovskite solar cells fabricated under ambient conditions with high relative humidity. Japanese Journal of Applied Physics. 2015;**54**(10):100305. DOI: 10.7567/jjap.54.100305

[56] Li C, Wang A, Xie L, Deng X, Liao K, Yang JA, et al. Secondary lateral growth of MAPbI3 grains for the fabrication of efficient perovskite solar cells. Journal of Materials Chemistry. C, Materials for Optical and Electronic Devices. 2020;**8**(9):3217-3225. DOI: 10.1039/c9tc06293a

[57] Bag S, Durstock MF. Large perovskite grain growth in low-temperature solution-processed planar p-i-n solar cells by sodium addition. ACS Applied Materials & Interfaces. 2020;**8**(8):5053-5057. DOI: 10.1021/acsami.5b11494

[58] Muscarella LA, Hutter EM, Sanchez S, Dieleman CD, Savenije TJ, Hagfeldt A, et al. Crystal orientation and grain size: Do they determine optoelectronic properties of MAPbI3 perovskite? The Journal of Physical Chemistry Letters. 2019;**10**(20):6010-6018. DOI: 10.1021/acs.jpclett.9b02757

[59] Krishna A, Gottis S, Nazeeruddin MK, Sauvage F. Mixed dimensional 2D/3D hybrid perovskite absorbers: The future of perovskite solar cells? Advanced Functional Materials. 2019;**29**(8):1806482. DOI: 10.1002/adfm.201806482

[60] Chen S, Shi G. Two-dimensional materials for Halide perovskite-based optoelectronic devices. Advanced Materials (Deerfield Beach, Fla.). 2017;**29**(24). DOI: 10.1002/adma.201605448

[61] Correa Baena JP, Saliba M, Buonassisi T, Grätzel M, Abate A, Tress W, et al. Promises and challenges of perovskite solar cells. Science (New York, N.Y.). 2017;**358**(6364):739-744. DOI: 10.1126/science.aam6323

[62] Liao HC, Guo P, Hsu CP, Lin M, Wang B, Zeng L, et al. Enhanced efficiency of hot-cast large-area planar perovskite solar cells/modules having controlled chloride incorporation. Advanced Energy Materials. 2017;**7**(8):1601660. DOI: 10.1002/aenm.201601660

[63] Smith IC, Hoke ET, Solis-Ibarra D, McGehee MD, Karunadasa HI. A layered hybrid perovskite solar-cell absorber with enhanced moisture stability. Angewandte Chemie (International Ed. in English). 2014;**53**(42):11232-11235. DOI: 10.1002/anie.201406466

[64] Lin Y, Bai Y, Fang Y, Chen Z, Yang S, Zheng X, et al. Enhanced thermal stability in perovskite solar cells by assembling 2D/3D stacking structures. The Journal of Physical Chemistry Letters. 2018;**9**(3):654-658. DOI: 10.1021/acs.jpclett.7b02679

[65] Li L, Zhou N, Chen Q, Shang Q, Zhang Q, Wang X, et al. Unraveling the growth of hierarchical quasi-2D/3D perovskite and carrier dynamics. The Journal of Physical Chemistry Letters. 2018;**9**(5):1124-1132. DOI: 10.1021/acs.jpclett.7b03294

[66] Liang K, Mitzi DB, Prikas MT. Synthesis and characterization of organic–inorganic perovskite thin films prepared using a versatile two-step dipping technique. Chemistry of Materials. 1998:403-411. DOI: 10.1021/cm970568f

[67] Zhang H, Mao J, He H, Zhang D, Zhu HL, Xie F, et al. A smooth CH3NH3PbI3 Film via a new approach for forming the PbI2 Nanostructure together with strategically high CH3NH3I concentration for high efficient planar-heterojunction solar cells. Advanced Energy Materials. 2015;**5**(23):1501354. DOI: 10.1002/aenm.201501354

[68] Fu Y, Meng F, Rowley MB, Thompson BJ, Shearer MJ, Ma D, et al. Solution growth of single crystal methylammonium lead halide perovskite nanostructures for optoelectronic and photovoltaic applications. Journal of the American Chemical Society. 2015;**137**(17):5810-5818. DOI: 10.1021/jacs.5b02651

[69] Yang S, Zheng YC, Hou Y, Chen X, Chen Y, Wang Y, et al. Formation mechanism of freestanding CH3NH3PbI3 functional crystals: In situ transformation vs dissolution–crystallization. Chemistry of Materials. 2014;**26**(23):6705-6710. DOI: 10.1021/cm5028817

[70] Im JH, Jang IH, Pellet N, Grätzel M, Park NG. Growth of CH3NH3PbI3 cuboids with controlled size for high-efficiency perovskite solar cells. Nature Nanotechnology. 2014;**9**(11):927-932. DOI: 10.1038/nnano.2014.181

[71] Xiao Z, Bi C, Shao Y, Dong Q, Wang Q, Yuan Y, et al. Efficient, high yield perovskite photovoltaic devices grown by interdiffusion of solution-processed precursor stacking layers. Energy & Environmental Science. 2014;**7**(8):2619-2623. DOI: 10.1039/c4ee01138d

[72] Chauhan M, Zhong Y, Schötz K, Tripathi B, Köhler A, Huettner S, et al. Investigating two-step MAPbI3 thin film formation during spin coating by simultaneous in situ absorption and photoluminescence spectroscopy. Journal of Materials Chemistry A, Materials for Energy and Sustainability. 2020;**8**(10):5086-5094. DOI: 10.1039/c9ta12409h

[73] Chen H. Two-step sequential deposition of organometal Halide perovskite for photovoltaic application. Advanced Functional Materials. 2017, 2017;**27**(8):1605654. DOI: 10.1002/adfm.201605654

[74] Wu Y, Islam A, Yang X, Qin C, Liu J, Zhang K, et al. Retarding the crystallization of PbI2for highly reproducible planar-structured perovskite solar cells via sequential deposition. Energy & Environmental Science. 2014;**7**(9):2934-2938. DOI: /10.1039/c4ee01624f

[75] Li J, Zhu Z, Han X, Yu T, Xu Y, Feng J, et al. Direct molecule substitution enabled rapid transformation of wet $PbBr_2$(DMF) precursor films to $CsPbBr_3$ perovskite. ACS Applied

Energy Materials. 2021;**4**(7):6414-6421. DOI: /10.1021/acsaem.1c00154

[76] Burschka J, Pellet N, Moon SJ, Humphry Baker R, Gao P, Nazeeruddin MK, et al. Sequential deposition as a route to high-performance perovskite-sensitized solar cells. Nature. 2013;**499**(7458):316-319. DOI: /10.1038/nature12340

[77] Cao XB, Li YH, Fang F, Cui X, Yao YW, Wei JQ. High quality perovskite films fabricated from Lewis acid–base adduct through molecular exchange. RSC Advances. 2016;**6**(75):70925-70931. DOI: /10.1039/c6ra15378j

[78] Liu X, Wu Z, Fu X, Tang L, Li J, Gong J, et al. Highly efficient wide-band-gap perovskite solar cells fabricated by sequential deposition method. Nano Energy. 2021;**86**(106114):106114. DOI: /10.1016/j.nanoen.2021.106114

[79] Lee DG, Kim DH, Lee JM, Kim BJ, Kim JY, Shin SS, et al. High efficiency perovskite solar cells exceeding 22% via a photo-assisted two-step sequential deposition. Advanced Functional Materials. 2021;**31**(9):2006718. DOI: /10.1002/adfm.202006718

[80] Xu G, Xue R, Chen W, Zhang J, Zhang M, Chen H, et al. New strategy for two-step sequential deposition: Incorporation of hydrophilic fullerene in second precursor for high-performance p-i-n planar perovskite solar cells. Advanced Energy Materials. 2018;**8**(12):1703054. DOI: /10.1002/aenm.201703054

[81] Li Q, Zhao Y, Fu R, Zhou W, Zhao Y, Liu X, et al. Efficient perovskite solar cells fabricated through CsCl-enhanced PbI2 precursor via sequential deposition. Advanced Materials (Deerfield Beach, Fla.). 2018;**30**(40):e1803095. DOI: /10.1002/adma.201803095

[82] Wu J, Xu X, Zhao Y, Shi J, Xu Y, Luo Y, et al. DMF as an additive in a two-step spin-coating method for 20% conversion efficiency in perovskite solar cells. ACS Applied Materials & Interfaces. 2017;**9**(32):26937-26947. DOI: /10.1021/acsami.7b08504

[83] Roy P, Kumar SN, Tiwari S, Khare A. A review on perovskite solar cells: Evolution of architecture, fabrication techniques, commercialization issues and status. Solar Energy (Phoenix, Ariz.). 2020;**198**:665-688. DOI: 10.1016/j.solener.2020.01.080

[84] Kim JH, Williams ST, Cho N, Chueh CC, Jen AKY. Enhanced environmental stability of planar heterojunction perovskite solar cells based on blade-coating. Advanced Energy Materials. 2015;5(4):1401229. DOI: 10.1002/aenm.201401229

[85] Yang Z, Chueh CC, Zuo F, Kim JH, Liang PW, Jen AKY. High-performance fully printable perovskite solar cells via blade-coating technique under the ambient condition. Advanced Energy Materials. 2015;**5**(13):1500328. DOI: 10.1002/aenm.201500328

[86] Le BM, Chen Y, Baigl D. From convective assembly to Landau-Levich deposition of multilayered phospholipid films of controlled thickness. Langmuir: The ACS Journal of Surfaces and Colloids. 2009;**25**(5):2554-2557. DOI: 10.1021/la803646e

[87] Fanton X, Cazabat AM. Spreading and instabilities induced by a solutal Marangoni effect. Langmuir: The ACS Journal of Surfaces and Colloids. 1998;**14**(9):2554-2561. DOI: 10.1021/la971292t

[88] Ruschak KJ. Limiting flow in a pre-metered coating device. Chemical Engineering

Science. 1976;**31**(11):1057-1060. DOI: 10.1016/0009-2509(76)87026-1

[89] Carvalho MS, Kheshgi HS. Low-flow limit in slot coating: Theory and experiments. American Institute of Chemical Engineers. 2000;**46**(10):1907-1917. DOI: 10.1002/aic.690461003

[90] Khim D, Han H, Baeg KJ, Kim J, Kwak SW, Kim DY, et al. Simple bar-coating process for large-area, high-performance organic field-effect transistors and ambipolar complementary integrated circuits. Advanced Materials (Deerfield Beach, Fla.). 2013;**25**(31):4302-4308. DOI: 10.1002/adma.201205330

[91] Jeong DN, Lee DK, Seo S, Lim SY, Zhang Y, Shin H, et al. Perovskite cluster-containing solution for scalable D-bar coating toward high-throughput perovskite solar cells. ACS Energy Letters. 2019;**4**(5):1189-1195. DOI: 10.1021/acsenergylett.9b00042

[92] Bishop JE, Routledge TJ, Lidzey DG. Advances in spray-cast perovskite solar cells. The Journal of Physical Chemistry Letters. 2018;**9**(8):1977-1984. DOI: 10.1021/acs.jpclett.8b00311

[93] Gittens GJ. Variation of surface tension of water with temperature. Journal of Colloid and Interface Science. 1969;**30**(3):406-412. DOI: 10.1016/0021-9797(69)90409-3

[94] Schackmar F, Eggers H, Frericks M, Richards BS, Lemmer U, Hernandez SG, et al. Perovskite solar cells with all-inkjet-printed absorber and charge transport layers. Advanced Materials Technologies. 2021;**6**(2):2000271. DOI: 10.1002/admt.202000271

[95] Shah MA, Lee DG, Lee BY, Hur S. Classifications and applications of inkjet printing technology: A review. IEEE Access: Practical Innovations, Open Solutions. 2021;**9**:140079-140102. DOI: 10.1109/access.2021.3119219

[96] Cummins G, Desmulliez MPY. Inkjet printing of conductive materials: A review. Circuit World. 2012;**38**(4):193-213. DOI: 10.1108/03056121211280413

[97] Rong Y, Ming Y, Ji W, Li D, Mei A, Hu Y, et al. Toward industrial-scale production of perovskite solar cells: Screen printing, slot-die coating, and emerging techniques. The Journal of Physical Chemistry Letters. 2018;**9**(10):2707-2713. DOI: 10.1021/acs.jpclett.8b00912

[98] Horwood RJ. Towards a better understanding of screen print thickness control. Electrocomponent Science and Technology. 1974;**1**(2):129-136. DOI: 10.1155/apec.1.129

[99] Deng Y, Zheng X, Bai Y, Wang Q, Zhao J, Huang J. Surfactant-controlled ink drying enables high-speed deposition of perovskite films for efficient photovoltaic modules. Nature Energy. 2018;**3**(7):560-566. DOI: 10.1038/s41560-018-0153-9

[100] Deng Y, Van Brackle CH, Dai X, Zhao J, Chen B, Huang J. Tailoring solvent coordination for high-speed, room-temperature blading of perovskite photovoltaic films. Science. Advances. 2019;**5**(12):eaax7537. DOI: 10.1126/sciadv.aax7537

[101] Matteocci F, Vesce L, Kosasih FU, Castriotta LA, Cacovich S, Palma AL, et al. Fabrication and morphological characterization of high-efficiency blade-coated perovskite solar modules. ACS Applied Materials & Interfaces. 2019;**11**(28):25195-25204. DOI: 10.1021/acsami.9b05730

[102] Yang Z, Zhang W, Wu S, Zhu H, Liu Z, Liu Z, et al. Slot-die coating

large-area formamidinium-cesium perovskite film for efficient and stable parallel solar module. Science Advances. 2021;**7**(18). DOI: 10.1126/sciadv.abg3749

[103] Rana PJS, Febriansyah B, Koh TM, Muhammad BT, Salim T, Hooper TJN, et al. Alkali additives enable efficient large area (>55 cm^2) slot-die coated perovskite solar modules. Advanced Functional Materials. 2022:2113026. DOI: 10.1002/adfm.202113026

[104] Di Giacomo F, Shanmugam S, Fledderus H, Bruijnaers BJ, Verhees WJH, Dorenkamper MS, et al. Up-scalable sheet-to-sheet production of high efficiency perovskite module and solar cells on 6-in. substrate using slot die coating. Solar Energy Materials and Solar Cells. 2018;**181**:53-59. DOI: 10.1016/j.solmat.2017.11.010

[105] Lee DK, Jeong DN, Ahn TK, Park NG. Precursor engineering for a large-area perovskite solar cell with >19% efficiency. ACS Energy Letters. 2019;**4**(10):2393-2401. DOI: 10.1021/acsenergylett.9b01735

[106] Lim KS, Lee DK, Lee JW, Park NG. 17% efficient perovskite solar mini-module via hexamethylphosphoramide (HMPA)-adduct-based large-area D-bar coating. Journal of Materials Chemistry A. 2020;**8**(18):9345-9354. DOI: 10.1039/d0ta02017f

[107] Lee DK, Lim KS, Lee JW, Park NG. Scalable perovskite coating via anti-solvent-free Lewis acid–base adduct engineering for efficient perovskite solar modules. Journal of Materials Chemistry. A. 2021;**9**(5):3018-3028. DOI: 10.1039/d0ta10366g

[108] Park M, Cho W, Lee G, Hong SC, Kim MC, Yoon J, et al. Highly reproducible large-area perovskite solar cell fabrication via continuous megasonic spray coating of CH3 NH3 PbI3. Small. 2019;**15**(1):e1804005. DOI: 10.1002/smll.201804005

[109] Yu X, Yan X, Xiao J, Ku Z, Zhong J, Li W, et al. Interface modification effect on the performance of CsxFA1-xPbIyBr3-y perovskite solar cells fabricated by evaporation/spray-coating method. The Journal of Chemical Physics. 2020;**153**(1):014706. DOI: 10.1063/5.0012803

[110] Heo JH, Lee MH, Jang MH, Im SH. Highly efficient CH3NH3PbI3−xClx mixed halide perovskite solar cells prepared by re-dissolution and crystal grain growth via spray coating. Journal of Materials Chemistry A. 2016;**4**(45):17636-17642. DOI: 10.1039/c6ta06718b

[111] Gao B, Meng J. Flexible CH3NH3PbI3 perovskite solar cells with high stability based on all inkjet printing. Solar Energy (Phoenix, Ariz.). 2021;**230**:598-604. DOI: 10.1016/j.solener.2021.10.072

[112] Liang C, Li P, Gu H, Zhang Y, Li F, Song Y, et al. One-step inkjet printed perovskite in air for efficient light harvesting. Solar RRL. 2018;**2**(2):1700217. DOI: 10.1002/solr.201700217

[113] Priyadarshi A, Haur LJ, Murray P, Fu D, Kulkarni S, Xing G, et al. A large area (70 cm^2) monolithic perovskite solar module with a high efficiency and stability. Energy & Environmental Science. 2016;**9**(12):3687-3692. DOI: 10.1039/c6ee02693a

[114] Hu Y, Si S, Mei A, Rong Y, Liu H, Li X, et al. Stable large-area (10 × 10 cm^2) printable mesoscopic perovskite module exceeding 10% efficiency. Solar RRL. 2017;**1**(2):1600019. DOI: 10.1002/solr.201600019

Chapter 9

Solar Solutions for the Future

David M. Mulati and Timonah Soita

Abstract

The energy conversion efficiency and limits of perovskite/silicon solar cells are investigated. The influence of a layered approach in preventing lead leakage in perovskite solar cells is discussed. The highest efficiency perovskite tandem to date was achieved by pairing a perovskite top cell with a Si bottom cell in a four-terminal configuration, yielding 26.4%. Perovskite cell integrated with crystalline silicon cell to form a tandem solar device has shown high performance above the single pn-junction silicon devices. Although sufficient work and different strategies have been applied to increase efficiency in these devices, the tandem application has achieved efficiency of 29% in a short period.

Keywords: photovoltaics, conversion efficiency, perovskite solar cells and tandem solar cells

1. Introduction

The growing demand of energy, has forced researchers to look for cheap alternative sources of energy. Among these cheap sources energy under investigations, photovoltaics is one of them. The International Energy Agency (IEA) has noted declining prices of photovoltaics and estimates solar systems in general to supply 5% of global electricity consumption in 2030. These estimates are likely to rising to 16% by 2050. This can only be achieved by increasing the global production of solar energy to 4600 GW by 2050 [1]. At the same time solar and wind energy is expected to contribute up to 50% of total energy generated. Reduced costs of energy generation and improvements in performance will lead to more penetration of solar energy across the globe. These coupled with more research, will make solar energy cheaper than fossil fuel in the near future. Using innovative new designs, researchers are continuously working to improve photovoltaic energy systems. This work is geared towards reducing the power generation cost by combining several materials in photovoltaic cells. The best cell efficiency of 39.2% has been demonstrated from multi-junction solar cells. However, it should be observed that this is applied in space technologies that have complex fabrication processes. Currently the theoretical efficiency of crystalline silicon (c-Si) based solar cells is approximately 26.7% and for thin film technology it is ~23.4%. These solar cell technologies outlined above are mature, and already in production. However, the complexed of the production process and energy consumption needs to be reduced.

Currently inexpensive and easy to fabricate solar cells are under investigation by researchers. Most of these solar cell technologies are categorized as emerging PV

 IntechOpen

technologies. Examples include; organic solar cells, dye-sensitized solar cells, quantum dot solar cells; perovskite solar cells (PSC). PSC shows very promising results and can favorably compete with c-Si solar cells in terms of efficiency. Perovskite was discovered in the Ural Mountain, in Russia, and basically describes compounds having crystal structures like calcium titanium oxide ($CaTiO_3$). The materials being used for newly developed PSC have their structure in the form of ABX structure where A, and B are cations and X is the bonding anion. In comparison, with traditional Si cells PSC absorbs sunlight with a hundred times thinner active layer. This ABX structure allows for variation in the energy gap by mixing and matching scenario where we can have multiple compounds by substituting any of the constituents. Within a decade of its inception, PSC has already overtaken thin film technology such as cadmium telluride (CdTe) or copper-indium-gallium-selenide (CIGS) and reached the level of c-Si solar cells. Already an efficiency of 25.2% for single junction PSC has been realized by a team led by Prof. Michael Saliba at KRICT2 whereas the theoretical efficiency for these PSC's are about 31%. When Perovskite solar cell is in tandem with other solar cell technologies, they give varying efficiencies. CIGS-perovskite tandem cells have shown efficiencies up to 21.5%, whereas, when combined with c-Si technology the perovskite-c-Si tandem cells have shown efficiencies up to 28% that shows better performance in comparison with a single pn-junction of silicon; and also single junction of PSC. The efficiency of the perovskite/c-Si tandem solar cells are in the range of 32.8% for gallium-indium-phosphide/gallium arsenide (GaInP/GaAs) tandem solar cells. This type of solar cells is realized through intensified processes and expensive manufacturing techniques. With more research the efficiency of the Perovskite/silicon tandem cell can get towards 30%. Although PV technology is not the most widely used energy source; due to low energy conversion efficiency and high initial system cost; in comparison to non-renewable energy sources; it is a fast growing energy source in the power sector [2–4]. So far, solar modules from c-Si single-junction solar cells have their conversion efficiency in the lab as 26.3% [5], while the upper theoretical energy conversion efficiency of a solar cell with a bandgap of 1.14 eV (e.g., silicon) is about 33.5% [6]. Multi-junction solar cell approach has been used to increase the theoretical limits of single-junction solar cells [7–13]. It has been shown that series connection of tandem solar cells can reach conversion efficiency in excess of 40% if proper material combination is selected for the top and bottom solar cell [6, 7]. Higher than 40% can be attained if the relationship of the cells is $E_{G\_top} = 0.5 \times E_{G\_bot} + 1.14$ eV, where E_{G_top} and E_{G_bot} are the bandgaps of the top and bottom diode absorbers. This equation holds true when the bottom cell bandgap is between 0.8.5 and 1.2 eV; hence several material combinations can be identified. One of the materials suitable as the bottom solar cell is c-Si with a bandgap of 1.14 eV. This has led to many activities focusing on the development of tandem solar cell using a c-Si bottom solar cell. With c-Si bottom solar cell; the highest conversion efficiency can be achieved if the bandgap of the top-cell is about 1.7 eV.

When combining well established c-Si solar cell technology with other material systems or fabrication process; a number of crucial aspects must be considered. Although amorphous silicon has a fav0urable bandgap of 1.7 eV, its tail states hinder it in being applied on the formation of tandem cell with high values of open circuit voltage. This is a condition for attaining high efficiency in a tandem structure [14–18]. Also silicon oxide/c-Si based quantum dot and quantum well have been researched for the top solar cell material. Un-successfully the two materials have shown comparatively low conversion efficiency. Although, compound semiconductors have been researched as potential top solar cell absorber material; high fabrication temperatures, the lattice mismatch between silicon and compound semiconductors, and the fabrication cost are the biggest hindrances to its application. In the last decade; the perovskite material system has attracted a lot of research either for single-junction solar cells or as material for perovskite/silicon tandem solar cells [18–25]. This has shown high energy conversion efficiencies, with open-circuit voltages close to the theoretical limit of silicon material [26–33]. In addition, a variety of deposition methods at low temperatures can be used in the fabrication of a perovskite top solar cell on a c-Si bottom solar cell. Conversion efficiencies above 20% have been realized with perovskite single-junction solar cells [34–38].

The perovskite/silicon tandem solar cells have both high and low band gap material in a single device; enabling the device to be active in both long and short wavelength regions of the electromagnetic spectrum, where each wavelength region can effectively be converted to electric power resulting in high efficiencies. Also perovskite SCs (solar cells) have unique properties like high absorption coefficient, variable band-gap, high defect tolerance, high open circuit voltage, abundant availability of its constituent elements and easy processability. Perovskite SCs can use the high energy blue and green light much more efficiently than silicon SCs. We note that perovskite/silicon tandem solar cells with high efficiencies is only possible if the perovskite top solar cell and the silicon bottom solar cell operate at a value very near to their theoretical limit. It is worth noting that, perovskite/silicon tandem solar cells with certified energy conversion efficiencies above 27% have been achieved [39]. The realization of solar cells with higher energy conversion efficiencies approaching or even exceeding 30% is feasible in the near future.

2. The basics of photovoltaic cells

A solar cell is an electrical device which converts the energy of sunlight directly into electricity by the photovoltaic effect, which is a physical and chemical phenomenon. The solar cells structure consists of either a p-n junction or p-i-n junction [4, 40]. Initially the incident radiation directed on the surface are absorbed resulting in the creation of electron/hole pairs. Then these photo generated electron/hole pairs are separated by the electric field and subsequently collected at the terminals. The charge collection of the photo generated charges occurs due to diffusion, drift or the combination of both transport processes to the contacts of the solar cell. **Figure 1** provides an overview of different solar cells. Note that schematic figures are not to scale.

Figure 1a above is a typical diagram of a c-Si homo-junction solar cell. It is assumed that absorption of high energy photons is done in the whole of p-n junction. Electron/hole pairs that are produced are predominantly transferred by charge diffusion process. Hetero junction solar cell that consists of a c-Si absorber and amorphous silicon contact layers is shown in **Figure 1b**. The hetero junction structure when compared with traditional homo-junction solar cells, there is a minimization of optical losses, particularly in the emitter. As a result we have high short-circuit current density and high open-circuit voltages. Generally, amorphous silicon p- and n-layers are used to form electrical contacts. The main charge transport mechanism in c-Si is charge diffusion; since it has high diffusion length. The p-i-n structure is currently applied in many thin-film solar cells. In this arrangement the intrinsic material layer is placed between the p-type material and n-type material. The charge collection process is done by a drifting mechanism of the electron/hole pairs to the contacts. **Figure 1c** shows a typical homo-junction of an amorphous silicon thin-film solar cell. A

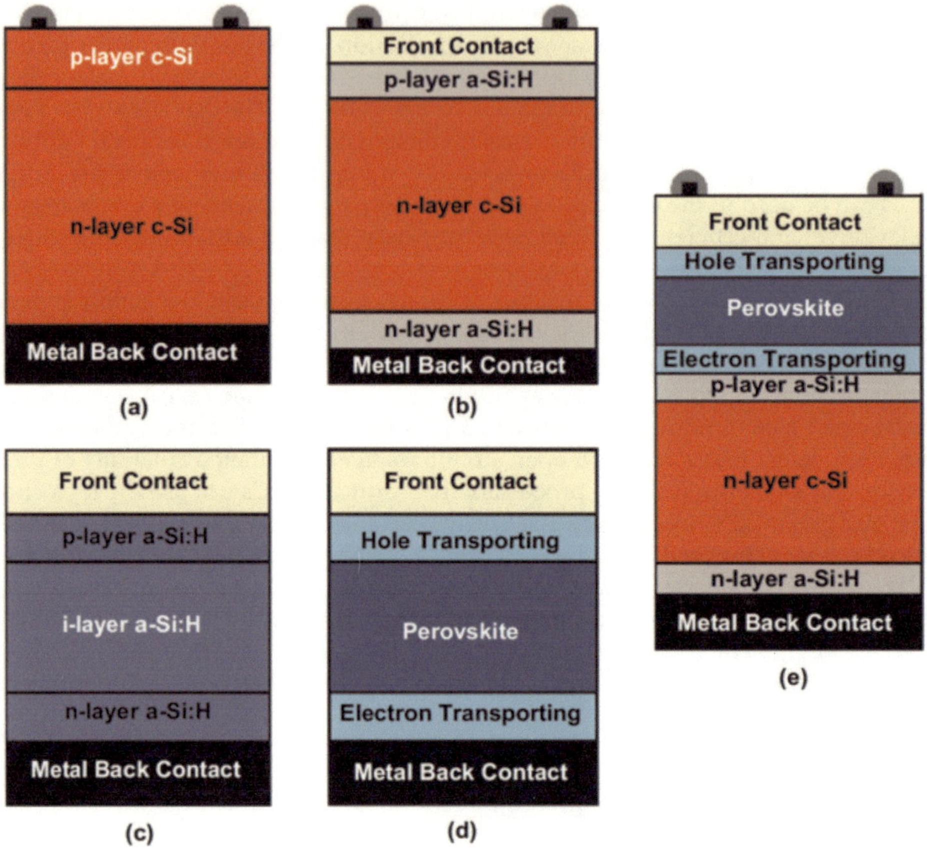

Figure 1.
Typical schematic diagrams of (a) homo-junction solar cell of c-Si, (b) hetero junction solar cell for a c-Si and amorphous silicon, (c) homo-junction thin-film solar cell for amorphous silicon, (d) hetero junction thin-film solar cell for perovskite, and (e) tandem solar cell of perovskite/silicon with layers showing different materials.

heterojunction thin-film solar cell is shown in **Figure 1d**. A perovskite layer is used as an absorber of the incident photon. Different kinds of materials for electron transporting/hole blocking layers are being researched for a possible contact layer. Similarly other materials are under investigation for hole transporting/electron blocking layers. For example, transparent conductive oxides (TCO) are being used as contact layers. Perovskites can be applied to conventional silicon, thus combining the strengths of both material classes: Silicon, in this case, utilizes sunlight in the red and infrared range of the solar spectrum efficiently, while perovskites are good at converting blue light. "If the materials, i.e. perovskites on silicon, are stacked on top of each other, the efficiencies of already commercial silicon cells can be increased considerably. This tandem idea has the potential to herald a solar revolution. Basically the critical parameter to characterizing a solar cell is the energy conversion efficiency. This is typically given by the ratio of the electrical output power density to the optical input power density. The standard optical input spectrum of air mass 1.5 is normally used [41]. The relationship of the efficiency with the physics of solar cell parameters involves short-circuit current density, fill factor, and open-circuit voltage. The short-circuit current density is obtained when the applied voltage is equal to zero, such that $J(V = 0) = J_{sc}$, while the open circuit voltage is also obtained when the current is equal to zero, i.e., $J(V = V_{oc}) = 0$. The maximum power density output for a typical solar cell is given by the product of V_{mp} and J_{mp} i.e., ($V_{mp} \times J_{mp}$), where V_{mp} and J_{mp} are the voltage and current density at the maximum power point (MPP). These two parameters are derived from the current–voltage characteristic, of a solar cell.

$$\eta = \frac{V_{mp} \times J_{mp}}{P_{in}} = \frac{V_{oc} \times J_{sc} \times FF}{P_{in}} \tag{1}$$

Hence the fill factor can be calculated by Eq. (2):

$$\mathrm{FF} = \frac{V_{mp} \times J_{mp}}{V_{oc} \times J_{sc}} \tag{2}$$

An ideal solar cell can be described by Eq. (3):

$$J(v) = J_o\left\{ \exp\left(\frac{qV}{KT_{cell}}\right) - 1\right\} - J_{sc} \tag{3}$$

From this Eq. (3); q is the elementary charge, V is the applied voltage, k is the Boltzmann constant, T_{cell} is the solar cell temperature, and J_0 is the saturation current density. The open-circuit voltage of the solar cell can be determined by using Eq. (4):

$$V_{oc} = \frac{kT_{cell}}{q} \ln\left(\frac{J_{sc}}{J_o} + 1\right) \cong \frac{kT_{cell}}{q} \ln\left(\frac{J_{sc}}{J_o}\right) \tag{4}$$

Understanding the fundamental limits in the energy conversion process of solar cells and determining a potential upper limit of the energy conversion efficiency is important in developing high-efficiency solar cells [42].

3. Solar cell conversion efficiency limit

The maximum conversion efficiency is the theoretical energy conversion limit of a semi-conductor –based solar cell. In deriving the limit we shall assume that the solar cell is described by a single-junction solar cell, which consists of a semiconductor with a constant bandgap. The light beam with photo energies equal or greater than the bandgap is absorbed, while photons with energies smaller than the bandgap are not absorbed. All the photo generated electron/hole pairs are assumed to be collected at contacts. Therefore the recombination of electron/hole pairs is not considered but only thermalization and absorption losses are taken into consideration. Thermalization losses occur for energies larger than the bandgap while absorption losses occur for photon energies smaller than the bandgap [42]. The absorbed photon flux density of the sun by the solar cell, is given by Eq. (5) [6, 42]:

$$F_{cell}(T = T_{sun}) = \frac{2\pi}{h^3c^2}\int_{E_g}^{\infty} \frac{E^2 \mathrm{d}E}{\exp\left(\frac{E}{kT_{sun}}\right) - 1} \tag{5}$$

where h, c, k, and E_g are Planck's constant, speed of light, Boltzmann constant, and energy bandgap of the photovoltaic material. The photon flux can be approximated by Eq. (6):

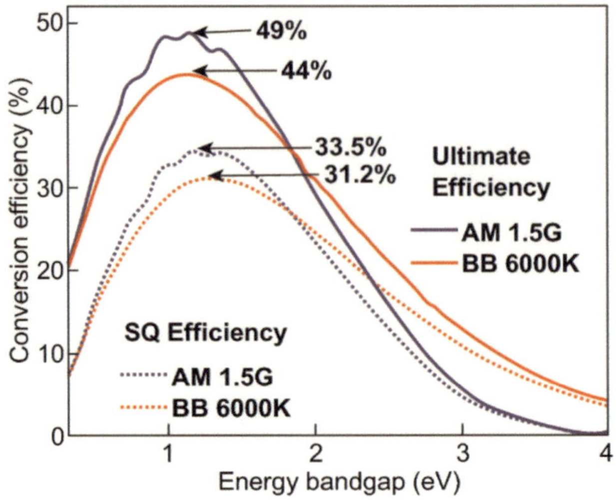

Figure 2.
Ultimate conversion efficiency and Shockley-Queisser limit of single-junction solar cells as a function of the bandgap. A blackbody spectrum at 6000 K and an AM 1.5G spectrum were used for the calculations.

$$F_{\text{cell}}(T = T_{\text{sun}}) = \frac{2\pi}{h^3c^2}\int_{E_g}^{\infty} \exp\left(-\frac{E}{kT_{\text{sun}}}\right)E^2\text{d}E \; = \int_{E_g}^{\infty} \varnothing_{\text{sun}}\text{d}E \tag{6}$$

where ϕ_{sun} is the blackbody radiation flux of the sun, which is given by Eq. (7):

$$\varnothing_{\text{sun}} = \frac{2\pi}{h^3c^2} \times E^2 \times \exp\left(\frac{-E}{kT_{\text{sun}}}\right) \tag{7}$$

The photocurrent density of the solar cell is given by $J = q \times F_{\text{cell}}\ (T = T_{\text{sun}})$. The electrical output power density of the solar cell is calculated by Eq. (8):

$$P_{\text{out}} = J \times V = q \times F_{\text{cell}}(T = T_{\text{sun}}) \times \frac{E_g}{q} = F_{\text{cell}}(T = T_{\text{sun}}) \times E_g \tag{8}$$

The input sun power density is given by Eq. (9) [42]:

$$P_{\text{in}} = \frac{2\pi}{h^3c^2}\int_{E_g}^{\infty} \frac{E^3\text{d}E}{\exp\left(\frac{E}{kT_{\text{sun}}}\right) - 1} \cong \frac{2\pi^5(kT_{\text{sun}})^4}{15h^3c^2} \tag{9}$$

From the above Eqs. (8) and (9), the energy conversion efficiency of a solar cell can be determined by $\eta = P_{\text{out}}/P_{\text{in}}$.

Using the blackbody spectrum at T = 6000 K and AM 1.5 global spectrum, the solar cell gives a maximum conversion efficiency of 44% and 49%, respectively, for an ideal bandgap of 1.1 eV as shown in **Figure 2**.

4. Methodology

Multi-level approach is employed for effective designing of tandem perovskite/silicon solar cell. This approach includes improving the performance of individual layers in each cell; examining the charge transport between each layer when they are stacked, and finally efficient light in-coupling between top and bottom cells (**Figure 3**).

4.1 Perovskite (PVSK) film processing

There are several methods in the fabrication of perovskite (PVSK) films employed as active layers in solar cells; e.g. spin coating, dip coating, gas-quenching (GQ), thermal co-evaporation or vapor phase conversion. Of all these methods, spin coating is most preferred, since it can control the production of consistent films with high quality when compared with the other techniques. It is easy to realize PVSK films on a flat substrate using the above methods generally. For textured substrate however the challenge remains, when using solution-processed spin coating. Consequently a combination of sequential co-evaporation and spin-coating was developed as a hybrid two-step deposition method. That forms a conformal PVSK layer on the micro-size pyramids of textured mono-crystalline silicon.

4.1.1 PVSK layer on flat substrates

Under PVSK processed on flat substrate substrates two techniques are discussed; GQ and solution spin-coating. GQ method was developed at the same time with the

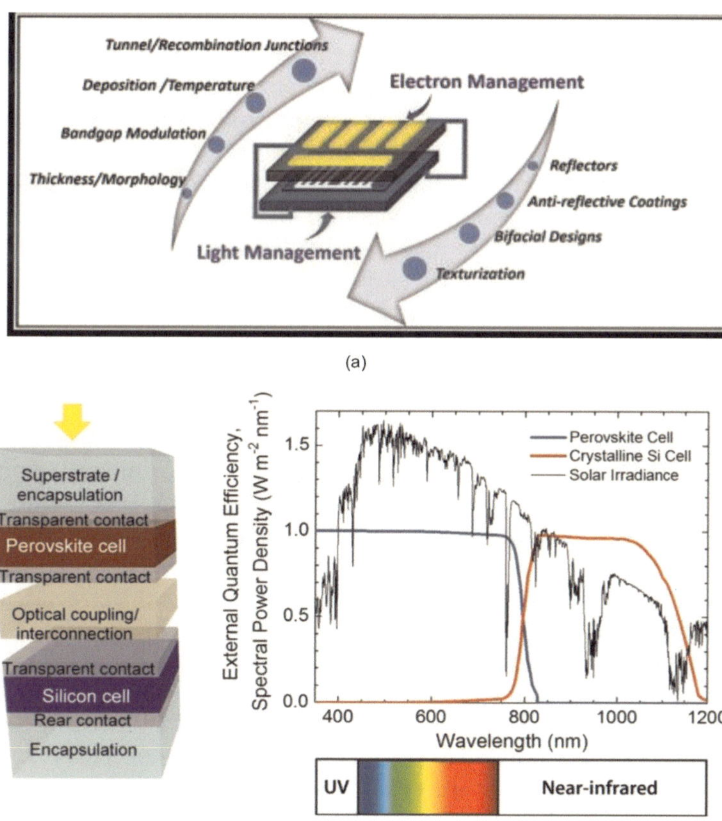

Figure 3.
a and b: (Left). Schematic of a perovskite-silicon tandem solar cell, together with the absorption spectrum of both perovskite.

solvent quenching method. Instead of using anti-solvent, in the GQ method, a flow of nitrogen gas was employed to facilitate the evaporation of precursor solution during one-step coating. By annealing, smooth films with densely packed grains are attained. Recently, many studies related to efficient multi-cation and multi-anion PVSK solar cells have been reported. Such a method has been widely used in fabricating PVSK layer in tandem devices. Adopting the GQ method, in 2020, McGehee and his team reported a 1.67 eV wide-band gap PVSK that consisted of triple-halide alloys of chlorine, bromine, and iodine. In addition the realizing the PVSK, the top cell showed improvement in carrier lifetime and charge-carrier mobility comparing to controlled ones. The reason for improvement was attributed to the enhanced solubility of chlorine by replacing iodine with bromine to shrink the lattice parameter. Of great importance, light-induced phase segregation in PVSK films was significantly suppressed. The conversion efficiency of 27% with an area of 1 cm^2, for this tandem cell was achieved in the laboratory. The cells had an improved stability with less than 4% degradation after 1000 h of MPP operation at around 60°C.

In the solution spin coating technique, a lot of work has also been done in the fabrication of PVSK films. By using this method, in 2016, a team led by Rech fabricated monolithic tandem cells with 18% conversion efficiency [43]. In 2019, Chen et al. combine two additives, MACl and MAH_2PO_2 in PVSK precursor, which significantly improve the morphology of the wide bandgap (1.74–1.70 eV) PVSK films, resulting in a high tandem V_{OC} of 1.80 V and improved conversion 25.4%. In 2020, Shin and his team used the solution spin coating method to develop stable PVSK solar cells with a band gap of about 1.7 eV and conversion efficiency of 20.7%. The fabricated solar cells were tested and found to be stable under extreme conditions. Those cells fabricated could retain nearly 80% of their initial efficiency after 1000 h under continuous illumination. In controlling both structural and electrical properties of the PVSK, anion engineering for materials is undertaken i.e. phenethylammonium (PEA)-based 2D additives is found to be important. Under this method high efficiency of 26.7% in a monolithic 2T wide gap PVSK/Si tandem solar cell was realized by combining spectral responses of the top and bottom diodes.

4.1.2 PVSK layer on textured c-Si

Combination of two-step deposition method: Currently, a single-side texturing arrangement of monolithic PVSK/Si tandem devices is predominantly common. Texturing the back side is done to enhance light trapping properties of the solar cell. In comparison with a double-side polished c-Si device, that has their front surface flat-polished in order to be conformable with the solution based PVSK manufacturing process; light trapping property in such an arrangement is not perfect. Consequently, there is a need to build high conversion efficient tandem cells by using double-side textured c-Si approach. This technique has been used by Ballif and his team to come up with a two-step deposition method, where sequential co-evaporation and spin-coating processes are applied. This resulted into conformal PVSK absorber layers on the micrometer-sized pyramids of textured monocrystalline Si. This type of arrangement resulted in a high current density of 19.5 mA cm^{-2}. The team achieved a conversion efficiency of 25.2% after texturing of the c-Si bottom cell in the micrometer range pyramids. This process reduced the primary reflection loss, enhancing light trapping properties in device.

PVSK on textured c-Si: Solution processed PVSK on textured c-Si has several limitations. These include uncovered Si peaks, shunt paths, and poor charge collection in films with variable thickness, etc. The covering of pyramidal peaks by PVSK of good quality has been reported by a team led by Sargent [44]. The improvement of drift and diffusion of photo-generated carriers in these films enhanced charge collection. In this approach, they used PVSK of wide-bandgap solar cells with a bottom cell of pyramidal-textured c-Si. This approach resulted into improvement of depletion width, in the PVSK and enhancement of carrier collection. In addition to increasing the carrier diffusion length, they used a passivator on the PVSK rough surfaces. And this passivation suppresses the undesired phase segregation. These attributes resulted into PVSK/c-Si cells achieving a conversion efficiency of 25.7% and good thermal stability at 85°C and MPP tracking at 40°C. Blade-coated PVSK on textured silicon with pyramids less than 1 μm in height has been reported by Huang–s group in 2020. Similarly a conformal hole transport layer and perovskite layer that fully covers the textured silicon solar cell were fabricated using nitrogen-assisted blading process. This perovskite/silicon tandem device achieved a conversion efficiency of 26% on textured silicon [45].

In conclusion, several deposition techniques for PVSK film have been advanced and widely researched on. A tandem cell of good quality of polycrystalline films can also be fabricated on both flat and textured substrates. In PVSK/Si tandem solar cells, there are no technical difficulties for flat c-Si. But for textured c-Si, it still appears to be challenging to get conformal films with uniform thickness via sequential co-evaporation and spin-coating methods, in particular for textured monocrystalline Si with large micrometer-sized pyramids. This calls for more investigation and development of more convenient and efficient deposition methods for PVSK films on the textured substrate.

5. Key results

As shown in the **Figure 4**; scanning electron microscopy (SEM) image of the semi-transparent cell; Absorption spectra of Rb-doped and Rb-free perovskite and external quantum efficiency (EQE) of the semi-transparent perovskite cell and filtered silicon cell put together with the absorption and transmittance of the semi-transparent perovskite cell. It is observed that the integrated current from the EQE of semi-transparent cell is 18.2 mA/cm^2, and the integrated current from the filtered silicon cell is 18.7 mA/cm^2. These values are good for marching the two cells. These results are similar with those of the currents determined from *J–V* characteristics. In the **Figure 4c**, *J–V* characteristics of the silicon cell with and without filter, and reverse, forward scan and steady state efficiency of the semi-transparent perovskite cell. The current density versus voltage (*J–V*) characteristic for this tandem solar cell is shown.

The application of detailed balance limit calculations that used in single junction solar cells can also be extended to tandem or multi-junction solar cells. This was first done on detailed balance calculation for tandem solar cells by De Vos [7]. Then latter Green gave a general description of the detailed balance theory for multi-junction solar cells [16]. For this work, this theory is applied to perovskite/silicon tandem solar cells. Typical tandem solar cell can be used either as a two terminal (2-T) or four-terminal (4-T) devices. Plots of tandem configurations (a) 2-T, (b) 4-T and (c) spectrum splitting and the energy conversion efficiency of two and four-terminal tandem solar cells are provided in **Figure 5a–c**.

4-T device is the case where the incoming radiation is split into two diodes which are electrically separate. The conversion efficiency of more than 40% can be realized

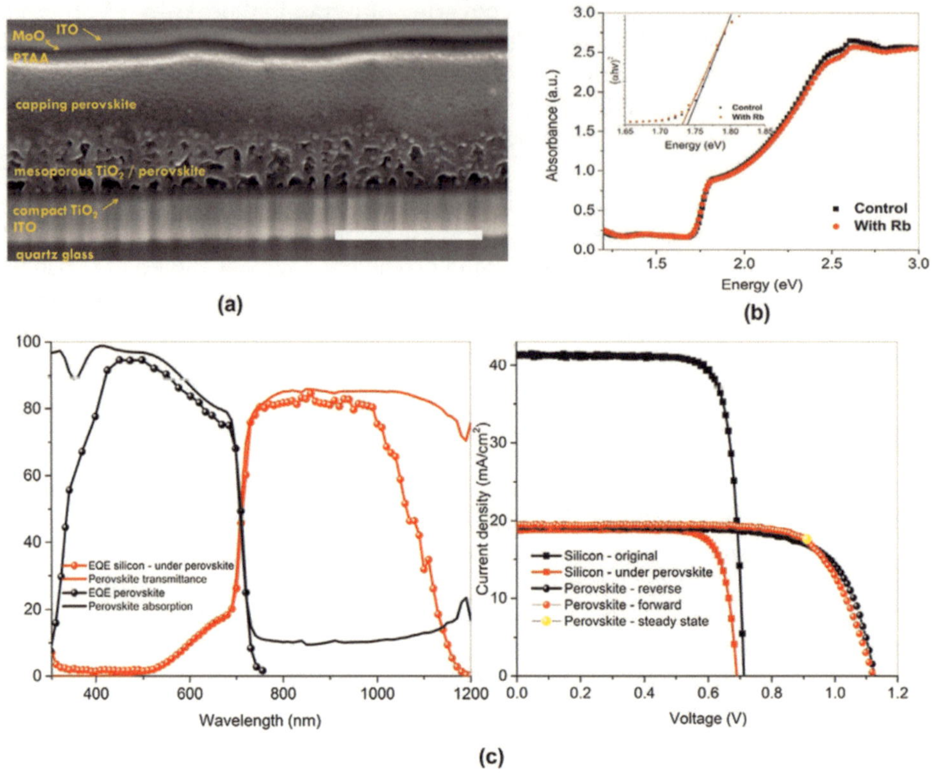

Figure 4.
(a) X-sectional (SEM) image of the semi-transparent tandem cell (b) Rb-doped and Rb-free perovskite absorption spectra. (c) EQE of the PVSK cell and filtered silicon cell and their J–V *characteristics [23].*

by several combinations of band-gaps in PVSK. The overall electrical energy conversion efficiency is determined from the sum of output power generated by both diodes independently. The incoming radiation is split among the two solar cells that form the tandem structure. In monolithic tandem solar cell tittled 2-Terminal (2-T), all the layers corresponding to the two sub-cells are assembled straight on top of another sub-cell i.e. **Figure 5a**. In four terminal (4-T), the sub-cell structures are constructed separately, and then mechanically stacked top cell onto bottom cell (**Figure 5b**). This configuration enable independent optimization of each sub-cell. The optical splitting tandem solar cell termed as 4-T optical operates with optical spectrum filter to split the light spectrum to each sub-cell. The sub-cells merely function independently without any integration, which makes the selection of sub-cells more flexible.
Figure 6a shows the conversion efficiency of a two-terminal device or a serial connected tandem solar cell that is determined by the current at zero applied voltage. The total short-circuit current is equal to the current at zero applied voltage of the bottom solar cell if the short-circuit current of the bottom cell is smaller than the short-circuit current of the top cell. The total short-circuit current is determined by the short-circuit current of the top cell if the short-circuit current is larger than the short-circuit current of the bottom cell. Ideally the short-circuit current of a tandem solar is said to matched if the short-circuit current of the top cell and the bottom cell is equal or almost equal. By matching the bandgaps of the bottom and top solar cells; the energy conversion efficiency of a tandem solar cell is maximized. With proper

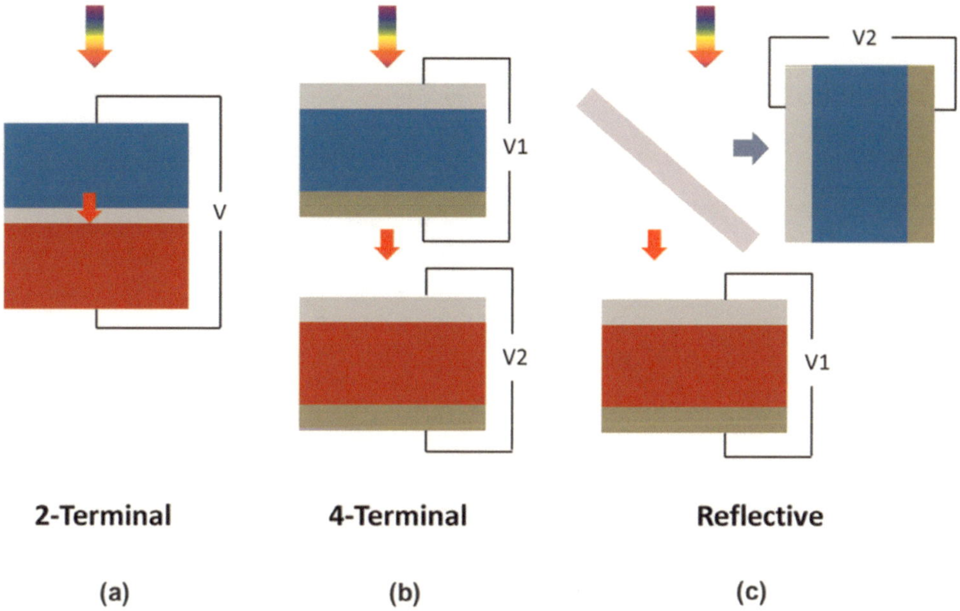

Figure 5.
Two-terminal and four-terminal configuration for tandem cells and light splitting.

combination of band gaps of the top cell; and bottom c-Si solar cell known, the two-terminal tandem solar cell's short-circuit current is matched. There is a possibility of the two-terminal tandem solar cells reaching the energy conversion efficiencies of the four terminal tandem solar cells. The relationship for maximum energy conversion efficiency is determined by the equation; $E_{G\_top} = 0.5 \times E_{G\_bot} + 1.14$ eV; where $E_{g\text{-}top}$ is energy gap for the top cell; $E_{g\text{-}bottom}$ is the energy gap for the bottom cell. For PVSK/c-Si tandem solar cell, maximum energy conversion efficiency is achieved when the bandgap of the top diode is about 1.725 eV. This gives the maximum energy conversion efficiency of about 43% for this arrangement. Using perovskite (MAPbI3) as the base absorber with a bandgap of about 1.6 eV; the maximum energy conversion efficiency of a perovskite/silicon tandem solar cell is approximately 33%. By using the bottom solar cell of bandgap 0.9 eV; and the top cell of perovskite with a bandgap of 1.6 eV, a higher energy conversion efficiency of approximately 44% is achieved.

6. Discussion

It is observed that perovskites have gained considerable attention as a photovoltaic material [26–28]. From its inception in 2009, the energy conversion efficiency of single-junction PSC has been increasing to over 22% [34–38]. It is true that, perovskites are a promising material system for the implementation of tandem or multi-junction solar cells. For the case of perovskite/c-Si tandem solar cells, there is a possibility of reaching high energy conversion efficiencies while potentially maintaining low fabrication and maintenance cost.

Perovskite solar cells are so named because they use a class of crystal structure similar to that found in the mineral known as perovskite. They are structured

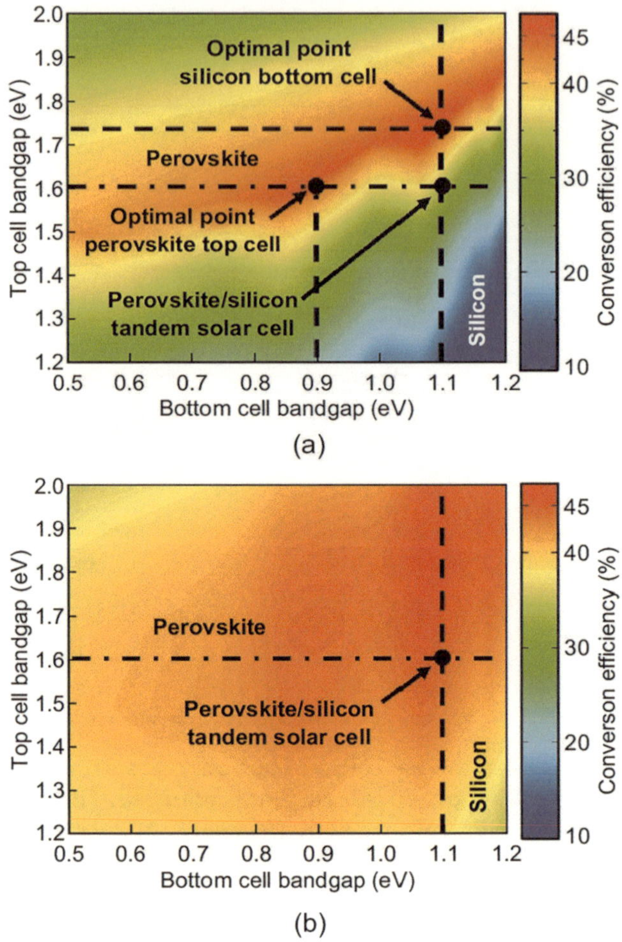

Figure 6.
The energy conversion efficiency of two and four-terminal tandem solar cells.

compounds, commonly hybrid organic-inorganic lead halide-based materials. A layered approach is used in preventing lead leakage in PSC. There is a recently developed process (on–device sequestration approach) that can be easily incorporated with the PSC configurations. Acting as anti-reflecting agent; a transparent lead absorbing film is applied to front conducting glass. The breakthrough in both device architecture and module manufacturing of Si-Perovskite tandem technology will lead to successful commercialization of these devices. Currently, the best devices in this technology with high efficiencies are developed using laboratory scale processes that include spin-coating and anti-solvent dropping methods for PVSK. These techniques are not strongly formed and economical for large scale manufacturing, since their substandard solvent coverage will lead to poor quality pinhole PVSK layers. The alternative to this challenge for perovskite solar cell, is a transition of fabrication processes to large scale deposition techniques such as blade coating, solution processes, printing and spray coating are recommended.

The recombination layers between Si and Perovskite cells should be impenetrable and free from pinholes which is challenging for solution processes involved. The blade coating process requires flat surfaces; but with the existence of textured structure of c-Si cell, this approach is quite complicated. In addition to these fabrication issues, the tandem technology currently shows higher material cost over the process cost. The use of expensive organic transport materials on perovskite cells is a long-standing moisture stability problem and needs an alternative as inorganic PSC. The wearing down of perovskite and leaching of ionic lead (Pb) out of the cell leads to serious environmental hazards and hinder their entry into commercial sector. Although a lot of investigation is concentrated towards replacing Pb^{2+} with dual cations, vacancies and all possible transition metal ions, these alternatives in solar cell devices have so far not been successful. These issues are hindrance for further advancement in the efficiency for the tandem solar cells.

7. Conclusion

Perovskite silicon tandem solar cell perhaps has a very high potential to reach low level cost of electricity.

To achieve highly efficient and reliable tandem perovskite/silicon solar cells, a multi-level approach is required. This includes improving the performance of individual layers in each cell. Examining the charge transport in each layer when they are stacked, and finally efficient-light in-coupling between top and bottom cells. One of the critical conditions for high device efficiency is the proper choice of the bandgap for the top perovskite cell in the tandem. Optical losses due to parasitic absorption losses resulting from both inefficient intermediate reflecting layers and inefficient absorption in the top cell have to be addressed. Low absorption coefficient of silicon bottom cell reduces the light absorption and reflects in efficiency of tandem solar cell. Stable and efficient light management is necessary for further improvement in the device performance.

The research on replacing highly pure expensive Si with recycled left-over multi-crystalline Si in tandem cells without much compromise in efficiency will upscale low-cost tandem cells. Si heterojunction solar cells still exist as the favorite bottom cell over homo-junction cells. Efficient doping with hydrogen passivates and stabilizes silicon and its heterojunction cells. Theoretically it is possible to achieve higher efficiency from the tandem cells of 31.2% at AM 1.5G and 44% out of space.

Author details

David M. Mulati* and Timonah Soita
Physics Department, Jomo Kenyatta University of Agriculture and Technology, Nairobi, Kenya

*Address all correspondence to: dmulati@fsc.jkuat.ac.ke

References

[1] Bos E, et al. World Population Projections, October 2021 According to United Nations estimates elaborated by World meter

[2] RDuran Sahin A, Dincer I, Rosen MA. Thermodynamic analysis of solar photovoltaic cell systems. Solar Energy Materials & Solar Cells. 2017;**91**:153-159. DOI: 10.1016/j.solmat.2006.07.015

[3] Rau U, Paetzold UW, Kirchartz T. Thermodynamics of light management in photovoltaic devices. Physical Review B. 2014;**90**:035211. DOI: 10.1103/Phys R evB.90.035211

[4] Alharbi FH, Kais S. Theoretical limits of photovoltaics efficiency and possible improvements by intuitive approaches learned from photosynthesis and quantum coherence. Renewable and Sustainable Energy Reviews. 2015;**43**: 1073-1089. DOI: 10.1016/j. rser.2014.11.101

[5] Yoshikawa K, Kawasaki H, Yoshida W, Irie T, Konishi K, et al. Silicon heterojunction solar cell with interdigitated back contacts for a photo conversion efficiency over 26%. Nature Energy. 2017;**2**:17032. DOI: 10.1038/ nener gy.2017.32

[6] Shockley W, Queisser HJ. Detailed balance limit of efficiency of p–n junction solar cells. Journal of Applied Physics. 1961;**32**:510-519. DOI: 10.1063/ 1.17360 34

[7] De Vos A. Detailed balance limit of the efficiency of tandem solar cells. Journal of Physics D: Applied Physics. 1980;**13**:839-846. DOI: 10.1088/ 0022-3727/13/5/018

[8] Bush KA, Palmstrom AF, Yu ZJ, Boccard M, Cheacharoen R, et al. 23.6%-efficient monolithic perovskite/ silicon tandem solar cells with improved stability. Nature Energy. 2017;**2**:17009. DOI: 10.1038/nenergy.2017.9

[9] Sahli F, Werner J, Kamino BA, Bräuninger M, Monnard R, et al. Fully textured monolithic perovskite/silicon tandem solar cells with 25.2% power conversion efficiency. Nature Materials. 2018;**17**:820-826. DOI: 10.1038/s4156 3-018-0115-4

[10] Leijtens T, Bush KA, Prasanna R, McGehee MD. Opportunities and challenges for tandem solar cells using metal halide perovskite semiconductors. Nature Energy. 2018;**3**:828-838. DOI: 10.1038/s4156 0-018-0190-4

[11] Zhao D, Wang C, Song Z, Yu Y, Chen C, Zhao X, et al. Four-terminal all-perovskite tandem solar cells achieving power conversion efficiencies exceeding 23%. ACS Energy Letters. 2018;**3**:305-306. DOI: 10.1021/acs energy lett.7b012 87

[12] Qarony W, Hossain MI, Salleo A, Knipp D, Tsang YH. Rough versus planar interfaces: how to maximize the short circuit current of perovskite single and tandem solar cells. Materials Today Energy. 2019;**11**:106-113. DOI: 10.1016/j. mtener.2018.10.001

[13] Werner J, Niesen B, Ballif C. Perovskite/silicon tandem solar cells: Marriage of convenience or true love story? An overview. Advanced Materials Interfaces. 2018;**5**:1700731. DOI: 10.1002/admi.20170 0731

[14] Meillaud F, Shah A, Droz C, Vallat-Sauvain E, Miazza C. Efficiency limits for single-junction and tandem solar cells. Solar Energy Materials & Solar Cells. 2006;**90**:2952-2959. DOI: 10.1016/ j.solma t.2006.06.002

[15] Shah A, Meier J, Vallat-Sauvain E, Droz C, Kroll U, Wyrsch N, et al. Microcrystalline silicon and 'micromorph' tandem solar cells. Thin Solid Films. 2002;**403–404**:179-187. DOI: 10.1016/S0040-6090(01)01658-3

[16] Meier J, Spitznagel J, Kroll U, Bucher C, Faÿ S, Moriarty T, et al. Potential of amorphous and microcrystalline silicon solar cells. Thin Solid Films. 2004;**451–452**:518-524. DOI: 10.1016/j.tsf.2003.11.014

[17] Green MA. Third Generation Photovoltaics: Advanced Solar Energy Conversion. Berlin: Springer; 2003. pp. 35-66

[18] Qarony W, Hossain MI, Hossain MK, Uddin MJ, Haque A, Saad AR, et al. Efficient amorphous silicon solar cells: characterization, optimization, and optical loss analysis. Results in Physics. 2017;**7**:4287-4293. DOI: 10.1016/j.rinp.2017.09.030

[19] Lopez-Delgado R, Higuera-Valenzuela HJ, Zazueta-Raynaud A, Ramos A, Pelayo JE, Berman D, et al. Enhancing the power conversion efficiency of solar cells employing down-shifting silicon quantum dots. Journal of Physics: Conference Series. 2016;**773**:012087. DOI: 10.1088/1742-6596/773/1/01208 7

[20] Conibeer G, Perez-Wurfl I, Hao X, Di D, Lin D. Si solid-state quantum dot-based materials for tandem solar cells. Nanoscale Research Letters. 2012;**7**:193. DOI: 10.1186/1556-276X-7-193

[21] Lopez-Delgado R, Higuera-Valenzuela HJ, Zazueta-Raynaud A, Ramos-Carrazco A, Pelayo JE, Berman-Mendoza D, et al. Solar cell efficiency improvement employing down-shifting silicon quantum dots. Microsystem Technologies. 2018;**24**:495-502. DOI: 10.1007/s0054 2-017-3405-x

[22] Pi X, Li Q, Li D, Yang D. Spin-coating silicon-quantum dot ink to improve solar cell efficiency. Solar Energy Materials & Solar Cells. 2011;**95**: 2941-2945. DOI: 10.1016/j.solmat.2011.06.010

[23] Wolff CM, Zu F, Paulke A, Toro LP, Koch N, Neher D. Reduced interface-mediated recombination for high open circuit voltages in CH3NH3PbI3 solar cells. Advanced Materials. 2017;**29**: 1700159. DOI: 10.1002/adma.20170 0159

[24] Leguy AMA, Hu Y, Campoy-Quiles M, Alonso MI, Weber OJ, et al. Reversible hydration of CH3NH3PbI3 in films, single crystals, and solar cells. Chemistry of Materials. 2015;**27**: 3397-3407. DOI: 10.1021/acs.chemmater.5b00660

[25] Yin W-J, Shi T, Yan Y. Unusual defect physics in CH3NH3PbI3 perovskite solar cell absorber. Applied Physics Letters. 2014;**104**:063903. DOI: 10.1063/1.4864778

[26] Zhang H, Wang H, Chen W, Jen AKY. CuGaO2: a promising inorganic hole-transporting material for highly efficient and stable perovskite solar cells. Advanced Materials. 2017;**29**:1604984. DOI: 10.1002/adma.20160 4984

[27] Liu M, Johnston MB, Snaith HJ. Efficient planar heterojunction perovskite solar cells by vapour deposition. Nature. 2013;**501**:395-398. DOI: 10.1038/nature12509

[28] Jeon NJ, Lee J, Noh JH, Nazeeruddin MK, Grätzel M, Il Seok S. Efficient inorganic-organic hybrid perovskite solar cells based on pyrene arylamine derivatives as hole transporting materials. Journal of the American Chemical Society. 2013; **135**:19087-19090. DOI: 10.1021/ja410 659k

[29] Fang Z, Liu L, Zhang Z, Yang S, Liu F, Liu M, et al. CsPbI2.25Br0.75 solar cells with 15.9% efficiency. Scientific Bulletin. 2019;**64**:507-510. DOI: 10.1016/j.scib.2019.04.013

[30] Zuo C, Scully AD, Vak D, Tan W, Jiao X, McNeill CR, et al. Self-assembled 2D perovskite layers for efficient printable solar cells. Advanced Energy Materials. 2019;**9**:1803258. DOI: 10.1002/aenm.20180 3258

[31] Zuo C, Vak D, Angmo D, Ding L, Gao M. One-step roll-to- roll air processed high efficiency perovskite solar cells. Nano Energy. 2018;**46**: 185-192. DOI: 10.1016/j. nanoen.2018.01.037

[32] Zuo C, Ding L. An 80.11% FF record achieved for perovskite solar cells by using the NH4Cl additive. Nanoscale. 2014;**6**: 9935. DOI: 10.1039/C4NR0 2425G

[33] Zuo C, Bolink HJ, Han H, Huang J, Cahen D, Ding L. Advances in perovskite solar cells. Advancement of Science. 2016;**3**:1500324. DOI: 10.1002/ advs.20150 0324

[34] Burschka J, Pellet N, Moon SJ, Humphry-Baker R, Gao P, Nazeeruddin MK, et al. Sequential deposition as a route to high-performance perovskite-sensitized solar cells. Nature. 2013;**499**:316-319. DOI: 10.1038/nature12340

[35] Park N-G. Perovskite solar cells: An emerging photovoltaic technology. Materials Today. 2015;**18**:65-72. DOI: 10.1016/j.mattod.2014.07.007

[36] Lee J-W, Seol D-J, Cho A-N, Park N-G. High-efficiency perovskite solar cells based on the black polymorph of HC(NH2)2PbI3. Advanced Materials. 2014;**26**:4991-4998. DOI: 10.1002/ adma.20140 1137

[37] Hao F, Stoumpos CC, Cao DH, Chang RPH, Kanatzidis MG. Lead-free solid-state organic–inorganic halide perovskite solar cells. Nature Photonics. 2014;**8**:489-494. DOI: 10.1038/n photon. 2014.82

[38] Green MA, Hishikawa Y, Dunlop ED, Levi DH, Hohl-Ebinger J, Yoshita M, et al. Solar cell efficiency tables (Version 53). Progress in Photovoltaics: Research and Applications. 2019;**27**:3-12. DOI: 10.1002/pip.3102

[39] Oxford PV-The Perovskite Company. Oxford PV Perovskite Solar Cell Achieves 28% Efficiency [Internet]. 2018. Available at: https://www.oxford pv.com/news/oxford-pv-pe rovskite-solar-cell-achieves-28- efficiency [Accessed 19 June 2019]

[40] Limpert S, Bremner S, Linke H. Reversible electron–hole separation in a hot carrier solar cell. New Journal of Physics. 2015;**17**:095004. DOI: 10.1088/ 1367-2630/17/9/09500 4

[41] Reference Solar Spectral Irradiance: ASTM G-173 [Internet]. 2018. Available from: http://rredc.nrel.gov/solar/spec tra/am1.5/astmg173/astmg173.html [Accessed 19 March 2018]

[42] Kosyachenko LA. Solar Cells: New Approaches and Reviews. Chernivtsi National University. Ukraine: InTech; 2015. p. 388. DOI: 10.5772/58490

[43] Albrecht S, Saliba M, Pablo Correa J, et al. Monolithic perovskite/silicon heterojunction tandem solar cells processed at low temperature. Energy & Environmental Science. 2016;**9**:81-88. DOI: 10,1039/c5EE02965A

[44] Hou Y, Chen B, Chen H, Sargent EH, et al. Efficient tandem solar cells with solution-processed perovskite on

textured crystalline silicon. Science 2020;**367**(6482):1135-1140. DOI: 10.1126/science aaz3691

[45] Young-Kim J, Lee JK, Shin H, et al. High-efficiency perovskite solar cells. Chemical Reviews. 2020;**120**(15): 7867-7918. DOI: 10.1021/acs.chemrev. oc00107

Chapter 10

Organic/Inorganic Halide Perovskites for Mechanical Energy Harvesting Applications

Venkatraju Jella, Swathi Ippili, Hyun You Kim, Hyun-Suk Kim, Chunjoong Kim, Tae-Youl Yang and Soon-Gil Yoon

Abstract

Organic/inorganic halide perovskites (OIHPs) have recently emerged as promising candidates for the creation of high-efficiency electronic and optoelectronic devices, having superior performance because of their unique features such as excellent optical and electronic properties, cost-effective fabrication, solution-processing, and simple device architecture. The noteworthy dielectric and ferro/piezoelectric properties of OIHPs have enabled the design of mechanical energy harvesters (MEHs). Considerable research has been conducted on using OIHPs in the field of piezoelectric and triboelectric nanogenerators. In this chapter, we describe the potential of OIHP materials, such as organic and inorganic halide perovskites, for harvesting ambient mechanical energy and convert it into electrical energy. Furthermore, the crystal structure of OIHPs along with their dielectric, piezoelectric, and ferroelectric properties are discussed in detail. Recent innovations in OIHP-based MEHs are also summarized. The role of OIHP-polymer composites in enhancing the performance and operational stability of nanogenerators is discussed. Certain issues and challenges facing contemporary OIHP-based MEHs are stated, and finally, some directions for future developments are suggested.

Keywords: OIHP, piezoelectricity, triboelectricity, mechanical energy, nanogenerator

1. Introduction

The rapid progress in artificial intelligence and internet-of-things technologies has increased the demand for portable and sustainable energy sources that can enable perpetual operation [1, 2]. Mechanical energy harvesters (MEHs) that convert abundant mechanical energy from the environment (wind, raindrops, water flow, and vibrations) as well as from human motions (walking, jogging, and running) into electricity, is considered a promising solution to alleviate the energy crisis for supplying power to low-power consumed portable electronics [3]. In particular, piezoelectric nanogenerators (PENGs) and triboelectric nanogenerators (TENGs) have received immense attention as they efficiently convert mechanical energy into electricity for powering portable and wearable electronics [4]. Furthermore, their simple structure, easy fabrication process, high energy-conversion efficiencies, flexibility, and mechanical robustness makes them well-suited for energy generation [4, 5]. PENGs transform mechanical energy into electricity by generating electric dipoles through the deformation of piezoelectric materials, while TENGs convert mechanical energy into electricity effectively through a coupling of contact electrification and electrostatic induction. In the past decade, several flexible MEHs have been demonstrated using diverse piezoelectric nanostructured materials including ceramics (e.g., $PbZr_xTi_{1-x}O_3$ (PZT) and $BaTiO_3$), semiconductors (e.g., ZnO and CdS), and polymers (PVDF and its derivatives) [6–9]. Among these materials, perovskite-structured ceramics (i.e., ferroelectric materials) are commonly used materials for constructing efficient PENGs and sensors because of their strong dielectric and ferroelectric/piezoelectric properties [10, 11].

In recent years, organic/inorganic halide perovskites (OIHPs) have emerged as promising materials for solar cells with extremely high-power conversion efficiencies over 25% because of their unique optical and electrical properties while having a simple solution process [12]. Besides the discovery of the intriguing ferroelectric and piezoelectric properties of OIHPs have accelerated their recent application in PENGs [13–15]. In 2016, the first thin-based PENG was reported based on the solution-processed ferroelectric $MAPbI_3$ thin films [14]. Subsequently, several flexible PENGs based on OIHP thin films and OIHP–polymer composite films have been developed [13]. OIHPs also exhibit impressive dielectric properties, which is one of the essential features for fabricating efficient TENGs. The first TENG based on $MAPbI_3$ displayed light-dependent triboelectric output characteristics with a moderate performance [15]. Later, a series of commonly used Cs-based perovskites were applied in TENGs owing to their stability compared to organic perovskites [16, 17]. The light-active nature along with the ferro/piezoelectric properties of OIHP harnesses the light-dependent output characters of PENGs as well as TENGs, which allows the nanogenerators to be used as bimodal sensors for sensing pressure and light [18].

In this chapter, we introduce the viability of OIHP materials for mechanical energy harvesting in the form of nanogenerators. The crystal structure and dimensionality of OIHPs along with their dielectric, piezoelectric, and ferroelectric properties are discussed in detail. In addition, the operating mechanisms of OIHP based-MEHs (PENG and TENG) are discussed. Furthermore, the recent progress of various MEHs based on a broad range of OIHP/OIHP–polymer composite materials is summarized. Finally, a brief glimpse into current challenges and future developments for OIHP-based nanogenerators is provided.

2. Structure and dimensionality of OIHPs

The term "perovskite" represents a class of materials originating from the mineral calcium titanate ($CaTiO_3$) and having a crystal structure of ABO_3, which were discovered in 1839 by Gustav Rose [19]. Oxide perovskites are widely used in various dielectric, ferroelectric, piezoelectric, and pyroelectric applications. However, OIHPs differ from inorganic ceramic perovskites by containing halide anions in place of oxide anions. Three-dimensional (3D) OIHPs also have the general crystal structure of ABX_3 (**Figure 1a**) [20], in which A represents an organic or inorganic monovalent cation (e.g., methylammonium ($MA^+ = CH_3NH_3^+$), formamidinium ($FA^+ = CH(NH_2)_2$), or

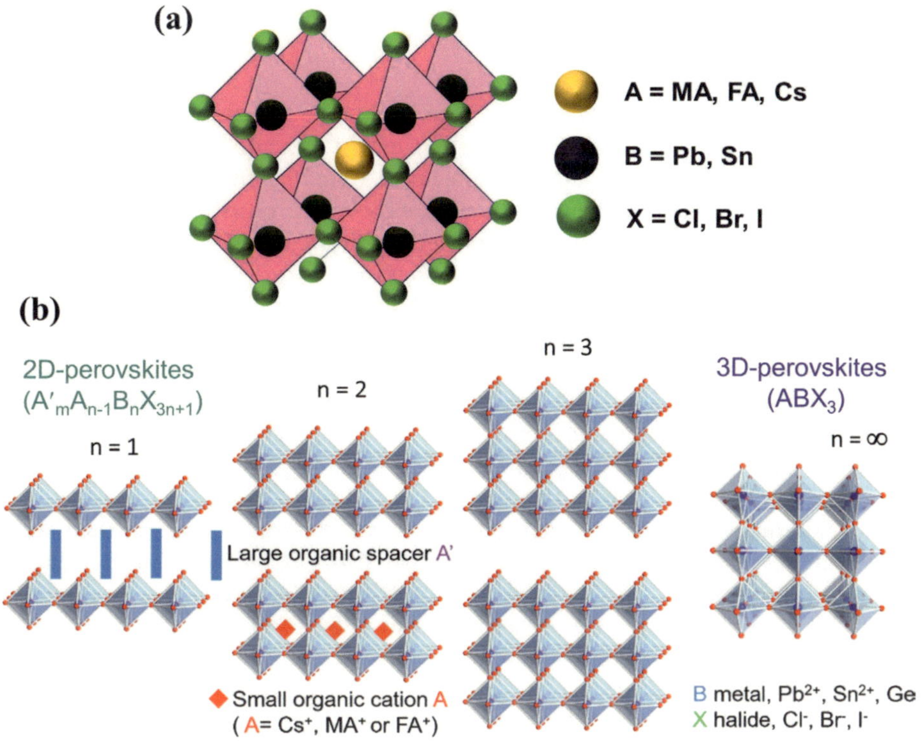

Figure 1.
Schematic representation of a) typical ABX_3 type 3D OIHP structure, b) structures of 2D OIHPs of $A'_mA_{n-1}Bn_{X3_{n+1}}$ (with the value of n increasing from n = 1 to n = ∞) [24].

cesium (Cs^+)), B denotes a divalent metal cation (e.g., Pb^{2+} or Sn^{2+}), and X indicates a halide anion (Cl, Br, and I). In the crystal structure, A-site cations are connected with 12 neighboring X, while B-site is coordinated by 6 X anions to form cuboctahedral and BX_6 octahedral geometries, respectively. The formation of the perovskite structure and its stability can be evaluated by the Goldschmidt tolerance factor (t) and the octahedral factor (μ) [21]. The tolerance factor is given by $t = (r_A + r_X)/\sqrt{2}(r_B + r_X)$, where r_A, r_B, and r_X are the ionic radii of A, B, and X, respectively. The octahedral factor is given by $\mu = r_B/r_X$, which is directly correlated with a BX_6 octahedron. The tolerance factor and octahedral factor values of OIHPs are expected to be in the range of $0.813 < t < 1.107$ and $0.44 < \mu < 0.90$ [13], respectively. OIHPs tend to form ideal cubic, orthorhombic, and hexagonal structures when $0.8 < t < 1.0$, $t < 0.8$, and $t > 1$, respectively [13, 22]. 2D perovskites or layered perovskites are formed by introducing large organic functional groups into the 3D structure (**Figure 1b**) and have received immense attention because of their excellent ambient stability [23]. These 2D perovskites can be prepared using a mixture of small cations that forms perovskite and a large organic cation that forms the layered metal halide. The general chemical formula for these layered perovskites is $A'_mA_{n-1}B_nX_{3n+1}$, where A′ is a monovalent (m = 2) or divalent (m = 1) long-chain organic cation (e.g., aromatic or aliphatic alkylammonium), which acts as a spacer. A, B, and X are cations and anions similar to the ones in 3D OIHPs, and n specifies the number of perovskite layers. Here, n = ∞ corresponds to a 3D structure, n = 1 represents a 2D structure, and other values of n denote a quasi-2D structure [24].

3. Structure and dimensionality of OIHPs

3.1 Dielectric properties

A dielectric is referred as an insulating material that is polarized under an applied external electric field. The response of a dielectric material to an applied field is expressed in terms of permittivity. The dielectric constant or relative permittivity (ε_r) of a material is usually obtained from the ratio of its permittivity to the permittivity of a vacuum (ε_0). Materials having a large dielectric constant have the ability to develop higher polarization for an applied electric field. In general, inorganic materials are well-known dielectric materials. In recent times, OIHPs are attracting extensive attention to be used as dielectric materials because of their simple low-temperature synthesis process. ABX_3-structured materials exhibit higher dielectric constant values owing to the ease of polarizing the cell structure. Specifically, distortion of the edge-sharing BX_6 octahedra in the ABX_3 structure can produce an electric dipole between the A and B sites. OIHPs demonstrate impressive dielectric properties analogous to ceramic perovskites, but the values are relatively lower owing to the existence of polar organic cations in the center of the perovskite structure, which can introduce orientational disorder and polarization. The dielectric properties of various lead (Pb) and lead-free OHIPs have been previously investigated experimentally [14, 25, 26]. Kim et al. measured the temperature-dependent dielectric properties of $MAPbI_3$ thin films and confirmed their tetragonal-cubic phase transition. $MAPbI_3$ revealed a dielectric constant value of 52 at 100 kHz [14]. Furthermore, structural tuning of $MAPbI_3$ also affects its dielectric properties. For example, with the partial incorporation of Cl into $MAPbI_3$, the dielectric constant of the resultant films increased to 90.9 at 100 kHz (**Figure 2a**), while Br-incorporated $MAPbI_3$ films exhibited a dielectric constant of 71.6 at 100 kHz [25]. In addition, a structural transition was also observed for partially incorporated Fe^{2+} into $MAPbI_3$ from the dielectric study [26]. The partial replacement of Pb^{2+} with Fe^{2+} ions exhibited a tetragonal–cubic phase transition

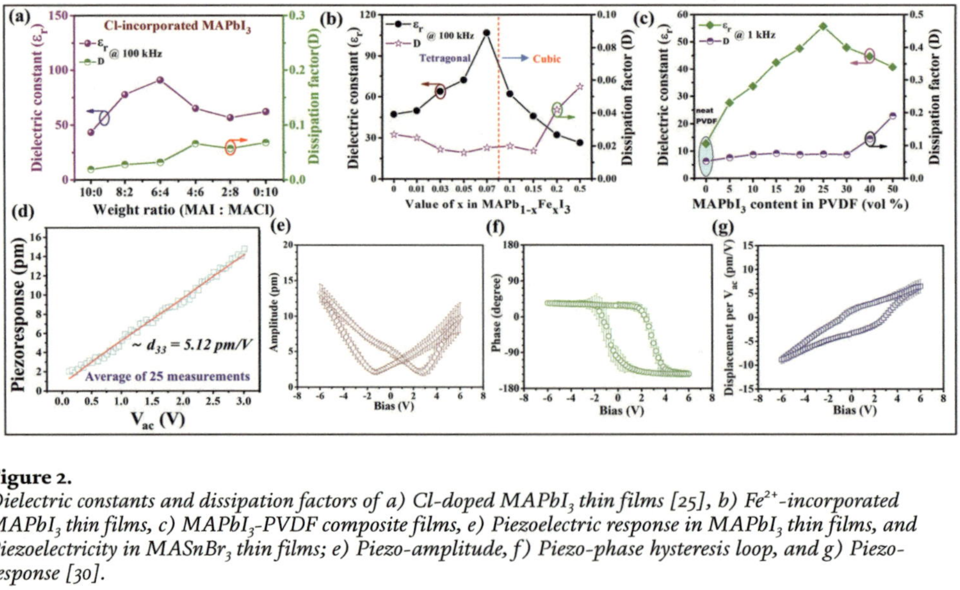

Figure 2.
Dielectric constants and dissipation factors of a) Cl-doped $MAPbI_3$ thin films [25], b) Fe^{2+}-incorporated $MAPbI_3$ thin films, c) $MAPbI_3$-PVDF composite films, e) Piezoelectric response in $MAPbI_3$ thin films, and Piezoelectricity in $MASnBr_3$ thin films; e) Piezo-amplitude, f) Piezo-phase hysteresis loop, and g) Piezo-response [30].

as discovered by the frequency dielectric study of Fe^{2+}-incorporated $MAPbI_3$ thin films (**Figure 2b**). The dielectric constant (ε_r) at 100 kHz of $MAPb_{1-x}Fe_xI_3$ films continuously increased as the Fe^{2+} content increased to x = 0.07, attaining a maximum value of 107. It then decreased for larger Fe^{2+} content, indicating the ferroelectric-to-paraelectric phase transition for x = 0.07. The dielectric properties of OHIP-polymer composites were also investigated by a few researchers. In general, the heterogeneous materials interfaces in polymer composite films can induce an interfacial or a Maxwell–Wagner–Sillars polarization that results in an abrupt change of the total dielectric constant [27, 28]. As the $MAPbI_3$ content in PVDF polymer increased, the dielectric constant of the $MAPbI_3$–PVDF composite rapidly increased because of large dipole–dipole interactions (**Figure 2c**) [29]. In addition, the interaction between the organic action of MA^+ and -CF_2- between the perovskite and PVDF leads to the self-orientation of polymer chains, enabling the nucleation of the electroactive phase that results in the formation of a spontaneous polar β-phase in the composite films. Similarly, the dielectric properties of lead-free $MASnBr_3$–PDMS composite films were also examined for various percentage weights of $MASnBr_3$ (0 to 25 wt.%) [30]. As the $MASnBr_3$ increased from 5 to 15 wt.% content, the dielectric constants of composite films progressively increased and achieved a maximum of 36.23 for 15 wt.% at 1 kHz. By contrast, the dielectric constant values of composite films decreased with a higher loading amount of $MASnBr_3$ owing to increased agglomeration of $MASnBr_3$ particles, which leads to the leaky nature of composite films. The tuning of the dielectric and piezoelectric properties of OIHPs by a simple solution process makes them suitable to be used to construct efficient mechanical energy harvesters.

3.2 Piezoelectric properties

The piezoelectric effect refers to the capability of certain materials to generate electric charges under applied mechanical stress, which is also known as the direct piezoelectric effect. Conversely, an electric field applied to the material induces mechanical strain, which is called the converse piezoelectric effect. This unique property of piezoelectric materials allows their use as sensors and actuators. The structural requirement for a material to exhibit piezoelectricity is non-centrosymmetric. Piezoelectricity was first demonstrated in 1880 by brothers Pierre and Jacques Curie. The direct piezoelectric effect is observed in many natural crystalline materials such as Rochelle salt, quartz, topaz, and human bone. In addition, many engineered materials, in particular, inorganic perovskite materials with a structure of ABO_3 including PZT, $BaTiO_3$ and $(K,Na)NbO_3$, exhibit a noticeable piezoelectric effect, and are widely studied for several applications [6]. The piezoelectric response or piezoelectric energy-harvesting capability of any piezoelectric material can be determined by its piezoelectric coefficient and is proportional to the dielectric constant and polarization (i.e., $d_{33} \alpha\, \varepsilon_r P_r$) [13].

OIHPs also exhibit relatively good piezoelectric properties similar to inorganic ceramic perovskites, but the values are comparatively lower. In recent years, some researchers have investigated the piezoelectric properties of OIHPs in order to determine their potential in various device applications. Kim et al. investigated the piezoelectric coefficient (d_{33}) of solution-processed polycrystalline $MAPbI_3$ films using piezoresponse force microscopy (PFM) and reported a d_{33} of 5.12 pm/V [14] (**Figure 2d**). A single-crystalline device may provide direct evidence regarding the piezoelectric properties of OIHPs, whereas studying the piezoelectric properties of polycrystalline films using PFM could face challenges owing to inaccurate estimation of tip contact area and other artifacts arising from surface topography and crystal orientation [27]. In this regard, Dong et al. verified the piezoelectric effect in single-crystalline $MAPbI_3$ by depositing two parallel facet gold electrodes [28]. They obtained a d_{33} of 2.7 pm/V along the (001) direction for single-crystal $MAPbI_3$ using the laser interferometry method. Compositional tuning of OIHPs also substantially altered their piezoelectric properties. The partial replacement of Pb with Fe in $MAPbI_3$ improved the d_{33} of 17.0 ± 6.0 pm/V for $MAPb_{1-x}Fe_xI_3$ (x = 0.07) [26]. Similarly, there was a considerable improvement in d_{33} of 20.8 pm/V observed for lead-free $MASnI_3$ by substituting Sn in the Pb-site in $MAPbI_3$ [31]. Likewise, lead-free $MASnBr_3$ perovskite displayed a d_{33} of 2.7 pm/V [30] (**Figure 2e–g**). The slanted butterfly shape in amplitude loop of the $MASnBr_3$ is caused from the electrochemical properties of defects, vacancies, and ions, which indicate that the $MASnBr_3$ possess both electrochemical and piezoelectric properties. According to Ding et al., replacing the A and X site in $MAPbI_3$ with FA and Br, respectively, significantly increased d_{33} (25 pm/V) for $FAPbBr_3$ having a particle size 50–80 nm, which is a five-fold enhancement over $MAPbI_3$ [32]. In addition, inorganic $CsPbBr_3$ films also revealed a higher d_{33} of 40.3 pm/V after poling than organic $MAPbI_3$ films [33]. As a subclass of 3D OIHPs, the piezoelectric properties of 2D perovskites (vacancy-ordered double perovskites) are also recently attracting significant research attention owing to their excellent ambient stability compared to 3D OIHPs. Although 2D OIHPs exhibit superior ambient stability, very few piezoelectric studies have been focused on the recently evolved 2D OIHPs. For instance, solution-processed $(ATHP)_2PbX_4$ displayed ferroelectric behavior with a large d_{33} of 76 pC/N and a giant piezoelectric voltage co-efficient (g_{33}) of 660.3×10^{-3} V.m/N [34]. In addition, most 2D OIHPs have ferroelectric natures, thus displaying superior piezoelectric properties as a subclass of piezoelectric materials.

3.3 Ferroelectric properties

Ferroelectric materials are a class of dielectric materials that exhibit ferroelectricity. Ferroelectricity is the ability of materials to possess spontaneous electric polarization and originates from a non-centrosymmetric crystal structure. The direction of spontaneous polarization can be reversed in accordance with an applied external electric field. Ferroelectric materials belonging to the perovskite family (ABX_3 crystal structure) are a subclass of pyroelectric and piezoelectric materials. Ferroelectric materials show ferroelectric behavior only below the Curie temperature (T_C). Above T_C, these materials display the paraelectric state (i.e., they are only polarized under an applied electric field). For example, a well-known inorganic perovskite, $BaTiO_3$, undergoes a structural transition from tetragonal to cubic above 393 K [35]. Recently, OIHPs have been studied specifically with a focus on their ferroelectric properties because of their structural transition and impressive dielectric properties. Although there is a debate on existing ferroelectricity in $MAPbI_3$, several researchers have conducted theoretical and experimental investigations to search for evidence of ferroelectricity in OIHPs. For instance, Kutes et al. provided the first experimental evidence of ferroelectricity in solution-processed $MAPbI_3$ thin films with a grain size of ~100 nm by directly observing the ferroelectric domains through a PFM study, which is a necessary tool to observe the ferroelectric domains at nanometer resolution. The reversible switching of those ferroelectric domains was also realized by electrical poling with a DC bias [36]. Rakita et al. conducted an experimental investigation for the existence of ferroelectricity in tetragonal $MAPbI_3$. They observed the polarization inversion under an external field, lack of inversion symmetry, and spontaneous polarization based on the measurements of a polarization-electric field (P-E) hysteresis loop, second harmonic generation signals, and pyroelectric response, respectively [37]. Kim et al. confirmed the tetragonal-to-cubic phase transition by measuring temperature-dependent dielectric properties [14]. These studies clearly indicate the experimental evidence for ferroelectricity in $MAPbI_3$ films. In addition, the PFM study upon lead-free $MASnI_3$ demonstrates its ferroelectric property (**Figure 2e–g**) [31]. From the PFM results, a well-defined butterfly-shaped hysteresis loop and the existence of 180° of domain switching validate ferroelectric polarization in $MASnI_3$ films. However, commonly used 3D OIHPs including $MAPbI_3$ and $MASnI_3$ exhibit relatively low T_C, limiting their wider applicability at high temperature. Pan et al. reported stable 3D (3-ammoniopyrrolidinium) $RbBr_3$[(AP)$RbBr_3$] perovskites synthesized by evaporation of the precursor's solution, which exhibited a ferroelectric nature at a high Tc = 440 K (**Figure 3**) [38].

Recently, researchers have been keen to study the ferroelectric properties of 2D OIHPs having advantageous characteristics including structural flexibility, diversity, and excellent moisture stability. The large asymmetric A-site cation

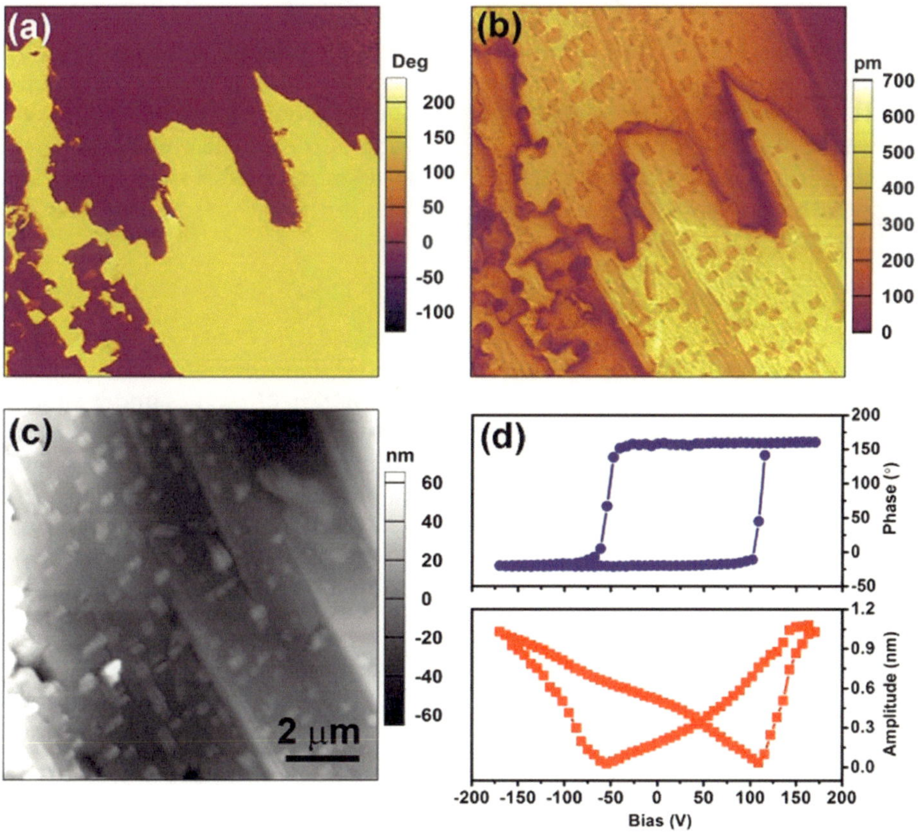

Figure 3.
PFM analysis of 2D layered (ATHP)$_2$PbBr$_4$ films; a) lateral PFM phase, b) amplitude, c) corresponding topography images and d) obtained local piezoelectric response phase hysteresis (top) and amplitude (bottom) loops under applied DC-bias [34].

provides an additional asymmetry to the 2D OIHP crystal. The orientation of such a large cation promotes ferroelectricity. Many excellent 2D OIHPs were designed and their ferroelectric properties investigated. For instance, 2D $(ATHP)_2PbBr_4$ was synthesized using the simple solution method and its ferroelectric property was investigated by measuring the P-E curve and lateral PFM [34]. The 180° contrast of domain orientation in the PFM phase image and separation of the adjacent domains by the domain walls in the PFM amplitude are a direct indictor of ferroelectricity for $(ATHP)_2PbBr_4$ (**Figure 4**). Another 2D OIHP [$(4,4\text{-DFHHA})_2PbI_4$ (4,4-DFHHA=4,4-difluorohexahydroazepine)] displayed ferroelectric properties with a spontaneous polarization of 1.1 $\mu C/cm^2$ at room temperature, with a T_C of 454 K [39].

4. MEHs based on OIHPs

A mechanical energy harvester that can produce electricity from mechanical vibrations is a very promising tool to realize sustainable energy generation in remote/indoor environs and even through human body movements. MEHs generally operate based on either the piezoelectric effect, triboelectric effect, or electromagnetic induction effects [13]. Amongst them, PENGs that operate based on piezoelectric effect and TENGs that operate based on triboelectric effect have attracted intense attention because of their direct power conversion ability and relatively easier fabrication processes. Intense research efforts have been conducted with regard to material and structural design, functionality, operation mechanism, and performance optimization of PENGs and TENGs [8, 13, 18]. More recently, the unique properties, low-temperature solution processing, and tunability of properties via compositional-tuning of OIHPs have highlighted their superior potential in the field of mechanical energy harvesters.

4.1 The working mechanism of a PENG

The working mechanism of a general piezoelectric nanogenerator having the metal–insulator–metal structure along with the formation of dipoles under applied force is schematically portrayed in **Figure 4**. In a non-poled piezoelectric material, the dipoles will be randomly orientated, while the orientation of those dipoles will be changed according to the applied electric field direction under poling. Initially, there

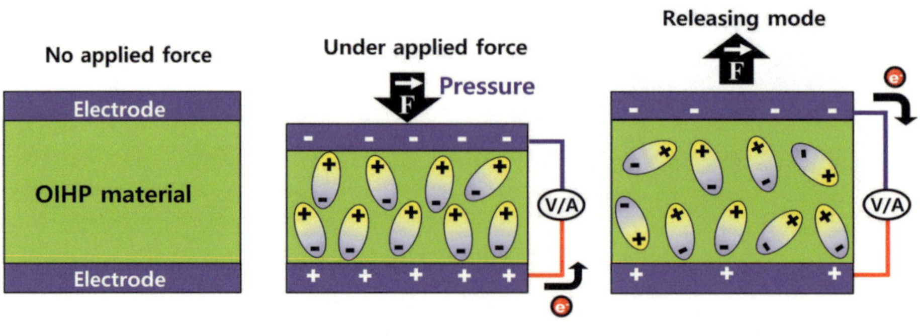

Figure 4.
Schematic illustration of working mechanism of OIHP-based PENG.

will be no generation of output from the PENG without any applied strain owing to the absence of potentials (or in equilibrium state) at the electrodes. However, when mechanical stress is applied normal to the PENG, the piezoelectric material undergoes compressive deformation leading to the generation of dipoles within the active material. The subsequent dipolar polarization results in a piezoelectric potential difference between the two electrodes of the nanogenerator. This, in turn, leads to the flow of charges from one electrode to the other electrode through the external circuit by producing an electrical output signal. When the applied force is withdrawn, the piezoelectric potentials vanish and the electrons flow back to the original position and generate an electrical output signal with the opposite polarity. This output generation is a cyclic process under applied cyclic pressures. In addition, the output of a PENG significantly depends on the material and also some external parameters like applied pressure and frequency.

4.2 The working mechanism of a TENG

A TENG can effectively transform the irregular and randomly distributed mechanical energy into usable electricity via coupling contact electrification with electrostatic induction [18]. In general, when two different materials are in contact with each other, chemical bonds will be formed between the interface of those materials, leading to charge (i.e., electrons, or ions, or molecules) transfer from one material to another because of the difference in their electron affinities [40]. When the two surfaces are separated from each other, the potential drop in the triboelectric charges induces charges into the electrodes via electrostatic induction effect. The potential difference between the electrodes drives electrons to flow between the two electrodes, thus generating triboelectricity form the devices. Based on this principle, four kinds of TENGs with different modes of operations have been developed, as shown in **Figure 5**. In the contact–separation mode, the first invented operation mode of TENG, two dielectric films are placed face to face, and metal electrodes are deposited on the opposite surfaces of the dielectric layers (**Figure 5a**). The TENG operates when the force is applied normal to the device. In the lateral sliding mode, the device structure is similar to that of contact–separation mode (**Figure 5b**), and the TENG operates when the two films keep sliding against one another. This sliding operation offers more efficient charge transfer compared to that offered by the contact–separation mode [41]. By contrast, single-electrode mode is designed to work independently and can be moved freely (**Figure 5c**). This mode is composed of a moving dielectric film and an electrode film connected to the ground. When the top dielectric film approaches towards and/or departs from the bottom electrode, the distribution of the local electrical field may change by generating a potential difference between electrode and ground. This leads to a flow of electrons between the ground and electrode and generates electricity. The freestanding triboelectric-layer mode consists of two symmetrical electrodes underneath a moving dielectric layer that has electrodes of similar sizes (**Figure 5d**). In this mode of operation, no direct physical contact between the two triboelectric layers can be realized, which tends to extend the lifetime of the TENG [42]. Among these modes, the OIHP-based TENGs developed so far were constructed and operated based on vertical contact–separation mode [43, 44]. In the OIHP-based TENGs, the OIHP film fabricated on the electrode-coated substrate is a triboelectric material and is naturally separated from the counter triboelectric material using spacers. Furthermore, when a piezoelectric material like OIHP is used to construct a TENG, the dipoles formed by mechanical deformation of a piezoelectric material under an applied force promote the

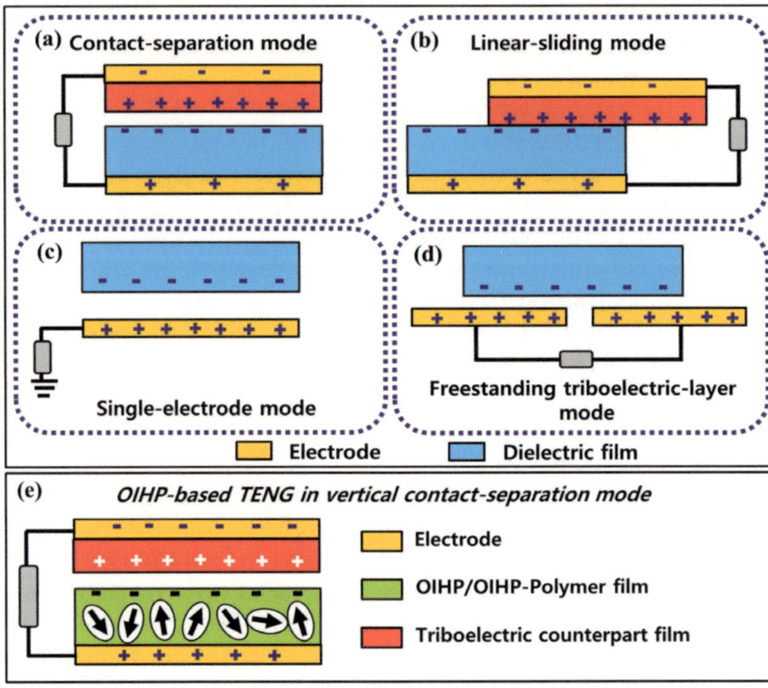

Figure 5.
Schematic depiction of a–d) the four fundamental operating modes of TENG, e) vertical contact–separation mode of OIHP/OIHP-polymer composite based TENG.

generation of more charges onto the surface of the OIHP film during TENG operation, as schematically portrayed in **Figure 5e** [8].

4.3 OIHP-based PENGs

OIHP materials have only been recently applied to PENGs because of their favorable characteristics, which include high piezoelectricity, flexibility, large-area fabrication and low-temperature synthesis along with the biocompatibility of lead-free OIHPs [30, 31]. In addition, various OIHP-polymer composite materials were developed to achieve flexible PENGs with improved mechanical and air stability [29]. Although the output power of OIHP PENGs is very moderate and is lower than OIHP solar cells, with increasing research efforts, the output power has enormously increased with polymer composites. OIHP-based PENGs having the typical metal–insulator–metal structures similar to other piezoelectric materials based devices and were constructed on flexible plastic substrates using simple solution methods [45, 46]. Yoon et al. reported the first OIHP PENG using solution-processed $MAPbI_3$ thin films as shown in **Figure 6a**. The PENG poled at an applied field of 80 kV/cm demonstrated an output voltage and current density of ~2.7 V and ~140 nA/cm^2 under a mechanical pressure of 0.5 MPa (**Figure 6b**) [14]. Later, many researchers focused on improving the output performance of OIHP PENGs [25, 26, 47]. For example, the lateral-structured PENG with inter digitated electrode (IDE) patterns using a $MAPbI_3$ active layer and ZnO & Cu_2O-charge transport layers achieved improved output current values [47]. The device was poled under a low (12 kV/cm) electric field for 10 min and was

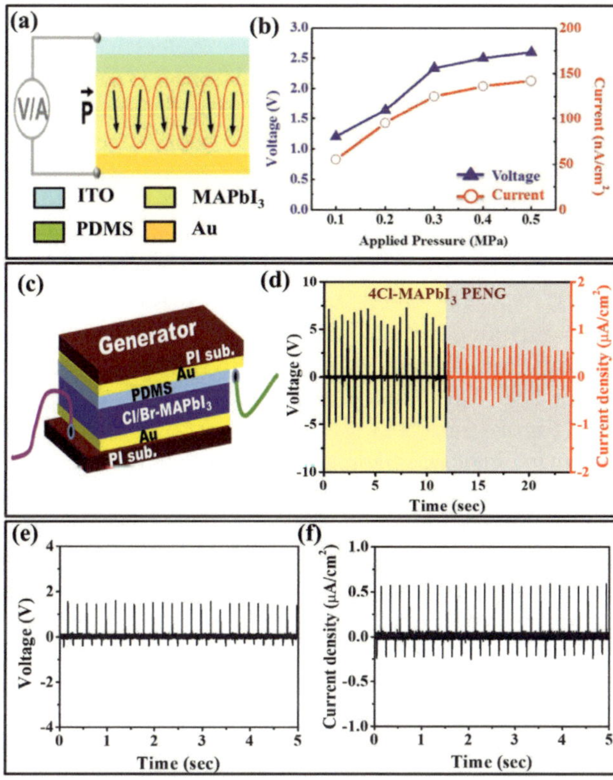

Figure 6.
a) Schematic picture of $MAPbI_3$-based PENG, and b) corresponding pressure-dependent piezoelectric output performance [14]. c) Schematic depiction of Cl/Br-doped $MAPbI_3$ PENG, and d) piezoelectric output performance of 4Cl-doped $MAPbI_3$ PENG [25]. Piezoelectric output performance of $MASnBr_3$ PENG: e) output voltage and f) current density signals [30].

able to generate a voltage of ~1.47 V and a current of ~0.56 μA under 0.2 MPa pressure. In addition, the output of OIHP PENGs can be further enhanced by controlling the dielectric and piezoelectric properties of perovskite material via the concept of functional-modification of perovskite. A high amount of Cl or Br doping into $MAPbI_3$ perovskite leads to enhanced dielectric and piezoelectric properties, which results in better piezoelectric output performance from halide doped-$MAPbI_3$ PENGs compared to the pure $MAPbI_3$ PENG (**Figure 6c**) [25]. As shown in **Figure 6d**, the poled 4Cl-$MAPbI_3$ PENG generated a particularly high output voltage and current density of ~5.9 V and ~0.61 μA/cm^2, respectively, because of the improved dielectric constant (ε_r = 90.9) and remanent polarization (P_r = 0.56 μC/cm^2) of perovskite film. Similarly, the partial incorporation of Fe^{2+} into the Pb^{2+} sites of $MAPbI_3$ perovskite using the simple solution method rapidly enhanced the piezoelectric output performances of PENGs [26]. As discussed earlier, with increasing Fe^{2+} content, the morphological and crystalline properties of the $MAPb_{1-x}Fe_xI_3$ thin films were improved, leading to improvement of dielectric and piezoelectric properties up to a doping amount of 7 at.% (x = 0.07). After 10 at.% (x = 0.10) doping, the $MAPb_{1-x}Fe_xI_3$ samples exhibited a structural transition from tetragonal to cubic; this was a paraelectric material and is unsuitable for PENG applications. However, as the Fe^{2+} concentration increased, the piezoelectric output performance of $MAPb_{1-x}Fe_xI_3$ thin-film PENGs linearly increased and achieved a maximum of 4.52 V for 7 at.%-doped PENG. In addition, the same 7 at.%-doped PENG demonstrated a much higher piezoelectric output of 7.29 V and 0.88 μA/cm^2 after poling at an applied electric field of 30 kV/cm.

However, the high toxicity of Pb makes it inappropriate for direct application in the human body or real environs. Researchers have been searching for other lead-free materials in an effort to develop alternatives to lead-based nanogenerators. One of the emerging lead-free OIHP materials is Sn-based perovskite, which is eco-friendly, biocompatible, and has a large piezoelectric coefficient comparable to that of ceramic $PbTiO_3$, which makes it a promising candidate for high-performance nanogenerators in the medical field [31, 48]. The poled lead-free $MASnI_3$ PENG produced an output voltage of 3.8 V and a current density of 0.35 μA/cm^2 under an applied pressure of 0.5 MPa [31]. Similarly, the lead-free $MASnBr_3$ PENG displayed an output voltage and current density of 1.56 V and 0.58 μA/cm^2, respectively, under the same applied pressure of 0.5 MPa (**Figure 6e** and **f**) [30]. The generated low output from the $MASnBr_3$ film is because of a lower piezoelectric coefficient of 2.7 pm/V compared to the $MASnI_3$ d_{33} value of 20.8 pm/V [30, 31]. In addition to organic–inorganic halide perovskites (OHPs), some researchers have also explored inorganic halide perovskite (IHP) materials for PENG applications because of their decent environment stability compared to OIHPs. In particular, $CsPbX_3$ has attracted considerable interest in device applications given its higher chemical stability than other perovskites. The $CsPbBr_3$ nanogenerator was developed on a plastic substrate with the structure of PET/ITO/PDMS/$CsPbBr_3$/ITO/PET and poled at an applied electric field of 25 kV/cm [33]. The PENG demonstrated better output performance with an output voltage and current of 16.4 V and 604 nA, respectively, after optimized poling conditions. The same device was further able to sense selective motions, such as eye-blinking, throat movements, and finger motions of a human body, highlighting the potential of $CsPbBr_3$ materials for physiological sensing applications.

Although many studies prove the potential of materials in harvesting mechanical energy for generating the electricity, the practical application of OIHP-based PENGs has not been realized so far because of their lower outputs. Furthermore, they are completely incompatible with irregular mechanical deformations. Hence, a key solution proposed was to create composite OIHP structures with polymer materials for the construction of high-performance and long-term air-stable nanogenerators that can withstand highly harsh environs. The first OIHP–PDMS composite-based PENG (PET/ITO/$FAPbBr_3$-PDMS/Al) was developed by incorporating ferroelectric $FAPbBr_3$ nanoparticles into a PDMS polymer, spin-coating the composite onto an indium tin oxide (ITO)-coated PET substrate, and integrating the film with Al foil acting as a top electrode [32]. This PENG demonstrated a maximum piezoelectric output voltage and current density of 8.5 V and 3.8 μA/cm^2, respectively, under pushing. Another group developed an eco-friendly PENG using lead-free $MASnBr_3$-PDMS composite material (**Figure 7a**), which displayed a high piezoelectric output voltage of 18.8 V, current density of 13.76 μA/cm^2, and power density of 74.52 μW/cm^2 under an applied pressure of 0.5 MPa (**Figure 7b**) [30]. In addition, the PENG exhibited enormous air-stability over 120 days and mechanical durability over more than 10,000 cycles. However, the non-uniform dispersion of perovskite materials in highly viscous polymers like PDMS may result in modest interactions between PDMS and perovskite crystals that could reduce the piezoelectric output performance of nanogenerators [13]. Soon after, researchers have made use of ferroelectric PVDF polymers to realize high-performance nanogenerators because of its ferroelectric nature [29, 49–52].

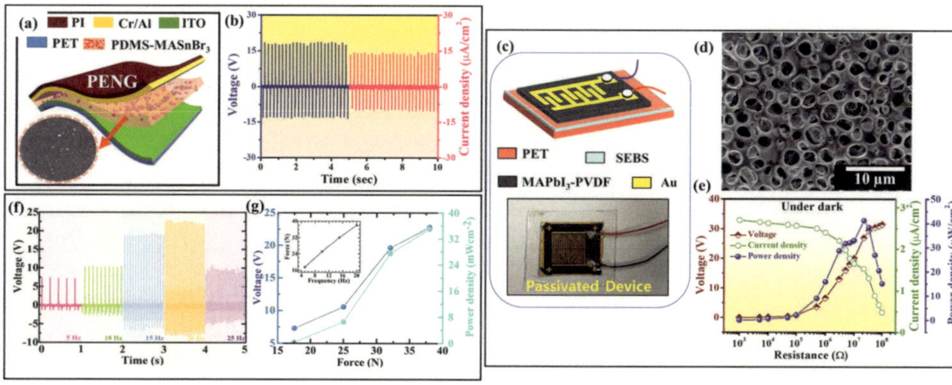

Figure 7.
a) Schematic representation of $MASnBr_3$-PDMS composite based PENG, and b) corresponding piezoelectric output performance [30]. c) Schematic diagram of $MAPbI_3$-PVDF composite based PENG (inset is fabricated device), d) SEM image of 25 vol% $MAPbI_3$-PVDF composite film, and e) corresponding load-resistance dependent piezoelectric output performance [52]. f) Frequency-dependent piezoelectric output voltage signals of $FASnI_3$-PVDF composite PENG and g) Force-dependent piezoelectric output voltage and power densities of $FASnI_3$-PVDF composite PENG and [50].

PVDF, a semi-crystalline ferroelectric polymer, is mainly available in four phases (α, β, γ, and δ). Among these, the β-phase is one of the polar phases and is a highly electroactive phase with superior piezoelectric properties [49]. Hence, the $MAPbI_3$ perovskite solution was mixed with PVDF solution and spin-coated onto the desired plastic substrates to construct PENGs (**Figure 7c**) [52]. The 25 vol% $MAPbI_3$-PVDF composite films showed porous-like morphology with good dispersion of $MAPbI_3$ nanoparticles into the PVDF matrix as shown in SEM image of **Figure 7d** [29, 52]. By increasing the volume fraction of $MAPbI_3$, the dielectric and ferroelectric properties of composite films improved remarkably owing to the enhanced β-phase content of the PVDF matrix caused by the strong polar interactions or hydrogen bonding between $MAPbI_3$ and PVDF. In addition, PVDF encapsulation significantly increased the air-stability of $MAPbI_3$ perovskite over 6 months [29]. The 25 vol% $MAPbI_3$-PVDF PENG, with the IDE-structure given in **Figure 7c**, generated a high piezoelectric output voltage, current density and power density of 33.6 V and 3.54 $\mu A/cm^2$, 41.18 $\mu W/cm^2$, respectively, at an applied pressure of 300 kPa, while demonstrating long-term operational stability and mechanical stability due to SEBS polymer passivation (**Figure 7e**) [52]. Furthermore, another eco-friendly PENG based on lead-free $FASnI_3$-PVDF nanocomposite materials that has a high piezoelectric coefficient of 73 pm/V demonstrated a piezoelectric output voltage of 23 V [50]. The piezoelectric output of the same PENG is highly influenced by the applied frequency and force as shown in **Figure 7f** and **g**, respectively. Furthermore, the developed lateral-structured PENGs based on highly uniform $CsPbBr_3$-PVDF composite fibers reveal a recordable piezoelectric output performance with an output voltage of 103 V and circuit current of 170 $\mu A/cm^2$, which is noticeably higher than many OIHP/OIHP-polymer materials [51]. The same composite PENG exhibits enhanced thermal/water/acid–base stabilities along with exceptional mechanical stability. These results open up a route for more simple and cost-effective production of high-performance PENGs using OIHPs and their polymer composite materials for mechanical energy harvesting and sensor applications.

4.4 OIHP-based TENGs

TENGs have been intensively utilized as flexible power sources and self-powered sensors [8]. Two dissimilar triboelectric nature materials lying at the extreme opposite ends of the triboelectric series are usually employed to fabricate TENGs in order to achieve higher output power. In particular, materials with high dielectric properties are suitable for realizing efficient TENGs, which are designed as part of capacitors [8]. Therefore, OIHPs are recognized as one of the most promising candidates for developing efficient TENGs because of their remarkable dielectric and piezoelectric properties along with low-temperature synthesis [52]. The first OIHP-based TENG developed using $MAPbI_3$ perovskite is operated as a self-powered photodetector based on the combined properties of photoelectric and triboelectric effects [15]. The TENG comprising of two triboelectric parts (Cu/PET as a negative triboelectric material and $MAPbI_3/TiO_2$/FTO as a positive triboelectric material) as illustrated in **Figure 8a** is operated in a fundamental vertical contact–separation mode. This TENG generated a triboelectric peak-to-peak output voltage of 8 V under mechanical pushing in darkness because of triboelectrification. The output is immediately decreased by nearly 37.5% (~5 V) under illumination with a light-intensity of 100 mW/cm^2, giving rise to a high responsivity of 7.5 V/W due to photogenerated charges in the light-active $MAPbI_3$ film (**Figure 8b**). Further, the compositional tuning and electrical poling of perovskite materials can significantly improve the triboelectric performance of TENGs, because compositional modification and ion migration under poling process both tend to alter the conductivity of the OIHP films, which in turn changes the surface potential and electron affinity of those films [43]. Clearly, as shown in **Figure 8c**, the conductivity of the $MAPbI_3$ perovskite film can noticeably change to either p-type (MAI rich) or n-type (PbI_2 rich) by regulating the MAI/PbI_2 ratio during the precursor synthesis [43]. This concept can further extend to TENG applications to realize high-performance TENGs. The composition-tuned $MAPbI_3$ perovskite is paired with PTFE and nylon (PA6) polymer films to develop TENGs (**Figure 8d**). Here, the

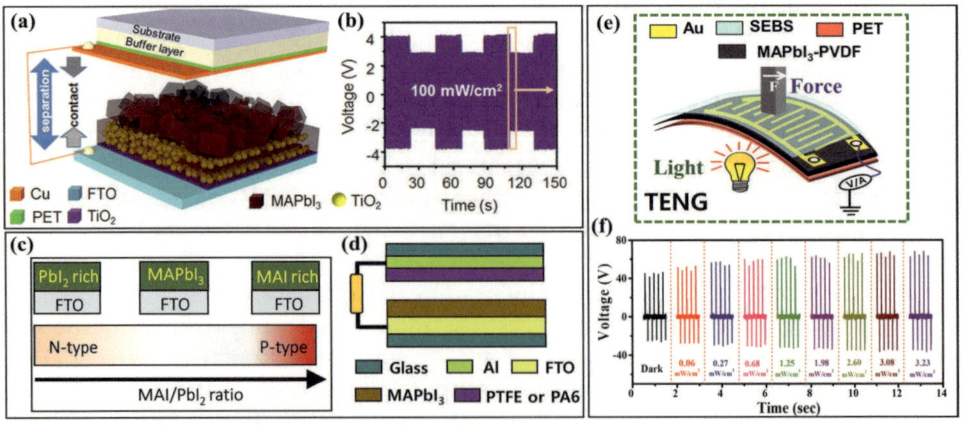

Figure 8.
a) Schematic representation of $MAPbI_3$-based TENG, and b) corresponding light-dependent triboelectric output performance [15]. c) Schematic illustration of n and p type conversion in $MAPbI_3$ films by controlling of MAI/PbI_2 ratio, and d) schematic diagram of $MAPbI_3$-based TENG with the counter triboelectric parts of PTFE/PA6. e) Schematic drawing of $MAPbI_3$-PVDF composite-based TENG in single-electrode mode operation, f) corresponding light-dependent triboelectric output voltage signals under constant pressure of 300 kPa [52].

500-nm-thick $MAPbI_3$ film acts as a triboelectric positive friction layer while pairing with the triboelectric negative PTFE film in PT-PVK TENG and generates a peak output current density of 61.25 mA/m^2. By contrast, the $MAPbI_3$ film acts as a triboelectric negative friction layer while pairing with the triboelectric positive PA6 in PA-PVK TENG and generates a peak output current density of 21.5 mA/m^2 with opposite polarity compared to that of PT-PVK TENG. The poling process further enhances the triboelectric output of PT-PVK TENG. The device generated a maximum output voltage, current density, and peak power density of 979 V, 106 mA/m^2, and 24 W/m^2 after an optimal compositional tuning (MAI/PbI_2 ratio of 2) and poling process (E_P = 4 V/μm). Similarly, the TENGs fabricated using composition tuned-Cs based perovskites demonstrate notable variations in their triboelectric output depending upon A-site or B-site, or halogen modification [44]. The TENG (glass/FTO/$CsPbBr_{3-y}Cl_y$//PVDF/Ag) demonstrates increasing output performance with increasing Cl content and reaches an output similar to that of only $CsPbCl_3$-based TENG owing to increased electron-donating ability with increasing doping amount. Here, perovskite acts as a triboelectric positive layer, while PVDF films act as a triboelectric negative layer.

As in the case of the OIHP–polymer PENGs, the OIHPs were further composited with polymer materials to improve the long-term operational stability along with air-stability of OIHP–polymer TENGs. To this end, several TENGs with different structures, materials, and modes of operations were developed and their feasibility to harvest the mechanical energy was demonstrated. The flexible single-structure multifunctional device with the structure of $MAPbI_3$-PVDF/Au-IDE/SEBS can harvest mechanical energy and simultaneously sense multiple external stimuli like light and pressure (**Figure 8e**) [52]. The TENG in a single-electrode mode generates an output voltage of ~44.7 V, a current density of ~4.34 $\mu A/cm^2$, and a power density of ~59.52 $\mu W/cm^2$ under cyclic contact–separations in darkness. Furthermore, the triboelectric output gradually increases with increasing light-intensity and reaches a maximum voltage of 67.9 V (**Figure 8f**), current density of 7.44 $\mu A/cm^2$, and power density of 158.34 $\mu W/cm^2$ at a high light intensity of 3.23 mW/cm^2. This significant enhancement in triboelectric output is because of the combined photoelectric and triboelectric properties of the $MAPbI_3$–PVDF active layer. Under mechanical pushing, when the pushing stack (Al_2O_3/Al-stack) touches the surface of the SEBS polymer, contact electrification results in the generation of charges with opposite polarities on the surfaces of the pushing stack as well as SEBS polymers. Concurrently, the active piezoelectric $MAPbI_3$–PVDF layer undergoes deformation, thus generating dipoles. Owing to the combined triboelectric and piezoelectric effects, the charges will be induced on the Au electrode, leading to a higher potential difference between the electrode and the ground. The resultant potential difference allows the flow of electrons through the external circuit to the ground, thus generating improved outputs. However, when the TENG is illuminated under the applied pressure, the induced triboelectric charge allows the rapid injection of photogenerated charge carriers from $MAPbI_3$ into the Au electrode. This results in a significantly higher triboelectric output under illumination compared to dark-state. Similarly, the fabricated $MAPbI_3$–PDMS composite e-skin-based TENG is highly capable of harvesting mechanical energy and producing neural-stimulating electrical signals without relying on an external power supply [53]. The triboelectric output performance of e-skin significantly increases as the bending radius increases and shows high output voltage and currents of 0.659 V and 8.94 nA, respectively, for a bending angle of 60°. In addition, the device displays strain-dependent and light-stimulated voltage variations, which enable the device to operate as a self-powered pressure and physiological sensor application. More recently, a stretchable, breathable, and long-term stable hybrid MEH has been developed based on eco-friendly, 2D layered lead-free $Cs_3Bi_2Br_9$, PVDF-HFP and SEBS composite (LPPS-NFC) nanofibers prepared via an electrospinning process [54]. Here, the strong electron-accepting nature of perovskite materials acts as a nucleating agent and improve the crystallinity and polar β-phase of PVDF polymers. The developed composite nanofibers can efficiently harvest the mechanical energy in piezoelectric as well as triboelectric modes. The LPPS-NFC stretchable device with the structure of Spandex/Ag-SEBS/LPPS-NFC//Al generates much larger peak-to-peak outputs with a voltage of 400 V, current density of 1.63 $\mu A/cm^2$, and power density of 2.34 W/m^2 in the hybrid mode based on the combined piezoelectric and triboelectric effects of composite film. Furthermore, the LPPS-NFC based MEHs reveal excellent stability and are able to produce stable outputs even under harsh mechanical deformations like washing, folding, and crumpling, indicating the superior potential of these LPPS-NFC-based MEHs for use in smart textile-based wearable devices. All these results demonstrate the high potential of hybrid perovskites as triboelectric materials, given their superior dielectric property and stepping forward for high-performance TENG platforms.

5. Current challenges and future prospects

Plenty of research effort has been expended in the study of OIHP nanogenerators to prove the potential of OIHPs as promising active materials for mechanical energy harvesting. The development of OIHP-based mechanical energy harvesters can substantially advance IoT and AI systems. OIHP nanogenerators can operate as sensors that have a wide range of utility in environment monitoring, health monitoring, motion detection, robotics, e-skin, and human-machine interactions. Furthermore, those nanogenerators can supply power to conventional batteries in small-scale and portable electronic devices. However, the key factors that need to be resolved currently in the field of OIHP-based MEHs are air-stability, encapsulation, toxicity, mechanical sturdiness, and moderate performances. Future developments in this field are likely to be focused on the following aspects. First, the output performance and energy conversion efficiency of OIHP-based MEHs should be improved to meet the requirements of small-scale/portable devices. Second, in order to be implemented in wearable devices, the OIHP nanogenerators should be highly flexible, stretchable, and lightweight, and must be able to withstand harsh environs. Third, the eco-friendly nature and low toxicity of OIHP-based devices is a key characteristic for use in health monitoring/biomedical devices; thus, the need of lead-free OIHPs for MEH applications is necessary. Systematic investigations of OIHPs having various dimensions can introduce a new platform for designing high-performance nanogenerators. Controlling the dielectric and ferro/piezoelectric properties of various OIHPs via compositional and structural engineering can also assist nanogenerators to improve energy conversion efficiencies. Furthermore, it has been recognized that layered 2D OIHPs have better piezo/ferroelectric properties along with decent moisture and air stability compared to the 3D OIHPs owing to the presence of long-chain organic cation molecules. It is expected that flexible and stretchable self-powered systems with dynamic sensing properties are the future direction of wearable electronic devices. Therefore, integrating OIHPs with flexible piezoelectric polymers will aid in the construction of air-stable, mechanically robust and high-performance nanogenerators.

6. Conclusions

In this chapter, we presented the ability of OIHP materials including organic and inorganic halide perovskite materials to produce electricity by harvesting the ambient abundant mechanical energy. The structural suitability of various OIHPs and OIHP–polymer composites for developing high-performance MEHs is discussed in detail along with their dielectric, piezoelectric, and ferroelectric properties. In addition, some significant works based on OIHP-MEHs in the form of PENGs and TENGs are summarized. Finally, the existing issues and challenges facing current research are stated and some future research directions for pursuing the commercialization of OIHP-based MEHs in wearable, portable electronic devices are suggested.

Acknowledgements

This work was supported by a National Research Foundation of Korea (NRF) grant funded by the Korean government (MSIP) (NRF-2021R1A2B5B03002016, NRF-2021R1A2C1010797, NRF-2021R1I1A1A01060012) and by the Basic Research Program through the National Research Foundation of Korea (NRF) funded by the Ministry of Education (Grant No. 2021R1A6A1A03043682).

Conflict of interest

The authors declare no conflict of interest.

Author details

Venkatraju Jella*†, Swathi Ippili*†, Hyun You Kim, Hyun-Suk Kim, Chunjoong Kim, Tae-Youl Yang and Soon-Gil Yoon*
Department of Materials Science and Engineering, Chungnam National University, Daejeon, Republic of Korea

*Address all correspondence to: venkatrajujella@gmail.com, iswathirs@gmail.com and sgyoon@cnu.ac.kr

† Both authors contributed equally to this work.

References

[1] Nižetić S, Šolić P, de Gonzáleze DL, Patrono L. Internet of Things (IoT): opportunities, issues and challenges towards a smart and sustainable future. Journal of Cleaner Production. 2020;**274**:122877. DOI: 10.1016/j.jclepro.2020.122877

[2] Shi Q, Dong B, He T, Sun Z, Zhu J, Zhang Z, et al. Progress in wearable electronics/photonics—moving toward the era of artificial intelligence and internet of things. InfoMat. 2020;**2**:1131-1162. DOI: 0.1002/inf2.12122

[3] Lee JH, Kim J, Kim TY, Al Hossain MS, Kim SW, Kim JH. All-in-one energy harvesting and storage devices. Journal of Materials Chemistry A. 2016;**4**:7983-7999. DOI: 10.1039/C6TA01229A

[4] Jin DW, Ko YJ, Ahn CW, Hur S, Lee TK, Jeong DG, Lee M, Kang C.-Y, Jung JH: Polarization- and electrode-optimized polyvinylidene fluoride films for harsh environmental piezoelectric nanogenerator applications. Small. 2021;17:2007289. DOI: 10.1002/smll.202007289

[5] Mao Y, Zhang N, Tang Y, Wang M, Chao M, Liang E. A paper triboelectric nanogenerator for self-powered electronic systems. Nanoscale. 2017;**9**:14499-14505. DOI: 10.1039/C7NR05222G

[6] Lee H, Kim H, Kim DY, Seo Y. Pure piezoelectricity generation by a flexible nanogenerator based on lead zirconate titanate nanofibers. ACS Omega. 2019;**4**:2610-2617. DOI: 10.1021/acsomega.8b03325

[7] Le AT, Ahmadipour M, Pung S-Y. A review on ZnO-based piezoelectric nanogenerators: Synthesis, characterization techniques, performance enhancement and applications. Journal of Alloys and Compounds. 2020;**844**:156172. DOI: 10.1016/j.jallcom.2020.156172

[8] Ippili S, Jella V, Thomas AM, Yoon C, Jung J-S, Yoon S-G. ZnAl–LDH-induced electroactive β-phase and controlled dielectrics of PVDF for a high-performance triboelectric nanogenerator for humidity and pressure sensing applications. Journal of Materials Chemistry A. 2021;**9**:15993-16005. DOI: 10.1039/D1TA02966E

[9] Lu L, Ding W, Liu J, Yang B. Flexible PVDF based piezoelectric nanogenerators. Nano Energy. 2020;**78**:105251. DOI: 10.1016/j.nanoen.2020.105251

[10] Luo Y, Szafraniak I, Zakharov ND, Nagarajan V, Steinhart M, Wehrspohn RB, et al. Nanoshell tubes of ferroelectric lead zirconate titanate and barium titanate. Applied Physics Letters. 2003;**83**:440-442. DOI: 10.1063/1.1592013

[11] Jiang W, Zhang R, Jiang B, Cao W. Characterization of piezoelectric materials with large piezoelectric and electromechanical coupling coefficients. Ultrasonics. 2003;**41**:55-63. DOI: 10.1016/S0041-624X(02)00436-5

[12] Jeong J, Kim M, Seo J, Cao W. Pseudo-halide anion engineering for α-FAPbI3 perovskite solar cells. Nature. 2021;**592**:381-385. DOI: 10.1038/s41586-021-03406-5

[13] Jella V, Ippili S, Eom J-H, Pammi SVN, Jung J-S, Tran V-D, et al. A comprehensive review of

flexible piezoelectric generators based on organic-inorganic metal halide perovskites. Nano Energy. 2019;**57**:74-93. DOI: 10.1016/j.nanoen.2018.12.038

[14] Kim YJ, Dang TV, Choi HJ, Park BJ, Eom JH, Song HA, et al. Piezoelectric properties of $CH_3NH_3PbI_3$ perovskite thin films and their applications in piezoelectric generators. Journal of Materials Chemistry A. 2016;**4**:756-763. DOI: 10.1039/C5TA09662F

[15] Su ZX, Zhao HY, Li J, Yuan ZL, Wang GZ, Cao G, et al. High-performance organolead halide perovskite-based self-powered triboelectric photodetector. ACS Nano. 2015;**9**:11310-11316. DOI: 10.1021/acsnano.5b04995

[16] Xie GX, Yang XY, Duan JL, Duan YY, Tang QW. Bulk Pt/CsPbBr3 Schottky junctions for charge boosting in robust triboelectric nanogenerators. Journal of Materials Chemistry A. 2020;**8**:11966-11975. DOI: 10.1039/D0TA04463F

[17] Wang YD, Yang XY, Xu WK, Yu XP, Duan JL, Duan YY, et al. Triboelectric behaviors of inorganic $Cs_{1-x}A_xPbBr_3$ halide perovskites toward enriching the triboelectric series. Journal of Materials Chemistry A. 2020;**8**:25696-25705. DOI: 10.1039/D0TA09982A

[18] Ippili S, Jella V, Thomas AM, Yoon S-G. The recent progress on halide perovskite-based self-powered sensors enabled by piezoelectric and triboelectric effects. Nanoenergy Advances. 2021;**1**:3-31. DOI: 10.3390/nanoenergyadv1010002

[19] Goldschmidt V. M Krystallbau und chemische zusammensetzung. zusammenfassender vortrag, gehalten vor der deutschen chemischen gesellschaftam. 1927;**60**:1263

[20] Yi Z, Ladi NH, Shai X, Li H, Shen Y, Wang M-K. Will organic–inorganic hybrid halide lead perovskites be eliminated from optoelectronic applications? Nanoscale Advances. 2019;**1**:1276-1289. DOI: 10.1039/C8NA00416A

[21] Park N-G. Perovskite solar cells: an emerging photovoltaic technology. Materials Today. 2015;**18**:65-72. DOI: 10.1016/j.mattod.2014.07.007

[22] Li Z, Yang M, Park J-S, Wei S-H, Berry J, Zhu K. Stabilizing perovskite structures by tuning tolerance factor: Formation of formamidinium and cesium lead iodide solid-state alloys. Chemistry of Materials. 2016;**28**:284-292. DOI: 10.1021/acs.chemmater.5b04107

[23] Mao L, Ke W, Pedesseau L, Wu Y, Katan C, Even J, et al. Hybrid Dion–Jacobson 2D lead iodide perovskites. Journal of the American Chemical Society. 2018;**140**:3775-3783. DOI: 10.1021/jacs.8b00542

[24] Mao L, Stoumpos CC, Kanatzidis MG. Two-dimensional hybrid halide perovskites: principles and promises. JAM Chemical Society. 2019;**141**:1171-1190. DOI: 0.1021/jacs.8b10851

[25] Jella V, Ippili S, Yoon S-G. Halide (Cl/Br)-incorporated organic–inorganic metal trihalide perovskite films: Study and investigation of dielectric properties and mechanical energy harvesting performance. ACS Applied Electronic Materials. 2020;**2**:2579-2590. DOI: 10.1021/acsaelm.0c00473

[26] Ippili S, Jella V, Kim J, Hong S, Yoon S-G. Enhanced piezoelectric output performance via control of dielectrics in Fe^{2+}-incorporated $MAPbI_3$ perovskite thin films: Flexible piezoelectric generators. Nano Energy.

2018;**49**:247-256. DOI: 10.1016/j.nanoen.2018.04.031

[27] Labuda A, Proksch R. Quantitative measurements of electromechanical response with a combined optical beam and interferometric atomic force microscope. Applied Physics Letters. 2015;**106**:253103. DOI: 10.1063/1.4922210

[28] Dong Q, Song J, Fang Y, Shao Y, Ducharme S, Huan J. Bioinspired ferroelectric polymer arrays as photodetectors with signal transmissible to neuron cells. Advanced Materials. 2018;**28**:2816. DOI: 10.1002/adma.201603618

[29] Jella V, Ippili S, Eom J-H, Choi J, Yoon S-G. Enhanced output performance of a flexible piezoelectric energy harvester based on stable $MAPbI_3$-PVDF composite films. Nano Energy. 2018;**53**:46-56. DOI: 10.1016/j.nanoen.2018.08.033

[30] Ippili S, Jella V, Kim J, Hong S, Yoon S-G. Unveiling predominant air-stable organotin bromide perovskite toward mechanical energy harvesting. ACS Applied Materials & Interfaces. 2020;**12**:16469-16480. DOI: 0.1021/acsami.0c01331

[31] Ippili S, Jella V, Eom J-H, Kim J, Hong S, Choi J-S, et al. An ecofriendly flexible piezoelectric energy harvester that delivers high output performance is based on lead-free $MASnI_3$ films and $MASnI_3$- PVDF composite films. Nano Energy. 2019;**57**:911-923. DOI: 1016/j.nanoen.2019.01.005

[32] Ding R, Liu H, Zhang X, Xiao J, Kishor R, Sun H, et al. Flexible piezoelectric nanocomposite generators based on formamidinium lead halide perovskite nanoparticles. Advanced Functional Materials. 2016;**26**:7708-7716. DOI: 10.1002/adfm.201602634

[33] Kim DB, Park KH, Cho YS. Origin of high piezoelectricity of inorganic halide perovskite thin films and their electromechanical energy-harvesting and physiological current-sensing characteristics. Energy & Environmental Science. 2020;**13**:2077-2086. DOI: 10.1039/C9EE03212F

[34] Chen XG, Song XJ, Zhang ZX, Li PF, Ge JZ, Tang YY, et al. Two-dimensional layered perovskite ferroelectric with giant piezoelectric voltage coefficient. Journal of the American Chemical Society. 2020;**142**:1077-1082. DOI: 10.1021/jacs.9b12368

[35] Smith MB, Page K, Siegrist T, Redmond PL, Walter EC, Seshadri R, et al. Crystal structure and the paraelectric-to-ferroelectric phase transition of nanoscale $BaTiO_3$. Journal of the American Chemical Society. 2008;**130**:6955-6963. DOI: 10.1021/ja0758436

[36] Kutes Y, Ye V, Zhou Y, Pang S, Huey BD, Padture NP. Direct observation of ferroelectric domains in solution-processed $CH_3NH_3PbI_3$ perovskite thin films. The Journal of Physical Chemistry Letters. 2014;**5**:3335-3339. DOI: 10.1021/jz501697b

[37] Rakita Y, Bar-Elli O, Meirzadeh E, Kaslasi H, Peleg Y, Hodes G, et al. Tetragonal $CH_3NH_3PbI_3$ is ferroelectric. Proceedings of the National Academy of Sciences of the United States of America. 2017;**114**:E5504. DOI: 10.1073/pnas.1702429114

[38] Pan Q, Liu Z-B, Tang Y-Y, Li P-F, Ma R-W, Wei R-Y, et al. A three-dimensional molecular perovskite ferroelectric: (3-ammoniopyrrolidinium) $RbBr_3$. Journal of the American Chemical Society. 2017;**139**:3954-3957. DOI: 10.1021/jacs.7b00492

[39] Chen X-G, Song X-J, Zhang Z-X, Zhang H-Y, Pan Q, Yao J, et al. Confinement-driven ferroelectricity in a two-dimensional hybrid lead iodide perovskite. Journal of the American Chemical Society. 2020;**142**(22):10212-10218. DOI: 10.1021/jacs.0c03710

[40] Feng H, Zhao C, Tan P, Liu R, Chen X, Li Z. Nanogenerator for biomedical applications. Advanced Healthcare Materials. 2018;**7**:1701298. DOI: 0.1002/adhm.201701298

[41] Lin L, Wang V, Xie Y, Jing Q, Niu S, Hu Y, et al. Sliding-triboelectric nanogenerators based on in-plane charge-separation mechanism. Nano Letters. 2013;**13**:2916-2233. DOI: 10.1021/nl400738p

[42] Wang S, Niu S, Yang J, Lin L, Wang ZL. Quantitative measurements of vibration amplitude using a contact-mode freestanding triboelectric nanogenerator. ACS Nano. 2014;**8**:12004-12013. DOI: 0.1021/nn5054365

[43] Huang S, Shi L, Zou T, Kuang H, Rajagopalan P, Xu H, et al. Controlling performance of organic-inorganic hybrid perovskite triboelectric nanogenerators via chemical composition modulation and electric field-induced ion migration. Advanced Energy Materials. 2020;**10**:2002470. DOI: 10.1002/aenm.202002470

[44] Yu XP, Wang YD, Zhang JH, Duan JL, Yang XY, Liu LM, et al. Halogen regulation of inorganic perovskites toward robust triboelectric nanogenerators and charging polarity series. Journal of Materials Chemistry A. 2020;**8**:14299. DOI: 10.1039/D0TA05531J

[45] Yoon C, Ippili S, Jella V, Thomas AM, Jung J-S, Han Y, et al. Synergistic contribution of flexoelectricity and piezoelectricity towards a stretchable robust nanogenerator for wearable electronics. Nano Energy. 2021;**91**:106691. DOI: 10.1016/j.nanoen.2021.106691

[46] Ding R, Zhang X, Chen G, Wang H, Kishor R, Xiao J, et al. High-performance piezoelectric nanogenerators composed of formamidinium lead halide perovskite nanoparticles and poly(vinylidene fluoride). Nano Energy. 2017;**37**:126-135. DOI: 10.1016/j.nanoen.2017.05.010

[47] Jella V, Ippili S, Eom, J.H, Kim YJ, Kim HJ, Yoo S-G: A novel approach to ambient energy (thermoelectric, piezoelectric and solar-TPS) harvesting: Realization of a single structured TPS-fusion energy device using $MAPbI_3$. Nano Energy. 2018;52:11-21. DOI: 10.1016/j.nanoen.2018.07.024

[48] Liu S, Zheng F, Grinberg I, Rappe AM. Photoferroelectric and photopiezoelectric properties of organometal halide perovskites. The Journal of Physical Chemistry Letters. 2016;7:1460-1465. DOI: 10.1021/acs.jpclett.6b00527

[49] Ueberschlag P. PVDF piezoelectric polymer. Sensor Review. 2001;**21**:118-126. DOI: 10.1108/02602280110388315

[50] Pandey R, Sb G, Grover S, Singh SK, Kadam A, Ogale S, et al. Microscopic origin of piezoelectricity in lead-free halide perovskite: Application in nanogenerator design. ACS Energy Letters. 2019;**4**:1004-1011. DOI: 10.1021/acsenergylett.9b00323

[51] Chen H, Zhou L, Fang Z, Wang S, Yang T, Zhu L, et al. Piezoelectric nanogenerator based on in situ growth all-inorganic $CsPbBr_3$ perovskite nanocrystals in PVDF fibers with long-term stability. Advanced Functional Materials. 2021;**31**:2011073. DOI: 10.1002/adfm.202011073

[52] Ippili S, Jella V, Eom S, Hong S, Yoon S-G. Light-driven piezo- and triboelectricity in organic–inorganic metal trihalide perovskite toward mechanical energy harvesting and self-powered sensor application. ACS Applied Materials & Interfaces. 2020;**12**:50472-50483. DOI: 10.1021/acsami.0c15009

[53] Guan H, Lv D, Zhong T, Dai Y, Xing L, Xue X, et al. Self-powered, wireless-control, neural-stimulating electronic skin for in vivo characterization of synaptic plasticity. Nano Energy. 2020;**67**:104182. DOI: 10.1016/j.nanoen.2019.104182

[54] Jiang F, Zhou X, Lv J, Chen J, Chen J, Kongcharoen H, et al. Stretchable, breathable, and stable lead-free perovskite/polymer nanofiber composite for hybrid triboelectric and piezoelectric energy harvesting. Advanced Materials. 2022;**2022**:2200042. DOI: 0.1002/adma.202200042

Chapter 11

Encapsulation against Extrinsic Degradation Factors and Stability Testing of Perovskite Solar Cells

Edwin Ramírez, Rafael Betancur, Juan F. Montoya, Esteban Velilla, Daniel Ramírez and Franklin Jaramillo

Abstract

Commercialization of perovskite solar technology depends on reaching a stable functioning of the devices. In this regard, both intrinsic (chemistry phenomena of the different device layers) and extrinsic factors (environmental) need to be considered. In this chapter, we report the state of the art of encapsulation techniques against extrinsic degradation mechanisms. Our analysis includes the most common encapsulation structures, materials employed and their by-products, standard methods to test the stability of the devices (accelerated testing, outdoor and degradation monitoring), and security requirements to prevent the health/environmental hazard of lead leakage.

Keywords: encapsulant materials, extrinsic degradation, stability measurements, perovskite, solar cells

1. Introduction

Commercialization of perovskite photovoltaic technology (PPT) relies on the golden triangle of solar cell performance whose vertices are lifetime, cost, and efficiency. In the last years, PPT has achieved photovoltaic conversion efficiency (PCE) up to 25.7% at lab scale, almost matching their silicon counterparts [1]. Moreover, the impressive advances in the fabrication of solar modules by scalable solution deposition techniques [2] such as doctor blade, slot-die, or ink-jet printing have enabled a rapid performance growth of large-area devices. According to NREL, the PCE of perovskite modules has increased from 11.8% to 17.9% in the last 4 years. Also the scale of devices has evolved from "Submodule" with an active area of 200–800 cm^2, to "Small module" with area ranging from 800 to 6500 cm^2 [3]. A recent techno-economic model established a cost range for solution processed perovskite modules of \$3.30/W–0.53 USD/W [4], which is competitive with silicon solar cells. The low cost projected for PPT is linked to the processability by scalable solution-based deposition techniques. Therefore, the figures of merit for cost and efficiency of perovskite solar cells (PSCs) are almost pairing, in few years—those achieved for silicon after decades of technological development. However, the recent record for stability of PSC reached

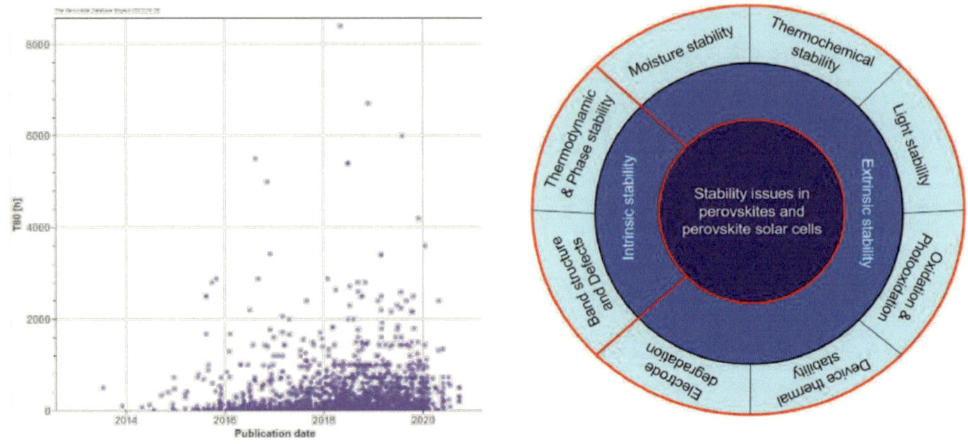

(a) (b)

Figure 1.
(a) State of the art of the stability of perovskite photovoltaic (PV) devices based on the data available in the open-access Perovskite Database [6], and (b) stability issues in perovskite solar cells. Reproduced with permission from references [6, 7].

9000 hours under operational tracking [5], which is far behind the proven lifetime of 25 years (>200,000 hours) reported for silicon solar cells.

In recent years, several research articles have reported stability measurements of perovskite solar cells. **Figure 1a** shows the historical evolution of T_{80} for 1833 PSC devices recorded in the Perovskite Database [6]. T_{80} is a figure of merit defined as the time taken to observe a drop of 20% in the initial power output of the solar cell. Before 2016, only few articles reported T_{80} because the main focus was related to improving efficiency and processability of PSC. Since then, an increasing number of scientific articles report the T_{80} of PSC devices achieving in most cases values up to 1000 hours. Noteworthy, less articles report 1000 hours $< T_{80} <$ 2000 hours and only a small proportion surpasses 2000 hours. This analysis of the scientific literature reveals the urgent necessity of increasing the PSC lifetime targeting to hundreds of thousands of hours.

A complex interaction of factors determines the stability of perovskite solar cells as shown in **Figure 1b**. Devices are degraded by multiple variables such as heat, light, electric load, moisture, and oxygen, which act simultaneously in real operation conditions. Such complexity explains why the progress in stability remains behind the rapid advancements in PCE and processability of PSC. Thus, understanding the degradation mechanisms is of crucial importance to overcome stability issues. Degradation mechanisms can be classified into *intrinsic*, which are related to the compositional and crystallographic structure of the perovskite material, and *extrinsic*, associated with the interaction of the PSC with external factors during their operational life.

2. Intrinsic and extrinsic degradation mechanisms

2.1 Intrinsic degradation mechanisms

Hybrid halide perovskites (HHPs) have been considered as "soft crystalline materials" due to their low formation energy and stability dictated by a delicate thermodynamic balance [8]. Active layers of solar cells have been obtained from HHPs with the general chemical formula ABX_3, where A is an organic or inorganic cation, B is a metallic cation, and X is a halogen anion. This structure has enabled the use of a wide range of atoms as A cations or X anions without significantly losing their high photovoltaic (PV) performance [9]. However, the structure can only accommodate a certain combination of ions due to restriction of sizes in order to have a stable cubic or tetragonal 3D structure. This restriction is often expressed by the Goldschmidt tolerance factor (t), which is calculated in terms of the ionic radii of the constituent ions. The 3D perovskite structure is favored by $0.8 < t < 1$ [8]. In halide perovskites used for PV, the A-site is a monovalent cation, commonly methylammonium ($CH_3NH_3^+/MA^+$), formamidinium ($CH(NH_2)_2^+/FA^+$), cesium (Cs^+), or a combination thereof. The B-site is a divalent metallic cation, mostly lead (Pb^{2+}) but in some cases can be tin (Sn^{2+}), and the X-site is a halide anion that is commonly iodide (I^-) or bromide (Br^-). A stable structure must preserve charge neutrality. Therefore, the valences of A and B should sum the charge of X multiplied by three. The mentioned restrictions in sizes and charges of the constituent ions determine that only certain perovskite formulations are stable at operational temperature ranges. For instance, the archetypical methylammonium lead iodide ($MAPbI_3$) has a tolerance factor of 0.911 forming a tetragonal structure at room temperature. This phase is stable up to 327.4 K turning into a cubic structure at higher temperatures [10]. During the fabrication of perovskite films, as well as during the solar cell operation, the material is subjected to temperature cycles that can promote the formation of unwanted PV inactive phases. This phenomenon has been observed for formamidinium lead iodide ($FAPbI_3$), which crystallizes into a non-perovskite hexagonal phase (yellowish phase) at room temperature due to its high tolerance factor ($t = 1.04$). This perovskite turns into a cubic structure after annealing over 150 °C [7]. $FAPbI_3$-based PSC devices have been developed motivated by the higher thermal stability of FA^+ compared with MA+ cation, but a careful material processing is required in order to avoid the formation of the non-active hexagonal phase of $FAPbI_3$. Although pure $FAPbI_3$ or $CsPbI_3$ perovskites yield higher thermal stability of the A-site cation, they are not the preferred choice due to thermodynamic phase stability issues. Alloys of A-site cations have been used to produce materials such as $FA_xCs_{1-x}PbI_3$ or $FA_{0.75}MA_{0.15}Cs_{0.1}Pb(I_{0.83}Br_{0.17})_3$ used as active layers in high performance devices [11]. Thus, compositional tuning is a suitable strategy to overcome thermodynamic stability issues. Note that the proportion between the A-site cations must be strictly controlled to achieve a suitable tolerance factor. In addition, since the perovskite bandgap depends in part on the energy level of the anion occupied p orbital, such compositional tuning also determines the resulting bandgap.

The benchmark $MAPbI_3$ perovskite has shown high defect tolerance preserving its opto-electronic properties even at high trap density (10^{14}–10^{16} cm^{-3}). In contrast, PV grade gallium arsenide (GaAs) must have a defect concentration as low as 10^7 cm^{-3} [12]. This high tolerance to defects enables the synthesis of HHPs by solution processing techniques. Density functional calculations show that the growth conditions of the perovskite correlate with the concentration and type of defects. Specifically, the I/Pb ratio determines the formation energy of defects. It was found that formation energy of deep trap states is very high, therefore the probability of having nonradiative recombination centers is low. As a result, shallow trap states found in the $MAPbI_3$ perovskite are not detrimental of the photovoltaic performance [13]. This outstanding defect tolerance of the $MAPbI_3$ perovskite derives from its exceptional band structure. Notably, related perovskites with compositional mixtures at the A-site cations or halides have also shown defect tolerance if they have an adequate Goldschmidt factor. Accordingly, a HHP with high structural stability is also defect tolerant.

In summary, compositional tuning is a suitable strategy to increase intrinsic stability, which means a thermodynamically stable structure with high defect tolerance. Both properties are of utmost importance for a photovoltaic material.

2.2 Extrinsic degradation mechanisms

Beyond compositional tuning, additional strategies must be developed to increase extrinsic stability. During solar cell operation, the perovskite active layer may interact with external factors such as heat, light, moisture, oxygen, electric bias, and other interface or external agents. Here, we briefly review some of the most important degradation mechanisms. Further information can be found elsewhere.

Water: When water interacts with perovskite, it can form hydrate or dehydrate phases according to reactions (1) and (2).

$$CH_3NH_3PbI_3(s) + H_2O(g) \longleftrightarrow CH_3NH_3PbI_3 \cdot H_2O \quad (1)$$

$$4[CH_3NH_3PbI_3 \cdot H_2O] \longleftrightarrow (CH_3NH_3)_4 PbI_6.2H_2O + 3PbI_2 + 2H_2O \quad (2)$$

These reactions are reversible. Thus, perovskite can be regenerated exposing it to an inert environment. However, some irreversibility can appear due to phase segregation. Water forms hydrogen bonds with the A-site cation weakening its interaction with the lead halide octahedra [14]. As a result, the perovskite becomes prone to degradation by external other stressors such as heat or electric bias. Once the perovskite is saturated by moisture, it fully decomposes to PbI_2 and MAI.

Heat: During solar cell operation, the device is subjected to temperature cycles. As mentioned before, PV active phases of perovskites are stable in a temperature range. Moreover, some A-site cations are volatile organic molecules, which can be converted into gas-phase products when the perovskite reaches some critical temperatures. For instance, the $MAPbI_3$ can be decomposed to PbI_2, ammonia and methyl iodide when it is heated to 85°C in an inert atmosphere [15]. Additionally, the materials commonly used in hole or electron transporting layers are organic molecules, which can also be degraded at some temperatures. The main strategies to avoid degradation by heat are compositional engineering of the A-site cation and encapsulation with materials with good heat dissipation properties.

Light: light-induced degradation has been attributed to the migration of vacancies in the perovskite layer [16]. UV degradation takes place in the absence of moisture and particularly in the device stack denoted as n-i-p, especially those containing TiO_2, which can induce photo-degradation. However, it has been demonstrated that UV-degraded devices can be subsequently recovered by 1-sun light soaking. The UV degradation/recovery phenomenon has been attributed to the free carriers generated by light soaking after neutralization of accumulated trap states and generated free charges [17].

Moreover, light can induce redistribution of halide and metal ions in the $MAPbI_3$ perovskite film. This phenomenon causes the increment of the photoluminescence under illumination due to the diffusion of I^- species [18]. Ionic migration leads to phase segregation in the perovskite layer with halide-rich or halide-deficient areas being the most commonly found. These defects are carrier trapping states with smaller bandgaps, which generate the increment in the photoluminescence. In the presence of moisture or oxygen, this phenomenon is more pronounced due to the passivation effect of superoxide molecules generated by the reaction of O_2 and H_2O with light [19]. In alloyed perovskites, light-induced A-site cation segregation has also been observed [20].

The main strategy against extrinsic degradation mechanisms is based on the encapsulation process, which is reviewed in the following.

3. Encapsulant structures

The encapsulation of solar cells has been highly guided by the industrial experience with commercial silicon solar cells. In general, the components for encapsulating a solar device [22] are presented in **Figure 2**. The general requirements that each encapsulation component needs to meet in order to protect the perovskite solar cells from external degradation factors are summarized in **Table 1**. These components include:

3.1 Polymeric encapsulant

It is the most critical component (also known as pottant or filling material). The list of its requirements includes: excellent adhesion properties, optical matching for better sunlight injection, electrical insulation, mechanical support, serves to physically separate cells and strings, UV stability, mechanical strength, thermal conduction, and stands thermal cycles. In addition, the selected encapsulant needs to adapt to specific processing conditions such as narrow temperature/pressure operational ranges and restrictions about the solvents and by-products that are considered "safe" to interact with the solar cell during the encapsulation process. Typical polymeric encapsulant materials are ethylene-vinyl acetate (EVA), PVB, TPU, silicone, ionomer, and UV-curable resin. However, it is important to mention that some devices, mostly at research level, do not include a polymeric encapsulant precisely to prevent device affectation after the interaction of the solar cell with the encapsulation material. In such cases, it is common to find in the filling space a desiccant material [22]. Finally, the standard encapsulation process for commercial PV devices, for instance, silicon ones, includes two layers of encapsulant fully surrounding the solar cell and its electrical contacts.

3.2 Substrate (bottom barrier)

It provides mechanical support and must provide a good thermal conductivity and a physical barrier against weathering. Typical materials include:

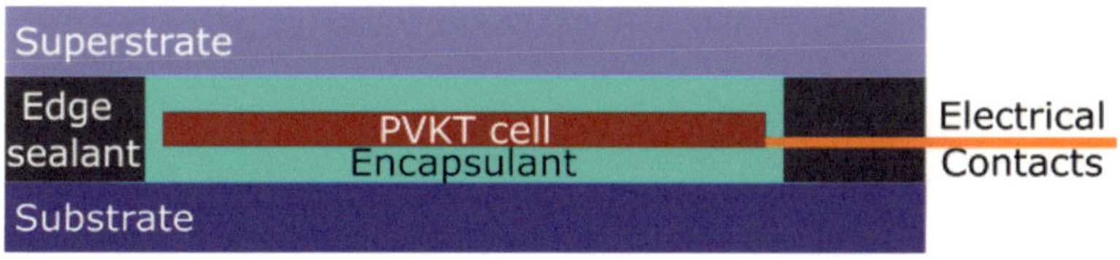

Figure 2.
Components used to encapsulate a perovskite solar cell.

General requirements		Encapsulation component			
		E	S	SP	ES
Mechanical	Resistance to break/tear		✓	✓	
	High adhesion strength between components	✓	✓	✓	✓
	Fixation of cells	✓			
Electrical	Insulation between cells/strings	✓	✓	✓	
	High dielectric breakdown	✓	✓	✓	✓
	High volumetric resistivity	✓	✓	✓	✓
Optical	High optical transmittance	✓		✓	
	Refractive index matching	✓		✓	
	Resistance to UV-induced yellowing	✓	✓	✓	✓
Physical protection	Low moisture absorption	✓	✓	✓	✓
	Low oxygen absorption		✓	✓	✓
Thermal	High thermal conductivity	✓	✓	✓	✓

E: Encapsulant; S: substrate; SP: superstrate; and ES: edge sealant.

Table 1.
General requirements of each encapsulation component.

- Glass
- Fluoropolymers-based: polyvinyl fluoride-based films such as Tedlar® and ethylene-tetrafluoroethylene-based films such as Tefzel™.
- Tedlar-based: TPT (Tedlar/PET/Tedlar), TPE (Tedlar/PET/EVA), TAT (Tedlar/aluminum foil/EVA), TPAT (Tedlar/PET/aluminum foil/Tedlar), TPOT (Tedlar/PET/oxide/Tedlar).
- Polyethylene terephthalate (PET) and polyethylene naphthalate (PEN) based: PAP (PEN/aluminum foil/PET), Teijin Teonex, BaSO4-filled PET

3.3 Superstrate (top barrier)

It is the front barrier of the solar cell. Accordingly, it provides physical protection against weathering and serves as optical coupling for the injection of sunlight. Frequently, the superstrate includes surface coatings such as UV protection and antireflective coatings (ARC). Typical materials include:

- Glass: low-iron and tempered.
- Flexible glass.
- Flexible polymer ultra-barriers [23] Amcor, Viewbarrier (Mitsubishi Plastic Inc), Opteria, Tera barrier.

<table>
<tr><th>Typical names</th><th>Architecture</th></tr>
<tr><td>Complete encapsulation [23]. Blanket encapsulation with edge sealing [2].</td><td>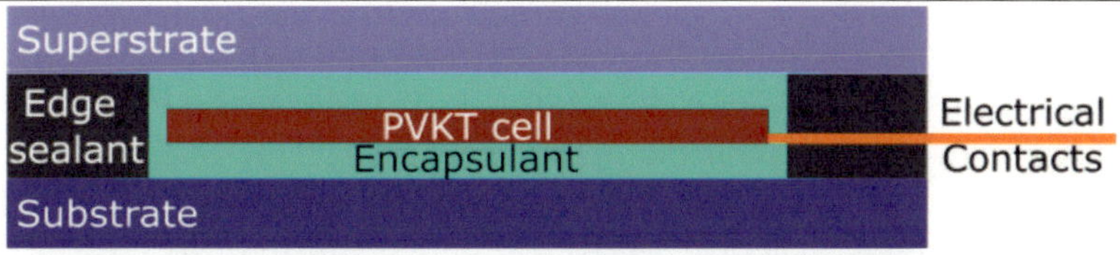
</td></tr>
<tr><td>Blanket encapsulation [22]</td><td>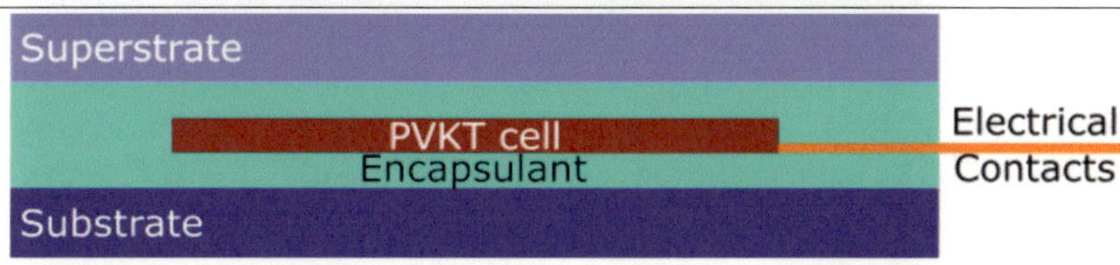
</td></tr>
<tr><td>Edged covered encapsulation [24]. The structure is filled with inert gas.</td><td>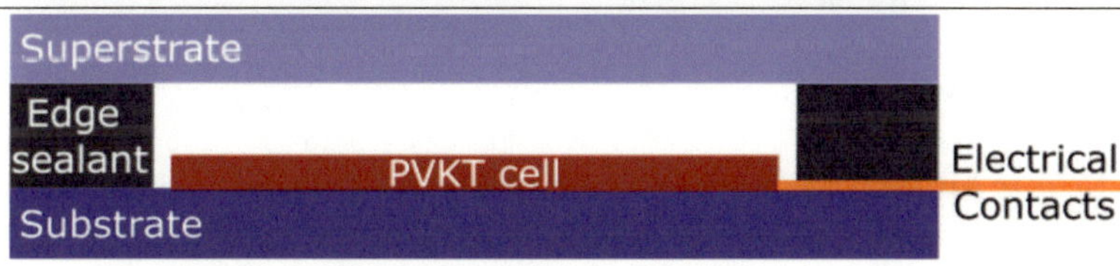
</td></tr>
<tr><td>Film encapsulation [24]</td><td>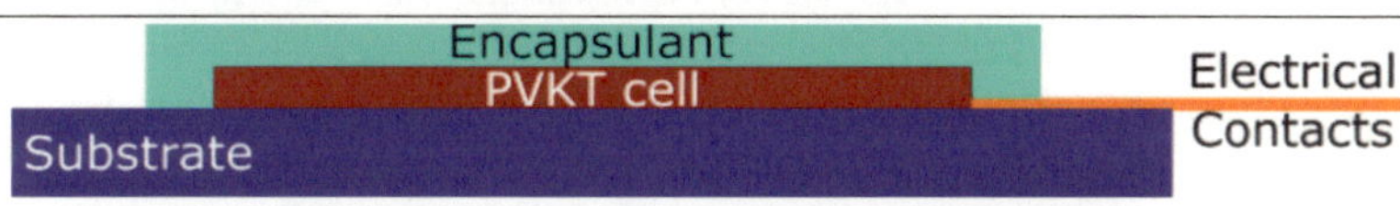
</td></tr>
<tr><td>Russian doll encapsulation [24]. The structure is filled with an inert gas.</td><td>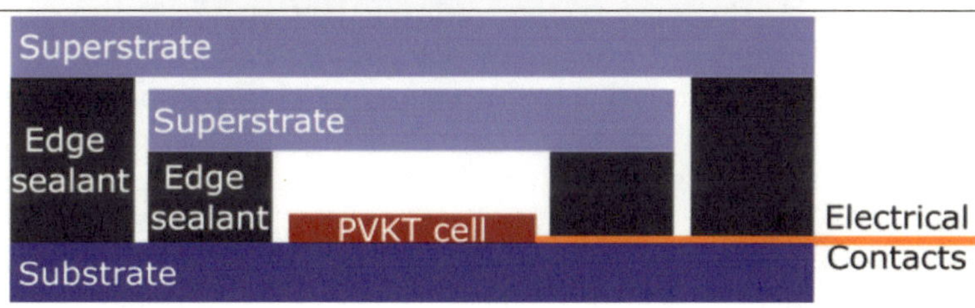
</td></tr>
</table>

Table 2.
Encapsulation architectures.

3.4 Edge sealant

Its primary function is blocking the lateral ingress of oxygen and moisture. Typical materials are polybutyl ("hot butyl"), silicones, and polyisobutylene (PIB). An interesting integration between edge sealant with desiccant materials was achieved by Quanex (SolarGain®).

Clearly, different configurations of the encapsulation components enable several encapsulation architectures. **Table 2** shows the most common architectures. Be advised that there is not a universal labeling of these structures:

4. Encapsulation materials and by-products

The previous section mentioned the encapsulation structures that have been used for perovskite-type solar cells (PSCs). For the manufacture of these structures, methods similar to those used for the encapsulation of silicon panels have been adopted, such as vacuum or roll-to-roll lamination processes [25]. These methods may include UV curing, high-pressure lamination, and temperatures between 80°C and 140°C to ensure good adhesion and avoid thermal degradation of the PSC [26]. The encapsulant materials must be chemically inert to the PSC layers and also serve as a barrier blocking extrinsic factors such as H_2O and O_2. In addition, the encapsulant materials must be stable under temperature, humidity, and illumination conditions, ensuring adequate electrical, optical, and mechanical properties of the resulting device (**Figure 3**). For instance, the materials used for FE (film encapsulant) and ES (edge sealant) must have a high volumetric resistance to offer electrical insulation, among these materials, EVA, Surlyn (ethylene methacrylic acid copolymer), and polyolefin (POE) have volume resistivities of 1×10^{14}, 6×10^{15} years 3×10^{16} Ω cm, respectively [27]. The main properties of the most studied materials used for FE and ES are presented below.

4.1 Optical properties

Mainly for FE, the materials used must present high optical transmittance since they are deposited on the active area of the solar cell and must guarantee that the light passes through. On the other hand, ES materials have fewer optical requirements since they are located in the non-active area of the cell.

EVA, Surlyn, polyvinyl butyral (PVB), UV-cured epoxy resins, polyurethane (PU), and POE have optical transmittance values around 90% [27]. However, the encapsulant EVA (the most common encapsulant used in commercial silicon cells) turns yellowish/brownish after a few years of operation leading to a decreased transmittance [28]. This is due to thermal stress of the copolymer, degradation by UV radiation, or a combination of both factors [28]. This outcome indicates that even for polymers that show high optical transmittance after the lamination process, it is necessary to perform accelerated aging testing under thermal and UV radiation conditions to confirm their durability. If after this kind of tests, the materials show a significant decay in transmittance, a possible solution is the incorporation of anti-oxidants, UV absorbing materials, or the use of Ce-doped glasses to absorb UV and prevent the aging of the encapsulant [29].

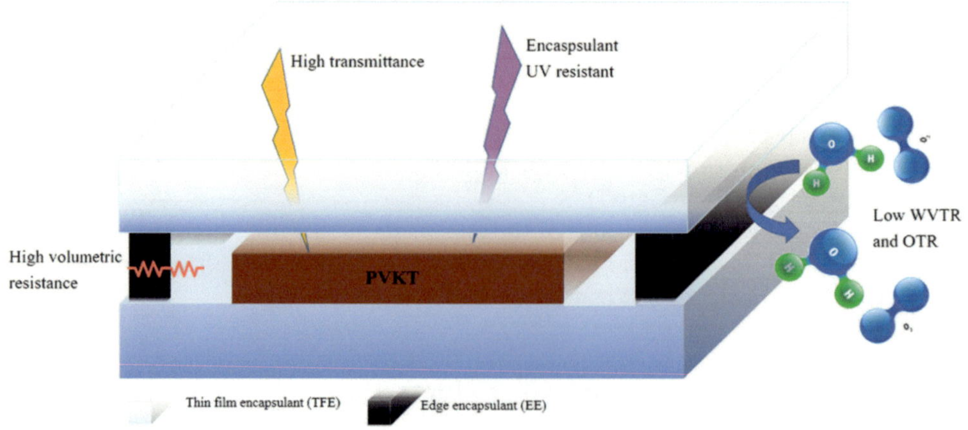

Figure 3.
Scheme of complete encapsulation system.

4.2 Mechanical properties

The materials used in FE and ES must have a low elastic modulus in order to relieve strain and avoid delamination processes. The copolymers EVA and POE, mainly used in FE, have an elastic modulus between 10 and 80 MPa [27]. On the other hand, Surlyn copolymer has an elastic modulus around 400 MPa. For this reason, it has been reported that due to the brittleness of this material, delamination occurs after several measurement cycles [30]. These polymeric materials with low elastic modulus that minimize the presence of cracks under applied stresses are promising candidates for encapsulating flexible perovskite-based devices.

Among the materials that have shown the best results as edge sealing against moisture and oxygen are PIB and glass frits [25, 31]. PIB is a more versatile material with a low modulus of elasticity (9 MPa), which makes it a candidate for rigid and flexible encapsulation systems, while glass frits and epoxy resins are an option for encapsulation in rigid substrates, due to their mechanical rigidity making them prone to cracking. In addition, as further limitations, glass frits require temperatures >100°C for curing and currently are costly [25].

Solar cells must operate in ambient conditions, and it is necessary to anticipate the mechanical behavior of the encapsulating materials in different temperatures. For this reason, they are usually subjected to a thermal cycling test between −40°C and 85°C [27]. The glass transition temperature (T_g) of the encapsulant should be low enough to prevent the encapsulant from embrittlement in low-temperature conditions and failure, resulting in water and oxygen ingress and subsequent degradation. Polymeric materials that have a lower T_g usually have a lower cross-link density, a more flexible structure, and a higher free volume. For instance, the nonpolar chains in Surlyn are grouped together, and the polar ionic groups attract each other. This condition allows the polymer to behave similar to cross-linked polymers being more rigid and less permeable (**Figure 4a** and **b**), whereas EVA has a polymer structure that results in a more flexible but more permeable structure (**Figure 4c**).

4.3 OTR, WVTR, and prevention of volatilization of internal decomposition products

Due to the chemical nature of perovskite solar cells, the main purpose of the encapsulation is to mitigate degradation by extrinsic factors H_2O and O_2. Water vapor transmission rate (WVTR) and oxygen transmission rate (OTR) are the parameters that allow to quantify the water vapor and oxygen that penetrate through an encapsulant film in specific conditions of temperature and relative humidity. Because water molecules are smaller than O_2 molecules, WVTR is used frequently to characterize the barrier properties of the encapsulant. It has been reported that the optimal

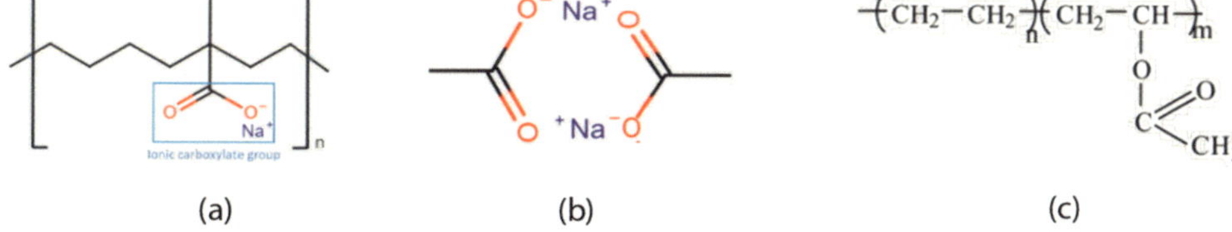

Figure 4.
Polymer molecules: (a) Surlyn, (b) ionic interaction in the Surlyn copolymer, and (c) EVA.

encapsulation materials should have WVTR between 10^{-3} and 10^{-6} gm^{-2} day^{-1} [24]. However, the WVTR varies greatly and depends on both the structure of the polymer and the polarity of the molecules. The nonpolar groups in the polymers are associated with low water affinity and result in surface contact angles >90 degrees. For instance, the thermoplastic polyurethane (TPU) has nonpolar groups and presents a contact angle around 150 degrees. This hydrophobicity allows the PSC modules encapsulated with this material retain 97.52% of the initial efficiency after 2136 hours under outdoor conditions [26].

In addition, to prevent extrinsic degradation, the materials used for FE can also contribute to reducing intrinsic degradation. Under stress factors, perovskite can decompose into volatile species such as HI, NH_3, and CH_3I [25]. However, in the low confined volume of FE/PVSC, the partial pressure of the degradation of volatile species starts to be high, up to the equilibrium point where the decomposition reactions are suppressed. For this reason, the materials used as FE are of special interest, but the reactivity with perovskite layers and the possible formation of by-products must be studied [32].

4.4 Chemical properties and by-products

Materials used in FE must be chemically inert to PSC under UV-Vis radiation, ambient temperature, and high humidity conditions, while those materials used in ES must not release substances that degrade perovskite during thermal, UV, or laser curing. On the other hand, organic encapsulants should have resistance to UV degradation and should not present hydrolysis reactions, for instance, PDMS, POE, PIB, and glass frits have not exhibited any reactions that promote the degradation of perovskite or the material itself in accelerated aging tests. On the other hand, materials such as EVA and Surlyn under prolonged illumination and thermal stress produce acetic acid and acid methacrylate as by-products, respectively [22]. PU has ester bonding (R–NH–COOR'), which in the presence of high humidity leads to hydrolysis and depolymerization [33]. Similarly, it has been reported that PVB, due to its chemical structure, is sensitive to hydrolysis reactions and should be combined with low WVTR edge sealants [34]. On the other hand, the components of UV curable epoxy resins might be inert with PSC but have traces of moisture among them that subsequently degasify and degrade the perovskite layer.

Finally, most of these encapsulation materials are compatible with silicon cells, including EVA with acetic acid as a by-product. However, for PSC encapsulation materials with higher stability are required. Among these, PDMS, polyolefin, and PIB are shown to be the main candidates for encapsulation. Besides, to mitigate intrinsic and extrinsic degradation, the most promising structure is complete encapsulation, with an FE material that has high compatibility with PSC and an edge sealant with the lowest WVRT.

5. Stability testing and characterization

The main parameter to evaluate the stability performance of solar panels is the maximum power (or the panel efficiency), which depends on environmental variables such as solar irradiance and panel temperature [35]. In this context, the failure of an individual device is defined as the time at which the output power drops to 20% below the initial rated power. This parameter corresponds to the standard definition of the lifetime of photovoltaic devices (T_{80}) used as a figure of merit and commonly estimated from regressions analysis using the maximum power as a function of time [36]. Moreover, T_{80} depends on different factors such as the materials and procedures used to fabricate the device, weather, and installation conditions, etc. [37]. Nevertheless, to improve the lifetime, the solar panel must include encapsulation/packaging materials to mitigate degradation processes, increase the electrical insulation, and provide mechanical and thermal support [38].

Delamination or adhesion loss of the encapsulant can be considered the most frequent and severe cause of module degradation, affecting the sunlight absorption and allowing the water-moisture penetration into the device [39]. In addition, most of the recognized failure models on PV silicon modules are related to the packaging materials [40]. **Figure 5** shows the relationship between the device degradation and some failure modes such as discoloration, hydrolysis, corrosion, current leak, encapsulant embrittlement, and delamination. Herein, the degradation occurs when the substrate (backsheet) or encapsulant (typically EVA) is mainly affected by UV, heat, or water ingress. Besides, **Figure 5** correlates the involved tests according to the failure modes. For example, visual inspection and/or thermographic analysis (IF image) for discoloration, delamination or cell crack, chemical degradation for hydrolysis, series resistance (R_s) for the corrosion process, insulation test for leaking currents, and validating the status of the packaging material as a dielectric, etc. Finally, the overview also highlights the importance of the I-V curve and the extracted parameters from this curve to track the degradation processes, such as fill factor (FF), series resistance (R_s), shunt resistance (R_{sh}), etc. [41–43].

In this context, PV manufacturers widely recognized international standards such as IEC 61215 to identify potential failures in silicon photovoltaic modules [44]. This qualification testing is based on three stress factors: light (irradiance and UV), heat, and moisture. Herein, the initial stabilization (exposing the modules to simulated sunlight) and characterization (visual inspection, performance, insulation, and leakage currents) are essential to verify manufacture label values (datasheet). The final stabilization and wet leakage current test are performed to determine the module degradation and evaluate the pass criterium. It is worth noting that the output power determination is performed after a defined cooling time. Thus, eight randomly selected modules are tested into five groups as shown in **Figure 6**.

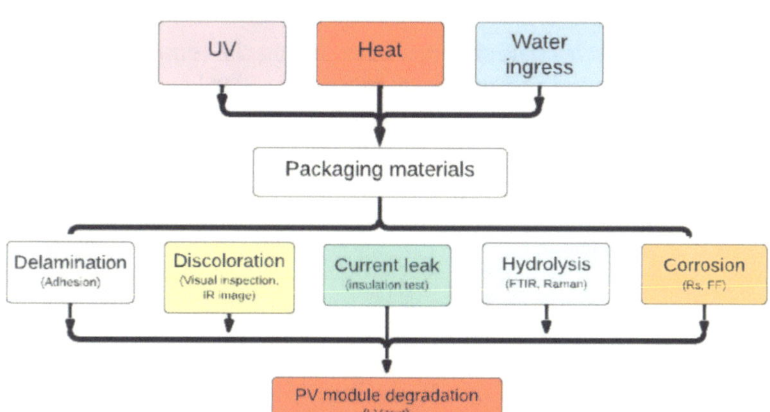

Figure 5.
Correlation of failure modes with the packaging materials used in PV modules. Adapted from [40].

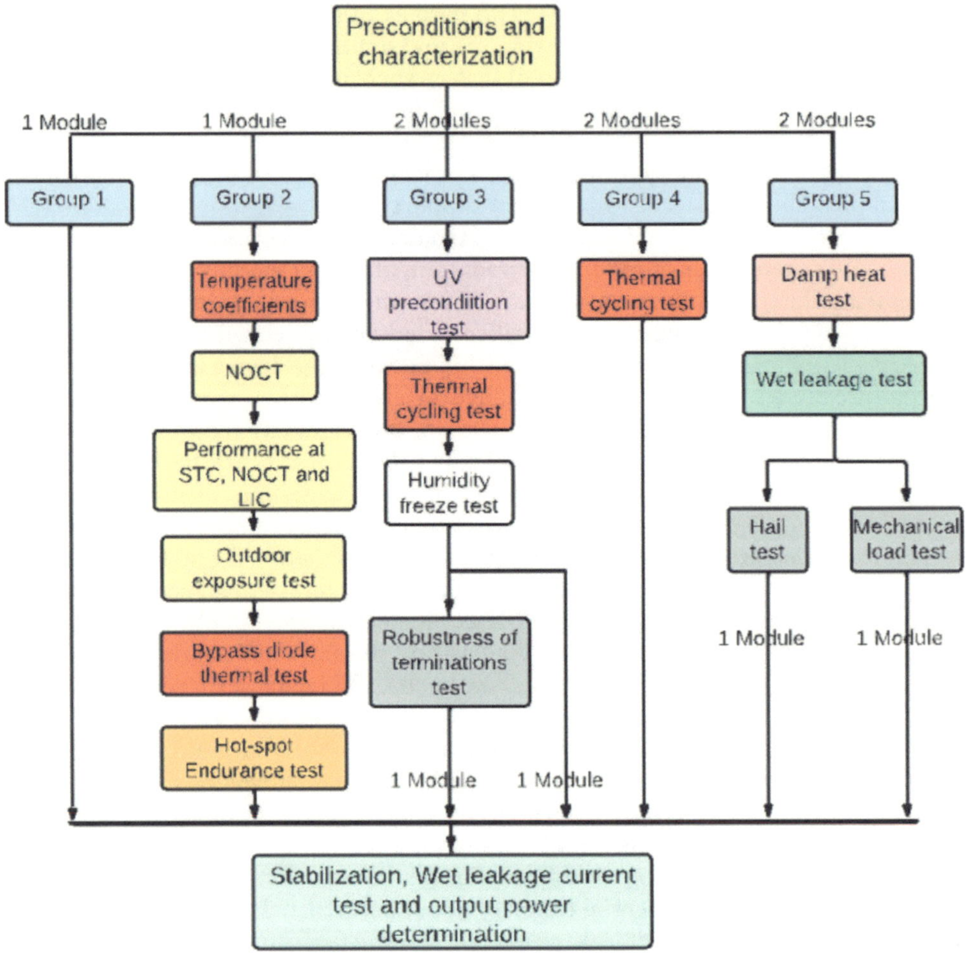

Figure 6.
Flowchart for design qualification of PV modules according to IEC 61215. Adapted from [45].

- *Group 1:* one module is taken as reference (control).
- *Group 2:* one module is characterized considering the temperature coefficients and different power rating conditions such as standard test conditions (STC, irradiance of 1000 W/m^2 and module temperature of 25°C), nominal operative cell temperature conditions (NOCT, irradiance of 800 W/m^2 and ambient temperature of 20°C), and low irradiance conditions (LIC, irradiance of 200 W/m^2 and module temperature of 25°C). Moreover, the bypass diode and hot-spot are tested.
- *Group 3:* two modules are tested in order to determine the ability to withstand the temperature (thermal cycling) and humidity effects (humidity freeze test). In both tests, the applied current and experiment setup must follow the standard suggestions.
- *Group 4:* two modules are tested in order to determine the ability to withstand temperature changes (cycling test).

- *Group 5:* two modules are tested in order to determine the ability to withstand the humidity penetration (damp heat test, 85°C/85%RH). Moreover, mechanical tests are included.

In the case of perovskite solar cells and modules, a broad range of efficiencies have been reported due to the diversity of structures used in the fabrication, highlighting the critical role of the protocols for obtaining reproducible devices [2]. Moreover, several protocols have been designed to evaluate the stability of perovskite devices focused in laboratory-scale cells. From these, the protocol most widely spread is the result from the International Summit on Organic Photovoltaic Stability (ISOS) [46].

Related to the deployment of perovskite technology, several features must be highlighted from the stability tests reported in the literature. Ethylene-vinyl acetate (EVA) as an encapsulant has been successfully tested following the temperature cycles test suggested by IEC 61215 [30]. Polyisobutylene as a barrier layer showed promising results in thermal cycles and damp heat tests [47]. Carbon layer as a barrier increased the long stability of devices up to 12,000 hours of exposure under continuous illumination [48]. A printable mesoscopic solar cell with carbon as the electrode and hot melt polyurethane as encapsulant passed the accelerated tests suggested by IEC 61215 [5]. Epoxy resin was used as encapsulant to evaluate the outdoor performance for minimodules following the international standard IEC 61853-1 [49]. The lifetime of minimodules encapsulated with EVA was estimated in outdoor tests concerning the depicted degradations patterns for the maximum power evolution and ideality factor providing insight concerning the degradation processes [43].

Although a lot of work has been reported on the stability of perovskite technology, the average lifetime (T_{80}) is still short and reaches just a few months [50]. This fact remarks the essential role of the packaging materials to protect the solar cells and electrodes from the environment guaranteeing lower degradation rates. In fact, the stability results suggest that there is still room for improvement, particularly outdoor test investigations to provide insights related to failure modes [51]. Nevertheless, it is worth noting that the qualification testing does not test for all failure mechanisms, and for that, it cannot be used to provide a prediction of the device lifetime [52]. Besides, the qualification testing of IEC 61215 is proper for modules (module level); thus, the scaling of the technology and some particularities of the perovskite technology must be considered to adapt or include other requirements for testing the device stability, as occurred with the light-soaking effects (power stabilization) for thin-film modules [44].

6. Security

In addition to the relevance of encapsulation as an important aspect to overcome the extrinsic degradation and improve the operational stability of the PSCs, health and environmental security are also pertinent toward commercialization of the technology. Like other photovoltaic modules, such as silicon, Perovskite modules can be damaged due to several uncontrollable causes that could include hailstones, fire during operation, or some other natural disaster. Therefore, a proper encapsulation can also contribute to prevent environmental issues associated with constituent materials leakage. Particularly, lead represents the most hazardous environmental contaminant among all the constituent materials in a PSC. It has been estimated that for a typical 400–550 nm thick perovskite layer, the unit area concentration of Pb ranges from 0.4

to 0.75 g/m^2 [53, 54], which is a high value, when compared with the amount of lead present in an automobile battery that contains 20 pounds (9,000,000 mg) on average.

If a PSC made from a $CH_3NH_3PbI_3$ is in direct contact with water, it immediately decomposes into PbI_2 and CH_3NH_3I [55]. When a broken device is exposed to simulated rain, it loses up to 72% of lead after 5 min of leaching and 100% after 72 hours. In order to prevent contamination by lead leakage, there are basically two alternatives. The first one is to make a long-lasting encapsulation that can be "indestructible," which is quite difficult to achieve, but somehow possible using a self-healing coating that could heal itself after any kind of scratch. The second strategy is focused on mitigating the leakage after the encapsulation has failed, by means of chemical lead sequestration, using lead adsorbents in the device structure. A schematic representation of these two strategies is shown in **Figure 7**.

In the case of lead leakage prevention using chemical absorbing materials, resins are the preferred choice. Among them, *P*,*P*′-di(2-ethylhexyl)methanediphosphonic acid (DMDP) and *N*,*N*,*N*′,*N*′-ethylenediaminetetrakis(methylenephosphonic acid) (EDTMP) as Pb-chelating agents, or sulfonic acid cation exchange resins, such as Amberlyst15TMH, can be used [56, 57]. **Figure 8** shows two examples of these materials, one chelating resin and one cation exchange resin. In the first case, the two phosphonic acid groups in each DMDP molecule can strongly bind with a Pb^{2+} with a binding energy of 252 kJ mol^{-1}. In the second case, the sulfonic acid groups act as adsorption sites for Pb^{2+} ions and have a surface area (~50 m^2 g^{-1}) due to its nanoscopically porous structure with nanoparticle sizes of ~40 nm. As the two materials act when lead is in its ionic state, they can be introduced as an external coating of the device, but also in combination with the solid state perovskite film, in some cases without decreasing the device performance [56].

The use of a self-healing coating to prevent lead leakage is shown in **Figure 9**. Resins with low glass transition temperature (T_g) are attractive for this purpose. Specifically, epoxy resins with T_g below 50°C can be sandwiched between the

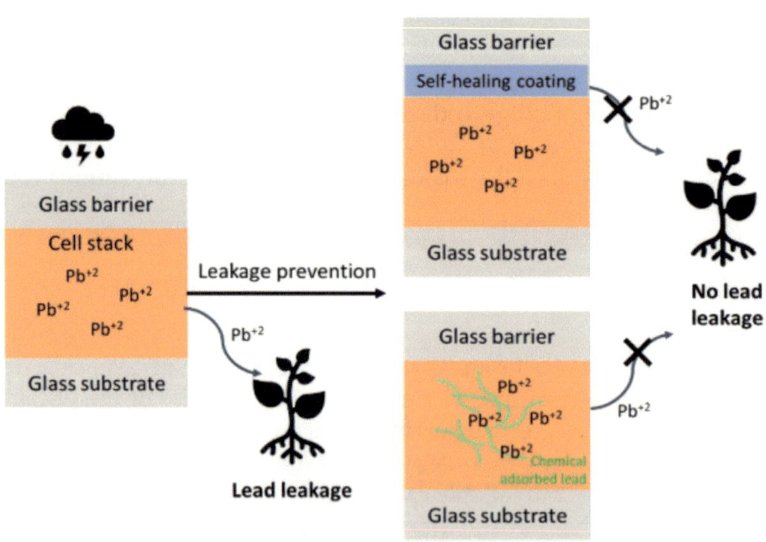

Figure 7.
If a solar cell is broken, the use of two strategies can prevent the lead leakage. The first one (top-right) is the use of a self-healing coating that can automatically repair if it is damaged, and the second one (bottom-right) is the use of an adsorbent chemical that can capture leached lead before it reaches the soil.

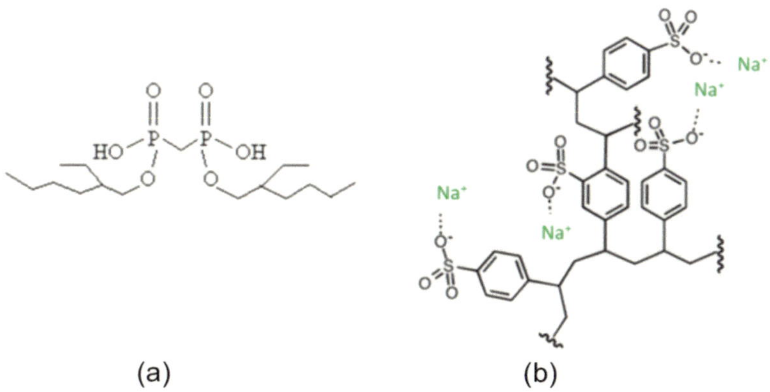

Figure 8.
Chemical structure of (a) DMDP resin, and (b) Amberlyst15TMH cation exchange resin.

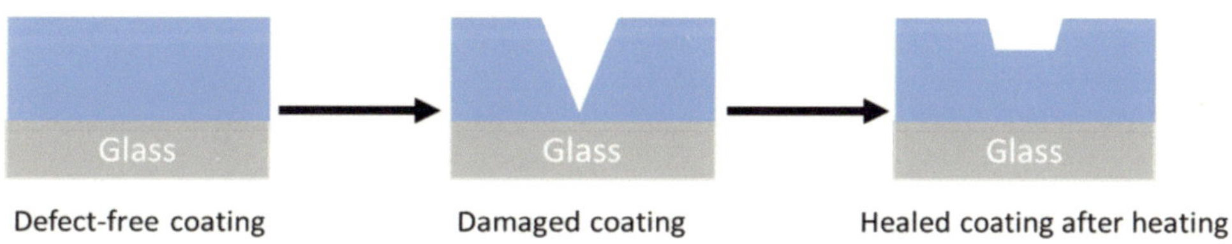

Figure 9.
Schematic representation of the self-healing process of a glass (gray) coated substrate that can be damaged and then recover after heating at low temperature (below 50°C).

perovskite solar module and the top glass cover, and when the glass and the coating are damaged, the heat caused by sunshine can increase the device temperature above T_g, leading to a softening of the epoxy resin and a structural accommodation that can fill the empty spaces (cracks of cut areas).

7. Conclusions

In line with the impressive advances in efficiency (25.7% at laboratory scale) and cost reduction associated with the fabrication of perovskite solar modules using scalable solution deposition techniques, the protection of perovskite solar cells against extrinsic degradation factors, such as moisture, oxygen, heat, and sunlight, is a matter of intense research toward reaching the golden triangle of solar cell performance (lifetime, cost, and efficiency). As reviewed in this chapter, the particularities to improve the lifetime of the perovskite technology have demanded three main aspects:

Innovation in encapsulation materials fully compatible with the perovskite device structure from the chemical and processing frameworks. Up to date, PDMS, polyolefin, and PIB are the main candidates as encapsulants, and the complete encapsulation with edge sealing is the most promising structure to prevent H_2O and O_2 ingress and decrease degradation into volatile components.

Adapting the testing methods to better characterize and predict the temporal performance evolution of the perovskite devices and their failure mechanisms.

Finding alternatives to mitigate the environmental hazard of lead leakage where self-healing encapsulation structures and lead sequestrants outstand.

By consolidating all of these aspects have the potential to reach a robust encapsulation solution that paves the way for commercial perovskite solar cells.

Acknowledgements

The authors gratefully acknowledge the financial support provided by the Colombia Scientific Program within the framework of the call *Ecosistema Científico* (contract no. FP44842—218-2018). Edwin Ramírez thanks the Ministry of Science, Technology, and Innovation of Colombia (MINCIENCIAS), for the national doctoral scholarship number 567-2012.

Author details

Edwin Ramírez, Rafael Betancur, Juan F. Montoya, Esteban Velilla, Daniel Ramírez and Franklin Jaramillo*
Center for Research, Innovation and Development of Materials—CIDEMAT, University of Antioquia UdeA, Medellín, Colombia

*Address all correspondence to: franklin.jaramillo@udea.edu.co

References

[1] Best Research-Cell Efficiency Chart | Photovoltaic Research | NREL. Available from: https://www.nrel.gov/pv/cell-efficiency.html [Accessed: April 30, 2022]

[2] Park NG, Zhu K. Scalable fabrication and coating methods for perovskite solar cells and solar modules. Nature Reviews Materials. 2020;**5**(5):333-350. DOI: 10.1038/s41578-019-0176-2

[3] Champion Photovoltaic Module Efficiency Chart | Photovoltaic Research | NREL. Available from: https://www.nrel.gov/pv/module-efficiency.html [Accessed: April 30, 2022]

[4] Mathews I et al. Economically sustainable growth of perovskite photovoltaics manufacturing. Joule. 2020;**4**(4):822-839. DOI: 10.1016/J.JOULE.2020.01.006

[5] Mei A et al. Stabilizing perovskite solar cells to IEC61215:2016 standards with over 9,000-h operational tracking. Joule. 2020;**4**(12):2646-2660. DOI: 10.1016/J.JOULE.2020.09.010

[6] Jacobsson TJ et al. An open-access database and analysis tool for perovskite solar cells based on the FAIR data principles. Nature Energy. 2021;**7**(1):107-115. DOI: 10.1038/s41560-021-00941-3

[7] Mazumdar S, Zhao Y, Zhang X. Stability of perovskite solar cells: degradation mechanisms and remedies. Frontiers in Electronics. 2021;**2**:712785. DOI: 10.3389/FELEC.2021.712785

[8] Dunlap-Shohl WA, Zhou Y, Padture NP, Mitzi DB. Synthetic approaches for halide perovskite thin films. Chemical Reviews. 2019;**119**(5):3193-3295. DOI: 10.1021/ACS.CHEMREV.8B00318/ASSET/IMAGES/MEDIUM/CR-2018-00318H_0058.GIF

[9] Vásquez-Montoya M, Montoya JF, Betancur R, Ramirez D, Jaramillo F. Perovskite solar cells: new precursors and challenges for scaling-up. In: Chemical Solution Synthesis for Materials Design and Thin Film Device Applications. Amsterdam, Netherlands: Elsevier; 2021. pp. 477-508

[10] Stoumpos CC, Malliakas CD, Kanatzidis MG. Semiconducting tin and lead iodide perovskites with organic cations: phase transitions, high mobilities, and near-infrared photoluminescent properties. Inorganic Chemistry. 2013;**52**(15):9019-9038. DOI: 10.1021/IC401215X/SUPPL_FILE/IC401215X_SI_004.CIF

[11] Saliba M et al. Cesium-containing triple cation perovskite solar cells: improved stability, reproducibility and high efficiency. Energy & Environmental Science. 2016;**9**(6):1989-1997. DOI: 10.1039/C5EE03874J

[12] Tennyson EM, Doherty TAS, Stranks SD. Heterogeneity at multiple length scales in halide perovskite semiconductors. Nature Reviews Materials. 2019;**4**(9):573-587. DOI: 10.1038/s41578-019-0125-0

[13] Yin WJ, Shi T, Yan Y. Unusual defect physics in CH3NH3PbI3 perovskite solar cell absorber. Applied Physics Letters. 2014;**104**(6):063903. DOI: 10.1063/1.4864778

[14] Leguy AMA et al. Reversible hydration of CH3NH3PbI3 in films, single crystals, and solar cells. Chemistry of Materials. 2015;**27**(9):3397-3407. DOI:

10.1021/ACS.CHEMMATER.5B00660/SUPPL_FILE/CM5B00660_SI_001.PDF

[15] Conings B et al. Intrinsic thermal instability of methylammonium lead trihalide perovskite. Advanced Energy Materials. 2015;**5**(15):1500477. DOI: 10.1002/AENM.201500477

[16] Shahivandi H, Vaezzadeh M, Saeidi M. Theory of light-induced degradation in perovskite solar cells. Solar Energy Materials & Solar Cells. 2020;**208**:110383. DOI: 10.1016/J.SOLMAT.2019.110383

[17] Lee SW et al. UV degradation and recovery of perovskite solar cells. Scientific Reports. 2016;**6**(1):1-10. DOI:10.1038/srep38150

[18] DeQuilettes DW et al. Photo-induced halide redistribution in organic–inorganic perovskite films. Nature Communications. 2016;**7**(1):1-9. DOI: 10.1038/ncomms11683

[19] Brenes R et al. Metal halide perovskite polycrystalline films exhibiting properties of single crystals. Joule. 2017;**1**(1):155-167. DOI: 10.1016/J.JOULE.2017.08.006

[20] Domanski K et al. Migration of cations induces reversible performance losses over day/night cycling in perovskite solar cells. Energy & Environmental Science. 2017;**10**(2):604-613. DOI: 10.1039/C6EE03352K

[21] Pern J. Module Encapsulation Materials, Processing and Testing (Presentation) [Online]. 2008. Available from: https://www.nrel.gov/docs/fy09osti/44666.pdf [Accessed: April 24, 2022]

[22] Wang Y et al. Encapsulation and stability testing of perovskite solar cells for real life applications. ACS Materials Au. 2.3. 2022:215-236. DOI: 10.1021/acsmaterialsau.1c00045

[23] Sutherland LJ, Weerasinghe HC, Simon GP. A review on emerging barrier materials and encapsulation strategies for flexible perovskite and organic photovoltaics. Advanced Energy Materials. 2021;**11**(34):2101383. DOI: 10.1002/AENM.202101383

[24] Lu Q et al. A review on encapsulation technology from organic light emitting diodes to organic and perovskite solar cells. Advanced Functional Materials. 2021;**31**(23):1-41. DOI: 10.1002/adfm.202100151

[25] Li J et al. Encapsulation of perovskite solar cells for enhanced stability: Structures, materials and characterization. Journal of Power Sources. 2021;**485**(June 2020):229313. DOI: 10.1016/j.jpowsour.2020.229313

[26] Fu Z et al. Encapsulation of printable mesoscopic perovskite solar cells enables high temperature and long-term outdoor stability. Advanced Functional Materials. 2019;**29**(16):1-7. DOI: 10.1002/adfm.201809129

[27] Boyd CC, Cheacharoen R, Leijtens T, McGehee MD. Understanding degradation mechanisms and improving stability of perovskite photovoltaics. Chemical Reviews. 2019;**119**(5):3418-3451. DOI: 10.1021/acs.chemrev.8b00336

[28] Akraa MA, Hasan AS, Kadhim MJH. Spectroscopy characterization of ethylene vinyl acetate degradation by different kinds of accelerated aging. Baghdad Science Journal. 2020;**17**(3):795-805. DOI: 10.21123/bsj.2020.17.3.0795

[29] McIntosh KR, Powell NE, Norris AW, Cotsell JN, Ketola BM. The effect of damp-heat and UV aging tests on the optical properties of silicone and EVA

encapsulants. Progress in Photovoltaics: Research and Applications. 2011;**19**(1):294-300. DOI: 10.1002/pip.1025

[30] Cheacharoen R, Rolston N, Harwood D, Bush KA, Dauskardt RH, McGehee MD. Design and understanding of encapsulated perovskite solar cells to withstand temperature cycling. Energy & Environmental Science. 2018;**11**(1):144-150. DOI: 10.1039/c7ee02564e

[31] Emery Q et al. Encapsulation and outdoor testing of perovskite solar cells: comparing industrially relevant process with a simplified lab procedure. ACS Applied Materials & Interfaces. 2022. DOI: 10.1021/acsami.1c14720

[32] Philippe B et al. Chemical and electronic structure characterization of lead halide perovskites and stability behavior under different exposures-A photoelectron spectroscopy investigation. Chemistry of Materials. 2015;**27**(5):1720-1731. DOI: 10.1021/acs.chemmater.5b00348

[33] Raman RK, Gurusamy Thangavelu SA, Venkataraj S, Krishnamoorthy A. Materials, methods and strategies for encapsulation of perovskite solar cells: from past to present. Renewable and Sustainable Energy Reviews. 2021;**151**(September 2020):111608. DOI: 10.1016/j.rser.2021.111608

[34] Peike C, Hülsmann P, Blüml M, Schmid P, Weiß K-A, Köhl M. Impact of permeation properties and backsheet-encapsulant interactions on the reliability of PV modules. ISRN Renewable Energy. 2012;**2012**:1-5. DOI: 10.5402/2012/459731

[35] IEC 61853-2:2016. Photovoltaic (PV) module performance testing and energy rating-Part 2: Spectral responsivity, incidence angle and module operating temperature measurements. Published by the International electrotechnical commission 2016. Edition 1.0 2016-09

[36] Meeker W, Hong Y, Escobar L. Degradation models and analyses. In: Encyclopedia of Statistical Sciences. Hoboken, NJ, USA: John Wiley & Sons, Inc.; 2011. pp. 1-23

[37] Phinikarides A, Kindyni N, Makrides G, Georghiou GE. Review of photovoltaic degradation rate methodologies. Renewable and Sustainable Energy Reviews. 2014;**40**:143-152. DOI: 10.1016/j.rser.2014.07.155

[38] Hasan O, Arif AFM. Performance and life prediction model for photovoltaic modules: effect of encapsulant constitutive behavior. Solar Energy Materials & Solar Cells. 2014;**122**:75-87. DOI: 10.1016/j.solmat.2013.11.016

[39] Ndiaye A, Charki A, Kobi A, Kébé CMF, Ndiaye PA, Sambou V. Degradations of silicon photovoltaic modules: a literature review. Solar Energy. 2013;**96**:140-151. DOI: 10.1016/j.solener.2013.07.005

[40] Wang E, Yang HE, Yen J, Chi S, Wang C. Failure modes evaluation of PV module via materials degradation approach. Energy Procedia. 2013;**33**:256-264. DOI: 10.1016/j.egypro.2013.05.066

[41] Velilla E, Cano J, Jimenez K, Valencia J, Ramirez D, Jaramillo F. Numerical analysis to determine reliable one-diode model parameters for perovskite solar cells. Energies. 2018;**11**(8):1963. DOI: 10.3390/en11081963

[42] Padilla A, Londoño C, Jaramillo F, Tovar I, Cano JB, Velilla E. Photovoltaic performance assess by correcting the I–V curves in outdoor tests. Solar

Energy. 2022;**237**:11-18. DOI: 10.1016/j.solener.2022.03.064

[43] Velilla E, Jaramillo F, Mora-Seró I. High-throughput analysis of the ideality factor to evaluate the outdoor performance of perovskite solar minimodules. Nature Energy. 2021;**6**(1):54-62. DOI: 10.1038/s41560-020-00747-9

[44] IEC 61215-1-3:2016. Terrestrial photovoltaic (PV) modules - Design qualification and type approval - Part 1-3: Special requirements for testing of thin-film amorphous silicon based photovoltaic (PV) modules. Published by the International electrotechnical commission 2016. Edition 1.0 2016-12

[45] Hu Y et al. Standardizing perovskite solar modules beyond cells. Joule. 2019;**3**(9):2076-2085. DOI: 10.1016/j.joule.2019.08.015

[46] Khenkin MV et al. Consensus statement for stability assessment and reporting for perovskite photovoltaics based on ISOS procedures. Nature Energy. 2020;**5**(1):35-49. DOI: 10.1038/s41560-019-0529-5

[47] Shi L et al. Accelerated lifetime testing of organic-inorganic perovskite solar cells encapsulated by polyisobutylene. ACS Applied Materials & Interfaces. 2017;**9**(30):25073-25081. DOI: 10.1021/acsami.7b07625

[48] Grancini G et al. One-year stable perovskite solar cells by 2D/3D interface engineering. Nature Communications. 2017;**8**(1):15684. DOI: 10.1038/ncomms15684

[49] Velilla E, Ramirez D, Uribe J-I, Montoya JF, Jaramillo F. Outdoor performance of perovskite solar technology: silicon comparison and competitive advantages at different irradiances. Solar Energy Materials & Solar Cells. 2019;**191**(October 2018):15-20. DOI: 10.1016/j.solmat.2018.10.018

[50] He S, Qiu L, Ono LK, Qi Y. How far are we from attaining 10-year lifetime for metal halide perovskite solar cells? Materials Science & Engineering R: Reports. 2020;**140**:100545. DOI: 10.1016/j.mser.2020.100545

[51] Velilla Hernández E, Bernardo Cano Quintero J, Felipe Montoya J, Mora-Seró I, Jaramillo IF. Outdoor performance of perovskite photovoltaic technology. In: Thin Films Photovoltaics. London: IntechOpen; 2022

[52] Osterwald CR, McMahon TJ. History of accelerated and qualification testing of terrestrial photovoltaic modules: A literature review. Progress in Photovoltaics: Research and Applications. 2009;**17**(1):11-33. DOI: 10.1002/pip.861

[53] Park NG, Grätzel M, Miyasaka T, Zhu K, Emery K. Towards stable and commercially available perovskite solar cells. Nature Energy. 2016;**1**(11):1-8. DOI: 10.1038/nenergy.2016.152

[54] Hailegnaw B, Kirmayer S, Edri E, Hodes G, Cahen D. Rain on methylammonium lead iodide based perovskites: possible environmental effects of perovskite solar cells. Journal of Physical Chemistry Letters. 2015;**6**(9):1543-1547. DOI: 10.1021/ACS.JPCLETT.5B00504/SUPPL_FILE/JZ5B00504_SI_001.PDF

[55] Berhe TA et al. Organometal halide perovskite solar cells: degradation and stability. Energy & Environmental Science. 2016;**9**(2):323-356. DOI: 10.1039/C5EE02733K

[56] Li X, Zhang F, He H, Berry JJ, Zhu K, Xu T. On-device lead sequestration

for perovskite solar cells. Nature. 2020;**578**(7796):555-558. DOI: 10.1038/s41586-020-2001-x

[57] Chen S, Deng Y, Xiao X, Xu S, Rudd PN, Huang J. Preventing lead leakage with built-in resin layers for sustainable perovskite solar cells. Nature Sustainability. 2021;**4**(7):636-643. DOI: 10.1038/s41893-021-00701-x

Chapter 12

Lead-Free Perovskite and Improved Processes and Techniques for Creating Future Photovoltaic Cell to Aid Green Mobility

Rira Kang, Tae-ho Jeong and Byunghong Lee

Abstract

Perovskites material is in the spotlight as photovoltaic device due to their optical and physical properties. In a short period of time, this organic-inorganic pevskite can achieve about energy conversion efficiencies of 25.6% by anti-solvent and spin-coating based process. In addition, ambipolar carrier transport properties of perovskite materials open up new directions for the high-efficiency thin-film solar cells. Despite its attractive properties in solar cell application, concerned about device stability and the use of lead compounds ($APbX_3$, A = a cation X = halide) with toxicity cause the potential risk for the human body and environment issue. Therefore, the use of a new classed strucutral materials with intrinsic stability and beneficial optoelectronic properties can be considered as a start of the next chapter in pervoksite device. This chapter is structured into two major parts: In section 1, we introduce more stable class of perovskite, A_2SnX_6, where Sn is in the 4+ oxidation state. A detailed discussion on the ramifications of material structure and chemistry-related challenges is presented for solution processing, along with careful characterization. In section 2, we talk about the direction of development for perovksite materials to be a next chapter of energy source for a green mobility.

Keywords: perovskite, AMX3, A2MX6, lead-free perovskite, CsSnI3, Cs2SnI6, coating technique, impedance analysis, vacuum evaporation

1. Introduction

1.1 Lead Free Perovskite Materials

The continual development in efficiency of lead-halide-based perovskites has yielded phenomenal success with efficiencies above 25% [1]. To date, however, there have been few reports of scalable solution or vapor processing techniques being applied to the deposition of perovskite films with nontoxicity and stability. It is apparent that the path toward commercialization of solution-processed perovskite solar cells requires the development of fabrication protocols compatible with high-volume roll-to-roll or sheet-fed processing techniques. Although the use of lead in perovskite is expected to have many human and environmental issues in the future, these factors are not currently being considered due to the battle for solar cell efficiency. This research trends make the future of perovskite solar cell darker. Therefore, the most important key to the commercialization of perovskite solar cells is the development of lead-free structured materials with competitive solar cell efficiency.

1.2 Tin-based perovskite

Tin of a group 14 element with comparable ionic radii is currently considered as one of the best candidates to replace lead compounds due to its meaningful electrical and optical properties [2]. Completely lead-free perovskite solar cells based on $CH_3NH_3SnI_3$ were reported in 2014 yielding efficiencies of 5.23% [3] and 6.4% [4] with the suppression of Sn^{4+} accomplished through ultrahigh purity starting materials, fastidious synthesis, and glove box device fabrication protocols. In terms of optical bandgap calculated by Shockley-Queisser for the highest efficiency, $CH_3NH_3SnI_3$ (Eg ~1.3 eV) is considered the ideal materials ($Eg_{,\ ideal}$ ~ 1.35 eV). However, relatively lower efficiency, the poor long-term stability, and low reproducibility of these films caused by the tendency for Sn to get oxidized are insufficient to replace this material with lead-based perovskite [4]. To improve power conversion efficiencies (PCEs) as inhibiting the oxidation of Sn^{2+} to Sn^{4+}, $CH_3NH_3Sn_xPb_{1-x}I_3$, which is mixture compounds of Sn and Pb, is reported [5]. Mixtures containing about x = 0.25 of Sn ($CH_3NH_3Sn_{0.25}Pb_{0.75}I_3$) showed ~7.4% of the best efficiency [6]. Further improvement (PCE ~ 10.1%) was demonstrated as adding Cl, which show a better film coverage, effective exciton dissociation, and charge transport [7].

1.3 $CsSnI_3$-typed perovskite

With Kanatzidis's group, we introduce p-typed $CsSnI_3$ semiconductor, which is a distorted three-dimensional perovskite structure that crystallizes in the orthorhombic *Pnma* space group at room temperature (RT) (**Figure 1(a)**) [11]. The Sn^{2+} center sits in a distorted octahedral environment with six I^- anions, resulting in stereochemically inactive so-called $6s^2$ lone pair of electrons. The $\{SnI_{6/2}\}^-$ octahedra condense to form a three-dimensional framework via corner-sharing with the Cs^+ countercations residing at 12 coordinate interstices within the network made by eight $\{SnI_{6/2}\}^-$ octahedra. This structure exhibits direct bandgap properties of ~1.3 eV and very high hole mobility of μ_h = 585 $cm^2\ V^{-1}\ s^{-1}$ with p-type conduction behavior at RT. Thermoelectric power measurements gave positive Seebeck coefficients over the entire temperature range with linear dependence on temperature, suggesting p-type conduction (**Figure 1(b)**). The hole mobility can be estimated from the electrical conductivity using the equation $\mu_h = \sigma p^{-1} e^{-1}$ (where μ_h is the hole mobility, σ is the electrical conductivity, p is the hole concentration, and e is the electronic charge). The calculated value is close to that obtained from the Hall effect measurements. Therefore, this compound can be considered as being an excellent candidate material for replacing lead-based perovskite. In 2012, a related inorganic $CsSnI_3$ perovskite was used as a hole conductor by combining the ruthenium dyes, reaching 8.5% efficiency [12]. From these properties and initial works, the field of p-typed perovskite material has attracted great attention. After these initial works, the field of perovskite-based solar

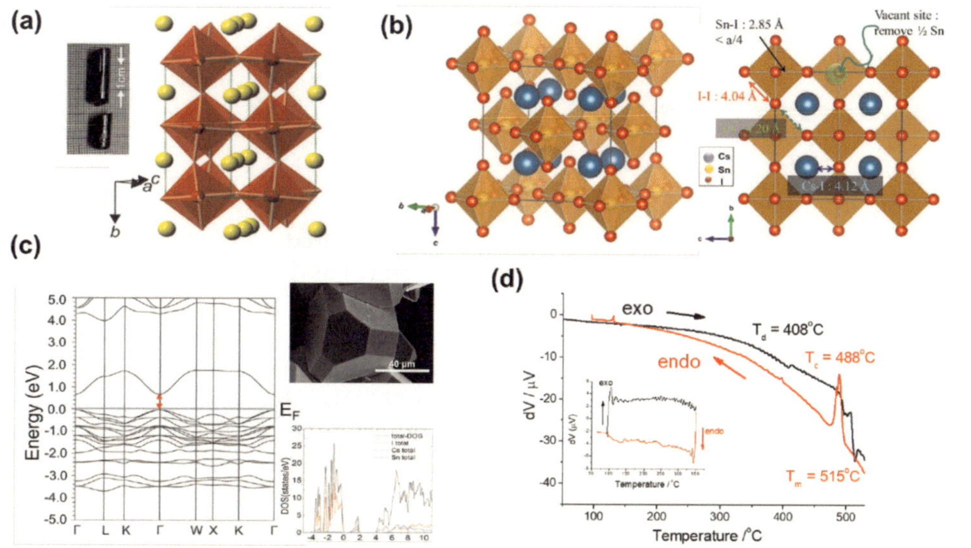

Figure 1.
(a) Distorted three-dimensional perovskite structure of $CsSnI_3$ at RT. red polyhedron, $\{SnI_{6/2}\}^-$: Yellow, Cs. (b) Crystal structure of Cs_2SnI_6 from the VESTA program. (c) DTA and graphs for Cs_2SnI_6 for temperature maxima of 400–600°C. reprinted from [8–10].

cells has literately exploded, with extremely exciting very recent results. $CsSnI_3$ successfully also was demonstrated as light absorber [13]. The high photocurrent densities of more than 22 $mAcm^{-2}$ can be attained by utilizing $CsSnI_3$ due to its favorable bandgap, optical properties, and low exciton binding energies (BEs) (18 meV) [14]. Nonetheless, an improvement of open-circuit voltage (~0.24 V) is still required [15]. Also, the fabrication of tin-based $ASnI_3$ perovskite cells is highly unstable in the ambient due to the tendency for Sn to get oxidized, and its easy oxidation creates Sn^{4+} that originates a metal-like behavior in the semiconductor which lowers the photovoltaic performance [5, 16]. Therefore, a clear improvement on the stability remains an objective.

1.4 Cs_2SnI_6-typed perovskite

A new class of perovskite variants A_2BX_6 would be attractive candidates because *B* in A_2BX_6 is expected to be the +4 oxidation state upon assumption of the A^+ and X ion states, which would lead to a stable structure in air and moisture [8]. It can be inferred that the electronegativity and ionic radii play an important role to be the stable structure, the lattice parameters of which are determined by the competition between the ionic Coulomb and exchange correlation contributions [17]. This typed crystals are *Fm-3 m* space group with four formula units in one unit cell, and it is chemically bonded from the Coulomb interactions between the particular ions [17]. In this crystalline structure, 12-fold of each monovalent cation A coordinates (in the unit of the lattice constant) are (0.25, 0.25, 0.25), each tetravalent cation B is sixfold coordinated by the halogen ions (x, 0, 0), where x is somewhat different (varying around 0.2) for different structures. From optical band structure, halogen ions are located in the upper valence band and the bottom of conduction bands is formed dominantly by the cationic *s*-states. Usually incorporation of the d-transition metal ions (in the case of doped crystals) gives additional localized *d*-states, which form the bottom of the conduction bands.

1.5 Characteristics of Cs_2SnI_6 perovskite

1.5.1 Structural properties

The α-phase of Cs_2SnI_6 crystalizes into the face-centered-cubic (fcc) K_2PtCl_6 type with the space group $Fm\overline{3}m$ (the antifluorite structure) and the lattice parameter a of 11.6276 (9) [16]. The unit cell is configured of four $\{SnI_6\}^{2-}$octahedra at the corners and the face centers and eight Cs^{2+} cations at the tetragonal interstitials. (see **Figure 1(b)**) Alternatively, Cs_2SnI_6 can be regarded as a defective variant of the AMX_3 structure type similar to those of $CsSnI_3$, $CH_3NH_3SnI_3$ and $CH_3NH_3PbI_3$ compounds, in which the $\{SnI_6\}$ octahedra connect to each other by sharing their corners. In $Cs_2SnI_{6,}$ half of the octahedral *Sn* atoms are missing creating discrete $\{SnI_6\}^{2-}$ octahedra. The compound is therefore a molecular salt and contains Sn^{4+} rather than Sn^{2+} in $CsSnI_3$. This accounts of the stability of the material. After the half of the Sn atoms are removed, the $\{SnI_6\}^{2-}$ octahedra shrink slightly, leading to the smaller Sn − I bond length (2.85 Å) [17] in Cs_2SnI_6 than that in $CsSnI_3$ (3.11 Å) [9] as well as the smaller intraoctahedral I − I bond length (4.04 Å) than that of interoctahedral I − I′ bond lengths (4.20 Å).

1.5.2 First-principles electronic band structure of Cs_2SnI_6

The self-consistent full-potential linearized augmented plane wave method (LAPW) within density functional theory (DFT) and the generalized gradient approximation (GGA) of Perdew, Burke, and Ernzerhof for the exchange and correlation potential and WIEN2k program is used for electronic structure calculations [18–23]. The modified Becke-Johnson exchange potential is also employed for bandgap correction [24]. From these calculations, Cs_2SnI_6 estimate a direct bandgap of ∼1.3 eV at the Γ point comprising filled I-5p orbitals and empty hybrid I-6p/Sn-5 s orbitals for the valence band maximum (VBM) and conduction band minimum (CBM), respectively (**Figure 1(c)**). The valence and conduction bands are surprisingly well dispersed in energy, for a molecular $\{SnI_6\}^{2-}$ salt compound, with ∼1 eV and ∼ 0.5 eV bandwidth, respectively. Such a band configuration appears to justify the remarkably high electron and hole mobility of pristine Cs_2SnI_6.

1.5.3 Electrical properties

From Hall effect measurements, electrical resistivity of a pressed polycrystalline pellet of Cs_2SnI_6 by annealing at 200°C shows a reasonably low value of ∼100 Ω·cm. The electron carrier concentration is measured to be on the order of $\sim 1 \times 10^{14}$ cm^{-3} by RT and $\approx -2.6 \times 10^3$ μV/K of the Seebeck coefficient. Plus, electron mobility of the pristine bulk material that behaves as an n-type semiconductor shows about 310 cm^2/V·s. Interestingly, with doping Sn^{2+} (as SnI_2) in Cs_2SnI_6, p-typed semiconductor behavior can be observed ($\sim 1 \times 10^{14}$ cm^{-3} of a nearly identical carrier concentration and $S \approx 1.9 \times 10^3$ μV/K of the Seebeck data). The hole mobility and resistivity of the p-type Cs_2SnI_6 show ∼42 cm^2/V·s and 780 Ω·cm, respectively. Although electron mobility level of p-type Cs_2SnI_6 is lower than that of the $CH_3NH_3PbI_3$ perovskite, it still preserves a considerable hole mobility [5, 16]. And, we can clearly understand the ambipolar nature of Cs_2SnI_6 from these characteristics of both n- and p-type behavior.

1.5.4 Thermal properties of Cs_2SnI_6

The thermal stability of Cs_2SnI_6 is assessed by differential thermal analysis (DTA). The temperature maximum in each scan was varied progressively by a 100°C increment between 400°C and 600°C (**Figure 1(d)**). The melting point T_m was determined to be at ~515°C, but X-ray diffraction suggests that the melting proceeds through partial decomposition to CsI (mp = 621°C, bp = 1277°C) and SnI_4 (mp = 143°C, bp = 348.5°C). It is not clear at which temperature the decomposition occurs, but it can be tentatively assigned to the reversible process occurring at $T_d \sim$ 410°C. The further thermal properties of Cs_2SnI_6 powder are analyzed by thermal gravimetric analysis (TGA) and differential scanning calorimetry (DSC) measurements. For the TGA data, the drastic weight loss onset occurs at 269°C, which is attributed to the beginning of the decomposition of Cs_2SnI_6 to SnI_4 and CsI from the perovskite frameworks. Nearly 44% weight loss between 270°C and 350°C of Cs_2SnI_6 is observed. In general, DSC measurement can be helped to understand a sublimation and a decomposition process for Cs_2SnI_6. The DSC curve exhibits two small endothermic peaks at 200°C and 308°C. Both TGA and DSC results also confirm that the Cs_2SnI_6 crystal is relatively good stability and nonexplosive character. In order to verify this behavior, three independent samples of Cs_2SnI_6 are prepared starting from pristine, solution precipitated material. The first sample consists of the 'as made' fresh material, whereas the other two have been annealed at 350°C and 550°C, respectively, in evacuated ampoules. In agreement with the DTA and TGA data, the material obtained from 350°C annealing remains unchanged, whereas the treatment at 550°C results in a molten solid ingot which is, however, contaminated with orange crystals of SnI_4. High-resolution powder X-ray diffraction using synchrotron radiation confirms the decomposition of Cs_2SnI_6 to SnI_4 and CsI above 410°C. This is accompanied by a relative loss of crystallinity, as judged by the loss of diffraction intensity and the relative broadening of the reflections.

1.6 Cs_2SnI_6 formation processes

Deposition of perovskite films by spin coating process with anti-solvent is a highly common method employed in perovskite photovoltaics research [25]. This method is a good way on a laboratory scale, but it is not suitable for large-area or mass production process. Here, we introduce a new two-step process (a CsI deposition in Step 1 by e-spraying process and vaporization of SnX_2 or SnX_4 in Step 2) method of Cs_2SnX_6 (X = halide) compounds film formation. In earlier research, a series of experiments describe how we have optimized our two-step solution processes for synthesizing iodosalt $Cs_2SnI_{6-x}Br_x$ thin films to achieve suitable properties as solar photon absorbers for light to charge conversion [10]. This paper is well explained what the importance of Step 1 is and how to apply this. As another approach for making better thin films, we adopted the method suggested by Saparov et al. [26] that Vapor SnI_4 treatment is a thermally activated chemical reaction and the reaction temperature is independent of the SnI_4 temperature needed to establish the vapor concentration. The schematic representation of the experimental vaporized step was illustrated in **Figure 2(a)**. The as-made CsI film prepared by e-spraying and excess SnI_4 powder was placed in a sealed glass container with the different positions (the center: CsI

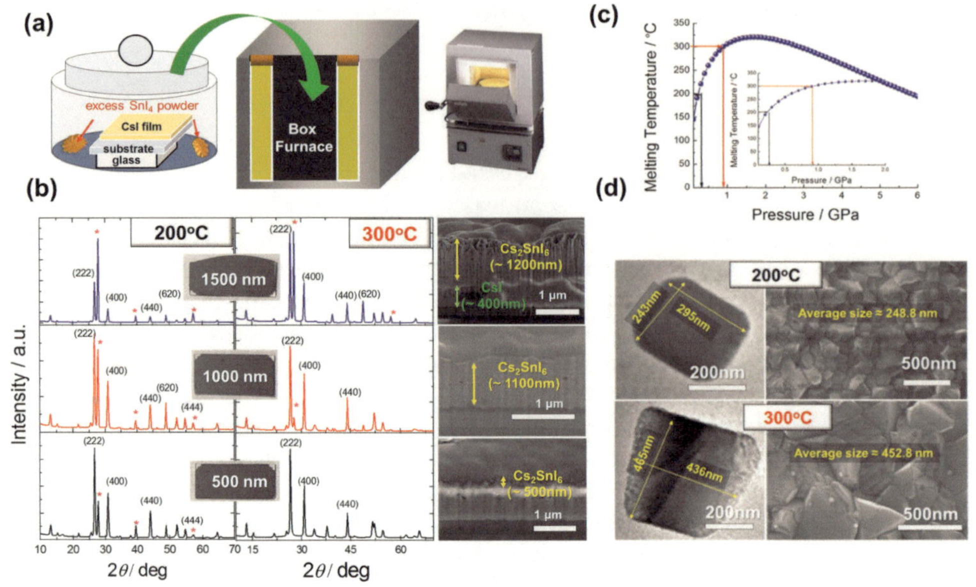

Figure 2.
Vaporized technique as a step 2 process: (a) illustration showing vaporizing procedure; (b) XRD analysis and cross-sectional image for the different film thickness; (c) vaporized condition for SnI_4 powder; and (d) morphological study from SEM and TEM analysis for 200°C and 300°C for 1 hour.

film, the edge: excess SnI_4 powder) and rapidly heated to the reaction temperature (25–200°C or 300°C) for 1 hour in a box furnace at a rate of 10°C/min. In our first observation, even if the films show very smooth surfaces, SnI_4 vaporization condition at 200°C for 60 min is not enough to fully convert the Cs_2SnI_6 crystal because of the existence of the main peaks of CsI film at all the thickness appears at $2\theta = 27.6^{o}$ (see **Figure 2(b)**). Therefore, we choose a different condition based on the ground of the work of Fuchizak [27, 28]. In more detail, at the initial stage of the experimental research into the melting behavior, Simon's equation is used:

$$\frac{p}{p_0} = \left(\frac{T_m}{T_0}\right) C_s{-}1 \qquad (1)$$

where T_0 and p_0 are the values at the reference state, T_m is the melting temperature at pressure p, and c_S is a constant. This substitution is believed not to cause any significant issues. Second, the Kechin melting curve equation is discussed. This equation was derived from

$$\frac{d\ln T_m}{dp} = \frac{\Delta V}{\Delta H} = \frac{\Gamma_m}{K_m} \qquad (2)$$

Here, ΔV and ΔH denote the changes in volume and enthalpy upon melting, and hence the first equality merely states the Clapeyron-Clausius relationship. K represents the isothermal bulk modulus [27]. The subscript "m" denotes that the quantities are calculated from a melting curve, or more precisely, along a solidus. Thus, Γ_m and K_m are considered as the asymptotic quantities of Γ and K evaluated in the solid state. Because only a solid state is involved in the assessment of the second equality, it was referred to as a "one-phase" approach in contrast to the first equality, called a "two-phase" approach. Γ is defined by:

$$\Gamma = \frac{-dlnT}{dp} = 2\left(\Upsilon - \frac{1}{3}\right) \tag{3}$$

where γ is Grüneisen's parameter, and the latter equality is evolved from Lindemann's melting law. The quite intriguing point in Kechin's treatment for Eq. (2) is to employ the Padé approximation to express the RHS and to obtain the solution:

$$\frac{T_m}{T_0} = \left(1 + \frac{p}{a}\right)^b e^{-cp} \tag{4}$$

when L = M = 1 was chosen in the Padé approximant. The constants, a, b, and c, are expressible in terms of the original thermodynamic quantities contained in Eq. (2) as follows:

$$a = \frac{\Delta H_0}{\Delta H_0'} = \frac{K_{mo}}{K_{m0}'}$$

$$b = \frac{\Delta V_0}{\Delta H_0'} + ac = \frac{\Gamma_{mo}}{K_{m0}'} + a \text{ and } c = \frac{\Delta V_0'}{\Delta H_0'} = -\frac{\Gamma_{m0}'}{K_{m0}'}$$

where a prime denotes a pressure derivative, and the subscript "0" means that the quantity is estimated at $p = p_0(\simeq 0)$. When c = 0, Eq. (3) can be simplified to Eq. (1). Eq. (3) is "almighty" in that it can capture an unusual melting curve with a maximum at $p_{max} = b/c - a$.

However, Eq. (3) was used only as a fitting guide, and no examination was attempted to demonstrate the fitted parameters on the basis of Eq. (4). Here, we are simply fitted to Eq. (3), treating a, b, and c as fitting parameters with "best-fit" value. The overall aspect of the fit is not bad, but the actual melting curve seems to break more abruptly near 1.5 GPa, beyond which it becomes almost flat, with a slight maximum at about 3 GPa (**Figure 2(c)**). Based on this information, we test a different sets of condition Cs_2SnI_6 film produced by vaporized SnI_4 treatment with the different thicknesses. XRD analysis and morphology study for the different thickness can be seen in **Figure 2(b)**. All of the diffraction peaks are indexed as Cs_2SnI_6 with the space group, Fm-3 m(225) (JCPDS #04–016-3227). This experimental study shows that the complete reacted 500-nm-thick Cs_2SnI_6 film can be obtained after SnI_4 vapor exposure at 300°C for 1 hour. However, when the thickness is over 500 nm, a CsI impurity peak started to appear. Thus, to remove unreacted CsI, a different treatment condition is needed (e.g. we can also confirm the completely converted Cs_2SnI_6 film at 300°C for 2 hour). Furthermore, we obtain the large Cs_2SnI_6 crystal for the 300°C cases. From SEM top-view images in **Figure 2(d)**, the 300°C treated sample has shown an increased grain size with diameters of 453 ± 35 nm, while the diameters of 200°C treated sample are in the range of 240 ± 58 nm. The TEM images are used to further examine the crystal size. The sizes for 200°C and 300°C treated Cs_2SnI_6 films are estimated to be 269 ± 24 nm and 450 ± 16 nm, respectively, which are consistent with the SEM observations. The grain size (D) of two samples is also independently calculated from XRD data using Scherrer formula [29]. The (222) peak at 2 θ = 26.5° is fitted to estimate the grain size of Cs_2SnI_6.. The 222 peak gives an estimate of the average crystallite size only in the *ab*-plane direction. As expected, the average grain size increases only slightly after 1 h of annealing at 300°C (D = 58.80 ± 0.2 nm) compared with the 200°C treated samples (D = 52.21 ± 0.6 nm). The increased grain size correlates with an improved film conductivity. This improvement is mainly provided by the carrier mobility being enhanced from 1.94 $cm^2/(V \cdot s)$ for the 200°C treated film to 11.24$cm^2/(V \cdot s)$ for the 300°C treated Cs_2SnI_6 film, and to a lesser extent by the carrier concentration that increases slightly from 1.57×10^{15} cm^{-3} to 4.89×10^{15} cm^{-3} at a film thickness of 1.5 μm. This experimental research indicates that bulk electrical conductivity reduces with the decreasing grain size. The electrical property change can be attributed to fewer boundaries impeding electron mobility in the 300°C treated films [30].

The optical properties of Cs_2SnI_6 film prepared by the vaporized technique can be also seen in **Figure 3(a)**. The film formed vaporized condition followed the same optical (1.6 eV) and electrical properties are quite similar to the literature results [26]. This reason can be understood by the following experiment: X-ray photoemission spectroscopy (XPS) and ultraviolet photoelectron spectroscopy (UPS) measurements of the Cs_2SnI_6 film produced by the different methods were performed for compositional and chemical states analysis. The presence of Cs, Sn, and I elements is clearly discernible (**Figure 3(b)**). The binding energies of 619.7 eV and 631.2 eV are indicative of I3d bonded to Cs3d at binding energies (BEs) of 725.1 and 739 eV, in good agreement of result for CsI peak [31]. In the case of the Sn compound, the main binding energies of $Sn3d_{5/2}$ and $Snd_{3/2}$ obtained from solution method are about 488 eV and 496 eV, respectively, attributed to Sn^{4+} state (see **Figure 3(b)** middle column). Interestingly, in the case of the vaporization method, the main binding energy at 487.3 eV can be assigned to Sn^{2+}, leading to intrinsic defects and instable form, with a nominal formula $Cs^{+}{}_2Sn^{2+}(I_6)^{4-}$ [26, 32–34]. The vaporization of a system of tetrahedral MX_4 groups linked through vertices (silica-like structure) is accompanied by a reduction in the coordination number of the metal, in some instances to polymeric species which dissociate to monomers at higher temperature [35]. Thus, the formation of the Sn^{2+} oxidation states can be explained by halogen transfer during the high-temperature process [36–38]. The position of energy levels for occupied states can be also determined by UPS analysis. (**Figure 3(b)** right column) The BE of the HOMO onset of Cs_2SnI_6 film prepared from the different method is determined by the intersect of the linear extrapolation of the leading edge of the HOMO peak and the straight background line. The Fermi level in most of the presented spectra is fixed at zero binding energy, and all the measured positions are

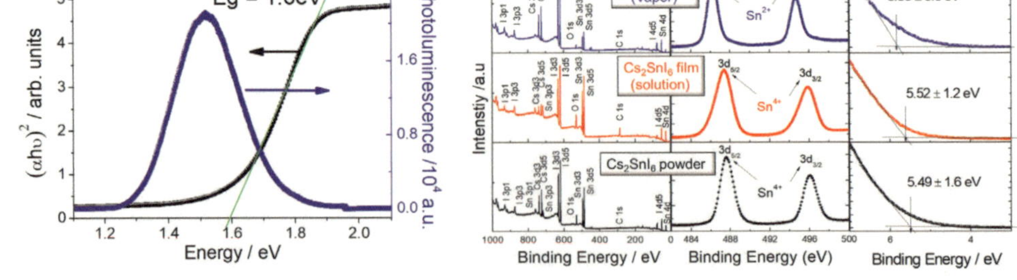

Figure 3.
(a) The optical absorption and PL analysis at 1000 nm thickness prepared by the vaporized technique at 300°C for 1 hour, and (b) XPS and UPS analysis for Cs_2SnI_6 prepared from the different technique.

referred with respect to the Fermi level. The HOMO onset of Cs_2SnI_6 powder and film from solution method are observed to be 5.52 ± 1.2 eV, while for the vapor method they are measured to be 5.58 ± 0.9 eV. The downshift of HOMO level with increasing bandgap can be considered by the distortion of crystal from the Sn^{2+} state. The different oxidation state can affect not only crystal structure account of the different ionic radius of Sn^{2+} (102 pm) and Sn^{4+} (0.69 pm) but also has a profound influence on a number of physical properties [39, 40]. The Sn^{2+}compounds are expected to have distortions of their bonded configurations because of the influence of their nonbonded pair of electrons, while the $5s^0p^0$ configuration of the Sn^{4+} ion should give regular octahedral coordination for tin in ionic lattices. The bond length of Cs_2SnI_6 can be seen in **Figure 1(b)** [17, 41]. The missed half of Sn atom brought the Sn-I length (2.89 Å) closer to the actual value (2.85 Å), while estimated Sn-I length of Sn^{2+} was about 3.22 Å. The incongruous Sn-I length leads to the distortion of Cs_2SnI_6. From Goldschmidt tolerance factor (t), we can simply understand the stability and distortion of Cs_2SnI_6 crystal with Sn^{2+} and Sn^{4+} oxidation state [42]. The tolerance factor is calculated from the ionic radius of the atoms [43]. A tolerance factor of 0.71–0.9 originates from a distorted perovskite structure with tilted octahedra. In the case of the tolerance factor is higher (>1) or lower (<0.71), perovskite phase cannot be formed. This rule made for oxide perovskite, but the trend is still valid for pero-halide perovskite materials structure. The calculated tolerance factor of Sn^{2+} oxidation state can be estimated by 0.9012 (from SPuDS software program), while the maximum t achievable of Sn^{4+} oxidation state is 0.998. The decreased t value of the Sn^{2+} oxidized state indicates that the network of corner-shared SnI_6 octahedral will tilt in order to fill space, causing a less stable structure.

Figure 4 illustrated the electronic structure of Sn^{2+} state $Cs^+{}_2Sn^{2+}I_6{}^{4-}$ and Sn^{4+} state $Cs^+{}_2Sn^{4+}I^-{}_6$ presented by Xiao et al. [33, 41]. The chemical bonding nature and the origin of the bandgap in Cs_2SnI_6 can be understood by DFT calculations for some hypothetic structure. The qualitatively arranged group of electronic structure on the energy scale can be seen in **Figure 4(a)**. The electron structure on an isolated [36] octahedron (i.e. $\{I_6\}^0$ cluster) is firstly displayed. The 18 I 5p orbitals of the $\{I_6\}$ octahedron are split to seven groups, following by the energy eigenvalues at the Γ point and the group theory. The six radial I 5p orbitals split into three groups of a_{1g} (I-I bonding) and e_g & t_{1u} (I-I antibonding). The 12 tangential I 5p orbitals form 4 triply

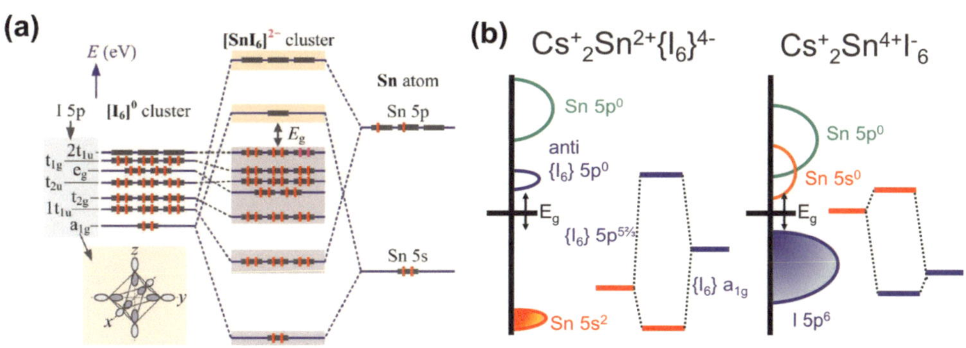

Figure 4.
(a) Total and projected DOSs of qualitative interaction diagram for the $\{I_6\}^0$ cluster, the $\{SnI_6\}$ cluster, and the $\{SnI_6\}$ sublattice models. The orbital are qualitatively arranged on the energy scale. A schematic illustration of $\{I_6\}$ a_{1u} orbital: (b) schematic electronic structure for the different state Cs_2SnI_6. Reprinted form [41].

degenerated groups of $1t_{1u}$ & t_{2g} (I-I bonding) and t_{2u} & t_{1g} (I-I antibonding). By adding a Sn atom and two electrons (transferred from the tow Cs atoms, which is ionized to Cs^+ in Cs_2SnI_6) into the $\{I_6\}$ octahedron, the electronic structure of a $\{SnI_6\}^{2-}$ octahedron cluster is obtained. Therefore, the main difference between the +2 and + 4 oxidation state is the Sn 5 s orbital position. The unoccupied Sn 5 s orbital at the $Cs^{+}{}_2Sn^{4+}I^{-}{}_6$ state is contributed to the conduction band maximum (CBM). However, the calculated $Cs^{+}{}_2Sn^{2+}I_6{}^{4-}$ state had the fully occupied Sn 5 s orbital and I 5p-Cs 6 s antibonding CBM state. The +2 oxidation state of Sn has deeper VBM, and it can be explained by its wide bandgap. Using the one-micron-thick Cs_2SnI_6 film produced by solution and vapor as a photosensitizer, we fabricated three different series of solar cells structure: (a) nanoporous TiO_2 /Cs_2SnI_6/Au; (b) nanoporous TiO_2 / Cs_2SnI_6/Spiro-OMeTAD/Au; and (c) nanoporous TiO_2/Cs_2SnI_6/LPAH/Au. As shown in the SEM cross section, a nanoporous TiO_2 layer with interpenetrating layer of Cs_2SnI_6 is placed as the next layer. A selected HTM layer is next deposited followed by the evaporation of a thin Au contact layer. Three different series of HTM layers prepared from solution and vapor processes are shown in **Figure 5(a)**. The bottom of SEM images is also displayed in their band diagrams. The *J-V* curves for each of the device structure are plotted in **Figure 5(b)**, along with a table showing their characteristics. For configuration a. and d., we observed a substantial short circuit in the device of both solution- and vapor-processed samples. For configuration e., the vaporized samples show the best performance in typed cells (η = 0.505%), while the solution-processed solar cell shows the best performance (η = 0.177%) in the configuration c.. The overall improved PCE at the vaporized process is attributed to their

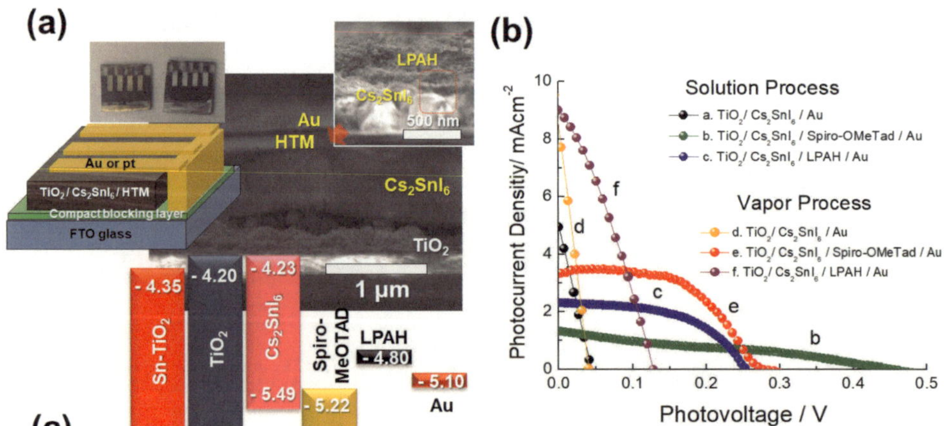

Process		Structure	V_{OC}(V)	J_{sc}(mA/cm²)	FF (%)	EFF (%)
Solution	a.	TiO_2 / Cs_2SnI_6/Au	0.043	5.04	25.4	0.055
	b.	TiO_2 / Cs_2SnI_6/ Spiro-OMeTAD/ Au	0.454	1.338	29.4	0.177
	c.	TiO_2 / Cs_2SnI_6/LPAH/ Au	0.253	2.20	52.3	0.305
Vapor	d.	TiO_2 / Cs_2SnI_6/Au	0.041	8.518	25.41	0.088
	e.	TiO_2 / Cs_2SnI_6/ Spiro-OMeTAD/ Au	0.278	3.308	55.19	0.505
	f.	TiO_2 / Cs_2SnI_6/LPAH/ Au	0.127	8.989	31.32	0.357

Figure 5.
(a) {, #558} and cross-sectional images of a Cs_2SnI_6 film device (bottom: band alignment diagram) (b) J-V curves for different structures and device performance as described in the table above.

smooth surface, but still a low efficiency at 0.5%. However, it should be noted that large-effective-surface-area polyaromatic hydrocarbon (LPAH) can be dispersed well in alcoholic solvent (such as ethanol and isopropanol) without any polymer binder or surfactants [44]. Interestingly, the LPAH suspension with isopropanol also shows long-term dispersion stability. We believe that our findings make it possible to use this unique carbon nanostructural material as HTM material. In order to prove the effectiveness of LPAH, we tested that the case of a methylammonium lead iodide ($MAPbI_3$) thin film can lead to high-efficiency device. (This book will not cover it).

Operating mechanisms of the Cs_2SnI_6-based solar cells have raised a number of questions. The optimization and further improvement of a new material require a deep knowledge of the working principles of this photovoltaic device. In order to understand the effectiveness for photosensitizer, the charge transfer process is studied by two different tools such as electrochemical impedance spectroscopy (EIS) and femtosecond transient optical spectroscopy (i.e. TAS and time-resolved PL (TRPL) spectroscopy) for measuring accumulation of a photogenerated charge and the diffusion length (L_D) [45–47].

1.6.1 Impedance analysis

In earlier reports, dye-sensitized solar cells (DSSCs) have been successfully modeled by equivalent circuit elements, which have helped to elucidate the roles of internal interfaces as well as device components [48, 49]. From simulated model, average charge carrier lifetime, electronic densities of states, and charge carrier concentrations can be calculated. However, in the single-electrode system like BHJ organic photovoltaic devices or perovskite solar cell, the different impedance model is applied due to their geometrical difference [50–52]. Herein, we consider the properties of the impedance associated with diffusion coupled with recombination [45, 46, 51]. As seen in **Figure 6(a)**, electron energy diagram of an electron-transporting materials (specially, nanostructured metal oxide, TiO2, SnO) in contact with a hole-conducting material (or redox medium), displaying the electrochemical potential of electrons E_{Fn} (Fermi level), when a voltage V is applied to the substrate, and assuming that conduction band energy (E_c), is stationary with respect to the redox level, E_{redox}. The equivalent circuit (transmission line model, TL) for a small periodic ac perturbation contains the resistance for electron transport throughout the metal oxide nanoparticles, r_{tr}; the resistance in the hole-transporting medium (hole or ion conduction) r_{HTM}; the recombination resistance at the metal oxide/HTM interface, r_{rec}; and the chemical capacitance for charge accumulation in the metal oxide particles, C_μ. The TL pattern related to the carrier transport at lower frequency which is due to a coupling of capacitance with recombination is demonstrated by a straight line. (The extension of the straight line cuts the semicircle at low frequencies).

The model corresponding to the reflecting boundary condition is shown in **Figure 6(a)** and contains three main elements. The first is the chemical capacitance (C_μ), which makes C_μ dominate the total capacitance at sufficient forward bias. The C_μ is related to the variation of the electron Fermi level in the TiO_2 caused by the variation of the electron density as a function of the voltage. The fitting of TL allows to separate the two resistive parameters, for an active film of area A and layer thickness L. The second is the recombination resistance, $R_{rec,}$ and the third is the transport resistance R_{tr}, that is reciprocal to the carrier conductivity, σ and the conductivity relates to the free electrons diffusion coefficient, as:

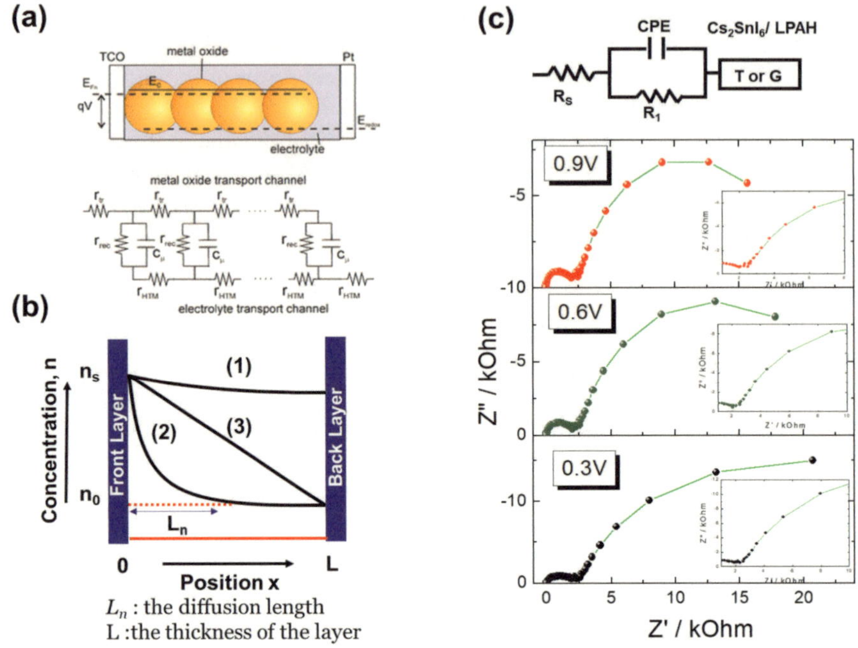

Figure 6.
(a) Electron energy diagram and general TL (b) concentration at the left boundary with concentration in the diffusion-recombination model. Curve represents the case (1) $L_n >> L$,(2) $L_n << L$ (3) long diffusion length with strong recombination at the back contact (c) the equivalent circuit used to fit the experimental data and impedance spectroscopy characterization. Reprinted from [46, 51].

$$R_{rec} = \frac{\tau_n}{C_\mu},\ R_{tr} = \frac{L}{A\sigma},\ \sigma = \frac{C_\mu D_n}{L.}$$

It is important to remark the following property:

$$R_{tr} = \left(\frac{L}{L_d}\right)^2 R_{rec} \text{ and } L_d = \left(\frac{R_{rec}}{R_{tr}}\right)^{1/2} L$$

The diffusion length (L_d) is also obtained from electron diffusion coefficient, D_n, and electron lifetime τ_n, as $L_d = \sqrt{D_n\tau_n}$ and indicates the average distance that generated or injected electrons travel before recombining. Influence of L_d of the carrier distribution in forward bias under dark conditions is illustrated in **Figure 6(b)**. For reflecting boundary (1) of long diffusion length, the carrier profile is nearly homogeneous. For short diffusion length (2), a gradient of carriers for the size of diffusion length is built from the injection point, and the rate of recombination at the back surface becomes another important factor. Finally, if the rate is large (3), excess carriers cannot remain at this boundary, and a gradient for the size of the semiconductor layer is built. **Figure 6(c)** shows a set of the characteristic impedance spectra pattern obtained for the Cs_2SnI_6/LPAH/Au at different applied voltages in the working conditions under 0.1 sun illumination. For all the spectra, an arc is observed at high frequencies related to the transport in Cs_2SnI_6/LPAH. At low frequencies, for samples, the classical feature of a transmission line, TL, discussed earlier is clearly visible. The TL pattern is defined by a straight line, associated with the carrier

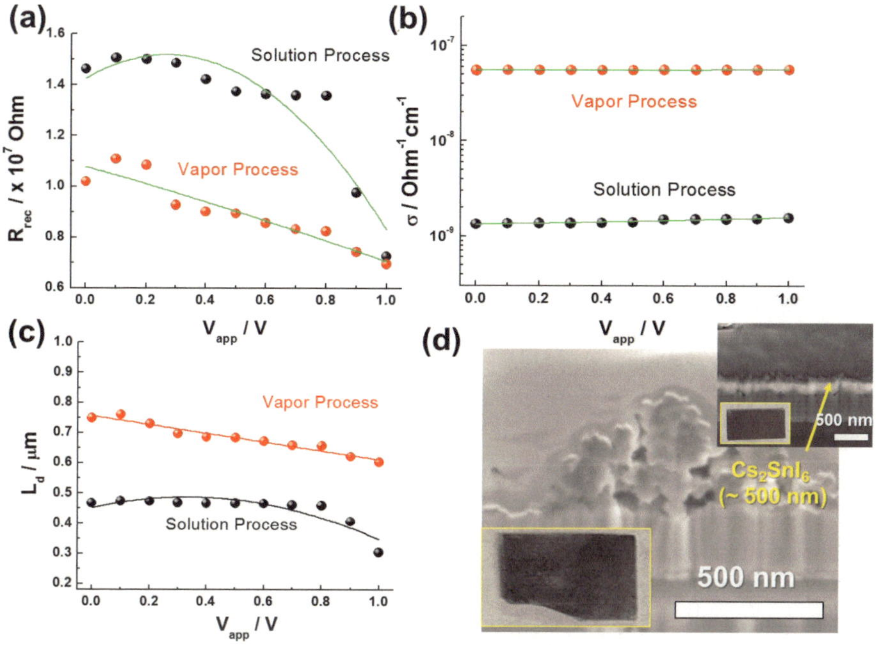

Figure 7.
Transport and recombination parameters vs. voltage: (a) recombination resistance, R_{rec}; (b) conductivity of active layer considering the geometric cell area; (c) diffusion length; and (d) cross-sectional SEM images of solution-processed film (inserted in (top) vapor processed film (bottom) photo image) for ~500 nm thick Cs_2SnI_6 film.

transport, followed by an arc at lower frequency, which is due to a coupling of capacitance with recombination. The fitting results are presented in **Figure 7**.

Important information about the recombination in the solar cell is contained in the recombination resistance, R_{rec}, and the transport rate is related to conductivity, σ. In the case of thin film (~500 nm), a Cs_2SnI_6 film produced by vaporized process exhibits two orders of higher conductivity, while it displays lower R_{rec} at comparable potentials (higher recombination rate) compared with Cs_2SnI_6 film from solution process. Therefore, diffusion length (L_d) of vapor-processed Cs_2SnI_6 film (~0.76 μm at low voltage) is increased by as much as 65.2%, compared with solution-processed film (~0.46 μm at low voltage). This reason can be considered as improved Cs_2SnI_6 film quality. For example, solution-processed Cs_2SnI_6 film shows the rough film with pinhole surface because of the loss during converting crystal formation such as being washed away by dropping SnI_4 alcoholic solvent (**Figure 7(d)**). The light absorption properties of Cs_2SnI_6 film can be determined by absorption coefficient. Materials with strong absorption coefficients more readily absorb photons, which excite electrons into CB. Thus, knowing the absorption coefficients of materials aids engineers in determining which material to use in their solar cell designs. The calculated coefficient (α) of 7.11×10^2 cm^{-1} at 550 nm measured for Cs_2SnI_6 film (at ~500 nm) is about two order lower than that of 1.32×10^4 cm^{-1} at 550 nm for $MAPbI_3$ film (at ~400 nm) [53]. Although a wider absorption spectrum of Cs_2SnI_6 film is benefit from light absorption, a low α must be improved to be the efficient photosensitizer. The higher α can be obtained by increasing film thickness developed. For example, α is estimated to be 1.7×10^3 cm^{-1} and 8.5×10^3 cm^{-1} at 550 nm for 1000 nm and 1500 nm Cs_2SnI_6 film, respectively, which indicates that α of thicker film is an order of magnitude higher than that of thinner film. However, from calculated diffusion length and the experimental limitation for a completely reacted Cs_2SnI_6 film, we conclude that about 1.0 μm thick is the best condition for the efficient solar cell. Unlike aforementioned results, the oppose behavior, i.e. increased transport rate and decreased recombination rate at thicker film, can be observed due to the existence of CsI impurity at vapor process. Consequentially, in the case of about 1.0 μm thick, solution-processed Cs_2SnI_6 film shows longer diffusion length (1.5 μm) than that of vapor process (0.78 μm). This result indicates solution process is more favorable technique for thicker layer film. However, vapor-processed Cs_2SnI_6 film shows a high photocurrent and increased performance regardless of CsI impurity (leading to decrease L_d) compared with solution method. Therefore, our group believes that a completely converted or single-crystalline Cs_2SnI_6 film over 1.5 μm thick under well-controlled vapor process leads to the outstanding solar performance.

1.6.2 Femtosecond time-resolved transient absorption spectroscopy

The further photo-induced charge transfer processes can be confirmed by the excited-state dynamics measured from femtosecond transient absorption (fs-TA) spectroscopy [54]. In the case of over 800 nm, the totally black colored Cs_2SnI_6 film has a problem to transmit light through the samples. Therefore, >500 nm of Cs_2SnI_6 film is used for this study. **Figure 8** presents the normalized ground-state fs-TA spectra of solution and vapor-processed Cs_2SnI_6 film on Al_2O_3 (<100 nm, Aldrich). Samples are excited with 600 nm, and 10 nJ laser pulses are used with the same

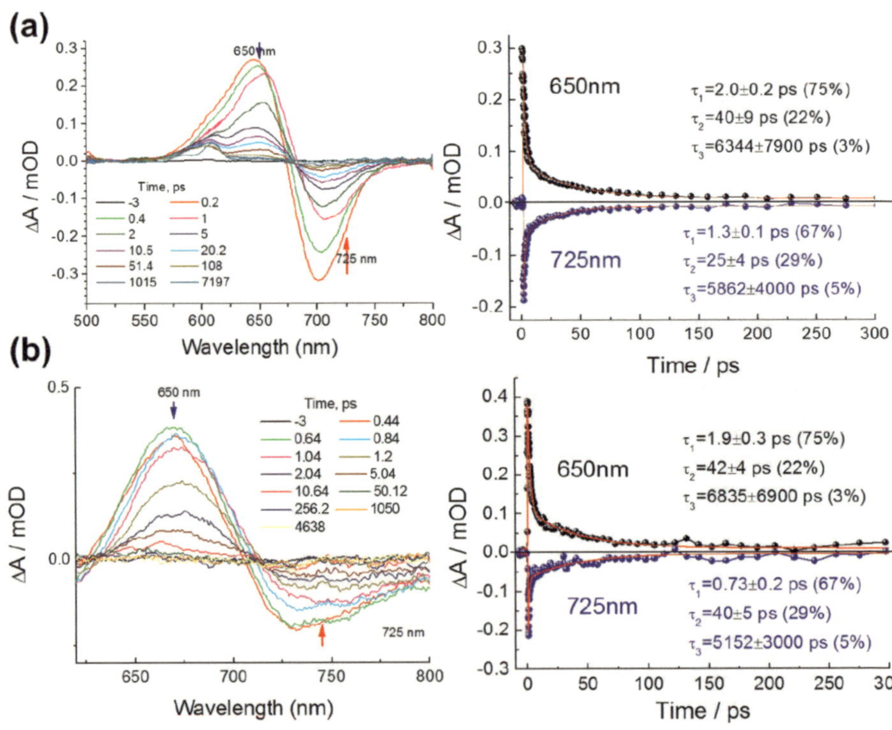

Figure 8.
Femtosecond transient absorbance spectra with white light continuum probe and pulsed fs laser excitation at 600 nm and 10 nJ laser pulses are used with the same spectrum spanning from 500 to 800 nm. Dynamics extracted at 650 nm (black) and 725 nm (blue) for (a) solution and (b) vapor process for a two-step Cs_2SnI_6 film.

spectrum spanning from 500 to 800 nm. Both samples show two main features: a positive band in the range of 600–680 nm and broad negative band peaked at 680–800 nm. The broad positive band is resulted in the superposition of a ground-state bleaching (GSB) and simulated emission (SE) due to the close resemblance to the spectra of steady absorption and photoluminescence; the negative spectral feature is assigned to photo-induced absorption of excited state (PIA) as bleaching transitions from valence band to a conduction band [55]. The peak at 590 nm in case of solution is considered a noise peak caused by the uncovered Cs_2SnI_6 layer film defect. In the global analysis procedure, the decay-associated spectra with time constant probing at 630 nm and 725 nm are plotted in **Figure 8(c)** and **(d)**.

The completely reacted and covered Cs_2SnI_6 film shows three time constants of 1.9 ps (74%), 42.4 ps (21%), and 6835 (5%) at 650 nm as well as 0.73 ps (73%), 40.5 ps (21%), and 5152 ps (6%) at 725 nm, while a solution-processed Cs_2SnI_6 film reveals three time constants of 2.02 ps (75%), 40.9 ps (22%), and 6344 ps (3%) at 650 nm and 1.31 ps (67%), 38.4 ps (29%), and 5862 ps (5%) at 725 nm. The fast component for these samples is accounted for charge carrier trapping at grain boundaries of perovskite. We assign the longer time component (τ_2) to electron injection into glass. This long-lifetime component, not resolved in this work, is most likely electron-hole recombination [56–59]. For 650 and 725 nm, the small differences observed between solution ($\tau_{2,650nm}$ = 40.9 nm and $\tau_{2,725nm}$ = 38.4 nm at solution process, while for vapor process, $\tau_{2,650nm}$ = 42.4 nm and $\tau_{2,725nm}$ = 40.5 nm). The small differences observed for both samples are not significative enough to draw any conclusion regarding the electron injection process. The further experiment for this measure did not progress by the difficulty in the sampling. For example, in electric field deposition system, a conductive substrate is prerequisite for making a continuous and homogeneous thin film. However, for fs-TA spectroscopy analysis, nonconductive substrate is favorable to clarify the electron transfer dynamics of material itself. Therefore, in this thesis, fs-TA is no longer used. In spite of possible charge dynamic properties charge injection and the similar diffusion length ($L_d \approx 0.8$ μm at ~500 nm) of Cs_2SnI_6 film compared with $MAPbI_3$-based solar cell ($L_d \approx 1$ μm), our initial finding showed very low device efficiencies. Nevertheless, we were encouraged because the device operated as a photosensitizer even when the conduction band energy level between TiO_2 and Cs_2SnI_6 layer is nearly the same. A possible solution to this problem can be sought in bandgap tuning, and this can be achieved i) through the modification of the electron-transporting layer and ii) using an appropriate $Cs_2SnI_{6-x}Br_x$ absorption layer. These two approaches are discussed in the next paper.

2. Toward scalable fabrication of perovskite solar cells for mobility

Over the last decade, various support policies for electric vehicles (EVs) have been established in key markets, facilitating a major expansion of EVs models. But the challenge remains enormous such as driving range improvement and electricity production for charging EVs. Therefore, there are many efforts to increase the mileage of EVs. The use of solar energy would theoretically be the ideal fuel for EVs. In addition, significant fiscal incentives were provided for cars equipped with solar roofs (see IRC Section 30D(g)(1)(A)). Thus, major manufacturers expressed great interest in vehicle-intergraded solar cells. Still, the limited coverage area and efficiency of crystalline silicon-based solar cells make it difficult to have been active in a car application. Recently, Sono Motors, Lightyear one, and Mercedes-Benz presented the concept of the electric concept sedan with a range of 620 miles (~1000 km) by increasing the solar panel area such as the roof, hood, and trunk. However, unlike stationary solar cells, mobile one requires higher standards (e.g. climatic tests such as temperature, humidity, and sun, and mechanical tests such as shock, drop, and scratch resistance are conducted for pedestrian safety.) Hyundai (Sonata and IoniQ5) and Toyota (Prius) successfully launched solar-roof-integrated cars from the lamination technique of the panoramic sunroof.

Hyundai Motor Group continues to study solar roofs, hoods, and trunks with different requirements. Although the development of next-generation solar cells has been underway for a long time, unfortunately, silicon is still a primary material for commercial solar cells. As mentioned earlier, the mobile car with a limited area requires a new material solar cell capable of higher efficiency (above 30%) and flexibility for easy application. Here, we deal with perovskite cells (PVCs) for mobility only from the point of view of the fabrication method. Perovskite is regarded as one of the next-generation materials that can overcome or create synergy with c-Si. Many researchers and company have suggested fabrication methods on a lab scale, not in the mass production stage. In this section, we show the scalable fabrication method of PSCs, considering that different integration technologies are required depending on the parts in the car.

2.1 Solution-based fabrication of perovskite solar cells

There are representative PSCs manufacturing companies. They present various fabrication methods and strategies for the commercialization of perovskite (**Table 1**). The latest trends in development in schools and companies, along with increasing interest in PSCs, are described on website [60]. Solution-based fabrication of PSCs is similar to organic solar cells (OSCs) and organic light-emitting diode (OLED). In the case of OSCs, relatively high efficiency (~18.2%) and mass production processes have been successfully researched in Kolon Industry [61]. Still, it has not been expanded due to the poor stability of organic materials. Moreover, the operating life issues of solution-processed OLED panels compared with the vacuum-processed device make it still an uncommon technique. Similar to these two devices, fabrication methods of PSCs have been developed. Similar to these two devices, PSCs are expected to have the same problem. To avoid these problems, we will briefly review the perovskite coating process.

Company	Manufacturing target	Technique.	Performance claims
Oxford PV (UK)	Rigid perovskite/Si tandem	PVD	29.8% @ 1.12cm^2 (tandem)
Microquanta (China)	Rigid glass-backed perovskite cells.	?	17.3% @ 17.3cm^2 (module).
Saul Tech. (Poland)	Printed flexible, lightweight, perovskite-only cells (niche application, e.g. IoT device)	Inkjet printing	25.5% with a small area
Panasonic (Japan)	Rigid perovskite cells. (comparable c-Si)	Inkjet printing	16.1% @ 802 cm^2 (module)
Toshiba (Japan)	lightweight modules for rooftops (LCOE of 7yen/kWh)	Meniscus coating	15.1%, @ 703. cm^2 (flexible)

Table 1.
Comparison of some companies commercializing perovskite.

2.1.1 Spin coating technique

It has been widely used to fabricate small cells less than 0.1 cm^2 and large module with area of 100 cm^2 as well [62]. The main advantage of this technique is simple, has high reproducibility, and is easy to control a nanoscale film thickness. For perovskite application, one-step and two-step of spin coating method are used. For one-step spin coating, the mixed precursor solution (organic halide salts and lead halide salts are dissolved in commonly used solvent such as DMF and DMSO) is spin-coated under layer (charge-transporting layer). In order to prevent incomplete crystallization during on-step process, the two-step method including so-called anti-solvent engineering was introduced [63] This method has been widely used to produce high-quality film because it rapidly induces supersaturation, but the anti-solvent method cannot guarantee uniformity over a large area as the wet film is washed out. Additionally, since this method causes materials loss and stains with large-area coating, it is necessary to develop a method to replace this technique.

2.1.2 Meniscus coating (slot die and blade) coating

The term "meniscus coating" is the translation of a meniscus across the surface of a substrate as the solution is spread through a coating head or blade. These coating methods include slot die coating and blade coating. These coating can be applied in both sheet-to-sheet (S2S) and roll-to-roll (R2R) (see **Figure 9(c)** and **(d)**). Using R2R process, the material loss rate is less than 1% and high productivity can be expected [64]. The film thickness of the meniscus coating can be governed by the precursor solution concentration, the coating speed, and the gap distance between the blade or coating head and the substrates. A blade or a coating head moves across a surface or vice versa in the case of R2R. The blade spread predispensed ink, and the slot-die spread ink through a microfluidic metal die. They form it into a wet film. Compared with the spin coating, the solvent evaporation rate of meniscus-coated perovskite film is relatively slow, which promotes the growth of larger crystals. Because of the difficulty in controlling crystallization of perovskite film only by the natural drying, additional mechanical (e.g. preheating substrate [65] and gas blowing [66]) and chemical treatment (e.g.. anti-solvent [67]) are required. The former two cases are

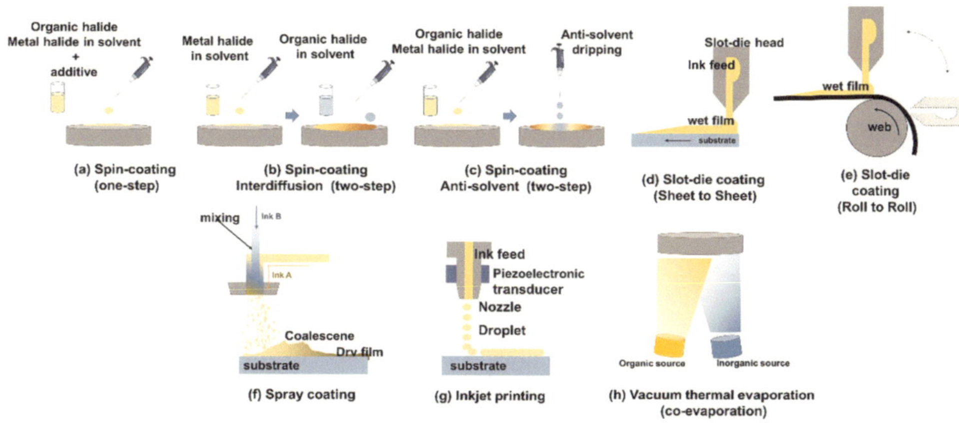

Figure 9.
Schematic of solution-based fabrication and vacuum thermal evaporation (co-evaporation).

possible through modified coating equipment. The latter, the two-step method with anti-solvent, has been commonly used in spin coating, as discussed before. However, this method is difficult to be transferred from spin coating to meniscus coating due to its narrow time window. In this regard, solvent engineering of perovskite precursor ink is required to enable a one-step method. Depending on the existence of an ink reservoir, the slot die coating is more useful than a blade coating for scale-up [68]. Regardless of substrate type, all slot-die coating is still difficult due to materials and device structure restrictions (see **Table 2**). Still, the current intermediate results of these coating are likely to be the first solid step toward future manufacturing of the PSCs cells.

2.1.3 Spray coating

As a low-temperature coating technology, spray coating with with low cost, high volume, rapid manufacturing, and low material loss rate is the most widely used technique. Spray coating uses relatively low-concentration "inks." This process can be divided into four stages: *i)* the generation of the ink droplets, *ii)* the transport of the droplets to the substrate, *iii)* the coalescence of the droplets into a wet film, and *iv)* the drying of the thin film (**Figure 9(f)**) [68]. In recent years, $\sim$ 15.5% of mini perovskite module (size$\sim$40 cm^2) produced by spraying process are reported [72]. Until now, efficiency of spray-coated PSCs still lags behind a meniscus-coated PSC due to nonuniform crystallization of perovskite and film coverage issue when one-step process is applied.

2.1.4 Inkjet printing coating

The inkjet printing method can reduce the procedure of the laser etching processing for module fabrication. For inkjet printing of the perovskite films, most inkjet-printed demonstrations utilize piezoelectric MEMs print heads, which provide controllable microfluidic jetting through a silicon-etched nozzle. To improve the reliability and the speed of printing, multiple jet nozzles are also devised [79]. Inkjet printing enables specific cell shapes for particular functions, such as small-scale utility power and building-integrated photovoltaics (BIPVs). Additionally, an essential advantage of continuous and other inkjet printing systems is that they do not require physical contact or critical gaps between the jet and the substrate, making them suitable for printing on uneven, curved, or pressure-sensitive surfaces [68]. It will also enable the printing of solar cells ideal for cars, especially esthetics, in the future. Saule Technologies uses this method for 1-m-wide perovskite in building-integrated photovoltaics (BIPVs) and EV charging ports. Thus, it is a very scalable technique for fabricating PSCs.

2.2 Solvent

Many papers have explained the growth mechanism of perovskite crystals in solvent [80, 81]. The interaction between solvents and perovskite compositions is significant in the nucleation and crystallization processes during the perovskite film formation. Noted that the choice of solvent in perovskite solution should consider how well solvent can dissolve precursors and crystallize them in a one-step method, as mentioned earlier. Polar aprotic solvents (DMF, DMSO, γ-butyrolactone (GBL), pyrrolidone (NMP), and acetonitrile (ACN)) are commonly used solvents. Organic

DOI: http://dx.doi.org/10.5772/intechopen.106256

Fabrication method	Structure	Type	Method (solvent)	Condition	Area	PCE (%)	Ref
Slot-die (all layers)	ITO/ SnO2/CsFAMAPbI3-xBrx/Spiro	S2S	One step (DMF/DMSO/ additives)	ambient	0.07 cm^2	14.5	[69]
Slot-die (only perovskite	ITO/TiO2/CH3NH3PbI3 − xCl/Spiro	S2S	One step (DMF)	glove box	168.75 cm^2	11.1	[70]
Slot-die (SnO2, perovskite)	PET/ITO/SnO2/ Cs0.15FA0.85.PbI2.85Br0.15/ Spiro	R2R	One step (DMSO, 2BE)	ambient	0.04cm^2 width 30 cm	13.5	[71]
Spray (TiO2, perovskite)	ITO/TiO2/CH3NH3PbI3 − xCl/PTAA	S2S	One step (DMF, GBL)	ambient	40 cm^2	15.5	[72]
Ultrasonic spray (all layers)	ITO/SnO2 Cs0.05FA0.81MA0.14PbI2.55Br0.45 l/ Spiro	S2S	One step (DMF/DMSO)	glove box	15.4mm^2	16.3	[73]
Inkjet printing (only PbI2)	FTO/TiO2/MAPbI3/Spiro/	S2S	Two step (DMF, DMSO)	ambient glove box	0.04 cm^2	18.3	[74]
Inkjet printing (all layers)	FTO/NiO/CsFAMAPbIBr/PCBM/	S2S	One step	glove box	10.5mm^2	16.5	[75]
Bar coating (only perovskite)	FTO/TiO2/(FAPbI3)0.95(MAPbBr3)0.05/ passivation/Spiro	S2S	One step (2ME/CHP)	ambient	31cm^2	17.53	[76]
Vacuum thermal evaporation (only perovskite)	FTO/TiO2/MA1-xCsxPbI3/Spiro	S2S	Co-evaporation (MAI treatment)	ambient	9 mm^2	20.13	[77]
Vacuum thermal evaporation (all layers)	ITO/Ca/C60/perovskite/TAPC/TAPC:MoO3/MoO	S2S	Sequentially evaporation	vacuum	0.1cm^2	17.6	[78]

Table 2.
Comparisons of some work by fabrication.

halides are relatively well soluble in organic solvents compared to metal halides (see **Table 2**). Many researchers prepare perovskite inks by dissolving organic halides into the DMF-based solvent before lead halides often decrease solvation time. The solvent has physical properties, such as Hansen's solubility parameters, dielectric constant, and Gutmann's donor number (DN). Solvents with dielectric constants more than 30 show generally good solubility for the perovskite precursors. In addition, halide in the solvents with low DN dominates the coordination with Pb^{2+}, and perovskites are quickly crystallized. In contrast, solvents with high DN compete for the coordination of I^- with Pb^{2+} slow down the crystallization of perovskites. Then higher DN solvents can be employed as solvent additives to control the crystallization of perovskite [82, 83]. However, by combining volatile non-coordinating solvents, i.e. 2-methoxy ethanol (2ME), and low volatile, coordinating solvent, i.e. N-cyclohexyl-2-pyrrolidone (CHP), Seok groups obtained bar-coated PSCs cells with a PCE of 20% with the area of 31 cm^2 [76]. 2ME has no or much weaker coordination capability. Thus, coordination ability and volatility (such as the boiling point and vapor pressure) can be critical parameters for perovskite crystallization. This solvent engineering will be a powerful technology for the scalability of printing. The toxicity of solvents, such as toxic DMF, skin-penetrating DMSO, or carcinogenic NMP, is also an important point, especially for operators who have direct contact with volatile solvents in the printing process [82]. It is difficult to determine whether 2ME is a commercially useful solvent or not due to the different regulations on industrial safety standards in country. On the other hand, DMSO is most environmentally friendly and least harmful to human health by systematically considering solvent production. Therefore, the search "green" solvent systems could also be critical for safe printing production [84].

2.3 Vacuum thermal evaporation-based fabrication of PSCs

Although vacuum thermal evaporation studies are relatively scarce, the following improvements led to PSCs with a PCE of 20%, not bad compared to the result of spin-coated PSCs (see **Table 2**) [77]. It was a significant result since there would be a feasible route to produce PSCs commercially, as proven by the CIGS and OLED industry. Next, we briefly summarize vacuum thermal evaporation-based fabrication methods.

2.3.1 Co-evaporation

This method is the most suitable vacuum-based process for many applications. The perovskite films are fabricated inside a high vacuum at a pressure of $10^{-5} \sim 10^{-6}$ bar. Each perovskite precursor is loaded into a crucible and heated to an appropriate sublimation temperature (**Figure 9(h)**). Removing the annealing step enables to deposit perovskite on any underlayer material [ref]. Perovskite films prepared in this way are more homogeneous, with better adherence to substrates than those prepared by spin coating. In addition, they are denser and pinhole-free, thus, more compatible with planar solar cells [85]. However, the most critical step is to control volatile organic halide (e.g. MAI) within the chamber, because it causes the compound to condense incompletely outside the evaporation cone region. The process is relatively slow and requires accurate periodic calibration to maintain deposition rates and precise stoichiometry.

2.3.2 Sequential evaporation

The method involves depositing several film layers sequentially on top of each other and then converting these multiple film layers through diffusion and recrystallization. The metal halide layer is typically deposited first and then converted by the organic halides. This method may not be ideal for optimal commercial scaling due to its throughput. By optimizing the system's pressure for each evaporation step, high efficiency of 17.6% was achieved in a small area. [78] This technique can be challenging to commercial scale as alternating evaporation can slow throughput and material utilization. When using a large vacuum chamber, the relatively long distance between the source and the substrate reduces the deposition rate and increases material waste. In addition, the vacuum-based fabrication of PSCs uses a variety of vapors, making control of the precursor stoichiometry challenging. The vacuum-based technology tends to be costly because of the sophisticated infrastructures required. However, major display companies such as Samsung and LG are converting their small- and medium-sized OLED panel manufacturing technology from the sixth generation (1500x1850 mm) to the eighth generation (2200 x 2500 mm). The evaporator and Fink Metal Mask (FMM) in a micrometer are the key to OLED panel production: The evaporator is critical equipment that forms red, green, and blue pixels, and FMM is a thin metal plate with small fine holes to induce the OLED material to be deposited on the substrate's required position. In addition, a curved display has also been developed. The curvature of curved TVs on the market today is roughly between 2000R and 4000R. The radius of curvature (R) is the reciprocal of the curvature. For example, 2000R TV refers to a curved TV radius of 2000 mm, and the lower R, the more pronounced curve, and the higher R, the more subtle curve. For mobile solar cell, the area of the panoramic sunroof in the vehicle is smaller than that of the 6Gen or 8Gen display. For example, the Hyundai YF sonata has a sunroof of 1000 x 1800 mm and a curvature of about 8125R (length) and 2512R (width). The vacuum deposition method applies to flexible substrates, metal foils, and ultrathin glass substrates with a thickness of 0.7 mm, so the range of solar application in a car is also wide

3. Summary

Perovskite solar cells have become one of today's most promising photovoltaic technologies. In this book, as the first section, we have introduced a lead-free $CsSnI_3$ and a stable molecular iodosalt, Cs_2SnI_6, and demonstrated that it is a vapor-processable semiconductor. Due to the instability properties of $CsSnI_3$, this research put more effort to develop Cs_2SnI_6. The use of a new class structural material with intrinsic stability and beneficial optoelectronic properties can be considered as a start of the next chapter in pervoksite photon devices. As the second section, large-area applicable perovskite coating technologies for commercialization were overhauled. PSCs based on the solution process can be manufactured at a much lower cost than conventional silicon cells as well as showing similar efficiency to silicon cells. However, to apply and commercialize this in real life, it needs a large-scale production technology while maintaining high efficiency. Until now, spin coating is the main technique. However, in general, this process has cracks or pinholes in the film when the coating area is expanded, resulting in poor density and uniformity. In addition, there is a limitation to reduce the amount of discarded solution, and it is difficult to do continuous (in-line) production. With increasing demand for mass production, production of M6-sized PSCs module has been many technological advances. Plus, improving stability is a problem that must be solved. For long-term reliability of PSCs, various encapsulation processes are also being studied. By overcoming many issues that perovskite can have, our company plans to integrate the PSCs to vehicles by 2025.

Acknowledgements

This work was supported by Hyundai Motor Group project (R-204696).

Author details

Rira Kang†, Tae-ho Jeong† and Byunghong Lee†*
Electronic Device Team, Hyundai Motor Group, Uiwang-si, Gyeonggi-do, Korea

*Address all correspondence to: redboho@gmail.com

† These authors contributed equally.

References

[1] NREL chart. Available from: https://www.nrel.gov/pv/cell-efficiency.html

[2] Mitzi DB et al. Conducting tin halides with a layered organic-based perovskite structure. Nature. 1994;**369**(6480): 467-469

[3] Hao F et al. Lead-free solid-state organic-inorganic halide perovskite solar cells. Nat Photon. 2014;**8**(6):489-494

[4] Noel NK et al. Lead-free organic-inorganic tin halide perovskites for photovoltaic applications. Energy & Environmental Science. 2014;**7**(9): 3061-3068

[5] Ogomi Y et al. CH3NH3SnxPb(1–x)I3 perovskite solar cells covering up to 1060 nm. The Journal of Physical Chemistry Letters. 2014;**5**(6):1004-1011

[6] Hao F et al. Anomalous band gap behavior in mixed Sn and Pb perovskites enables broadening of absorption Spectrum in solar cells. Journal of the American Chemical Society. 2014; **136**(22):8094-8099

[7] Zuo F et al. Binary-metal perovskites toward high-performance planar-heterojunction hybrid solar cells. Advanced Materials. 2014;**26**(37): 6454-6460

[8] Lee B et al. Air-stable molecular semiconducting Iodosalts for solar cell applications: Cs2SnI6 as a hole conductor. Journal of the American Chemical Society. 2014;**136**(43): 15379-15385

[9] Chung I et al. CsSnI3: Semiconductor or metal? High electrical conductivity and strong near-infrared photoluminescence from a single material. High hole mobility and phase-transitions. Journal of the American Chemical Society. 2012;**134**(20): 8579-8587

[10] Lee B et al. Solution processing of air-stable molecular semiconducting iodosalts, Cs2SnI6-xBrx, for potential solar cell applications. Sustainable Energy & Fuels. 2017;**1**(4): 710-724

[11] Shum K et al. Synthesis and characterization of $CsSnI_3$ thin films. Applied Physics Letters. 2010;**96**(22): 221903

[12] Fraiwan LF, et al. A wireless home safety gas leakage detection system. In: Biomedical Engineering (MECBME), 2011 1st Middle East Conference on 2011. 2011

[13] Kumar MH et al. Lead-free halide perovskite solar cells with high photocurrents realized through vacancy modulation. Advanced Materials. 2014; **26**(41):7122-7127

[14] Chen Z et al. Photoluminescence study of polycrystalline CsSnI3 thin films: Determination of exciton binding energy. Journal of Luminescence. 2012; **132**(2):345-349

[15] Sabba D et al. Impact of anionic Br– Substitution on open circuit voltage in Lead free perovskite (CsSnI3-xBrx) solar cells. The Journal of Physical Chemistry C. 2015;**119**(4):1763-1767

[16] Stoumpos CC, Malliakas CD, Kanatzidis MG. Semiconducting tin and Lead iodide perovskites with organic cations: Phase transitions, high Mobilities, and near-infrared Photoluminescent properties. Inorganic Chemistry. 2013;**52**(15):9019-9038

[17] Brik MG, Kityk IV. Modeling of lattice constant and their relations with ionic radii and electronegativity of constituting ions of A2XY6 cubic crystals (a=K, Cs, Rb, Tl; X=tetravalent cation, Y=F, Cl, Br, I). Journal of Physics and Chemistry of Solids. 2011;**72**(11): 1256-1260

[18] Singh D. Planewaves, Pseudopotentials, and the LAPW Method. Boston, MA: Kluwer Academic; 1994

[19] Kohn W, Sham LJ. Self-consistent equations including exchange and correlation effects. Physical Review. 1965;**140**(4A):A1133-A1138

[20] Hohenberg P, Kohn W. Inhomogeneous Electron gas. Physical Review. 1964;**136**(3B):B864-B871

[21] Perdew JP, Burke K, Ernzerhof M. Generalized gradient approximation made simple. Physical Review Letters. 1996;**77**(18):3865-3868

[22] Koelling DD, Harmon BN. A technique for relativistic spin-polarised calculations. Journal of Physics C: Solid State Physics. 1977;**10**(16):3107

[23] Blaha P et al. An Augmented Plane Wave + Local Orbitals Program for Calculating Crystal Properties. Austria: Universität Wien; 2001, Wien, Austria: Karlheinz Schwarz, Techn

[24] Tran F, Blaha P. Accurate band gaps of semiconductors and insulators with a Semilocal exchange-correlation potential. Physical Review Letters. 2009; **102**(22):226401

[25] Jeon NJ et al. Solvent engineering for high-performance inorganic–organic hybrid perovskite solar cells. Nature Materials. 2014;**13**(9): 897-903

[26] Saparov B et al. Thin-film deposition and characterization of a Sn-deficient perovskite derivative Cs2SnI6. Chemistry of Materials. 2016;**28**(7): 2315-2322

[27] Fuchizaki K et al. Polyamorphism in tin tetraiodide. The Journal of Chemical Physics. 2009;**130**(12):121101

[28] Fuchizaki K. Melting behavior of SnI4 reexamined. The Journal of Chemical Physics. 2013;**139**(24):244503

[29] Patterson AL. The Scherrer formula for X-ray particle size determination. Physical Review. 1939;**56**(10):978-982

[30] Xiao Z et al. Solvent annealing of perovskite-induced crystal growth for photovoltaic-device efficiency enhancement. Advanced Materials. 2014;**26**(37):6503-6509

[31] Qian W et al. Improving field emission by constructing CsI-AlN hybrid nanostructures. Journal of Materials Chemistry. 2012;**22**(35):18578-18582

[32] Weiss M et al. Preparation and characterization of methylammonium tin iodide layers as photovoltaic absorbers. Physica status solidi (a). 2016; **213**(4):975-981

[33] Xiao Z et al. Intrinsic defects in a photovoltaic perovskite variant Cs2SnI6. Physical Chemistry Chemical Physics. 2015;**17**(29):18900-18903

[34] Dang Y et al. Formation of hybrid perovskite tin iodide single crystals by top-seeded solution growth. Angewandte Chemie. 2016;**128**(10): 3508-3511

[35] Wells AF. Structural Inorganic Chemistry. Oxford University Press; 2012

[36] Apostolico L et al. The reaction of tin (iv) iodide with phosphines: Formation of new halotin anions. Dalton Transactions. 2009;**47**:10486-10494

[37] Apostolico L et al. The reaction of GeCl4 with primary and secondary phosphines. Dalton Transactions. 2004; **3**:470-475

[38] Guan J, Tang Z, Guloy AM. [H3N (CH2)7NH3]8(CH3NH3)2Sn(iv)Sn(ii) 12I46 - a mixed-valent hybrid compound with a uniquely templated defect-perovskite structure. Chemical Communications. 2005;**1**:48-50

[39] De Trizio L et al. Sn cation Valency dependence in cation exchange reactions involving Cu2-xSe nanocrystals. Journal of the American Chemical Society. 2014; **136**(46):16277-16284

[40] Lufaso MW, Woodward PM. Prediction of the crystal structures of perovskites using the software program SPuDS. Acta Crystallographica Section B. 2001;**57**(6):725-738

[41] Xiao Z. First-Principles Study on Electronic Structure and Doping for Novel Compound Semiconductors, BaZn2As2, SnS, and Cs2SnI6. 2015

[42] Goldschmidt VM. Die Gesetze der Krystallochemie. Naturwissenschaften. 1926;**14**(21):477-485

[43] Li Z et al. Stabilizing perovskite structures by tuning tolerance factor: Formation of Formamidinium and cesium Lead iodide solid-state alloys. Chemistry of Materials. 2016;**28**(1): 284-292

[44] Lee B, Buchholz DB, Chang RPH. An all carbon counter electrode for dye sensitized solar cells. Energy & Environmental Science. 2012;**5**(5): 6941-6952

[45] Kim H-S et al. Mechanism of carrier accumulation in perovskite thin-absorber solar cells. Nature Communications. 2013;**4**

[46] Gonzalez-Pedro V et al. General working principles of CH3NH3PbX3 perovskite solar cells. Nano Letters. 2014;**14**(2):888-893

[47] Xing G et al. Long-range balanced Electron- and hole-transport lengths in organic-inorganic CH3NH3PbI3. Science. 2013;**342**(6156):344-347

[48] Lee B et al. Materials, interfaces, and photon confinement in dye-sensitized solar cells. The Journal of Physical Chemistry B. 2010;**114**(45):14582-14591

[49] Lee B et al. Optimizing the performance of a plastic dye-sensitized solar cell. The Journal of Physical Chemistry C. 2011;**115**(19):9787-9796

[50] Garcia-Belmonte G et al. Charge carrier mobility and lifetime of organic bulk heterojunctions analyzed by impedance spectroscopy. Organic Electronics. 2008;**9**(5):847-851

[51] Fabregat-Santiago F et al. Characterization of nanostructured hybrid and organic solar cells by impedance spectroscopy. Physical Chemistry Chemical Physics. 2011; **13**(20):9083-9118

[52] Leever BJ et al. In situ characterization of lifetime and morphology in operating bulk heterojunction organic photovoltaic devices by impedance spectroscopy. Advanced Energy Materials. 2012;**2**(1): 120-128

[53] Nagpal P, Klimov VI. Role of mid-gap states in charge transport and photoconductivity in semiconductor

nanocrystal films. Nature Communications. 2011;**2**:486

[54] Margulies EA et al. Enabling singlet fission by controlling intramolecular charge transfer in π-stacked covalent terrylenediimide dimers. Nature Chemistry. 2016 advance online publication

[55] Liu Q et al. Exciton relaxation dynamics in photo-excited CsPbI3 perovskite nanocrystals. Scientific Reports. 2016;**6**:29442

[56] Wang L et al. Femtosecond time-resolved transient absorption spectroscopy of CH3NH3PbI3 perovskite films: Evidence for passivation effect of PbI2. Journal of the American Chemical Society. 2014;**136**(35):12205-12208

[57] Makarov NS et al. Spectral and dynamical properties of single excitons, Biexcitons, and Trions in cesium–Lead-halide perovskite quantum dots. Nano Letters. 2016;**16**(4):2349-2362

[58] Piatkowski P et al. Direct monitoring of ultrafast electron and hole dynamics in perovskite solar cells. Physical Chemistry Chemical Physics. 2015; **17**(22):14674-14684

[59] Cherepy NJ et al. Ultrafast Electron injection: Implications for a Photoelectrochemical cell utilizing an anthocyanin dye-sensitized TiO2 Nanocrystalline electrode. The Journal of Physical Chemistry B. 1997;**101**(45): 9342-9351

[60] Perovskite info. Available from: https://www.perovskite-info.com/companies

[61] Liu F et al. Organic solar cells with 18% efficiency enabled by an alloy acceptor: A two-in-one strategy. Advanced Materials. 2021;**33**(27): 2100830

[62] Zhao Y et al. Research progress in large-area perovskite solar cells. Photonics Research. 2020;**8**(7):A1-A15

[63] Li J et al. Optimization of anti-solvent engineering toward high performance perovskite solar cells. Journal of Materials Research. 2019; **34**(14):2416-2424

[64] Gu X et al. The meniscus-guided deposition of semiconducting polymers. Nature Communications. 2018;**9**(1):534

[65] Dai X et al. Meniscus fabrication of halide perovskite thin films at high throughput for large area and low-cost solar panels. International Journal of Extreme Manufacturing. 2019

[66] Deng Y et al. Tailoring solvent coordination for high-speed, room-temperature blading of perovskite photovoltaic films. Science. Advances. 2019;**5**(12):eaax7537

[67] Kim YY et al. Roll-to-roll gravure-printed flexible perovskite solar cells using eco-friendly antisolvent bathing with wide processing window. Nature Communications. 2020;**11**(1):5146

[68] Hamukwaya SL et al. A review of recent developments in preparation methods for large-area perovskite solar cells. Coatings. 2022;**12**(2):252

[69] Xu Z et al. Nanocrystal-enabled front-surface bandgap gradient for the reduction of surface recombination in inverted perovskite solar cells. Solar RRL. 2021;**5**(10):2100489

[70] Di Giacomo F et al. Up-scalable sheet-to-sheet production of high efficiency perovskite module and solar cells on 6-in. Substrate using slot die

coating. Solar Energy Materials and Solar Cells. 2018;**181**:53-59

[71] Galagan Y et al. Roll-to-roll slot die coated perovskite for efficient flexible solar cells. Advanced Energy Materials. 2018;**8**(32):1801935

[72] Heo JH et al. Highly efficient CH3NH3PbI3−xClx mixed halide perovskite solar cells prepared by re-dissolution and crystal grain growth via spray coating. Journal of Materials Chemistry A. 2016;**4**(45):17636-17642

[73] Bishop JE et al. Fully spray-coated triple-cation perovskite solar cells. Scientific Reports. 2020;**10**(1):6610

[74] Li Z et al. Ink engineering of inkjet printing perovskite. ACS Applied Materials & Interfaces. 2020;**12**(35): 39082-39091

[75] Schackmar F et al. Perovskite solar cells with all-inkjet-printed absorber and charge transport layers. Advanced Materials Technologies. 2021;**6**(2): 2000271

[76] Yoo JW et al. Efficient perovskite solar mini-modules fabricated via bar-coating using 2-methoxyethanol-based formamidinium lead tri-iodide precursor solution. Joule. 2021;**5**(9): 2420-2436

[77] Zhu X et al. Superior stability for perovskite solar cells with 20% efficiency using vacuum co-evaporation. Nanoscale. 2017;**9**(34):12316-12323

[78] Hsiao S-Y et al. Efficient all-vacuum deposited perovskite solar cells by controlling reagent partial pressure in high vacuum. Advanced Materials. 2016; **28**(32):7013-7019

[79] Swartwout R, Hoerantner MT, Bulović V. Scalable deposition methods for large-area production of perovskite thin films. Energy & Environmental Materials. 2019;**2**(2):119-145

[80] Nayak PK et al. Mechanism for rapid growth of organic–inorganic halide perovskite crystals. Nature Communications. 2016;**7**(1):13303

[81] Ono LK, Juarez-Perez EJ, Qi Y. Progress on perovskite materials and solar cells with mixed cations and halide anions. ACS Applied Materials & Interfaces. 2017;**9**(36):30197-30246

[82] Wang Y et al. Printing strategies for scaling-up perovskite solar cells. National Science Review. 2021;**8**(8)

[83] Hamill JC, Schwartz J, Loo Y-L. Influence of solvent coordination on hybrid organic–inorganic perovskite formation. ACS Energy Letters. 2018; **3**(1):92-97

[84] Park SH, Jin IS, Jung JW. Green solvent engineering for environment-friendly fabrication of high-performance perovskite solar cells. Chemical Engineering Journal. 2021;**425**:131475

[85] Hamukwaya SL et al. A review of recent developments in preparation methods for large-area perovskite solar cells. Coatings. 2022

Section 3

Multifunctional Materials

Chapter 13

Tunable Multifuctionality in Heusler Alloys by Extreme Conditions

Devarajan Uthiran and Arumugam Sonachalam

Abstract

The multifunctional materials have demonstrated various properties such as shape memory effect (SME), magneto caloric effect (MCE), magneto resistance (MR), piezoresistance (PR), exchange bias (EB), half metallic ferromagnetism (HMF), and spin polarization. Among many Heusler compounds, Ni-Mn-Ga alloys provide SME, MCE, PR, and MR behaviors. These properties can be tuned by some external/internal perturbations such as pressure, magnetic field, and chemical composition. These alloys are prepared using an arc melting furnace under by melting the high-purity starting elements (99.99%). The aim of the book chapter is to enhance the multicaloric properties (MCE and PR) nearer to ambient temperature by the application of some external parameters. Hence, we have chosen few Heusler alloys. These materials are investigated under extreme conditions (hydrostatic pressure, high magnetic field, and low temperature). All the doped and undoped Ni-Mn-Ga alloy series alloys exhibit conventional MCE. The application of external magnetic field increases the magnetization for both alloys. The hydrostatic pressure influences Ms and broadens the hysteresis width in both the samples. The observed metamagnetic transition at ambient pressure gets suppressed at higher pressure. Also, high pressure induces larger magneto crystalline anisotropy. The effect of pressure on MCE is decreased for both $Ni_{2-x}Mn_{1+x}Ga$ (x = 0 and 0.15) alloys. These alloys exhibit –ve PR (x=0 @ 30 kbar) and +ve PR (x = 0.15@ 28 kbar) when subjected to hydrostatic pressure. The rate of change of T and resistivity with respect to pressure are calculated and show positive values for both the samples. The residual resistivity and electron-electron scattering factor are found to be decreased with pressure for x = 0, and it exhibits metallic behavior. However, both parameters increase for x = 0.15 alloy, and it may be related to static disorder effects and spin fluctuations.

Keywords: Heusler alloys, NiMnGa, magneto caloric effect, magnetoresistance, piezoresistance

1. Introduction

1.1 Investigation of the influence of hydrostatic pressure on the magnetic magnetocaloric properties, electronic transport, and piezoresistivity of $Ni_{2-x}Mn_{1+x}Ga$ (x = 0, 0.15) Heusler alloys

The discovery of new phase transformations can be performed through high-pressure studies. The hydrostatic pressure effect gives insistently to recognize the physical stuff of ingredients adjacent to the phase transitions with correlating phenomenon. Within various stages happening in scheme of magnetic shape memory, they exhibit phases at high and low temperature, high (austenite) and low (austenite). The self-generated lattice strain is brought by martensite transformation in Heusler alloys at low temperature. Freshly, the series of Ni-Mn alloys have been fascinating for numerous uses because of the concerning multifunctional assets such as shape memory effect (SME) [1, 2], magneto caloric effect (MCE) [3, 4], exchange bias (EB) behavior [5], and magneto resistance (MR) [6–10]. In the foresaid credible tenders, the resources having MCE demonstrations predict for custom in condensed matter cooling system. Within the diverse magnetic shape memory alloys, Ni-Mn-Ga has premeditated widely; afterward, enormous strain was developed by magnetic field. The samples endure first-order structural alteration from extraordinary thermal austenite to low thermal martensite and consume prominant magnetocrystalline anisotropy in the lower martensite in addition to low twinning stress, which provides magnetic field–induced strain. The change in composition considerately impacts the phase transformations, crystallographic structures, and magnetic properties in NiMnGa alloys [11–16].

Hydrostatic pressures are known to play a significant role on the magnetic and structural properties of these systems [17–19]. The relative stability of the high-temperature cubic phase and the low-temperature martensite phase can be influenced by pressure. Magnetism in these alloys mainly arises from RKKY exchange interaction, within the Mn atoms [20]. The exchange interaction between Mn-Mn atoms in Heusler alloys powerfully influences on the distance and is reformed by either chemical substitutions (or) hydrostatic pressure. These consequences suggest the observation of positive pressure coefficient of J_{Mn-Mn} [21]. Recently, the effect of pressure on some of the NiMn-based systems has been reported [22–26]. Hydrostatic pressure effect on magnetic and martensitic transition shifts toward higher temperature with decrease of ΔS_M in NiMnIn magnetic superelastic alloys [22]. Albertini *et al* reported the pressure effects on MCE in Mn-rich and Ni-rich Ni_2MnGa alloy and found that the MCE decreases (increases) for Ni (Mn)-rich alloy [23]. Kamarad *et al* [27] have studied the effect of hydrostatic pressure on magnetization of Ni-rich $Ni_{2+x}Mn_{1-x}Ga$ (x = 0, 0.15) compounds and reported that pressure decreases $\Delta S_M{}^{max}$ at T_M. Further, Esakki Muthu *et al* [26] have reported the effect of hydrostatic pressure on M_s and ΔS_M in $Ni_{50-x}Mn_{37+x}Sn_{13}$ (x = 2, 3) alloys. Nayak *et al* [24] reported on NiCoMnSb alloy that the pressure enhances the stability of the martensite phase and decreases $\Delta S_M{}^{max}$ besides an upward shift in T_M. By the hydrostatic pressure effect on the magnetic and magneto caloric property of the Heusler alloys. At the same time, up to now few reports are available on the pressure need of MCE in Mn fertile Ni-Mn-Ga alloys [23].

Band ferromagnetism is a significant phenomenon developed in several transition-metal compounds. High-pressure studies can provide valuable information on electronic structure and electron-electron interactions in intermetallics. Among many of these studies, hydrostatic pressure [24] has been recognized as an effective tool to change physical and chemical properties in solids. Nevertheless, resistivity, piezoresistance (PR) [25], and pressure-induced phase transition [26, 27] are renowned as important phenomena that occur in Heusler alloys. PR is the change in resistance provoked by pressure effect was first discovered by Smith [28] in semiconductors (Si, Ge) possessing anisotropic energy with wide band structures. Further, they are being broadly used as stress and strain sensors [29]. Numerous kinds of Heusler alloy systems such as Ni-Mn-X (X = Sn, Ga, Si, Sb) have been extensively studied [9, 30–48]. Among them, Ni_2MnGa, a potentially well-known magnetic shape memory alloy (MSMA) [1, 31, 32], has been considerably investigated material for both scientific and technological purposes and is known to exhibit various phenomena such as MCE, EB [32], MR, and magnetic-field–induced strain (MFIS) [33–36]. It undergoes structural transformation from cubic austenite (high temperature, high symmetry phase) to low-temperature martensite phase at T_M = 202 K [37] and second-order paramagnetic to FM phase transition at T_C = 376 K [27]. Additionally, the increased rate of T_M with pressure is 0.55 K/kbar. On the divergent, for $Ni_{1.85}Mn_{1.15}Ga$ (T_M = 138 K), T_M is diluted at the rate of −1.08 K/kbar with fixed hydrostatic pressure and magnetic field [38, 39]. The temperature-reliant electrical resistivity ρ(T), whichever frequently employed to separate metals from insulators, band-gap insulators, Anderson localized insulators, has been studied for Ni-Mn-Ga alloys, for instance, $Ni_2MnGa_{1-x}B_x$ (x = 0.03,0.05) [40], $Ni_2MnGa_{1-x}In_x$ (x = 0.05–0.15) [41], $Ni_{2.16}Mn_{0.84}Ga$ alloy [42], $Ni_{2+x}Mn_{1-x}Ga$ (x = 0-–.2) [9, 43], and $Ni_{49.5}Mn_{25.4}Ga_{25.1}$, $Ni_{51.1}Mn_{24.9}Ga_{24}$ alloys [44]. Furthermore, the time-dependent and field-induced hump in the resistivity was also observed in Ga doped Ni-Mn-Sn alloy [45]. Clear view of magneto structural transition has been investigated in $Ni_{2.18}Mn_{0.82}Ga$ [46]. Temperature dependence of electrical resistivity under various hydrostatic pressures has been studied for $Ni_{2.14}Mn_{0.84}Ga_{1.02}$, $Ni_{2.14}Mn_{0.92}Ga_{0.94}$, and Ni_2MnGa single crystals [47]. Large value of piezoresistance and magnetoresistance under uniaxial stress has been observed in $Ni_{45}Co_5Mn_{37.5}In_{12.5}$ [48]. In this work, we investigate the effect of hydrostatic pressure on the resistivity and piezoresistivity (PR) of $Ni_{2-x}Mn_{1+x}Ga$ (x = 0, 0.15) magnetic shape memory alloys. We have also determined the ρ_0 and (A) for the alloys.

1.2 Experimental techniques

The samples have been prepared by standard arc melting technique [15, 49]. The elemental ratio is influenced using energy-dispersive analysis of X-rays, which provide the actual chemical ratio as $Ni_{1.99}$ $Mn_{1.01}Ga_{1.00}$ and $Ni_{1.9}Mn_{1.15}Ga_{0.95}$ for x = 0 (Ni_2MnGa) and 0.15 ($Ni_{1.85}Mn_{1.15}Ga$), respectively. The magnetic studies are accomplished at different pressures by 9 Tesla Physical Property Measurement System (PPMS-9T)-Vibrating Sample Magnetometer (VSM) (Quantum design, USA) module furnished with the Cu-Be clamp-type pressure cell with 10 kbar upper pressure [28]. The thermomagnetic data are recorded with VSM for both cooling and heating mode in the temperature range of 2 K–300 K under ambient and high pressures up to 7.4 kbar for $Ni_{1.85}Mn_{1.15}Ga$ (x = 0.15) and 9.69 kbar for Ni_2MnGa (x = 0) nominal compositions at a constant magnetic field of 0.01 T. The isothermal magnetization M (H) is recorded up to 5 T at ambient and high pressure at different temperatures (180–250 K: x = 0 & 130–160 K: x = 0.15) across T_M. The materials preparation, characterization, and other studies such as isothermal magnetoresistance, magnetization on $Ni_{2-x}Mn_{1+x}Ga$ (x = 0, 0.15) alloys have been discussed [15, 49–51].

High-pressure electrical transport measurements are performed by four-probe resistivity method. Throughout the measurements, the samples with four-probe contacts are immersed in a Teflon capsuled Daphane (#7074) and kept inside the Be-Cu clamp-type hybrid pressure cell (30 kbar). Usual persistent DC current of 80 mA is provided by programmable fixed current source (224, Keithely, USA), voltage is restrained with nanovoltmeter (34420A, Agilent, USA), and temperature is assured through temperature controller (Lakeshore, USA) and is computerized with LABVIEW software. The external pressure is applied to sample using 20 Ton hydraulic press. The clamped pressure cell is externally loaded inside the closed-cycle refrigerator (CCR-VTI). The pressure is standardized using Bi-resistive transitions of Bi I–II (25.5 kbar) and II–III (27 kbar) at ambient as well as least temperature was reached (300– K) employing CCR-VTI. Expending with turbo molecular pump, the cavities have been vacuumized up to 10^{-6} mbar. A sufficient He gas is entered into container of CCR-VTI when the alloy is cooled, and the temperature of the alloy is suppressed by the compressor. An external mechanical refrigerator removes the warm helium vapor.

2. Results and discussion

The temperature dependence of magnetization M (T) is measured for $Ni_{2-x}Mn_{1+x}$ Ga (x = 0, 0.15) alloys at 0.01 and 5 T (**Figure 1**). While cooling from ambient temperature, from x = 0, a rapid decrease in magnetization happens at 209 K in the magnetic sources of 0.01 T, which point out the martensitic begins (M_s = 209 K). Advanced cooling at less than 188 K (martensite end (M_f = 188 K)) results in persistent magnetization up to 5 K. The reduction in magnetization nearby T_M is caused by prominent magneto crystalline anisotropy in the lower-order phase [1, 52]. Here, the T_M is calculated using the relation $T_M=(M_s+A_f)/2$. The hysteresis is noticed between cooling and warming modes, which describes the first-order structural transition. Comparable property is found in the M (T) curve for x = 0.15 at a field of 0.01 T (**Figure 1b**), which is in agreement with our earlier M (T) measurement at 0.01 T in $Ni_{1.84}Mn_{1.17}Ga$ in the limited temperature range [50]. The observed M_s values for x = 0 and 0.15 are consistent with reported literature [50, 53, 54]. The various characteristic

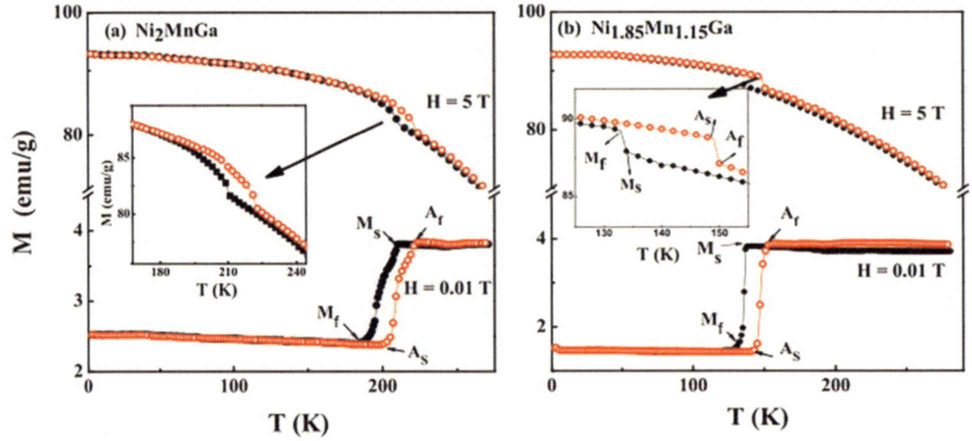

Figure 1.
Temperature dependence of magnetization at H=(0.01 & 5 T) for (a) Ni_2MnGa *and (b)* $Ni_{1.85}Mn_{1.15}Ga$.

transition temperatures obtained from our M (T) curves are shown in **Tables 1** and **2** for x = 0, 0.15. The characteristic transformations decrease with increasing Mn compositions at normal pressure. The closed view of **Figure 1** shows the enlargement of magneto thermal curves information approximate T_M measured. The H = 5 T enhances the magnitude of magnetization equated to the lower magnetic field (0.01 T) for both x = 0 and 0.15 (**Figure 1**). Besides, the magnetization of the low-stability phase is more than that of high-stability phase at 5 T for both alloys. This is owing to the alignment of magnetic domains in the oriented axis, it destroys the magneto crystalline anisotropy at H = 5 T [55, 56]. Similar increase in magnetization with the application of magnetic field has also been observed in $Ni_{1.84}Mn_{1.17}Ga$, $Ni_{50+x}Mn_{25-x}Ga$ (x = 0, 2, 3, 5), and $Ni_{1.75}Mn_{1.25}Ga$ [50, 57, 58].

Figure 2 indicates thermo-magnetization with cool and warm modes at different hydrostatic pressures up to 9.69 kbar (x = 0) and 7.4 kbar (x = 0.15) at H = 0.01 T. Notification from **Figure 2a** that M_s increases slightly with pressure, because of hybridization among Ni-3d and Mn/Ga atom. **Table 3** displays pressure derived function of M_s and T_M equated to existing reports. The hysteresis found in both cool and warm curvatures appears to be extended with enhancement of hydrostatic pressure. The possibility of owing to the variation in magneto crystalline anisotropy at minimum temperature in lower-order phase. Related pressure reliance has been described in x = 0 [54], where the heating and cooling curves show opposite trend compared with the present results.

Figure 2b demonstrates the M (T) curve at several hydrostatic pressures (0–7.4 kbar) for $Ni_{2-x}Mn_{1+x}Ga$ (x = 0.15). It is originated that M_S and T_M suppress as the pressure enhances (dM_s/dP = −2.027 K/kbar); dT_M/dP = −1.081 (K/kbar). At the same time, Albertini *et al* [23] noticed that martensitic transformation temperatures raise by pressure for $Ni_{1.9}Mn_{1.3}Ga_{1.8}$. The negative alteration of M_S graces the stability of cubic higher-order phase representing that martensite phase is low stability with

Pressure (kbar)	Transition temperatures (K)				
	M_s	M_f	A_s	A_f	$T_M = (M_s+A_f)/2$
0	209	188	205	222	216
1.37	209	185	208	227	218
6.06	210	209	212	230	220
9.69	210	212	213	233	222

Table 1.
Transition temperatures with several hydrostatic pressures of Ni_2MnGa.

Pressure (kbar)	Transition temperatures (K)				
	M_s	M_f	A_s	A_f	$T_M = (M_s+A_f)/2$
0	137	128	142	152	144.5
3.0	133	117	141	146	139.5
5.8	128	113	140	147	137.5
7.4	122	101	136	151	136.5

Table 2.
Transition temperatures with different hydrostatic pressures of $Ni_{1.85}Mn_{1.15}Ga$.

Sample	dT_M/dP (K/kbar)	dM_S/dP (K/kbar)
Ni_2MnGa	0.619	0.103
$Ni_{1.85}Mn_{1.15}Ga$	−1.081	−2.027
$Ni_{2.15}Mn_{0.85}Ga$ [23]	0.58	—
$Ni_{1.9}Mn_{1.3}Ga_{1.8}$ [23]	1.7	—
$Ni_{2.151}Mn_{0.771}Ga$ [59]	0.6	—
$Ni_{48}Mn_{39}Sn_{13}$ [60]	—	3.16
$Ni_{47}Mn_{40}Sn_{13}$ [60]	—	0.51

Table 3.
Pressure derivative values of T_M and M_S for various NiMnGa (Sn) systems.

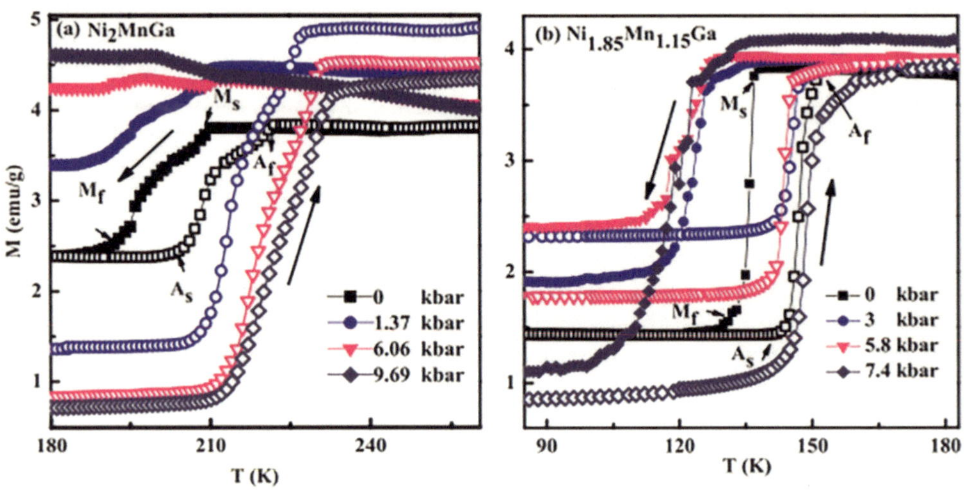

Figure 2.
Thermomagnetic curves (both cool and warm) of $Ni_{2-x}Mn_{1+x}Ga$ (x = 0, 0.15) alloys with various applied pressures at H = 0.01 T.

pressure. Furthermore, substantial alterations happen in electronic structure under pressure and may extend to various transformation temperatures. Pressure steadies the lower-order phase in further FM shape memory alloys (FSMAs) [61]. In addition, a large width of the hysteresis is noticed on increasing the pressure that is attributed to the enhancement of magneto elastic coupling with lattice strain, twin boundary motion, and first-order transition [61].

Pressure habituation of transformation in x=0, 0.15 is represented in **Figure 3**, which confirmed that transformation rises with pressure in x=0, while transformation suppresses in x = 0.15. This occurrence can be described in following ways. Enhancement of pressure will decrease the crystallographic volume, which affects characteristic transition temperatures. Furthermore, these alloys show that lower volume change with pressure results in slight change in M_s. Kim *et al,* described the pressure reliance of T_M in x = 0 and $Ni_{2.14}Mn_{0.84}Ga_{1.02}$ and established high volume change [39]. However, the volume change in our present alloys is lower than the reported $Ni_{50-x}Mn_{37+x}Sn_{13}$ (x = 2, 3) [60] and $Ni_{50}Mn_{34}In_{16}$ [22]. The characteristic

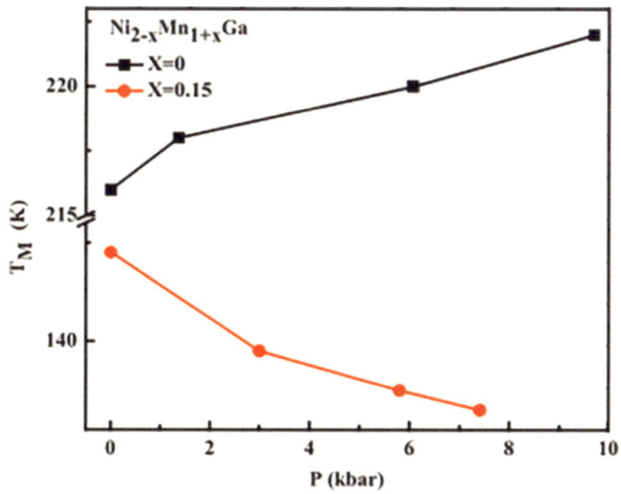

Figure 3.
Pressure reliance of T_M in $Ni_{2-x}Mn_{1+x}Ga$ (x = 0, 0.15) evaluated at H = 0.01 T under different hydrostatic pressures.

transition temperatures under different pressure are given in **Tables 1** and **2** for both alloys respectively.

The isothermal curves have restrained in x = 0 around T_M (180 K–250 K) at ambient, high pressures (**Figure 4a** and **b**). The magnetization work is absorbed into two ways: magnetic field is improved 0–5 and reduced 5–0 T. It is observed that magnetic field prompted first-order metamagnetic transition arises at 220 K, ambient pressure for x = 0 (**Figure 4a**). The enclosure of **Figure 4a**) displays the transition clearly. However, this transition disappears at 9.69 kbar. Related actions have been detected in $Ni_{2.208}Mn_{0.737}Ga$ [59]. From **Figure 4b**, the crossover in magnetization is noticed in the M (H) curve approximately 190–235 K. The magnetization is hard to impregnate for temperatures 190 and 196 K specifying a martensite phase, although at austenite phase at 223 K and 235 K the magnetization is relaxed to saturate. This is because of the firm magnetocrystalline anisotropy around transformations in x = 0 at high pressure. **Figure 4c** and **d** shows the isothermal curve of x = 0.15 at P = 0, 7.4 kbar. This M (H) is assessed with rate of temperature 130 K–160 K in gradual rise of 2 K; even though only picked out readings are drawn in **Figure 4c** and **d** for the clear visibility. At normal pressure, the field brought metamagnetic transition is seen in low field (0.68 T) at 146 K insert (**Figure 4c**). But, this transition is blocked at 7.4 kbar (**Figure 4d**). Similar to x = 0, the cusp in magnetization has been perceived between martensite and austenite temperature for x = 0.15 at P = 7.4 kbar, as viewed in (**Figure 4d**). Therefore, the subjection of pressure develops large magnetocrystalline anisotropy in both alloys.

The entropy change in magnetization (ΔS_M) at various pressures is evaluated using the Maxwell's relation

$$\Delta S_M = \int_0^H \left(\frac{\partial M(H,T)}{\partial T}\right)_H dH \tag{1}$$

ΔS_M is computed from the over equation using numerical integration of the isothermal curves. ΔS_M with temperature for various pressures is represented in

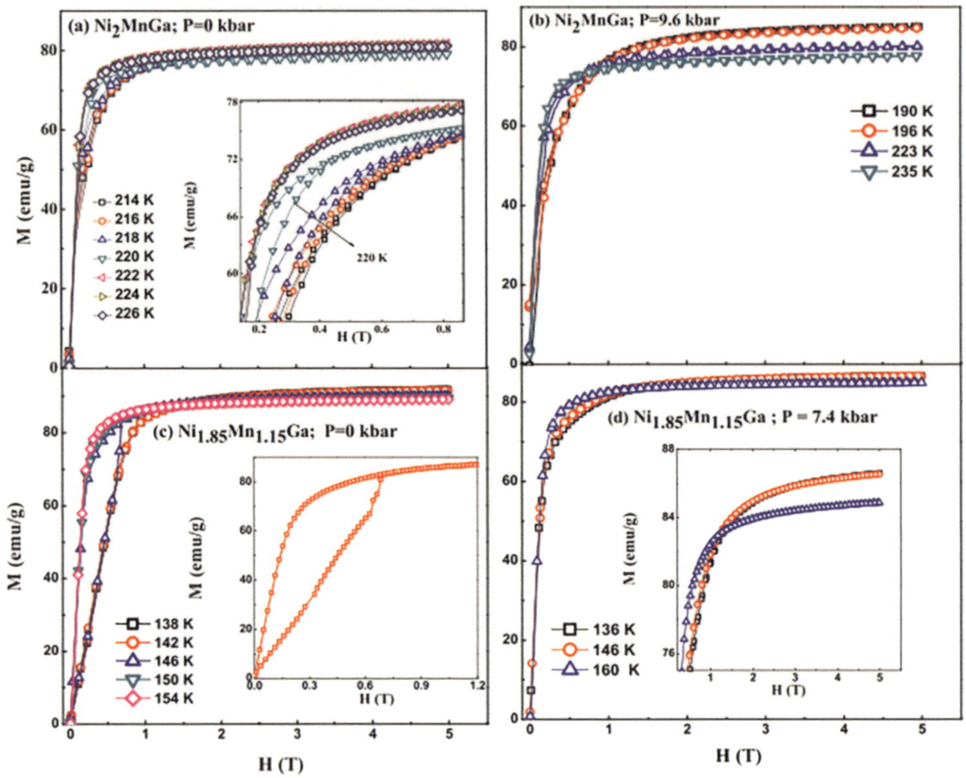

Figure 4.
Isothermal curves of x = 0 (a) ambient and (b) 9.69 kbar pressure, x = 0.15 (c) ambient and (d) 7.4 kbar pressure.

Figure 5. The usage of pressure declines the ΔS_M is 19.21 $Jkg^{-1}K^{-1}$ (P = 0 kbar) to 6.04 $Jkg^{-1}K^{-1}$ (P = 9.69 kbar) for x = 0. Nevertheless, the peak temperature of ΔS_M gains in the direction of higher temperature with pressure. Enhancement of the maximum temperature of ΔS_M near room temperature with the application of higher pressure. The degree of ΔS_M suppresses from 8.9 $Jkg^{-1}K^{-1}$ (P = 0) to 1.27 $Jkg^{-1}K^{-1}$ (P = 7.4 kbar) in $Ni_{1.85}Mn_{1.15}Ga$ and the highest temperature of ΔS_M moves near to lower temperature. Likewise, for Ni-rich $Ni_{2.208}Mn_{0.737}Ga$, ΔS_M shrinkages from 96 $Jkg^{-1}K^{-1}$ (P = 0) to 86 $Jkg^{-1}K^{-1}$ (P = 8 kbar) [59]. Additionally, in $Ni_{2.15}Mn_{0.85}Ga$, ΔS_M reduces from 24 $Jkg^{-1}K^{-1}$ (P = 0) to 20 $Jkg^{-1}K^{-1}$ (P = 11.7 kbar); however, in Mn-rich $Ni_{1.9}Mn_{1.3}Ga_{0.8}$, ΔS_M promotes from 4.5 $Jkg^{-1}K^{-1}$ (P = 0) to 6 $Jkg^{-1}K^{-1}$ (P = 12.2 kbar) [23]. Based on the earlier reports, it is concluded that pressure stimulates additional magnetic entropy in prominent additional Mn in $Ni_{2-x}Mn_{1+x}Ga$ compounds equated to the x = 0.15 sample analyzed and excess Ni in $Ni_{2-}Mn_{1+x}Ga$.

Temperature dependence of resistivity ρ (T) is measured for polycrystalline $Ni_{2-x}Mn_{1+x}Ga$ (x = 0, 0.15) at several hydrostatic pressures both in cool and warm sequences. In order to see the clear features, the heating cycle graph has presented for pressure (**Figure 6**) and the cool-warm lines are prearranged as inside view of **Figure 6** at least and maximized pressure. At normal pressure, ρ falls with decrement of temperature in x = 0 (**Figure 6a**) and exhibits least thermal hysteresis near martensite transition. This specifies that structural transitions are specified in the closed view of **Figure 6a**. Comparable ρ (T) is identified for x = 0 [47]. These results are perceived as hysteresis curves at ambient and P = 7 kbar. The fascinating report at 25

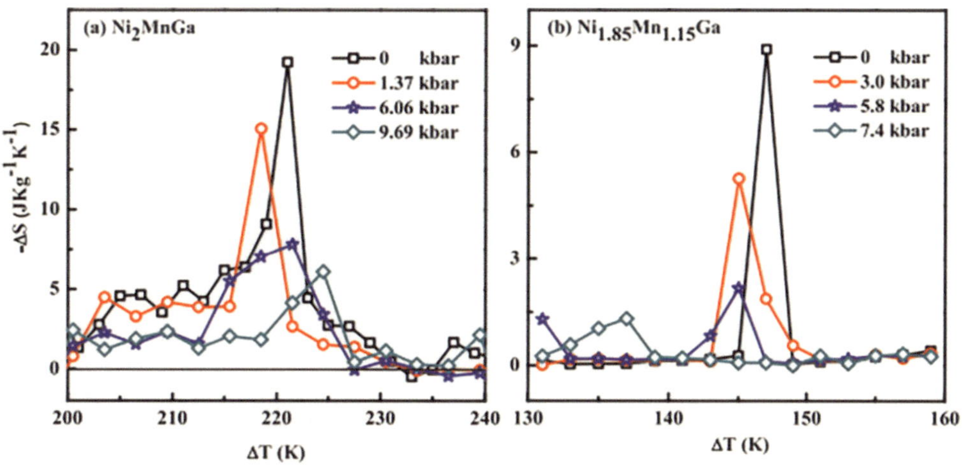

Figure 5.
Temperature desirable of magnetic entropy change in $Ni_{2-x}Mn_{1+x}Ga$ (x = 0, 0.15) alloys at different hydrostatic pressures.

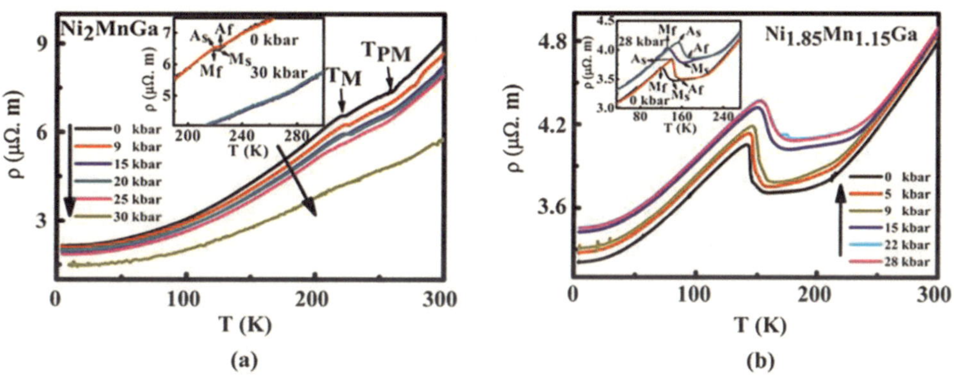

Figure 6.
Temperature dependence of resistivity for (a) Ni_2MnGa, (b) $Ni_{1.85}Mn_{1.15}Ga$ alloys at various pressures. Inset: shows ρ vs T in the temperature region of 180–300 K for the pressure of 0 kbar and 30 kbar.

kbar pressure, the hysteresis graph gets repressed in x = 0, at 30 kbar, the hysteresis completely be wiped out, and it can be evidently seen in inset of **Figure 6a**. As increasing pressure, its ρ decreases entire 4–300 K and boosts the metallic nature in the alloy due to enriched hybridization of valence band by applied pressure [40, 62, 63]. A mark of the pre-martensite transition (T_{PM}) detected at 260 K at ambient pressure for x=0. The cross of the T_{PM} is not clearly witnessed at higher pressure (30 kbar) (**Figure 6a**). The magnetic field and hydrostatic pressure are significantly affected on T_M and T_{PM} of Ni_2MnGa system [39, 64]. Similar ρ (T) has been found for x = 0.15 (**Figure 6b**), where the application of pressure increases the ρ due to the pressure-induced phase transformation, which indicates the phase changes from pre-martensite to martensite [46, 47]. Still, total ρ for x = 0.15 is lower than x =0. Besides, in demarcation to x = 0, considerable thermal hysteresis is seen for entire pressure. This may be related to electronic structure as compared with magnetic actions [19, 63, 65]. The application of pressure shifts the transformation temperature to higher values, and the clear view of transformations for x = 0.15 is illustrated from the inset **Figure 6b**.

Figure 7 displays the pressure addiction of T_{av}

$$[T_{av} = (A_s + A_f)/2] \quad (2)$$

Martensite to austenite transition for x = 0 and 0.15, which rises with improvement of pressure. The derivative pressures of T_{av} are 0.15 K/kbar and 0.82 K/kbar for x = 0 and 0.15, respectively. This directs the change in volume with pressure for $Ni_{1.85}Mn_{1.15}Ga$ matched to Ni_2MnGa. The additional Mn atoms will fill the Ga and Ni sites consequential in hybridization between Ni and Mn/Ga states in NiMnGa alloy [63]. The application of pressure increases the exchange interactions between Mn-Mn ions, which enhance the hybridization between Ni and Mn/Ga. This needs more thermal energy to drive martensite transition, which in turn increases the characteristic transformation temperature [66, 67]. An important aspect is that width of the hysteresis (given by the difference of $(A_s+A_f)/2$ and $(M_s+M_f)/2$) [68] decreases. This implies larger mobility of the twin boundaries with the application of hydrostatic pressure.

The temperature dependence of piezoresistivity (PR) of $Ni_{2-x}Mn_{1+x}Ga$ (x = 0, 0.15) alloys at various pressures is computed using the relation

$$PR(T) = [\rho_P(T)) - \rho_A(T))/\rho_A(T)] \quad (3)$$

where ρ_P and ρ_A are resistivity at pressure (P) and ambient pressure. Fascinatingly, PR has negative and positive sign for x = 0 and 0.15 (**Figure 8**). The highest PR is found at the martensite transition (near T_M) for all pressures. Inset of **Figure 8a** exposes that for x = 0, negative PR gradually enhances with pressure and peak value of –PR is 34% mentioned at 232 K for P = 30 kbar. For x = 0.15, a prominent peak of PR is remarked at T_M and the application of pressure increases the +PR (inset of **Figure 8b**). The maximum +PR of 17% is noticed at 154 K for P = 28 kbar. The observation of negative and positive PR is stimulated by decrement and increment in resistivity (ρ) with pressure for x = 0 and x0.15 presented in inset of **Figure 8a** and **b**. Therefore, it affirms that the PR of x = 0 responds with pressure. The PR activities are watched in single crystalline $Ni_{45}Co_5Mn_{37.5}In_{12.5}$ under uniaxial stress and attained peak value of PR 122% [48]. Hereafter, the Ni-Mn-Ga compounds are hardly prospective applications in subject of spintronics by revealing changes of PR.

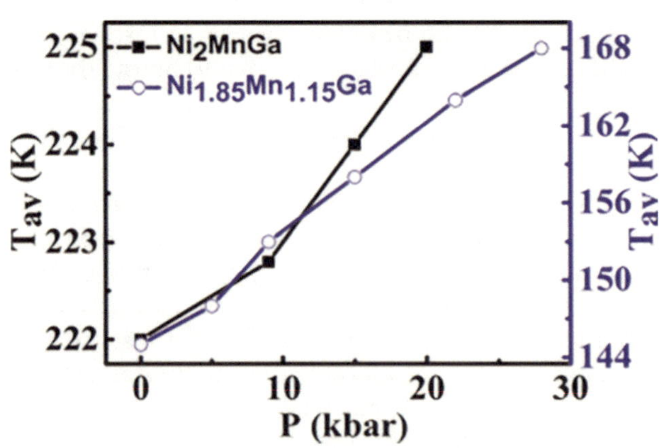

Figure 7.
Pressure dependence of T_{av} for $Ni_{2-x}Mn_{1+x}Ga$ (x = 0, 0.15) alloys.

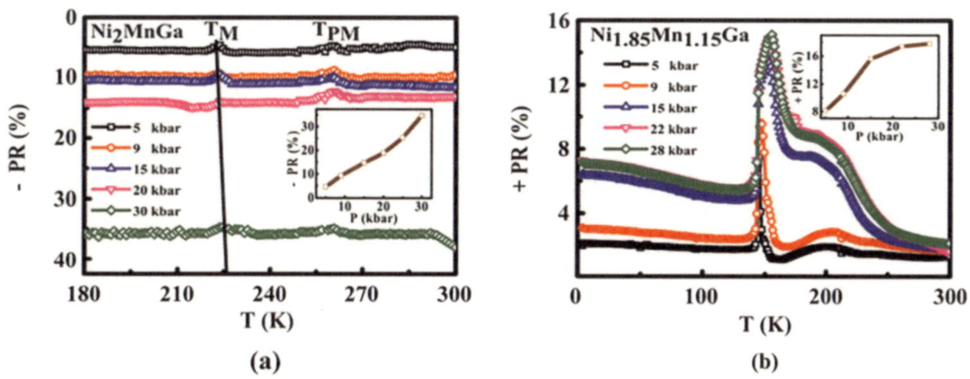

Figure 8.
Temperature habituation of PR at various hydrostatic pressures for (a) x = 0, (b) x = 0.15 alloys. Pressure variant of PR for x = 0 and x = 0.15 is exposed in insets of a and b.

Figure 9a and **b** shows variation of both residual resistivity (ρ_0) and electron scattering factor (A) with pressure for x = 0 and 0.15. The ρ_0 and (A) are obtained by fitting the simple electrical resistivity equation,

$$\rho = \rho_0 + \left(AT^2\right) \tag{4}$$

From the low thermal region 4–200 K (x = 0) and 4–130 K (x = 0.15) and the curved fitting are displayed in the inset of **Figure 9a** and **b**. The ρ_0 happens in these alloys due to impurities (or) imperfections. The (A) value at lower thermal region denotes electron-electron scattering [69]. From **Figure 9a**, the rate of ρ_0 and A reductions with pressure for x = 0. In demarcation, for x = 0.15 mutually ρ_0 and (A) gains with subjected pressure **Figure 9b**. The divergent changes of (A) involve that the subjection of pressure suppresses and enhances the electron-electron scattering for x = 0 and x = 0.15. In the case of x = 0.15, the fixed disorder on because of additional Mn that lodges the Ni locations possibly liable for rises in both the ρ_0 and (A) with pressure [7]. Another factor that might increase (A) could be related to the spin fluctuations due to Fermi surface nesting under pressure [70–72].

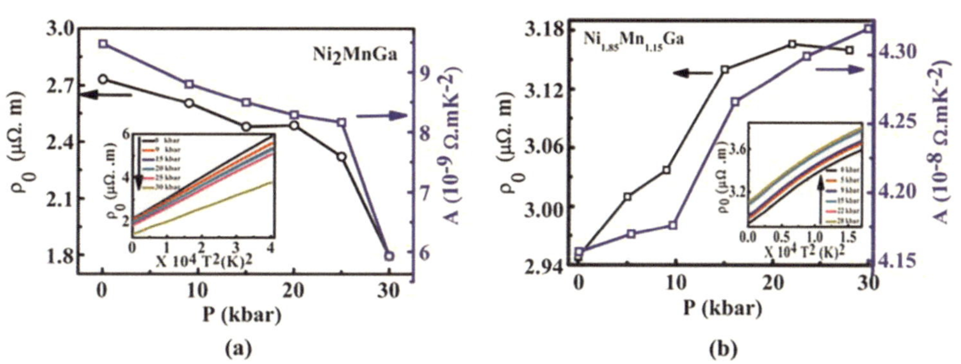

Figure 9.
Pressure discrepancy of residual resistivity (ρ_o) and electron scattering factor (A) for (a) Ni_2MnGa (b) $Ni_{1.85}Mn_{1.15}Ga$. The insets of a and b reveals ρ_o dependence T^2 for Ni_2MnGa and $Ni_{1.85}Mn_{1.15}Ga$.

3. Conclusion

The usage of outward magnetic field raises the magnetization for Ni_2MnGa and $Ni_{1.85}Mn_{1.15}Ga$ alloys. By applying hydrostatic pressure, which includes M_s and expands the hysteresis width in both alloys. The metamagnetic transition is detected at ambient pressure and suppressed at maximum pressure. The high pressure encourages greater magnetocrystalline anisotropy. Finally, the pressure impact on MCE is reduced in ΔS_M for both alloys. In compact of the work, Ni_2MnGa and $Ni_{1.85}Mn_{1.15}Ga$ alloys reveal negative and positive PR, when applied to hydrostatic pressure of 30 kbar and 28 kbar. The rate of variation of T_M and resistivity pertaining to pressure has been computed and exhibits positive numbers for the two alloys. The ρ_0 and (A) are originated to be reduced with pressure for Ni_2MnGa, which exhibit metallic property, merely both rise for $Ni_{1.85}Mn_{1.15}Ga$, and it may be associated to stationary disorder effects and spin oscillations. These materials have incredible potential interest in sensors and spintronics.

Acknowledgements

The author U.D acknowledges DST-SERB-NPDF, UGC-RFSMS-DSKPDF, and S.A and also acknowledges, DST-SERB, BRNS, DRDO, CEFIPRA, DMRL-Hyderabad, and UGC-DAE-CSR-Indore for their financial support and sample preparation, and characterizations.

Conflict of interest

The authors declare no conflict of interest.

Author details

Devarajan Uthiran[1,2,3]* and Arumugam Sonachalam[1]

1 Centre for High Pressure Research, School of Physics, Bharathidasan University, Tiruchirappalli, India

2 Department of Nuclear Physics, University of Madras, Chennai, India

3 Vel Tech Multi Tech Dr. Rangarajan Dr. Sakunthala Engineering College, Chennai, India

*Address all correspondence to: devaraman@gmail.com

References

[1] Ullakko K, Huang JK, Kantner C, Handley RCO, Kokorin VV. Applied Physics Letters. 1996;**69**:1966

[2] Kainuma R, Imano Y, Ito W, Sutou Y, Morito H, Okamoto S, et al. Nature (London). 2006;**493**:957

[3] Krenke T, Duman E, Acet M, Wassermann EF, Moya X, Mañosa L, et al. Nature. 2005;**4**:450

[4] Du J, Zheng Q, Ren WJ, Feng WJ, Liu XG, Zhang ZD. Journal of Physics D: Applied Physics. 2007;**40**:5523

[5] Esakki Muthu S, Rama Rao NV, Sridhara Rao DV, Manivel Raja M, Devarajan U, Arumugam S. Journal of Applied Physics. 2011;**110**:023904

[6] Sharma VK, Chattopadhyay MK, Shaeb KHB, Chouhan A, Roy SB. Applied Physics Letters. 2006;**89**: 222509

[7] Singh S, Rawat R, Esakki Muthu S, D'Souza SW, Suard E, Senyshyn A, et al. Physical Review Letters. 2012;**109**: 246601

[8] Yu SY, Liu ZH, Liu GD, Chen JL, Cao ZX, Wu GH, et al. Applied Physics Letters. 2006;**89**:162503

[9] https://journals.aps.org/prb/abstract/10.1103/PhysRevB.80.10440444

[10] Pathak AK, Gautam BR, Dubenko I, Khan M, Stadler S, Ali N. Journal of Applied Physics. 2008;**103**:07F315

[11] Chernenko VA. Scripta Materialia. 1999;**40**:523

[12] Pons J, Chernenko VA, Santamarta R, Cesari E. Acta Materialia. 2000;**48**: 3027

[13] K. Tsuchiya, H. Nakamura, D. Ohtoyo, H. Nakayama, H. Ohtsuka and M. Umemoto, (2001 Interscience Enterprises Ltd, Switzerland, 409.

[14] Chernenko V, Lvov V, Cesari E, Pons J, Portier R, Zagorodnyuk S. Materials Transactions, JIM. 2002;**43**:856

[15] Banik S, Singh S, Rawat R, Mukhopadhyay PK, Ahuja BL, Awasthi AM, et al. Journal of Applied Physics. 2009;**106**:103919

[16] Banik S, Mukhopadhyay PK, Awasthi AM, Barman SR. Advanced Materials Research. 2008;**52**:109

[17] Kanomata T, Shirakawa K, Kaneko T. Journal of Magnetism and Magnetic Materials. 1987;**65**:76

[18] Kanomata T. Journal of Magnetism and Magnetic Materials. 2002;**65**:1987

[19] Sasioglu E, Sandratskii LM, Bruno P. Physical Review B. 2005;**71**:214412

[20] Kubler J, Williams AR, Sommers CB. Physical Review B. 1983;**28**:1751

[21] Kyuji S, Endo S, Kanomata T, Ono F. Physica B: Condensed Matter. 1997; **237**:523

[22] Mañosa L, Moya X, Planes A, Gutfleisch O, Lyubina J, Barrio M, et al. Applied Physics Letters. 2008; **92**:012515

[23] Albertini F, Kamarád J, Arnold Z, Pareti L, Villa E, Righi L. Journal of Magnetism and Magnetic Materials. 2007;**316**:364

[24] Sagar A, Miller RC. Journal of Applied Physics. 1961;**32**:2073

[25] Bourgault D, Porcar L, Bruyère C, Jacquet P, Courtois P. The Review of Scientific Instruments. 2013;**84**:013905

[26] Chandra Shekar NV, K. Govinda rajan, Bull. Materials Science. 2001;**24**:1

[27] Khan M, Dubenko I, Stadler S, Ali N. Journal of Physics: Condensed Matter. 2004;**16**:5259

[28] Smith CS. Physics Review. 1954;**94**:1

[29] Suhling JC, Jaeger RC. IEEE Sensors Journal. 2001;**1**:14

[30] Soderberg O, Ge Y, Sozinov A, Hannula SP, Lindroos VK. Smart Materials and Structures. 2005;**14**:S223

[31] Jakob G, Elmers HJ. Journal of Magnetism and Magnetic Materials. 2007;**310**:2779

[32] Wang B, Liu Y. Meta. 2013;**3**:69

[33] Biswas C, Rawat R, Barman SR. Applied Physics Letters. 2005;**86**:202508

[34] Murray SJ, Marioni M, Allen SM, O'Handley RC, Lograsso TA. Applied Physics Letters. 2000;**77**:886

[35] Liang T, Jiang CB, Xu HB, Liu ZH, Zhang M, Cui YT, et al. Journal of Magnetism and Magnetic Materials. 2004;**268**:29

[36] Sozinov A, Lanska N, Soroka A, Zou W. Applied Physics Letters. 2013;**102**: 021902

[37] Webster PJ, Ziebeck KRA, Town SL, Peak MS. Philosophical Magazine B. 1984;**49**:295

[38] Chernenko VA, Lvov VA. Philosophical Magazine A. 1996;**73**:999

[39] Devarajan U, Esakki Muthu S, Arumugam S, Singh S, Barman SR. Journal of Applied Physics. 2013; **114**:053906

[40] Gautam BR, Dubenko I, Pathak AK, Stadler S, Ali N. Journal of Physics: Condensed Materials. 2008;**20**:465209

[41] Kourov NI, Pushin VG, Knyazev YV, Korolev AV. Physics of the Solid State. 2007;**49**:1773

[42] Bozhko AD, Vasilev AN, Khovailo VV, Dikshtein IE, Koledov VV, Seletskii SM, et al. Journal of Experimetal and Theoretical Physics. 1999;**88**:954

[43] Vasilev AN, Bozhko AD, Khovailo VV, Dikshtein IE, Shavrov VG, Buchelnikov VD, et al. Physical Review B. 1999;**59**:1113

[44] Chatterjee S, Giri S, De SK, Majumdar S. Physical Review B. 2010; **81**:214441

[45] Khovailo VV, Takagi T, Tani J, Levitin RZ, Cherechukin AA, Matsumoto M, et al. Physical Review B. 2002;**65**:092410

[46] Kim JH, Taniguchi T, Fukuda T, Kakeshita T. Materials Transactions. 2005;**46**:1928

[47] Maeda H, Fukudaa T, Kakeshita T. Journal of Alloys and Compounds. 2011; **509**:7840

[48] Porcar L, Bourgault D, Courtois P. Applied Physics Letters. 2012;**100**: 152405

[49] Singh S, Nayak J, Rai A, Rajput P, Hill AH, Barman SR, et al. Journal of Physics: Condensed Materials. 2013; **25**:212203

[50] Ahuja BL, Gulzar Ahmed S, Banik M, Itou YS, Barman SR. Physical Review B. 2009;**79**:214403

[51] Singh S, Petricek V, Rajput P, Hill AH, Suard E, Barman SR, et al. Physical Review B. 2014;**90**:014109

[52] Straka L, Heczko O. Journal of Applied Physics. 2003;**93**:10

[53] Banik S, Chakrabarti A, Kumar U, Mukhopadhyay PK, Awasthi AM, Ranjan R, et al. Physical Review B. 2006; **74**:085110

[54] Kamarad J, Albertini F, Arnold Z, Casoli F, Pareti L, Paoluzi A. Journal of Magnetism and Magnetic Materials. 2005;**290**:669

[55] Kira T, Mutata K, Inoue S, Koterazawa K, Jeong SJ, Yang GS, et al. Materials Transactions. 2004;**45**:1895

[56] Kharis Mulyukov Y, Musabirov I. Journal of Electromagnetic Analysis and Applications. 2010;**2**:431

[57] Babita I, Manivel Raja M, Gopalan R, Chandrasekaran V, Ram S. Journal of Alloys and Compounds. 2007;**432**:23

[58] Banik S, Rawat R, Mukhopadhyay PK, Ahuja BL, Chakrabarti A, Paulose PL, et al. Physical Review B. 2008;**77**: 224417

[59] Mandal K, Pal D, Scheerbaum N, Lyubina J, Gutfleisch O. Journal of Applied Physics. 2009;**105**:073509

[60] Esakki Muthu S, Rama Rao NV, Manivel Raja M, Arumugam S, Matsubayasi K, Uwatoko Y. Journal of Applied Physics. 2011;**110**:083902

[61] Ma SC, Xuan HC, Zhang CL, Wang LY, Cao QQ, Wang DH, et al. Applied Physics Letters. 2010;**97**:052506

[62] Nayak AK, Suresh KG, Nigam AK, Coelho AA, Gama S. Journal of Applied Physics. 2009;**106**:053901

[63] Roy S, Blackburn E, Valvidares SM, Fitzsimmons MR, Vogel SC, Khan M, et al. Physical Review B. 2009;**79**:235127

[64] Ma Y, Awaji S, Watanabe K, Matsumoto M, Kobayashi N. Applied Physics Letters. 2000;**76**(37):65

[65] Chakrabarti A, Biswas C, Banik S, Dhaka RS, Shukla AK, Barman SR. Physical Review B. 2005;**72**:073103

[66] Elliott RS, Shaw JA, Triantafyllidis N. International Journal of Solids and Structures. 2002;**39**:3845

[67] Fabbrici S, Porcari G, Cugini F, Solzi M, Kamarad J, Arnold Z, et al. Entropy. 2014;**16**:2204

[68] Ranjan R, Banik S, Barman SR, Kumar U, Mukhopadhyay PK, Pandey D. Physical Review B. 2006;**74**:224443

[69] Gratz E, Zuckermann HJ. Journal of Magnetism and Magnetic Materials. 1982;**29**:181

[70] Lee Y, Rhee JY, Harmon BN. Physical Review B. 2002;**66**:054424

[71] Burgaro C, Rabe KM, Dal Corso A. Physical Review B. 2003;**68**:134104

[72] https://journals.aps.org/prb/abstract/10.1103/PhysRevB.79.092410

Chapter 14

Metal Halide Hybrid Perovskites

Fency Sunny, Linda Maria Varghese, Nandakumar Kalarikkal and Kurukkal Balakrishnan Subila

Abstract

Halide Perovskites have gained much attention in the past decade owing to their impressive optical and electrical properties like direct tunable bandgaps, strong light absorption, high photoluminescence quantum yield, and defect resistance shown by them. These materials find application in numerous fields including photovoltaics, optoelectronics, catalysis, and lasing applications. Multidimensional hybrid perovskites have been extensively researched as these structures lead to superior results. They combine the properties of three-dimensional variant along with the stability of the two-dimensional perovskite. This chapter focuses on the unique properties of metal halide perovskites including the crystal structure, optical, electronic, and electrical properties. The different techniques followed for the synthesis of metal-halide nanostructures and 2D/3D hybrids are also included focusing on the changes in physical properties and the structure of these materials.

Keywords: metal halide perovskites, hybrid organic–inorganic perovskites, 2D/3D perovskites, structure, property enhancement

1. Introduction

Perovskite, commonly referring to as a calcium titanium oxide mineral, with the chemical formula of $CaTiO_3$, was discovered by Gustav Rose in the Ural Mountains of Russia in 1839 and is named after Russian mineralogist Lev Alexeievitch Perovski. Later on, the name perovskite structure has been given to any material, which has the same crystallographic structure as calcium titanium oxide ($CaTiO_3$). The general chemical formula of perovskite compound is ABX_3, where "A" and "B" are cations in which An atom is bigger in size than B atom, and "X" is an anion that binds to both cations and can be either halide [halide perovskite] or oxygen [oxide perovskite]. The ideal cubic symmetric perovskite structure has the B cation in 6-fold coordination, situated in the center surrounded by an octahedron of anion, the halogen atoms are in the faces center, and the A cation in 12-fold cubo octahedral coordination [1]. The A and B sites, both can accommodate inorganic cations resulting in the formation of inorganic perovskite. In the same way, if we replace inorganic cation A with small organic cations that leads to organic–inorganic hybrid materials [2].

Many physical properties of perovskite materials particularly electronic, magnetic, and dielectric properties depend on the crystal structure of perovskite, and the possibility for cations is limited by the stability of the structure, which can also produce several distorted structures. These distorted structures as the result of varying cations can be used to tune the properties of perovskite materials. Electro neutrality and ionic radii are the two important factors that determine the stability of perovskite materials. The stability of the perovskite can be estimated by the Goldschmidt tolerance factor and an octahedral factor. Even though these two factors predict the structure of perovskite, it is not easy to predict the type of distortion occurring in the structure, due to both the octahedral and Goldschmidt factors do not account for the various molecular interactions in the compound including ionic, covalent, or hydrogen bonding. Although the chemical formula and coordination number remain the same, these distortions reduce the symmetry of the perovskite to lower symmetry crystal structures like orthorhombic, rhombohedral, hexagonal, and tetragonal forms [3]. Different phases of the perovskite materials mainly depend on the temperature changes, that is at 100 K the perovskite shows a stable orthorhombic phase, but at 160 K, the tetragonal phase replaces the orthorhombic phase. At a higher temperature of about 330 K, the most stable cubic phase starts to appear and replace the tetragonal phase [4].

Compared with any other semiconductor materials, perovskite materials maintain a high crystallinity, and this enables the formation of versatile forms of perovskite materials from nanocrystals to macroscopic single crystals [5]. And one of the amazing facts about the perovskite material is the simplicity of its preparations; however, often simple methods create interesting chemistry and mechanisms that give the material unique properties and applications [3]. Different synthetic methods have their own special features; and hence, it is very important to choose the correct method based on the targeted compound and applications. Although perovskite materials are seen to have excellent properties and applications, there are still many properties that need to be explored. In this chapter, we discuss the unique properties of perovskite materials and their synthesis methods along with the effect of varying doping materials.

2. Types of perovskite halides

2.1 Inorganic perovskite

Material having the general formula ABX_3 is called perovskite material. Inorganic halide perovskite consisting of inorganic A site cations such as Cs^+, Rb^+, and B site with metal ion and X with halogen ion, have been demonstrated with improved stability toward moisture, light, and heat as compared to organic–inorganic hybrid perovskite and embrace unique structural features. These unique properties of inorganic halide perovskite are used in optical and photovoltaic applications. Meanwhile, inorganic halide perovskite shows dramatic change in phase transition temperature creating stability issues, and causing serious deformation, since most of the properties depend on chemical composition and crystal structure. According to environmental temperature, inorganic perovskite shows rapid phase changes, the favorable phase transition can be achieved by partial substitution of different halogen in inorganic halide perovskite. Incorporation of bromine to iodine, the visible light absorption range decreases and increases the band gap. The tin-based perovskite, $CsSnX_3$, shows a red shift in optical spectra compared to inorganic lead perovskite and $CsGeX_3$ shows similar optical spectra as $CsPbX_3$. In addition to the 3D inorganic halide perovskite ABX_3, heterovalent perovskites with +4 oxidation state in the B cation with chemical formula A_2BX_6 are also synthesized. Cs_2PdBr_6 synthesized via solution process shows excellent stability to moisture and suitable for optoelectronic application. $Cs_3Sb_2I_9$ is another inorganic halide perovskite posse's similar property to hybrid halide perovskite. The lead-free germanium-based perovskite $CsGeX_3$ posseses properties including high dielectric constants, photoabsorption coefficients, effective masses of charge carriers, exciton binding energies, and electronic band structures. Along with Cs ion, Rb ion can also be used as A-site cation in inorganic perovskite. Some of the Rb halide perovskites are reported. Among the A_2BX_6perovskite, Cs_2PdBr_6, Cs_2SnI_6, Cs_2TiBr_6 have been utilized in photovoltaic devices [6].

Compared with hybrid halide perovskite, inorganic halides show much more stability, therefore, a lot of studies have been focused on the synthesis of inorganic halide perovskite nanostructures using different strategies. Inorganic halide perovskite in nano size have shown superior optical, electrical, and optoelectronic properties which offer excellent platforms for distinct fundamental research and further development for future applications [7].

2.2 Organic inorganic perovskite

In the organic–inorganic hybrid perovskite, at least one of the A or B sites are occupied with organic ions, typically A sites are occupied with organic ion, B site with metal ion and X site with halogen ion. Methylammonium [MA] and formamidium [FA] with chemical formula CH_3NH_3 and $CH\ (NH_2)_2$, respectively are the most common organic ion occupied in A site of hybrid perovskite material. The possibilities of the formation of various perovskite materials are determined by various stability parameters, which determine whether a set of A, B or X ion may adopt the perovskite structure. In the case of organic ions, it is not easy to assign the ionic size, especially for nonspherically symmetric and charged complexes. Hence by considering the assumption of the molecule being free to rotate about its center of mass, the effective radii for organic cations are determined. The size restrictions, as outlined by the tolerance factor in the 3D structure can be lifted by slicing the perovskite structure. In the case of 2D layered derivative of perovskite structure, there is no restriction in the A cation length, and in the case of zero-dimension, size restrictions are not even considered as limitation in synthesis. This structural flexibility of structure in lower dimension structure provides a platform for preparation of varying structural material and tunable applications. When A site occupied with organic cation, it is necessary that it must contain a terminal functional group that interact with inorganic material, but the remaining part should not interfere with B or X components. Mono or diammonium cations are important factor in 2D layered perovskite structure, for example $(RNH_3)_2BX_4$ or $(NH_3RNH_3)BX_4$, in which R group denote organic functional group. The main advantage of this ammonium cation is that they form H-bonding with inorganic ions which give stability as well as the orientation of organic cations. The metal ion in site B can also influence the organic cation and H-bonding, especially when they show some kind of distortion in structure. Along with size and fit, the charge balance requirement is also important in respect to inorganic cation. In short both organic and inorganic cations are related to each other, allowing certain degree of influence of structure and properties of perovskite material.

The choice of organic cation and stoichiometry are two important factors on the orientation of resultant inorganic framework. In addition to the structural flexibility, the properties of both organic and inorganic cation combined together to give the structural properties of hybrid material. The energy transfer between organic and inorganic ions are studied and reported, include naphtelene and pyrene ion in Pb halide framework, azobenzene in $PbBr_4$, 3–2-(aminoethyl) indole in $CuCl_4$ are some of the examples, in which electronic tunability of both ions to create unique features.

In addition to the monomeric cation, incorporation of polymerizable moieties is also reported. The polyacetylene derivatives on lead bromide framework that have been reported is an example of polymerized structure. The structure shows enhanced air and moisture stability obtained from the protective polymer structure. Generally longer alkyl chains provide more degrees of freedom resulting more structural phase transition. Ferroelectric transitions of hybrid halide perovskite are reported; providing interesting possibilities for material design to ferroelectric random access memory and magnetic data storage. The hybrid perovskites can overcome some of the limitations of organic and inorganic quantum dot LED, including the issues of high cost, poor color purity, and high ionization energy. In total inorganic–organic hybrid halide perovskite found application in optical, electrical, and magnetic fields.

Even though many studies on hybrid perovskite are going on, the continued study of structure–property relationships in hybrid organic–inorganic halide perovskite is an important step for bringing reproducibility and predictability to the diverse and interdisciplinary fields [8].

2.3 2D/3D hybrid Heterojunction perovskite

Perovskite solar cells based on 2D/3D heterostructure have attracted researcher's attention in the past few years due to their promising photovoltaic application, and their amazing properties such as high absorption coefficient, tunable direct band gap and long exciton diffusion, and stability. The 3D halide perovskite attains a power conversion efficiency of 25%, which makes them attractive materials in the field of photovoltaic. Despite this efficiency, perovskite suffers stability issues including phase conversion of perovskite film, degradation in moisture, heat and irradiation, hinders the further development. Researchers have made so much effort to enhance the stability of 3D perovskite, among them introduction of 2D perovskite is one of the potential strategies for the stability improvement. The structural features especially the hydrophobic nature of the spacing as well as packing crystal structure in the 2D perovskite material makes them stable in the ambient temperature by protecting perovskite from the direct contact of moisture. However pure 2D crystals are not desirable, because of wide band gap and non-preferred crystallographic orientation [9]. Cation exchange has been used as a plausible way for transformation between 3D and 2D phases of halide perovskites [10]. Building on the combined benefit of both 2D and 3D perovskite material a new type of 2D/3D heterostructure is synthesized in the field of solar cell, in which 2D perovskite were incorporated onto the surface of 3D component as capping layer to increase the stability of the 3D perovskite phase without changing the device performance. That is in 3D\2D heterojunction architecture only the surface of the light absorbing perovskite layer is altered while the optoelectronic properties of the bulk film are remain intact. The 2D/3D hierarchical structural was first constructed for 3D $MAPbI_3$, the structure is $(PEA)_2(MA)_4Pb_5I_{16}$/$MAPbI_3$ (PEA = phenylethylammonium). This 2D/3D structure used in solar cell exhibit a strong moisture hindrance. Another structure $(BA)_2PbI_4$/$MAPbI_3$ (BA = butylammonium) shows much enhanced thermal stability and high photo conversion efficiency of 19.8%. Recently the 2D/3D FA (FA = formamidinium) structure also synthesized. For neat 2D perovskite, much study has been takes place and achieved in determining the nature of films, including thickness distribution, and charge transport. But to obtain a clear-cut understanding of the interfacial mechanism at the 2D/3D heterojunction, we need more information about the ligand dependence of 2D/3D heterojunction and its influence on charge collection. The ligand chemistry is very importance in knowing the thickness distribution and orientation of 2D perovskite, which is expected to play a major role in charge transfer at the heterojunction and solar cell. The ambient phase stability was studied for all the films and found that 2D/3D heterojunction assisted phase shows excellent stability especially for 3D $FAPbI_3$perovskite, particularly FPEA--based film maintains 84% of its value after ambient exposure for long time [11].

2.4 Double perovskite

In order to overcome the structural instabilities of halide perovskite especially lead halide perovskite, and the hybrid perovskite, researchers has sought in search of alternate crystal structure, found a promising material called double perovskite. ABX_3 is the general formula for an optimal-perovskite materials in which A and B are cations of different size and X is an anion similar in size to B cation. But in the case of double perovskite either A or B site can be occupied by two different types of cations, giving the formula $A'A''B_2X_6$ or $A_2B'B''X_6$. Among these two structures, the double B site perovskite is the preferred one because the physical properties of perovskite mainly depend on B site cation. The crystal structure of $A_2B'B''X_6$ depends on the arrangement of B′ and B″ cations in the sub lattices, which are mainly focusing on reducing the madelung [strain] energy of double perovskite due to the charge difference between these two B cations. Based on the charge difference between B′ and B″ there are three kinds of B cation sub lattices, called random, rock salt and layered structure [12]. The compound having random type sub lattices generally possess cubic or orthorhombic unit cell, and rock salt type sub lattices usually crystallize in cubic or monoclinic unit cell. When the B′ and B″ cations can alternate in one direction, the layered type is formed and possess a monoclinic unit cell [13]. The introduction of vacancy in the B site of $A_2B'B''X_6$ creates another type of double perovskite called vacancy ordered double perovskite. Cs_2SnI_6, $Cs_3Bi_2Br_9$, and Cs_2TeI_6 belong to this group which is also called as a defect tolerant semiconductor. **Figure 1** represents the crystal structure of double perovskite [14].

The experimental synthesis of $Cs_2BiAgCl_6$ and $Cs_2BiAgBr_6$ are the first documented double perovskite in the halide double perovskite avenue. Both these perovskites are crystallized in face centred cubic double structure, called elpasolite. The two B cations need a convenient oxidation state to form the perovskite phase and provide a combined charge of 4+; this charge can be equally or unequally divided. In the unequal distribution, the replacement of divalent cation by a pair of monovalent-trivalent cations maintains the charge neutrality. In search of trivalent cation, researchers reached the nitrogen family and found bismuth and antimony as the most suitable ones. And for monovalent cations, noble metals like copper, silver, or gold the ones with excellent electrical conductivity and optical properties are chosen. So far, among the many reported double perovskites, $Cs_2BiAgBr_6$ is the only double perovskite found application in any active devices [15]. Also, $Cs_2BiAgBr_6$ was successfully applied for X-ray detection and is of great importance in medical diagnosis, industrial application, and scientific researche. Along with hybrid double perovskite, rare earth metal containing double halide perovskite like MA_2KYCl_6, MA_2KGdCl_6 synthesized via solution-evaporation method have also found considerable importance [16]. The $Cs_2AgBiBr_6$nanocrystals exhibited impressive photo conversion of CO_2 into solar fuels, along with $Cs_2AgBiBr_6$, $Cs_2AgBiCl_6$, $Cs_2AgSbBr_6$, $Cs_2AgInCl_6$ found application in water splitting. The hybrid halide double perovskite is also synthesized, following are some of the reported perovskite MA_2KBiCl_6, $MA_2TlBiBr_6$, $MA_2AgBiBrl_6$. MA_2KBiCl_6 is the first synthesized hybrid double perovskite shows high resistivity and superior magnetic properties. Properties of doped double perovskite are also a

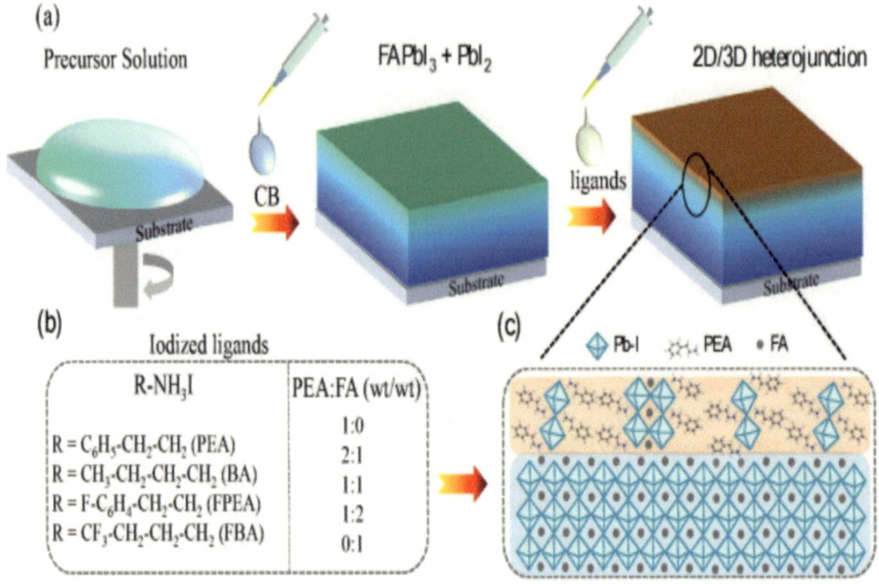

Figure 1.
Schematic representation of 2D/3D hierarchical structure: (a) fabrication of 2D/3D heterojunction; (b) spacers with different chemical structure and compositional engineering; (c) schematic model of 2D/3D hierarchical structure. (Source: Niu et al. [11]. Copyright 2019 American Chemical Society. Reprinted with permission).

topic discussion and Bi doped Cs_2SnCl_6 showed great potential as blue phosphor and LED exhibit white light emission. Recently, $Cs_2AgInCl_6$ doped with Na^+ cation also reported with efficient white emission via radiative recombination.

Compared to halide perovskite, the main issue associated with double perovskite is the much lower defect tolerance which sometime reduces the efficiency of perovskite materials. Based on the recent report, the double perovskite $Cs_2AgInCl_6$ shows much less PL efficiency than corresponding bulk structure. However, $Cs_2AgInCl_6$ could be used in LED, UV photo detector and scintillators [17]. There is plenty of double halides that can be synthesized according to the theoretical assumption. However, so far only few double perovskites have been synthesized. In order to produce more stable halide double perovskite, one should know some important points such as suitable composition, the properties of the material to be used, and the limitation to the synthetic procedure. Due to the low decomposition temperature of organic starting materials, the hybrid halide double perovskite is somewhat difficult to synthesis [18]. The major challenge in the synthesis of halide double perovskite is the high temperature requirement. The most studied double perovskite $Cs_2AgBiBr_6$ require high annealing temperature up to 285°C for obtaining high quality thin films. To some extent, this high temperature requirement puts limits on the synthesis and application of double halide perovskite [14].

3. Properties

3.1 Crystal structure

Perovskites are one of the oldest families of materials. Perovskites are normally formed by three molecules in the stoichiometric form ABX_3. The B site occupant is a divalent cation like lead, bismuth, tin, germanium, etc. and X is halide ion. In organic –inorganic perovskites, the A site is occupied by a monovalent cation such as methyl ammonium, or formamidinium. Inorganic perovskite halides have cesium, rubidium etc. The stability of perovskite structure is determined by the Goldschmidt tolerance factor, t, which is given by the equation $t = \frac{r_A + r_X}{\sqrt{2}(r_B + r_X)}$ based on the ionic radii of the component ions [19]. The different cations that can occupy the place in a stable perovskite should have ionic radii satisfying t with values between 0.76–1.13 constraining the options of possible cations [20]. In HOIPs, a notable number of tolerance factors are found to lie in the range of ~0.8–1.0 for stable perovskite structure. In the ideal case, perovskites take the cubic space group Pm 3 m with no variable parameters in the structure [20].

Apart from the original ABX_3 phase, hybrid perovskites also stabilize in a variety of structural phases. Hybrid organic–inorganic perovskites (HOIP) can be categorized from a structural perspective as ABX_3perovskites, $A_2BB'X_6$ double perovskites and A_3BX antiperovskite subclasses [20]. Halide double perovskites changes the simple perovskite ABX_3 structure to a 2 × 2 × 2 supercell, with the general formula $A_2B^IB^{III}X_6$. Here, the two bivalent cations B^{2+} are exchanged by a combination of one monovalent cation B^+ (e.g., Au^+, Cu^+, Ag^+, In^+) and one trivalent cation B^{3+} (e.g., Bi^{3+}, Sb^{3+}) [21].

The lower-dimensional 2D perovskites is composed of alternating layers of organic and inorganic phases called Ruddlesden–Popper (RP) phases having general formula $A_2A'_{n-1}B_nX_{3n+1}$. The size of the organic cation controls the 3D to 2D structural transition, in particular when it exceeds the critical size of Goldschmidt's tolerance factor, [22]. The 2D perovskites thus generated have alternating organic and inorganic sheets that are along [001] direction. This leads to the formation of layered structures. Though these phase results in excellent properties, it is affected by stability issues. In order to enhance the stability a combination of 3D and 2D phases are synthesized. The layered structure encompasses alternate 3D and 2D phases connected by the organic cation spacer (**Figure 2**) [22].

3.2 Electronic properties

Halide perovskites find numerous applications in myriad of fields owing to the exceptional electronic properties shown. The bandgaps of these materials are mainly determined by the halide ion due to the strong contribution of the 2p orbital of the halide ion. The bandgap can be tuned by changing the B-site ion as well [20].

HOIP are direct-bandgap semiconductors, with experimentally observed bandgaps ranging from ~1.2 to 2.8 eV. In HOIPs, the A-site cation also affects the band gap as amine cations could distort the anionic framework through hydrogen bonding and van der Waals interactions upon thermal and pressure perturbation. The electronic structure has been studied using computational methods. For a common HOIP structure like that of $MAPbI_3$, the conduction band minimum originates mainly from the 6p states of Pb, hybridized with a small amount of the 5p states of I, while the valence band maximum is mainly formed from the 5p states of I, mixed with a certain percentage of the 6 s states of Pb [20].

Double perovskites are seen to be materials with predominantly indirect band gaps. $Cs_2AgIn_xBi_{1-x}Cl_6$ NCs are the first double perovskite nanocrystals (NC) to have direct bad gap. Alloying the ratio of In to Br resulted in the shift of the band gap from indirect to direct. The direct band gap product showed better photoluminescence quantum yield (PLQY) [24]. Vacancy ordered double perovskites have been observed

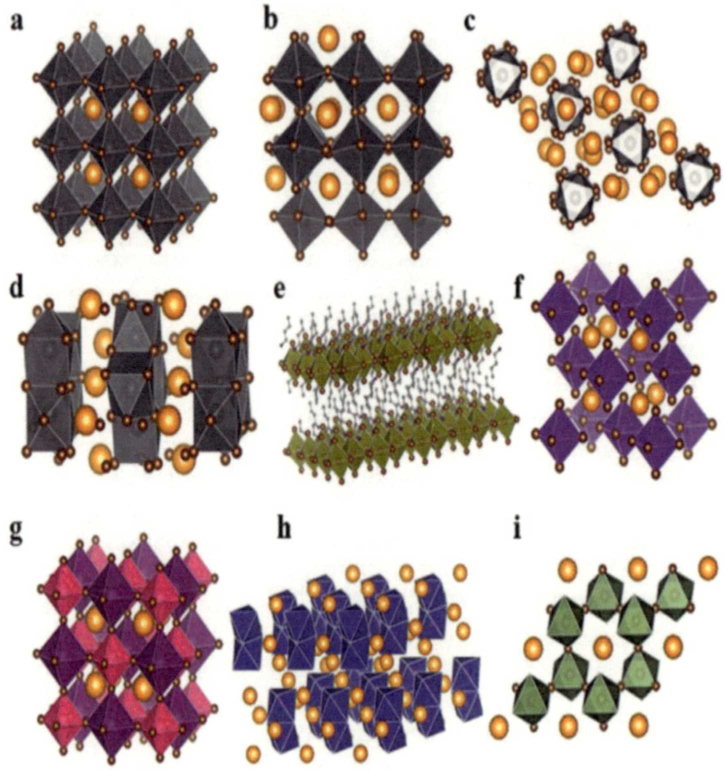

Figure 2.
Schematic representation of different metal halide structures: (a) cubic-phase ABX_3 (3D); (b) pseudocubic ABX_3 (3D); (c) A_4BX_6 (0D); (d) AB_2X_5 (2D); (e) A_2BX_4 (2D); (f) A_2BX_6 (0D); (g) $A_2B^+B^{3+}X_6$ (3D); and (h and i) $A_3B_2X_9$ (2D). (source: Shamsi et al. [23]. Copyright 2019 American Chemical Society. Reprinted with permission.)

to show direct bandgaps. Hybrid double perovskites that are isoelectronic to common HOIP such as $CH_3NH_3PbBr_3$ was synthesized to obtain direct bandgap materials [25].

3.3 Optical properties

HOIPs have been studied to show efficient broadband tunable optical properties. In the bulk form, HOIP-based LEDs are constructed as simple multilayered devices. Depending on the halide composition, these were found to be near-infrared, green and red-light emitters at room temperature with bright electroluminescence owing to the efficient radiative recombination of injected electrons and holes. However, the poor morphology of HOIP thin layers leads to lower efficiencies of the HOIP LEDs compared to conventional organic and quantum-dot LEDs [20].

The absorption measurements of Cs_2AgBiX_6 NCs with different halide compositions revealed the tunable exciton peaks ranging from 367 to 500 nm with the corresponding PL peaks varying from 395 to 575 nm [24]. The study of optical properties in double perovskites has shown that ligand passivation of surface defects can increase the PLQY (a 100-time increase). The colloidal NCs of $Cs_2AgBiBr_6$ were observed to exhibit dual absorption peak at 427 nm and 380 nm, the former from direct Bi s–p transition, while the latter may be assigned to the isolated octahedral

$BiBr_6^{3+}$ complex in the colloidal solution. The thin films of the double perovskites have shown stable single peak emissions with better PLQY [24].

In case of 2D/3D hybrid perovskites, the Ruddlesden–Popper structure of layered perovskites results in the variation of charge carrier dynamics and the optoelectronic properties with a varying value of n. As the value of n increases the optoelectronic properties tend to be that of the bulk material with broadened transient absorption features. For low values of n, on the other hand, the excitonic peak is enhanced and the material shows an increased monomolecular recombination rate [26].

3.4 Electrical properties

Metal halide perovskites show great carrier transport properties combined with long carrier lifetimes results in them being effectively used in photovoltaic applications. The long transport length helps in the perovskite thin films being used effectively for long distance transport of charges. This is achieved by the initial photoexcited carriers recombining away from the excitation spot which results in regeneration of photons that is reabsorbed. This forms charge carriers at significant distances away from the initial excitation point [20].

For efficient charge–carrier separation, the exciton binding energy (EB), which defines the lowest energy required to dissociate an exciton (electron–hole pair), must be small. In HOIPs, it has been shown experimentally that photo excitations directly generate free electrons and holes, rather than bound excitons. This is due to the fact that the EB is low enough ($\leq$25 meV) to allow charge separation at room temperature [20].

The combination of 2D and 3D perovskite structures in 2D–3D hybrid structures has yielded solar cells achieving high PCEs and excellent stability over thousands of hours. Engineering of the 2D/3D heterostructures has helped in achieving better performing solar cells [26].

3.5 Ferroelectric properties

Study of the existence of ferroelectric domains in perovskite materials are of great interest as they will enhance the electron–hole separation in the materials, resulting in better photovoltaic performance. Halide HOIPs have cations such as FA or MA that are of polar nature and shows order–disorder transitions across phase transitions. $MAPbI_3$ was the first to show potential ferroelectric behavior with the observation of a hysteresis in the current voltage plot. Further experiments using piezoelectric force microscopy have since proved that $MAPbI_3$ is indeed ferroelectric. Density functional theory (DFT) calculations along with symmetry mode analysis have shown that the origin of the spontaneous polarization as a combined effect of the relative movement of MA and the relaxation of the framework, which are coupled through hydrogen bonding [20].

3.6 Other properties

In HIOP the framework stiffness is seen to be proportional to the Pb–X bond strength, tolerance factor, as well as the electronegativity of the halogen atoms. The Young's moduli have been measured to be about ~10–20 GPa. This value correlates well with chemical and structural differences. The hardness properties show an inverted trend with respect to the halide ion (I > Br > Cl). The least rigid system was seen to be $MAPbI_3$ which also exhibited the highest resistance to plastic deformation. The thermal expansion behavior of these halide HOIPs had significant thermoelastic response (~30–40 × 10^{-6} K^{-1}), due to octahedral flexing. This could contribute to self-helaing property of possible point defects in device applications. At the same time, relatively low hardness properties indicate ease of plastic deformation, which could affect the cyclability of flexible cells and devices [20].

4. Synthesis methods

4.1 Synthesis of nanostructures

Nanostructures of halide perovskites have attracted great deal of attention due to the interesting properties shown in the nano scale. Various methods are followed for the synthesis of halide perovskites.

4.1.1 Ligand assisted synthesis

Colloidal halide perovskite nanostructures are synthesized using ligand assisted method. Long chain amines and a combination of short chain and long chain amines have been used as ligands for the synthesis of stable halide perovskites [27–29]. The surface adsorption and desorption of the ligands leads to the formation of stable nanostructures at relatively low temperature [29]. This helps in the more efficient application-based synthesis of perovskites. Quantum dots, nanocrystals (NCs), nanosheets (NS), and nanorods of halide perovskites have been synthesized using the colloidal method.

The colloidal hot-injection method has been used to successfully synthesize lead based, lead-free pure inorganic and organic–inorganic hybrid perovskite NPs having excellent structural and optical properties [30]. The ligands and the solvents used and the ratio of the ligand to precursors affect the morphology of the synthesized materials. In a typical method, the synthesis of $CsPbX_3$ (X = Cl, Br, I) is achieved by "controlled arrested precipitation of Cs^+, Pb^{2+}, and X^- ions into $CsPbX_3$ NCs is obtained by reacting Cs-oleate with a Pb(II)-halide in a high boiling solvent (octadecene) at 140–200°C" [31]. Hot injection method in the original and modified form has been used to synthesize double perovskites. $Cs_2AgBiBr_6$ double perovskite NCs with high crystallinity were synthesized via hot-injection approach. These NCs were capable of maintaining their structural stability in a set of varied environments such as low polarity solutions, 55% relative humidity and temperature of 100°C [24].

Colloidal synthesis of HOIP was first done using medium-length alkyl chain organic ammonium cations (bromides of octylammonium and octadecylammonium) as capping ligands to obtain luminescent NCs via the solvent-induced reprecipitation method. The change in the organic cation involved in the process led to particles with better emission and properties (**Figure 3**) [32, 33].

Ligand assisted reprecipitation (LARP) method is one of the common low temperature methods widely carried out to produce highly crystalline HOIPs with different morphologies. The nature of the capping ligand and the ratio of precursors to solvent are the controlling factors determining the shape, size and the formation mechanism of the HOIPs in this method. The commonly used solvents are Octylamine (OTAm, $C_8H_{17}NH_2$), methylamine (MA, CH_3NH_2), toluene ($C_6H_5CH_3$), N,N′-dimethyl formamide (DMF), γ-butyrolactone, oleylamine (OLAm), oleic acid (OLA), 1-octadecene (1-ODE). It is the interaction between the ligand and the added antisolvent

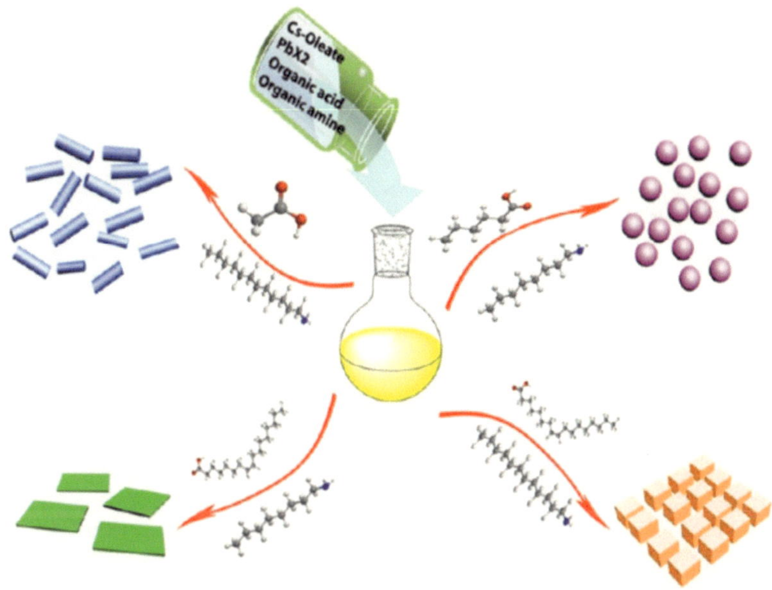

Figure 3.
A schematic of synthesis of different dimensional perovskite nanostructures using different organic ligands: Acetate acid and dodecylamine for nanorods; hexanoic acid and octylamine for spherical quantum dots; oleic acid and octylamine for few-unit-cell-thick nanoplatelets; oleic acid and dodecylamine for nanocubes. Reprinted with permission from [32]. Copyright 2016 American Chemical Society.

(e.g., Toluene) that results in the formation of crystalline or amorphous products [30]. Dual-ligand-assisted re-precipitation method under the synergistic effect of *n*-octanoic acid (OTAC), oleylamine (OLA), and (3-aminopropyl) triethoxysilane (APTES) was used for the synthesis of enhanced α-$CsPbI_3$ NCs [34].

4.1.2 Heating-up synthesis

Direct heating of the precursors in a suitable solvent (octadecene) has been followed for the easier synthesis of high-quality perovskite nanocrystals. The one pot approach results in a relatively simple, reproducible, easily scalable and tunable method [35]. The method has been extensively used in the synthesis of lead-free inorganic halide perovskites, wherein the precursors are added together into the solvent with or without the ligands and heated to a definite temperature. The reaction is allowed to take place for a specified time after which it is cooled to room temperature naturally. On centrifuging and washing the nanocrystals are obtained. A combination of hot injection and heating up method has been used for the production of double perovskite such as $Cs_2AgBi(Br/I)_6$ exhibiting better performance [36].

4.1.3 Hydrothermal synthesis

Hydrothermal method proves to be an efficient, simple and straight forward method for the synthesis of inorganic halide perovskites. In the method the precursors dissolved in halide acids or DMF are placed in Teflon liner inside an autoclave for the corresponding time [37, 38]. The reaction is allowed to take place at a constant temperature for time as low as 30 minutes. Double perovskites have been studied to be extremely sensitive to impurities. Therefore, the cleaning of the Teflon liner is especially important in this type of synthesis. Alloyed double perovskite materials such as $Cs_2AgInCl_6$ have been synthesized by the hydrothermal method. 10 M HCl solution was used as the solvent in this case. The product obtained was seen to be extremely stable with efficient white light emission on alloying sodium [38].

4.2 Thin film synthesis

Thin film preparation is done basically by spin coating the precursor solution dissolved in the required molar quantity. The coating is done usually onto substrates fit for the proposed application. Single crystal thin films are fabricated aimed at better efficiency and performance.

Thin films of 2D/3D hybrid structure of $BA_x(FA_{0.83}Cs_{0.17})_{1-x}Pb(I_{0.6}Br_{0.4})_3$ were prepared by the spin-coating of a blend of $FA_{0.83}Cs_{0.17}Pb(I_{0.6}Br_{0.4})_3$ and $BAPb(I_{0.6}Br_{0.4})_3$ in N,N-dimethylformamide, DMF. The coated film is then dried at 70°C for 60 s in nitrogen filled atmosphere followed by transfer into an oven where they were annealed in air at 175°C for 80 min to obtain the smooth films with the 2D crystallites at the boundaries of 3D grains [26]. Hybrid double perovskite has also been synthesized easily using the method of spin coating. Two step method has been followed for the synthesis of 2D/3D double perovskite wherein the organic cation such as PEA is spin coated onto the annealed films of the double perovskite [39]. Thin films of inorganic and double perovskites are easily synthesized normally by spin coating solution of the precursors in solvents such as DMF or dimethyl sulfoxide (DMSO) onto the suitable substrate for the appropriate applications in photovoltaics and optoelectronics [25].

4.3 Synthesis of single crystals

Single crystal nanostructures of halide perovskites are synthesized for improved application as semiconductors in electronics, optoelectronics, and photovoltaics. These structures provide improved photophysical properties and are therefore carry much importance [40].

The anisotropic growth rates of the crystals which depend on the feed ratios of the precursors, mineralizers, and solvents helps in the fabrication of the single-crystal thin films. Inverse temperature, temperature lowering, solvent evaporation and anti-solvent assisted crystallization methods are utilized for the process. Vapor phase epitaxial growth of thin films has been extensively used in inorganic halide perovskite monocrystal fabrication. Single crystals of HOIPs such as $FAPbI_3$ and $MAPbI_3$ have been reportedly sliced into wafers from their bulk counterpart in a top-down strategy for single crystal thin film strategy [41]. Dissolution-recrystallization pathway in solution synthesis from lead iodide (or lead acetate) films coated on substrates was used to grow single crystal nanowires, nanorods, and nanoplates of methylammonium lead halide perovskites ($CH_3NH_3PbI_3$ and $CH_3NH_3PbBr_3$). These single crystals showed increased photoluminescence and long carrier lifetimes [40].

HOIP single crystals have also been synthesized using bottom seeded solution growth and top seeded solution growth. The former method employs seed crystals that are fixed at the middle of a designated tray are rotated by the electric motor and the saturates solution was cooled to obtain the single crystals. The latter method employs seed crystals placed on a silicon substrate on top of a solution which facilitates the dissolution of the lower crystals due to the temperature discrepancy between the bottom and top of the solution that induced super-saturation to form the single crystals. 2D perovskite single crystals have been synthesized by a combination of vapor phase and solution processes. The process involves spin casting one of the precursors at an elevated temperature followed by chemical vapor deposition of the other one [42].

5. Property enhancement by doping

Doping, in general is an effective approach to enhance the properties of semiconductor materials by intentionally introducing impurity or heteroatom into the target lattices. The doping with anions and cations of different charge is also known to modify their micro structural, electrical, and magnetic properties. Considering the fact that halide materials possess interesting properties, researchers recognized that introduction of other materials into the halide perovskite could potentially lead to relevant discoveries and applications, with this aim, over the past few years, different kind of dopant materials are introduced. Among them most relevant ones are discussed here [43].

5.1 Metal ion

Due to the high abundance in the earth's crust, varying oxidation state and mostly nontoxic nature, doping by transition metals are the best choice to enhance the properties of perovskite materials. There are two methods to doping by metal ions in perovskite materials, alloying to partially replace the metal frame and inserting a small amount of transition metal in the lattices. Among these two insertions of small amount of transition metal ion into the perovskite is the preferred one, cause the energy transfer and charge carrier transfer processes between the dopant and perovskite crystals [44].

So far a large number of reports were being available on $CsPbX_3$ perovskite nanocrystals doped with divalent transition metals like Mn^{2+}, Co^{2+}, Cu^{2+} and Zn^{2+}. These metal ions play a role in eliminating the defects and distortions of perovskite crystals and exhibit dual color emission and efficient charge transfer. Among various methods available incorporating metal ion into perovskite material through hot injection is the feasible method, moreover, transition metals are economical and eco-friendly compared to other toxic materials and shows excellent properties without destroying the crystal structure [45]. Among these transition metals, Mn^{2+} need special attention due to its excellent properties. Mn-doping in nanocrystals increased the exciton luminescence and attribute the efficient energy transfer between exciton and host. Thus, for Mn doped $CsPbX_3$ achieved up to 60% luminescence quantum yield [16]. $MAPbBr_3$ the one of the most studied organic–inorganic perovskites show structural instabilities, which can be improved by doping with Zn metal ions. The Zn doped $MAPbBr_3$ reveals excellent optoelectronic properties and environmental stability due to increased lattice strain which leads to the improved interaction between bonds [46]. $Cs_2SbAgCl_6$ and its Cu^{2+} doped perovskite shows well-ordered double perovskite cubic structure with excellent conductivity. The researchers proposed that Cu^{2+} doping creates cation defect, which leads to increased conductivity of double perovskite. In conclusion, the antimony–silver based double perovskite doped with copper ion exhibit desirable properties comparison to the bare perovskite in greater bandgap tunability and stability and has impact on their morphological, optical, electronic behaviors [47].

In addition to the transition metals, alkali and alkaline earth metals, other group metals are also used as dopant in perovskite materials. The possibility of doping with alkali metal ion like Rb was explored and the result shows that Rb ion doping suppress the forming of impurity phases, and also increase the lifetime of perovskite materials [43]. In search of other metal doping, it has been found that Na^+, in bulk, $Cs_2SbInCl_6$ bring an increase in PL emission by three order magnitude compared to pure double halide material with an optimum Na content of 40%. This is where explained as the result of improved crystal quality and increased rate of radiative recombination [17]. There is also a reported result of increased hole concentration and mobility by Na doping and promoted electron injection in devices by Li doping in solar cell application [43]. Alkaline earth metals, Ca, Sr., and Ba, are proposed to be able to enhance the properties of perovskite with doping and has remarkable impact on their morphological, optical, and electronic behaviors. Besides alkali and alkaline earth metals, metals like Al^{3+} can provide tremendous morphological control to improve the properties of halide perovskite. The incorporation of indium (In3+) has also been reported to influence morphology by facilitating preferential growth of grains in several orientations [48]. In addition to that, Bi-doped bulk crystals underwent a significant band gap narrowing and shows improved stability and excellent optoelectronic and magnetic properties.

5.2 Lanthanides

To date, several successful doping of inorganic or hybrid perovskite by metal ions have been reported. However, the emissions for transition metal ions are broad band and confined to specific wavelength region limiting their application to limited energy structure. Hence lanthanide ions or called rare earth elements would be the most suitable candidate for energy and optical applications, because they possess rich and unique optical properties and emissions are in wide range with sharp line from UV to infrared region. In addition to that moving from Ce to Lu, a gradual decrease in ionic radii of lanthanides provide varying electrical, magnetic and chemical features provide the opportunities to study the changes of doping. Various lanthanide doped halide perovskite are studied and successful doping of Ce^{3+},Sm^{3+}, Eu^{3+}, Tb^{3+}, Dy^{3+}, Er^{3+} and Yb^{3+} into the $CsPbCl_3$ perovskite through hot injection methods are reported. Lanthanide doped double halide perovskite of the type $Cs_2AgInCl_6$ are also reported. For the lanthanide doped perovskite nanocrystals stable and tunable multi-color emission from visible to NIR regions are obtained [23]. However, few RE metals have been reported so far make it an attractive field. Eu^{3+} is reported to stabilize $CsPbI_3$ thin films.

Various materials like main group metals, transition metals and rare earth metals have been successfully doped into perovskite nanocrystals, single crystal, and polycrystalline film, giving rise to enhanced properties of perovskite materials. The various properties obtained through doping technology including improved stability, improved quality of thin films with reduced defect and enlarged grain size found application in optoelectronic devices including LEDs and solar cells. It is evident that dopant engineering has emerged as one of the excellent tools to enhance the properties of perovskite. Despite the fact that doping improves the properties our understanding about the mechanism related to doped halide perovskite is still limited. And number of scientific issues in terms of synthesis, doping methods, and structure and property relation are remaining unsolved. To fully exploit the possibilities of doping, there are certain questions that are needed to be answered, such as the role of metal ions in crystallization, their doping capability, the true position of metal ions, and how these differ with altered perovskite compositions.

Although 3D halide perovskite has been the prime target of doping, lower dimensional perovskite including 2D halide perovskite could be the next frontier for doping study, since they provide a rich space to study the interaction of electrically-, optically-, or magnetically-active do pants with quantum confinement effects [43].

5.3 Small molecules

Generally, in semiconductor devices, small molecules are used as dopant to enhance the properties of materials. The mechanism behind these doping is surface and interface charge transfer pathway. The small molecules like HATCN (Hexaazatriphenylenehexacarbonitrile), F_4TCNQ (2,3,5,6-tetrafluoro-7,7,8,8-tetracyanoquinodimethane) are used as dopant in low concentration in 2D perovskite materials. Also, molecules like cobaltocene and zethrene was used in inorganic perovskite materials and shows strong charge transfer from dopant to perovskite [44]. Even though doping with small molecule provide enhancement in property, formation of composite with small molecule is considered as more effective way to improve the properties of perovskite materials. The development of low cost, largest surface to volume ratio, low resistivity and high sensitivity along with ecofriendly perovskite-graphene based composite attain widespread attention in electrochemistry, sensing and optoelectrical applications. For example, $CsPbBr_3$ QDs/GO composite, that is cesium perovskite quantum dots and graphene oxide composite were used as photo catalyst for artificial CO_2 reduction. Quantum dot photo catalyst found wide attention, because of their large surface area and charge transfer mechanism. Moreover, the quantum confinement effect causes the shift in band position provides sufficient energy for photochemical reaction [49]. A perovskite and dual additive composite are reported in recent years, the PEO (polyethylene oxide) and TPBi (2,2′,2″-(1,3,5-Benzinetriyl)-tris(1-phenyl-1-H-benzimidazole)) blended perovskite act as the barrier for the excitonic dissociation at interface, which contribute to the increase in PL intensity. The possible reason for the change in PL is the smaller grain and higher surface coverage provided by the dual additive perovskite material, which can be possibly used as LEDs with high luminance and current efficiency [50].

5.4 Polymer

Perovskite-polymer composite received great attention in recent years due to their combination of properties from polymers and perovskite. The technique involves the formation of perovskite and polymer matrix in one pot reaction, to avoid the complexity of separate preparation. Compared with other technique like coating and ligand cross linking, forming composite with polymers is relatively easy to handle. In addition to the stability, polymers provide other advantages like convenient device fabrication mechanical performance, and enhanced luminescent properties. Recently poly (methyl methacrylate) (PMMA) and perovskite composite based solar cell provide a power conversion efficiency of 22.1%. Along with this, a lot of composites with excellent stability is reported. Even though excellent applications are reported, some serious issues are also found with polymers blending with perovskite. The blending of perovskite with polymers may result into serious aggregation. The large polarity difference between polymers and perovskite material results the aggregation of perovskite material, which reduce the efficiency of perovskite material. Another issue is the stability, the preparation and storage of perovskite material required high attention, to optimize the property of composite material. The time-consuming blending

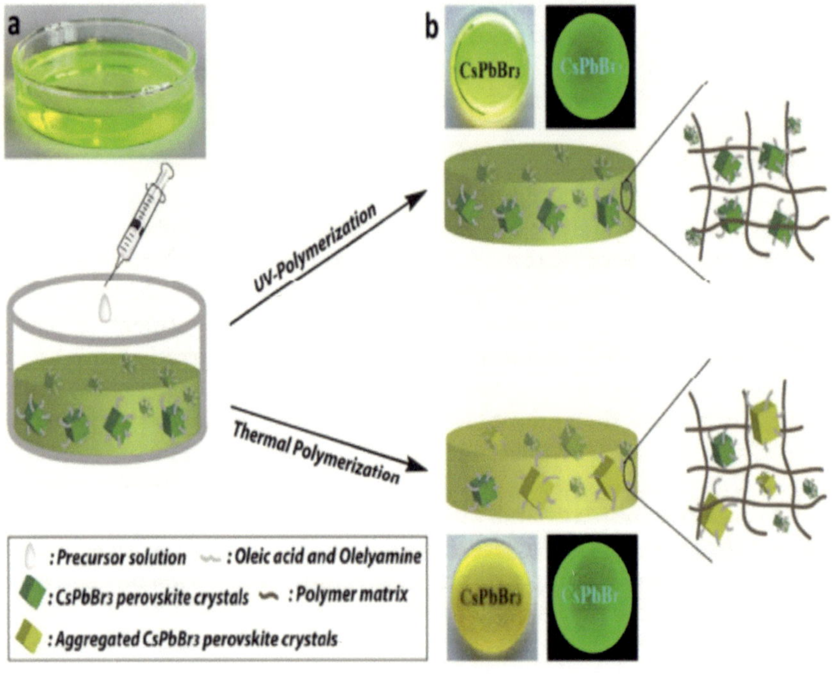

Figure 4.
Illustration of one-pot strategy to prepare perovskite-polymer composites ($CsPbBr_3$-polymer or $CH_3NH_3PbBr_3$-polymer). a) Formation of perovskite crystals in bulk monomers. The photo taken under room light is for the emissive bulk styrene after adding precursors. b) the UV- or thermal- polymerized perovskite-polymer composites. Representative disks (under room and UV light) are shown in photos reprinted with permission from [51]. Copyright 2018 American Chemical Society.

process, especially with high molecular weight polymer, result the decomposition of perovskite by moisture. The organic solvent used for dissolve polymers should be anhydrous to avoid the aggregation, and these solvents are not environment friendly and of high cost. Also needed to consider the fact that physical properties of polymers are different, hence it is necessary to know the characters of polymers before using as a composite with perovskite material. Commercially available polymers like PMMA, polystyrene (PS), polyethylenimine (PEI), polyvinylpyrrolidone (PVP), and poly (butyl methacrylate) (PBMA) are used with $CSPbBr_3$ perovskite material to form composite. The one pot synthesis of perovskite-polymer composite is illustrated in **Figure 4** [51]. The PMMA and perovskite material have difference in polarity make them less efficient, hence butyl methylacrylate with long alkyl chain of similar polarities was selected to prepare $CsPbBr_3$-PBMA shows deep green color than those of PMMA, indicating more stability. The organic halide perovskite-polymer composite of $MAPbBr_3$-PMMA could also synthesize via injection of precursor solution into the bulk MMA. The PL intensity of both $CsPbBr_3$-PMMA, $CsPbBr_3$-PBMA could remain high as 70% and 78% for month indicating stability. A white LED device was prepared based on the green emissive composite with phosphor of red emission. Also, the diverse selection of monomer provides controlled mechanical properties and flexibility to the composite also enables the preparation of device in large area. Recently, $CsPbBr_3$-PMMA was successfully prepared without using organic solvent provides a new direction to the perovskite-polymer composite synthesis and will broaden the use of polymer in perovskite science [51].

6. Conclusion

Metal halide hybrid perovskites are one of the most researched materials in the recent years owing to the fascinating properties they exhibit. These materials, therefore, find application in various fields. Inorganic, hybrid, organic–inorganic, 2D/3D mixed dimensional are all different types of halide perovskites that find applications in optoelectronics, photovoltaics, catalysis, etc. today. These variations come from the changes in the compositional elements in the basic structure family. The different phases exhibit peculiar properties that make them multifunctional. The crystal structure variation itself leads to changes in the properties. The changes in the electronic, optical, and electrical properties of the different types of metal halides are explored in the chapter. Ferroelectric property is mainly shown by the organic–inorganic counterparts owing to the polar nature of the organic cation. The synthesis techniques followed for the nanostructure, thin film and single crystals are varied and important as the growth of the materials and enhancement of its properties can be achieved by making changes in the synthesis procedures. The property enhancement of these materials is easily achievable by the addition of dopants or by fabricating polymer composites. These methods also help in enhancing the stability along with the properties making the extant of application of perovskite materials much wider.

Author details

Fency Sunny[1], Linda Maria Varghese[2], Nandakumar Kalarikkal[1,3,4] and Kurukkal Balakrishnan Subila[2]*

1 International and Inter University Centre for Nanoscience and Nanotechnology, Mahatma Gandhi University, Kottayam, Kerala, India

2 School of Chemical Sciences, Mahatma Gandhi University, Kottayam, Kerala, India

3 School of Pure and Applied Physics, Mahatma Gandhi University, Kottayam, Kerala, India

4 School of Nanoscience and Nanotechnology, Mahatma Gandhi University, Kottayam, Kerala, India

*Address all correspondence to: subilakb@gmail.com

References

[1] Gao P, Grätzel M, Nazeeruddin MK. Organohalide Lead perovskites for photovoltaic applications. Energy & Environmental Science. 2014;7(8): 2448-2463. DOI: 10.1039/C4EE00942H

[2] Cui J, Yuan H, Li J, Xu X, Shen Y, Lin H, et al. Focus issue review recent Progress in efficient hybrid Lead halide perovskite solar cells. Science and Technology of Advanced Materials. 2015;**16**:036004. DOI: 10.1088/1468-6996/16/3/036004

[3] Brittman S, Adhyaksa GWP, Garnett EC. The expanding world of hybrid perovskites: Materials properties and emerging applications. MRS Communications. 2015;**5**(1):7-26. DOI: 10.1557/MRC.2015.6

[4] Shi Z, Jayatissa AH. Materials perovskites-based solar cells: A review of recent Progress. Materials and Processing Methods. 2018;**11**(5):729-763. DOI: 10.3390/ma11050729

[5] Zhang W, Eperon GE, Snaith HJ. Metal halide perovskites for energy applications. Nature Energy. Nature Publishing Group. 2016;**1**(6):1038-1046. DOI: 10.1038/nenergy.2016.48

[6] Li B, Fu L, Li S, Li H, Pan L, Wang L, et al. Pathways toward high-performance inorganic perovskite solar cells: Challenges and strategies. Journal of Materials Chemistry A. 2019;7(36):20494-20518. DOI: 10.1039/C9TA04114A

[7] Huang J, Lai M, Lin J, Yang P. Rich chemistry in inorganic halide perovskite nanostructures. Advanced Materials. 2018;**30**(48):856-867. DOI: 10.1002/ADMA.201802856

[8] Saparov B, Mitzi DB. Organic-inorganic perovskites: Structural versatility for functional materials design. Chemical Reviews. 2016;**116**(7):4558-4596. DOI: 10.1021/ACS.CHEMREV.5B00715

[9] Ge C, Xue YZB, Li L, Tang B, Hu H. Recent Progress in 2D/3D multidimensional metal halide perovskites solar cells. Frontiers in Materials. 2020;7(9):601179-601188. DOI: 10.3389/FMATS.2020.601179

[10] Huang W, Wang Y, Balakrishnan SK. Controllable transformation between 3D and 2D perovskites through cation exchange. Chemical Communications. 2018;**54**(57):7944-7947. DOI: 10.1039/C8CC04261F

[11] Niu T, Lu J, Jia X, Xu Z, Tang MC, Barrit D, et al. Interfacial engineering at the 2D/3D heterojunction for high-performance perovskite solar cells. Nano Letters. 2019;**19**(10):7181-7190. DOI: 10.1021/ACS.NANOLETT.9B02781

[12] Yin W-J, Weng B, Ge J, Sun Q, Li Z, Yan Y. Oxide perovskites, double perovskites and derivatives for Electrocatalysis, Photocatalysis, and photovoltaics. Energy & Environmental Science. 2019;**12**:442. DOI: 10.1039/c8ee01574k

[13] Afroze S, Karim A, Cheok Q, Eriksson S, Azad AK. Latest development of double perovskite electrode materials for solid oxide fuel cells: A review. Frontiers in Energy. 2019;**13**(4):770-797. DOI: 10.1007/s11708-019-0651-x

[14] Igbari F, Wang Z-K, Liao L-S. Progress of lead-free halide double perovskites. Advanced Energy Materials. 2019;**9**(12):1803150. DOI: 10.1002/aenm.201803150

[15] Chatterjee S, Pal AJ. Influence of metal substitution on hybrid halide perovskites: Towards lead-free perovskite solar cells. Journal of Materials Chemistry A. 2018;**6**:3793-3823. DOI: 10.1039/ c7ta09943f

[16] Luo J, Bolisetty S, Peydayesh M, Mezzenga R, Li SL, Sun Z, et al. Featuring work from the research Group of Professor Chem Soc rev Sustainable Technologies for water purification from heavy metals: Review and analysis rational chemical doping of metal halide perovskites. Chemical Society Reviews. 2019;**48**:517. DOI: 10.1039/c8cs00563j

[17] Liu Y, Nag A, Iberato Manna L, Xia Z. Lead-free double perovskite Cs 2 AgInCl 6. Angewandte Chemie - International Edition. 2020;**113**(21):11696-11707. DOI: 10.1002/anie.202011833

[18] Meyer E, Mutukwa D, Zingwe N, Taziwa R. Metals Lead-free halide double perovskites: A review of the structural, optical, and stability properties as well as their viability to replace Lead halide perovskites. Metals. 2018;**8**(6):667-680. DOI: 10.3390/met8090667

[19] Manser JS, Christians JA, Kamat PV. Intriguing optoelectronic properties of metal halide perovskites. Chemical Reviews. 2016;**116**(21):12956-13008. DOI: 10.1021/ACS.CHEMREV.6B00136

[20] Li W, Wang Z, Deschler F, Gao S, Friend RH, Cheetham AK. Chemically diverse and multifunctional hybrid organic–inorganic perovskites. Nature Reviews Materials. 2017;**2**(3):1-18. DOI: 10.1038/natrevmats.2016.99

[21] Zhao XG, Yang JH, Fu Y, Yang D, Xu Q, Yu L, et al. Design of Lead-Free Inorganic Halide Perovskites for solar cells via cation-transmutation. Journal of the American Chemical Society. 2017;**139**(7):2630-2638. DOI: 10.1021/ JACS.6B09645

[22] Krishna A, Gottis S, Nazeeruddin MK, Sauvage F. Mixed dimensional 2D/3D hybrid perovskite absorbers: The future of perovskite solar cells? Advanced Functional Materials. 2019;**29**(8):1806482. DOI: 10.1002/ ADFM.201806482

[23] Shamsi J, Urban AS, Imran M, De Trizio L, Manna L. Metal halide perovskite nanocrystals: Synthesis, post-synthesis modifications, and their optical properties. Chemical Reviews. 2019;**119**(5):3296-3348. DOI: 10.1021/ ACS.CHEMREV.8B00644

[24] Fan Q, Biesold-McGee GV, Ma J, Xu Q, Pan S, Peng J, et al. Lead-free halide perovskite nanocrystals: Crystal structures, synthesis, stabilities, and optical properties. Angewandte Chemie International Edition. 2020;**59**(3):1030-1046. DOI: 10.1002/ANIE.201904862

[25] Chu L, Ahmad W, Liu W, Yang J, Zhang R, Sun Y, et al. Lead-free halide double perovskite materials: A new superstar toward green and stable optoelectronic applications. Nano-Micro Letters. 2019;**11**(1):1-18. DOI: 10.1007/ S40820-019-0244-6

[26] Buizza LRV, Crothers TW, Wang Z, Patel JB, Milot RL, Snaith HJ, et al. Charge-carrier dynamics, Mobilities, and diffusion lengths of 2D–3D hybrid Butylammonium–Cesium–Formamidinium Lead halide perovskites. Advanced Functional Materials. 2019;**29**(35):1902656. DOI: 10.1002/ ADFM.201902656

[27] Yun S, Kirakosyan A, Yoon SG, Choi J. Scalable synthesis of exfoliated Organometal halide perovskite nanocrystals by ligand-assisted ball milling. ACS Sustainable Chemistry & Engineering. 2018;**6**(3):3733-3738. DOI: 10.1021/ ACSSUSCHEMENG.7B04092/SUPPL_ FILE/SC7B04092_SI_001.PDF

[28] Chen K, Zhong Q, Chen W, Sang B, Wang Y, Yang T, et al. Short-chain ligand-passivated stable α-CsPbI3 quantum dot for all-inorganic perovskite solar cells. Advanced Functional Materials. 2019;**29**(24):1900991. DOI: 10.1002/ADFM.201900991

[29] Li H, Dong W, Shen X, Ge C, Song Y, Wang Z, et al. Enhancing the efficiency and stability of CsPbI3Nanocrystal-based light-emitting diodes through ligand engineering with Octylamine. Journal of Physical Chemistry C. 2022;**126**(2):1085-1093. DOI: 10.1021/ACS.JPCC.1C10115/SUPPL_FILE/JP1C10115_SI_001.PDF

[30] Ananthakumar S, Moorthy Babu S. Progress on synthesis and applications of hybrid perovskite semiconductor nanomaterials—A review. Synthetic Metals. 2018;**246**:64-95. DOI: 10.1016/J.SYNTHMET.2018.10.003

[31] Protesescu L, Yakunin S, Bodnarchuk MI, Krieg F, Caputo R, Hendon CH, et al. Nanocrystals of Cesium Lead halide perovskites (CsPbX3, X = Cl, Br, and I): Novel optoelectronic materials showing bright emission with wide color gamut. Nano Letters. 2015;**15**(6):3692-3696. DOI: 10.1021/NL5048779/SUPPL_FILE/NL5048779_SI_001.PDF

[32] Sun S, Yuan D, Xu Y, Wang A, Deng Z. Ligand-mediated synthesis of shape-controlled Cesium Lead halide perovskite nanocrystals via Reprecipitation process at room temperature. ACS Nano. 2016;**10**(3):3648-3657. DOI: 10.1021/ACSNANO.5B08193/SUPPL_FILE/NN5B08193_SI_001.PDF

[33] Huang H, Polavarapu L, Sichert JA, Susha AS, Urban AS, Rogach AL. Colloidal Lead halide perovskite nanocrystals: Synthesis, optical properties and applications. NPG Asia Materials. 2016;**8**(11):e328-e328. DOI: 10.1038/am.2016.167

[34] Chen X, Chen Y, Huang J, Li C, Ni Z, Zhang K, et al. Phase regulation and surface passivation of stable α-CsPbI3Nanocrystals with dual-mode luminescence via synergistic effects of ligands. Journal of Physical Chemistry C. 2022;**126**(11):5233-5243. DOI: 10.1021/ACS.JPCC.1C10040/SUPPL_FILE/JP1C10040_SI_001.PDF

[35] Chen X, Peng L, Huang K, Shi Z, Xie R, Yang W. Non-injection gram-scale synthesis of Cesium Lead halide perovskite quantum dots with controllable size and composition. Nano Research. 2016;**9**:1994-2006. DOI: 10.1007/s12274-016-1090-1

[36] Liu Z, Yang H, Wang J, Yuan Y, Hills-Kimball K, Cai T, et al. Synthesis of Lead-free Cs2AgBix6 (X = Cl, Br, I) double perovskite Nanoplatelets and their application in CO2 photocatalytic reduction. Nano Letters. 2021;**21**(4):1620-1627. DOI: 10.1021/ACS.NANOLETT.0C04148/SUPPL_FILE/NL0C04148_SI_001.PDF

[37] Murugadoss G, Thangamuthu R, Senthil Kumar SM, Anandhan N, Rajesh Kumar M, Rathishkumar A. Synthesis of ligand-free, large scale with high quality all-inorganic CsPbI3 and CsPb2Br5 nanocrystals and fabrication of all-inorganic perovskite solar cells. Journal of Alloys and Compounds. 2019;**787**:17-26. DOI: 10.1016/J.JALLCOM.2019.02.018

[38] Luo J, Wang X, Li S, Liu J, Guo Y, Niu G, et al. Efficient and stable emission of warm-white light from Lead-free halide double perovskites. Nature. 2018;**563**(7732):541-545. DOI: 10.1038/s41586-018-0691-0

[39] Sirtl MT, Hooijer R, Armer M, Ebadi FG, Mohammadi M, Maheu C,

et al. 2D/3D hybrid Cs2AgBiBr6 double perovskite solar cells: Improved energy level alignment for higher contact-selectivity and large open circuit voltage. Advanced Energy Materials. 2022;**12**(7):2103215. DOI: 10.1002/AENM.202103215

[40] Fu Y, Meng F, Rowley MB, Thompson BJ, Shearer MJ, Ma D, et al. Solution growth of single crystal Methylammonium Lead halide perovskite nanostructures for optoelectronic and photovoltaic applications. Journal of the American Chemical Society. 2015;**137**(17):5810-5818. DOI: 10.1021/JACS.5B02651

[41] Wang XD, Li WG, Liao JF, Kuang DB. Recent advances in halide perovskite single-crystal thin films: Fabrication methods and optoelectronic applications. Sol RRL. 2019;**3**(4):1800294. DOI: 10.1002/SOLR.201800294

[42] Dang Y, Ju D, Wang L, Tao X. Recent Progress in the synthesis of hybrid halide perovskite single crystals. CrystEngComm. 2016;**18**(24):4476-4484. DOI: 10.1039/C6CE00655H

[43] Zhou Y, Chen J, Bakr OM, Sun H-T. Metal-doped Lead halide perovskites: Synthesis, properties, and optoelectronic applications. Chemistry of Materials. 2018;**30**:6589-6613. DOI: 10.1021/acs.chemmater.8b02989

[44] Ou Q, Bao X, Zhang Y, Shao H, Xing G, Li X, et al. Band structure engineering in metal halide perovskite nanostructures for optoelectronic applications. Nano Materials Science. 2019;**1**(4):268-287. DOI: 10.1016/J.NANOMS.2019.10.004

[45] Naresh V, Lee N. Zn(II)-doped Cesium Lead halide perovskite nanocrystals with high quantum yield and wide color Tunability for color-conversion light-emitting displays. ACS Applied Nano Materials. 2020;**3**(8):7621-7632. DOI: 10.1021/ACSANM.0C01254

[46] Li R, Chen S, Li X, Yin G, Gong Y, Yu J, et al. Zn doped MAPbBr3 single crystal with advanced structural and optical stability achieved by strain compensation. Nanoscale. 2020;**12**(6):3692-3700. DOI: 10.1039/C9NR09657D

[47] Karmakar A, Dodd MS, Agnihotri S, Ravera E, Michaelis VK. Cu(II)-doped Cs2SbAgCl6 double perovskite: A Lead-free, low-bandgap material. Chemistry of Materials. 2018;**30**(22):8280-8290. DOI: 10.1021/ACS.CHEMMATER.8B03755

[48] Pan G, Bai X, Yang D, Chen X, Jing P, Qu S, et al. Doping lanthanide into perovskite nanocrystals: Highly improved and expanded optical properties. Nano Letters. 2017;**17**(12):8005-8011. DOI: 10.1021/ACS.NANOLETT.7B04575

[49] Macdonald TJ, Batmunkh M, Lin C-T, Kim J, Tune DD, Ambroz F, et al. Origin of performance enhancement in TiO_2-carbon nanotube composite perovskite solar cells. Small Methods. 2019;**3**(10):1900164. DOI: 10.1002/smtd.201900164

[50] Huang M-Y, Veeramuthu L, Kuo C-C, Liao Y-C, Jiang D-H, Liang F-C, et al. Improving performance of Cs-based perovskite light-emitting diodes by dual additives consisting of polar polymer and n-type small molecule. Organic Electronics. 2019;**67**:294-301. DOI: 10.1016/j.orgel.2018.12.042

[51] Xin Y, Zhao H, Zhang J. Highly stable and luminescent perovskite-polymer composites from a convenient and universal strategy. ACS Applied Materials & Interfaces. 2018;**10**(5):4971-4980. DOI: 10.1021/ACSAMI.7B16442

Chapter 15

The Mystery of Dimensional Effects in Ferroelectricity

Rolly Verma and Sanjeeb Kumar Rout

Abstract

The dimensional effect on ferroelectricity is a subject of long-understanding fundamental interest. While the low-dimensional finite ferroelectric structures are committed to the potential increase in electronics miniaturization, these anticipated benefits hinged on the existence of stable ferroelectric states in low-dimensional structures. This phenomenon can be understood from the point of basic physics. This chapter reviews the literature on the finite-size effects in ferroelectrics, emphasizing perovskite and polyvinylidene-based polymer ferroelectrics having technological importance. The reviewed data revealed that despite critical dimensionality being predicted in ferroelectrics, polarization switching phenomenon is possible in as thin as one monolayer film, at least in the case of P(VDF-TrFE) Langmuir–Blodgett thin film with stabilized functional properties. The roles of the depolarization field, electrode interfaces, domain wall motion, etc. in controlling the measured ferroelectric properties have been discussed. Further, the observed deviation from the bulk properties is explained based on both experimental and theoretical modeling.

Keywords: perovskite ferroelectrics, boundary conditions, dimensional confinement, polarization switching kinetics

1. Introduction

Ferroelectric materials have been recognized as one of the focal points in condensed matter physics and material science for over 50 years. This is the most exciting material used in the electronics industry possessing switchable spontaneous polarization with the direction of applied field stress. These ferroelectrics exhibit substantial piezoelectricity as well. Accordingly, these materials are widely exploited as ultrasonic devices, sensors, actuators, energy storage, memory components, and noticeably more consumer electronics products. At the next level up, modern electronics have taken the charge of electronics miniaturization with the nano-dimensional system including thin film and ultra-thin films precisely placed in the electronics circuit [1]. In the last few decades, the advancement in voltage-modulated scanning probe microscopy techniques, exemplified by piezoresponse force microscopy (PFM) and associated spectroscopies, opened a driveway to make use of ferroelectrics on a single-digit nanometer level. Current research in the United States and other nations is pushing the limits of miniaturization to the point that structures only hundreds of atom-thick will be commonly manufactured [2]. This high-precision microelectronics assembly is achieved by scaling down the materials in accord. Nevertheless, the performance of the ferroelectric material is related to the way they are structurally confined undoubtedly due to structure–property alliance. Whilst the dimensional downscaling of the ferroelectric materials from bulk to nanoscale boost the possibilities to endure the boxing up of increased numbers of components into single electronics integrated circuit, the functional properties are suppressed as the material goes down to the critical dimension. The theoretical studies on the nano-dimensional system including thin films and ultra-thin films have shown that ferroelectricity persists down to the nanoscale. However, the experimental approach at this scale revealed the disappearance of the ferroelectric switching phenomena as the critical size of the crystal in the ferroelectric system is reached. For example, 80% of the dielectric and piezoelectric properties of perovskite ceramics are suppressed compared to their bulk counterpart as the material is scaled down to $\sim$10 nm [3]. A bulk-like ferroelectricity with finite-size modifications has been observed in nanocrystals as thin as 25 Å crystalline ferroelectric polymer films [4–6], 100 Å perovskite films [7] and as small as 250 Å in diameter ultrafine nanoparticles [8]. These outcomes can be elucidated as the bulk ferroelectricity is stamped out by surface depolarization energies and inferred that the bulk transition is limited by minimum critical dimension. This is noted as the scaling effect. It occupies a prominent place in the research area as our limited intuition for the nanoworld and comprehensive knowledge of structure–property relations often lag behind technological advances. Since nanostructuring of ferroelectric materials ends up with the appearance of their critical size limit, below which the essential ferroelectric parameters cannot be sustained, a completely contrasting behavior has been observed in hafnium based thin films which displayed an unconventional form of ferroelectricity in thin films with a thickness of only a few nanometers. This allows the construction of nanometer-sized memories and logic devices. Until now, however, it is an unsolved mystery how ferroelectricity could turn-out at this scale. A study reported by scientists at the University of Groningen, Netherland revealed that migrating oxygen atoms (or vacancies) are supposed to be responsible for the distinguished polarization switching phenomena in a hafnium-based capacitor [9]. Likewise, Bune et al. [10] have reported the near-absence of finite-size effect in two monolayer crystalline Langmuir–Blodgett film of P(VDF-TrFE) ferroelectric polymer. This contrasting behavior of ferroelectrics increased the curiosity of the scientific community in this stream. Although, well-developed theories exist for bulk materials, the extrapolation of these theories to thin films and nanostructures is frequently ambiguous. Hence understanding the dimensional system and going into the issues with scaling and size effect is crucial and is the central challenge for the ferroelectrics-based electronics community.

The chapter is aspired to understand the fundamental mechanism underlying ferroelectric behavioral patterns in polymer and ceramics systems as it is scaled down to a critical dimensional range attractive for a variety of technological applications. This knowledge would be beneficial for the current ferroelectric materials as well as for designing new materials with even a cut above electroactive property. The chapter is divaricated into six sections. Section 1 introduces the topic of our discussion. Section 2 talks about the theoretical framework for the scaling effect in the ferroelectric system. Section 3 discusses about how material functional properties are depleted in nano-confined perovskite ferroelectric system including phase transition temperatures, spontaneous polarization, coercive field and piezoelectric coefficient. Next are the possible causes for the observed scaling effect. Section 5 explores the scaling effect in ferroelectric polymer thin films with special emphasis on PVDF and its copolymers.

The fundamental ferroelectric polarization switching mechanism for nanostructures is introduced and the models for thin films at the nanoscale are reviewed in Section 6. The nucleation-limited-switching (NLS) model based on region-to-region switching kinetics for polymer thin films will be highlighted. Finally, the observed results will be summarized and the future outlook for ferroelectric nanostructures are discussed. We clarify here that the goal of this chapter is not to review all the work in the vast field of ferroelectrics but rather to provide a scholastic presentation for the readers through the use of select case studies and authors experience in the field.

2. Theoretical framework

The more is the challenge for developing nano-scaled devices, the more is the challenge to sustain their ferroelectricity at this scale. To capture the comprehensive knowledge in the versatility of ferroelectricity as the material is scaled down, particularly at the nanoscale, a theoretical framework is exceedingly advantageous. The first-principle density functional theory (DFT)-based modeling and simulations plays a significant role as the fundamental properties could be envisioned and act as guidelines in the design of ferroelectric nanostructures. For the last decade, it has been successfully implied to various ferroelectric bulk crystals as well as nanostructures. According to first-principle density functional theory, ferroelectricity is analyzed in two possible ways [11]: (a) calculation of total energy by solving ground state problem for a given potential, (b) computation of linear response (LR). This is done by discovering the lowest order changes in ground state energy as the potential changes. The former provides the knowledge about the parameters which is the first derivative of total energy such as stress or electric polarization while the latter computes the properties corresponding to the second and third derivatives of total energy such as phonons, dielectric, piezoelectric and other compliances. In the perovskite ferroelectrics oxides, the transition metal is in d^0 state, therefore the effect of electronic interaction is rather weak on the ground state electrons. Hence the first-principle calculation can be quite useful in their studies. In ferroelectric oxides, it is very unlikely to have electronic excitations due to the presence of large band insulators with unsettled d-states of transition metal (B). DFT calculation ascertains the crystal structure through energy minimization such as phonons, Raman tensors, dielectric, piezoelectric and other compliances. For example, DFT calculation provides subtle information about which structural distortions can destabilize the cubic structure in perovskite ferroelectrics [12]. Further, DFT calculation explains that the temperature dependence ferroelectricity arises from the phonon contribution and these operations hold sway over the interesting piezoelectric response as well. Even so, it has some limitations, firstly these simulations are relevant for the material properties at $T = 0$ K (or at low temperature). Secondly, DFT theory could simulate no more than 150 atoms (for a short time scale ~ 100 ps) and have definable size errors in the approximation of thermodynamic properties of ferroelectrics. However, this dereliction is compensated in more intuitive way through an effective Hamiltonian methodology which dealt with finite temperatures along with large-scale simulations of ferroelectrics. This approach remains unaltered for the bulk ferroelectric but for thin film or at the nano-scale, effects of surrounding (appropriate boundary conditions) are captured as the estimated properties of nanostructures below the "critical dimension" depends on the length-scale measurement, that is on the ambient conditions not on the volume of a cluster [11]. Two important boundary conditions have been reported.

First is the mechanical boundary condition specially for thin film developed epitaxially on a substrate. For this, the required in-plane strain component by the lattice constant of the substrate are frozen to constant value while in thick films, all the strain component are free to fluctuate. Second is the electrical boundary condition that creates depolarization field arising due to bound charges at the surface partially recompensated by free carriers assembled at the electrode. This interesting finding is summarized here in the context of $BaTiO_3$. Using the first-principle calculations, Junquera and Ghosez explained that a favorable polar state can be realized only for $BaTiO_3$ film as thin as six-unit cells ($\sim$24 Å) and attributed this extraordinary ferroelectric stability to the depolarizing electrostatic field at the ferroelectric-metal electrode interface [13]. Since the depolarization field is responsible for diminishing ferroelectricity at the finite size, the author theoretically explained here that the electrons at the metal interface tend to screen the surface charge. As a result, the dipoles with the similar polarity appeared at the metal-ferroelectric interface and stabilized the ferroelectricity. The fusion of first-principle density functional (DFT) calculations with an effective Hamiltonian offers a multiscale driveway to analyze the various functional properties of ferroelectric oxides. It also provides the possibility to directly couple the properties to the atomic arrangements and the boundary conditions. Another important theory is Landau Devonshire theory which uses spatial inhomogeneity to show the smearing of phase transitions in ferroelectric nanostructures [14]. This theory explains that the inhomogeneity between a ferroelectric material and an electrode is a result of domain structure in ferroelectric thin films. A dead layer is formed between the film and the electrode. The reduced dead layer softens the domain structure contributing large dielectric response of the film. Landau–Devonshire theory is a free-energy-based phenomenological perspective for continuum mechanics ferroelectric functioning. This theory is very helpful in analyzing diverse phases in complex phase diagrams, microstructures as well as device simulations [15]. However, parameters obtained from Landau free energy are based on material-specific information. Therefore, it is highly desirable to link first-principle calculations with Landau-like theories at nonzero temperature so that analysis could be done at all length-scales fairly based on the information obtained from first principle calculations. The latter pushed the limit of fabrication of perovskite ferroelectrics below $\sim$ 15 nm [16] and as thin as 1 nm in ferroelectric polymer system [10].

3. Critical dimensional range for perovskite ferroelectrics

On the edge of the ferroelectrics class is the ABO_3 oxides (where 'A' and 'B' are two cations, often of different sizes, and O is the oxygen atom that bonds to both ions) occurring in the perovskite structure. Typical materials that crystallize in the perovskite structure having technological importance are ferroelectric $BaTiO_3$, $PbTiO_3$, piezoelectric $Pb(Zr, Ti)O_3$, electrostrictive $Pb(Mg, Nb)O_3$, multiferroic $BiFeO_3$ etc. Ferroelectricity is a cooperative phenomenon of the orchestration of the charged dipoles within the crystal structure. In perovskite, it is governed by the existence of long-range ordering of elemental dipoles up to a distance ranging from millimeter to a few microns. The lower-dimensional confinement of the perovskite ferroelectric material especially in the nano to sub Å level, strongly perturbs the long-range ferroelectric order as the fraction of surface/interface atoms is increased. As the ferroelectric particle goes down to the nano-range, there is a greater probability of the arrangement of constituent atoms at the surface of the particles, thereby ratio of surface area to volume ratio is increased that changes the free energy of the crystal, triggering immense changes in the functional parameters of the material [17] such as the abnormal lowering of ferroelectric to paraelectric phase transition temperature (T_c), suppression of remnant polarization (P_r), and increase in the coercive field (E_c). Few literature reports evidenced the shifting of T_c toward room temperature when the particle size is lowered down to 200 nm or below [18–20]. It has been suggested that the surface charge layer and the depolarization field played a significant role in this scaling effect as the depolarization effect breaks the material into small domains of different polarization to minimize the macroscopic charge generated on the surface as it is cooled through T_c [21]. Ivan et al. [22] also mentioned the role of depolarizing field in nanoconfined perovskite material using an ab initio derived Hamiltonian. Daniel and his research colleague well-articulated the nature of ferroelectric phase transition temperature (T_c) on downscaling the barium titanate (BTO) nanocrystals using the surface plasmon technique [23]. They proposed that the behavior of surface ferroelectricity seems to be different from the volume ferroelectricity and is characterized by very long relaxation time scales. For nanoscale ferroelectrics, the surface and the volume of the crystals are well-tuned due to the dominance of the surface over the whole nanocrystals. Therefore, the volume T_c may probably close to the bulk-like but for nanocrystals, it decreases significantly relative to the bulk value. The BTO crystal size > 0.1 μm exhibited bulk like properties with a phase transition temperature $T_c \sim 130°C$, while a continuous shift in the temperature range ~50–90°C has been observed for the crystals with dimensions <50 nm. This behavior may be the consequence of barium titanate nanocrystalline size distribution [23]. For lead titanate ($PbTiO_3$) crystals, size effects were found to be applicable below 100 nm. The T_c decreases from 500 to 486°C as the particle size decreases from 80 nm to 30 nm respectively with a more diffused peak in the lower dimension and the phase transition peak completely disappeared after 26 nm [24]. This scaling effect on T_c, typically implied by the relation:

$$\delta T = \frac{[T_c(\infty) - T_c(d)]}{T_c(\infty)} = Ad^{-\bar{\lambda}} \tag{1}$$

where $T_c(\infty)$ and $T_c(d)$ are phase transition temperature of bulk crystal and thin film of thickness '*d*' respectively, too deviates at the ultralow-dimensional scale as reported by Emad et al. [25]. Genesta et al. [26] reported the disappearance of ferroelectric switching in barium titanate nanowire below a critical size of about 1.2 nm. The author explained that the global contraction of the unit cell at the wire surface is attributed to the disappearance of ferroelectricity. Vincenzo and Randall [17] have provided a very good discussion about the size and scaling effect in the barium titanate ferroelectric system. Even though discrepancies on size limit still persist as ferroelectricity not only depends on the absolute critical size of the material but the preparation route to achieve the limit. For example, Ishikawa et al. illustrated that sol–gel-prepared $PbTiO_3$ nanoparticles exhibited a critical dimensional limit of ~10 nm at 300 K which was later defied by Fong et al. [27] who suggested the stable ferroelectric phase in $PbTiO_3$ thin films down to the thickness of 3-unit cells (1.2 nm) at room temperature. Recently Hao et al. [28] demonstrated the structural and polarization switching behavior of 4.5 nm $BaTiO_3$ ultrafine nanoparticles. The author attributed the switchable polarization to the presence of local spatial coherent asymmetric nanoparticles with discernable Ti-distortion and paved the way for the construction of high-density memory devices. This finding evidenced that the absence of ferroelectricity reported literature may not be inherent to the system. The abnormal response of phase transition temperature on downscaling the perovskite ferroelectrics extends to other ordered parameters as well. Daopei et al. [29] theoretically demonstrated three types of equilibrium polarization patterns based on various sizes and material parameters combination, i.e., monodomain, vortex-like, and multidomain, in isolated $BaTiO_3$ or $PbTiO_3$ octahedral nanoparticles embedded in a dielectric medium, like $SrTiO_3$ (ST, high dielectric permittivity) and amorphous silica (a-SiO_2, low dielectric permittivity) using a time-dependent Landau–Ginzburg method with coupled-physics finite-element-method-based simulations. The author further discussed the existence of. The critical particle size below which ferroelectricity vanishes in their calculations was 2.5 and 3.6 nm for $PbTiO_3$ octahedral nanoparticles for high- and low-permittivity matrix materials respectively. However, this size was unalike for $BaTiO_3$ octahedral nanoparticles (~3.6 nm) for all that of the matrix materials. Yan et al. [30] synthesized barium titanate nanoparticle by high-gravity reactive precipitation (HGRP) method and found that crystal with the size of 30 nm exhibited a completely paraelectric cubic phase which changes to tetragonal ferroelectric phase at 70 nm confirmed by XRD and Raman spectral analysis. Nuraje et al. [31] confirmed the tetragonal $BaTiO_3$ nanoparticles (~6–12 nm) at room temperature by electrostatic force microscopy (EFM). Besides, coercive field (E_c), the field of negligible polarization, an important functional parameter pertains to the scaling effect in perovskite ferroelectrics as well. According to Janovec–Ka–Dunn (JKD) law, the scaling dimension (~thickness 'd') of ferroelectric thin-film and the coercive field is given by semiempirical relation:

$$E_c \propto d^{-2/3} \tag{2}$$

Following the JKD scaling theory, Xu et al. [32] investigated the ferroelectric properties in 20–330 nm of (0 0 1)- and (1 1 1)-oriented $PbZr_{0.2}Ti_{0.8}O_3$ ceramics system. The change in the spontaneous polarization and the coercive field by lowering the dimension of thin PZT thin-film is delineated in **Figure 1**. Likewise, Venkata et al. [33], Hong et al. [34] also confirmed the falling of field-induced polarization behavior with the downscaling in perovskite polycrystals and ferroelectric nano-thin films respectively (**Figure 1**). It has been observed that (0 0 1)-oriented PZT film followed the JKD scaling while (1 1 1)-oriented heterostructures (~<165 nm) deviated from the expected scaling. The first principle DFT calculation attributed this deviation to the formation of a lower energy barrier phase for switching which eventually reduces the domain-wall energy and exacerbates the deviation.

However, defying the general hypothesis on the scaling effect in perovskite ceramics, an increase in long-range ferroelectric order is observed in $NaNbO_3$ by Juriji et al. [35] in 2017 as the material was scaled down below 0.27 μm which was attributed to the existence of intra-granular stresses induced during the formation of non-180° domain walls as the grain dimension is reduced. Recently, Lorenzo et al. [36] successfully developed an unusual ferroelectric orthorhombic phase (*Pmma*) in 24 nm crystal of $NaNbO_3$ using a microwave synthesis route. Further, the exceptional property of ferroelectricity's appearance in antiferroelectric $PbZrO_3$ ceramics as the material attained its critical dimension ~400–500 nm [37]. This unique result provided the possibility among the research community to stabilize ferroelectricity in lower dimensions which was not observed in other ferroic-system. The disappearance of ferroelectricity below the critical nano-dimension was long thought of the past. In recent years, advanced characterization techniques enabled the fundamental size

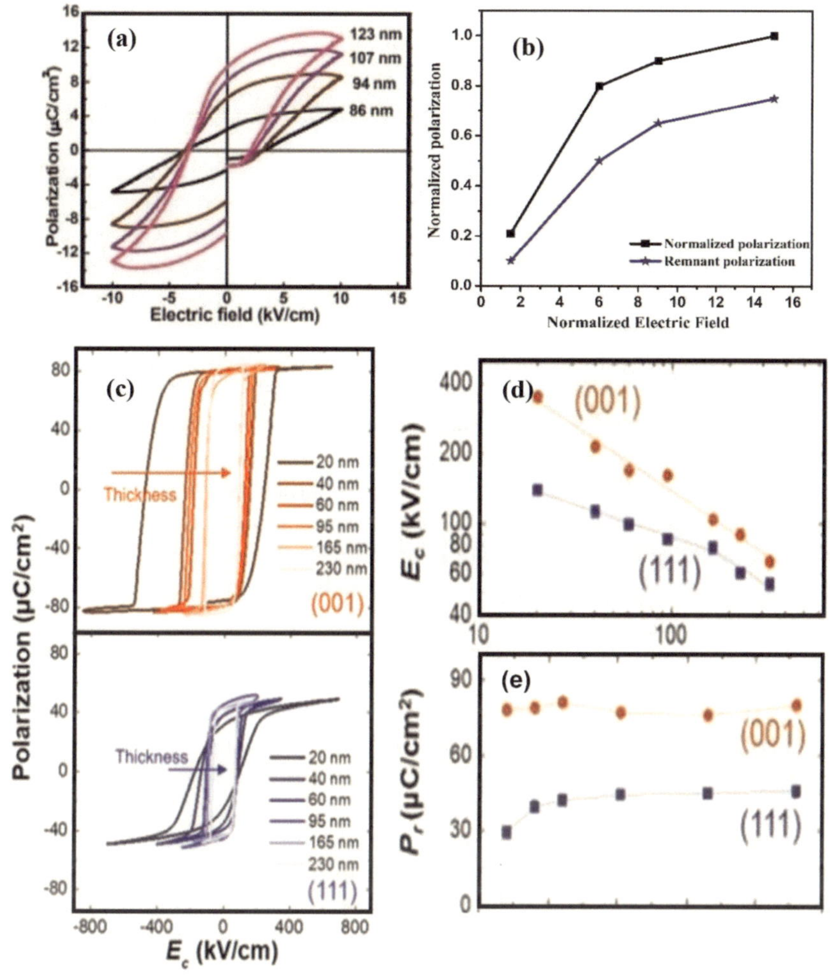

Figure 1.
Variation of (a) perovskite polycrystals [33] (open access) (b) ferroelectric nano-thin films [34] (c) polarization hysteresis, (d) Coercive field, (e) Remnant polarization (Pr) for (001)- and (111)- oriented PZT thin-films [32].

effect at the sub-Å level (~6 unit cells) in perovskite ferroelectric systems [13] that lowered the critical dimension for the existence of ferroelectricity in thin films by orders of magnitude.

4. Genesis of scaling effect in perovskite ferroelectrics

Ferroelectric instability is a consequence of a delicate balance between short-range and long-range dipolar interactions. These interactions are definitely perturbed in nanostructures. With the downscaling of ferroelectrics to nanoscale, the surface to volume ratio is changed, the short-range forces are altered at the surfaces and interfaces while long-range character is influenced by the limitation in finite sizes of the material. One of the critical issues in downscaling the perovskite ferroelectrics is the distortion of the ferroelectric phase such as orthorhombic or tetragonality (c/a) in crystals. It is noted that tetragonality in $PbTiO_3$ crystals were rapidly decreased to 1 as it was scaled down to 7 nm [24] following the relation:

$$\frac{c}{a} \approx 1 - exp^{-\sim d}, (where\ d\ is\ the\ grain\ size\ of\ the\ material). \quad (3)$$

However, this explanation was not appropriate as the theoretical calculations pushed the limit of fabrication of perovskite ferroelectrics as thin as ~15 nm [10, 16]. The origination of scaling and size effect is still not realized although two important explanations were suggested: (a) the distinctive intrinsic properties of nanoparticles smaller than critical dimension, (b) generation of local depolarization field due to the surface ions arresting the ferroelectric phase. The development of the depolarization field is a consequence of extrinsic effects such as electrical boundary conditions and electrode screening effect. It is the key issue in analyzing the ferroelectric domain structures, Further, the strain and the electrical polarization in ferroelectrics are coupled phenomena, therefore any misfit strain affairs change the spontaneous polarization of the material. Hence the materials are responsive to mechanical boundary conditions as well. The boundary conditions cognate with the contact situation between the surface of the ferroelectric film and the electrode, play a prominent role in the scaling effect of thin-films. The suppression of spontaneous polarization by instigating the surface and interfacial charges offsetting the normal component of the polarization, creates a depolarization field [38]. In a few cases, the depolarization electrical energy guided the retention of polar crystals by electrode screening effect [39]. The latter is associated with the perfect screening of the electrode and depolarization phase, thereby stabilizing the ferroelectric phase and its resulting properties [13]. While in other cases, it is completely considered for destabilizing the ferroelectric domains [17, 21, 40]. The size of the ferroelectric crystals strongly influences the magnitude of the depolarization field. The scaling of ferroelectrics to their critical dimensional range, being the surface charge remains constant, increases the voltage developed per unit length which induces the depolarization-field-induced scaling effect. The latter is eminent in thin films, when present strongly influences the ferroelectric domains. Further, with the reduced film thickness, rational growth is promoted that leads to strong mechanical boundary conditions, contributes to the scaling effect in ferroelectrics. Factors such as lattice mismatch in epitaxial grown thin films, the difference in the properties of the substrate and the ferroelectric film or growth-related strain generated during the fabrication process creates mechanical boundary conditions. It is associated with the substrate-induced stress/strain that is not only coupled with the spontaneous polarization but strongly influences the array of ferroelastic domains, if present. For example, if the polarization vector switches ferroelastically between [0 0 1] and [1 0 0] directions, then biaxial compression perpendicular to the polar axis will stabilize that orientation and increases the phase transition temperature. However, when these strain effects are overlaid on the scaling effect, the process is supposed to be reversed [41, 42]. Therefore, it is notable that mechanical boundary condition functions along with the intrinsic scaling effect [3]. In bulk ceramics, mechanical boundary conditions are created at the grain boundaries and developed a spontaneous dipole. Apart from surface/ferroelectric film interfaces, the other factors that strongly influence the scaling effect in ferroelectrics are the volume of domain walls and grain boundaries in the lower-dimensional scale of ferroelectric system. The extreme reduction in thin-films/grain size lessen the number of stable domain configurations and eventually mobility of domain boundaries decreases which resulted in low permittivity of the system [3]. Besides, crystal imperfections, doping effect, grain boundaries, microstructures, etc., are interlinked to the processing condition [43] may influence the scaling effect in perovskite ferroelectrics and requires independent assessment.

5. Scaling effect in ferroelectric polymers

Ferroelectric polymers such as poly(vinylidene fluoride) and its copolymer systems have evinced the distinguishing properties in lower-dimensional structures. Their nanostructures are emphasized as electrospun nanofibers [44], anodic aluminum oxide-templated nanotubes [45] and the 2D Langmuir–Blodgett (LB) nanofilm [46]. Few reports have also described the PVDF-nanosphere [47]. For example, Zhengguo et al. [48] reported the formation of P(VDF-TrFE) nanoparticles with sizes of 60–100 nm using a solution method with the successful application in low band-gap polymer photovoltaic devices. Mostly, these polymers are analyzed in the form of thin films [49–51]. Unlike ferroelectric ceramics, the polymer ferroelectrics are semi-crystalline (amorphous and crystal parts are intertwined) in nature, therefore the ferroelectricity in the polymer is strongly affected by the interaction between the crystalline and amorphous interface. This is known as the nanoconfinement effect [52], according to which the dipole switching in polymer ferroelectrics largely depends on the local electric field in the crystals. Definitely, these interactions are perturbed as the dimensionality of the polymer ferroelectrics goes down to the lowest possible range. As a consequence, the crystal orientations are varied that eventually influences the functional properties of the material. In the bulk form, P(VDF-TrFE, 70:30) exhibited the first-order ferroelectric to paraelectric phase transition temperature $T_c \sim 100°C$ and a spontaneous polarization of $P_s \sim 0.1\ C/m^2$ at room temperature [53]. A maximum polarization of 12 $\mu C/cm^2$ at 4 V has been observed for 100 nm thick P(VDF-TrFE) film which is attributed to the presence of crystalline β-phase (a type of crystal orientation) [54]. Similarly, Xu et al. [55] suggested the preferential crystal orientation for the maximum polarization of 10 $\mu C/cm^2$ and apparent coercive field ~6 MV/m in 500 nm thick PVDF film at a very low switching voltage of 3 V. The study of ferroelectric polymer in their ultra-low dimensions were not possible until the discovery of Langmuir–Blodgett (LB) [56] technique of monolayer formation as the thin films constructed by the conventional route of synthesis such as uniaxial or biaxial drawing [57], solvent casting [58], uniaxial stretching [59] or spin coating limited the thickness as thin as ~ 60 nm only [50]. Langmuir–Blodgett (LB) monolayer transfer technique produces high-quality ferroelectric polymer ultrathin films which are few monolayers thick and can be switched at 1 V, permitting precise control of the film nanostructures [5]. In 1993 ferroelectricity was first discovered in 30 monolayers (15 nm) LB films of P(VDF-TrFE) random copolymer. Later, in 1998, using this method, Bune et al. [4, 10] reported the ultrathin ferroelectric film of PVDF-TrFE copolymer with a thickness of 1 nm. This film was prepared using a horizontal Langmuir–Blodgett (LB) technique, known as Langmuir–Schaefer (LS) technique. This gave the recognition of two-dimensional ferroelectric polymer thin film system implying that the state of ferroelectricity may be achieved by coupling only within the plane of the film and unlocked a new frontier in polarization switching development in ultrathin-single crystal films [4, 5, 60–62]. However, the larger interfacial effect may arrest the ferroelectric switching even in PVDF-based Langmuir–Blodgett (LB) nanofilms [63]. The P(VDF-TrFE) ferroelectric LB films displayed complete polarization reversal in samples for the thickness ranging from 30 to 100 monolayers. Also, the partial reversal has been observed at eight monolayers thickness, the thinnest possible ferroelectric films made to date [64]. The 30-layer ferroelectric LB films (~15 nm) exhibited the phase transition temperature (T_c) in the range ~70–90°C lower than the typical values ~90–110°C for spun films of P(VDF-TrFE) [65]. The decrease in T_c is typically attributed to the depolarization interfaces. Zhu et al. [66] demonstrated the lowering of the spontaneous polarization to 5 μC/cm^2 at a very high electric field of 700 MV/m for 18 nm thick P(VDF-TrFE) LB film even with 80% of crystallinity. The impression of reduced ferroelectric response reaches out to piezoelectric responses as well. The piezoelectric coefficient of $|d_{33}|$ = 5 pm/V for a 30-layer ferroelectric LB film was measured using an interferometric method as compared to the bulk P(VDF-TrFE) film ~−41 pm/V and pure PVDF film ~−26 pm/V. Further, a large coercive field of 1.2 ± 0.3 V/m has been observed which is approximately 20 times larger than a bulk counterpart [64, 67]. To a great degree, the increase in coercive field as the film dimension is lowered is explained by power law ($E \sim d^{-0.7}$, d is the thickness of the ferroelectric film) [68, 69]. Nevertheless, the advancement in characterization techniques for the nanostructures further decreases the dimensionality with stable ferroelectric state. Recently, the single monolayer (0.5 nm) of P(VDF-TrFE) LB film surprisingly exhibited the ferroelectric switching calculated theoretically by Fridkin [70] as shown in **Figure 2**. Earlier the near-absence of finite-size effect was reported for the P(VDF-TrFE) LB film as thin as 2 monolayer (~ 10 Å) crystalline film [10]. A schematic representation showing the polarization switching in 1 and 10 monolayers of P(VDF-TrFE) thin films is delineated in **Figure 2**. Hence, it is noteworthy that there is no critical size thickness for exhibiting ferroelectric switching phenomena in ferroelectric

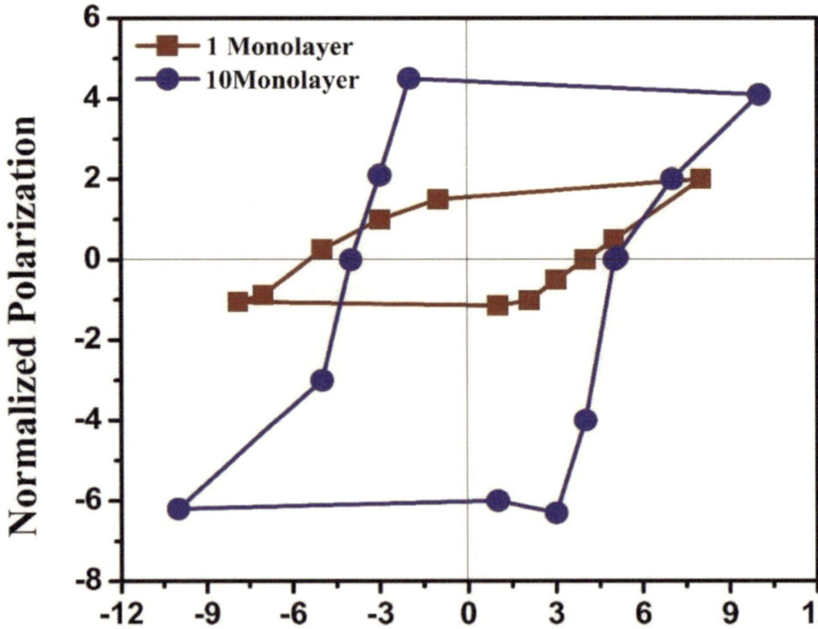

Figure 2.
The schematic representation of polarization switching in one and ten monolayer of P(VDF-TrFE) LB film (replotted taking the Ref. [10, 71]).

polymer P(VDF-TrFE) thin films. These outstanding results vitalized the search for the critical dimensions in other ferroelectrics.

6. Polarization switching kinetics for nanoscale ferroelectrics

Electric polarization is the first-order framework of ferroelectric transitions, whose non-zero value apprehends the ferroelectric phase from the paraelectric one. The phenomenon of macroscopic polarization reversal with the external field stress is termed polarization switching. The kinetics for the same was contrasting for the lower-dimensional system compared to its bulk counterpart. Ferroelectric materials including bulk ceramics, spin-coated epitaxial oxide thin film or the Langmuir–Blodgett polymer thin films, consist of widely distributed domains. Earlier studies have shown that polarization switching is a complex inhomogeneous phenomenon involving domain nucleation and growth. This process can be realized in terms of Kolmogorov–Avrami framework of inhomogeneous phase transformation, [72] where polarization is associated with the lower energy phase. At the macroscopic level, typically two frameworks have been observed in partially polarized ferroelectric materials: (a) the whole material may experience an identical polarization or (b) the presence of spatial inhomogeneous polarization. The second situation is practically observed in ferroelectric P(VDF-TrFE) thin films. Devonshire was the first scientist to develop a theory on polarization switching on barium titanate ceramics system based on Landau mean-field phase transition [73]. Later on, the theory was improved with the consideration of Ginzburg spatial inhomogeneity framework and termed as Landua–Ginzburg–Devonshire (LGD) theory [5, 74]. According to this theory, the free energy for macroscopic polarization which is considered as order parameter is expanded as Eq. (4).

$$F(P) = \frac{1}{2}\alpha p^2 + \frac{1}{2}\beta p^4 + \frac{1}{6}\gamma p^6 - EP \tag{4}$$

where α, β and γ are the Landau coefficients and E is the electric field within the ferroelectric material. The term EP of Eq. (1) defines the polarization alignment in the direction of the field to lower the free energy. The calculated P–E relation for P(VDF-TrFE) using Landau–Devonshire theory is shown in **Figure 3**.

In the computed P–E relation (**Figure 3**), the author theoretically explained an unstable region between point a and b and proposed that the polarization switching as a consequence of lowering the free energy of the system. Nevertheless, a gap always persists between the theoretical and experimental values. For example, the field for minimal polarization was computed in the order of magnitude in GV/m while it is typically 50 MV/m, as verified experimentally. The explanation of polarization switching based on nucleation and multidomain [75, 76], is labeled as *extrinsic switching*. This process involves the recasting of free energy of the crystal system due to the presence of sporadic dipolar defects, thereby lowering the energy barrier for local dipole reversal, thus creates a nucleation center for emerging ferroelectric switching domains. Likewise, the Monte-Carlo simulations unduly confirmed the non-collective polarization switching phenomenon mediated by the formation and development of domains as well.

However, the nanosized polymer ferroelectric P(VDF-TrFE) LB thin films (within the critical thickness) exhibited a critical behavior, a homogeneous non-domain

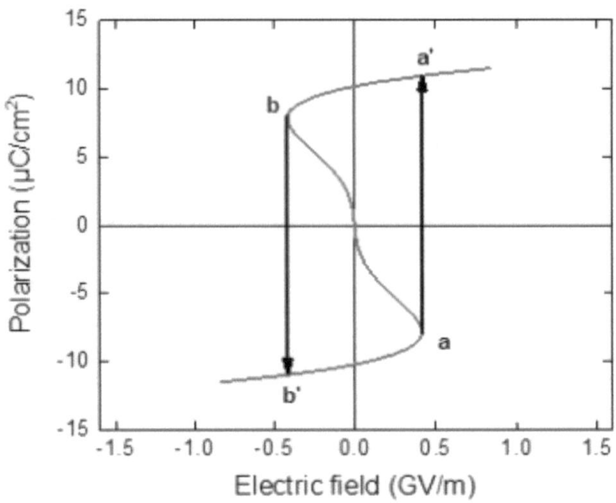

Figure 3.
The computed P-E relation for P(VDF-TrFE) using Landau-Devonshire theory [75, 76].

switching of polarization is observed [5]. Gaynutdinov et al. [71] demonstrated that polarization switching kinetics for 54 nm thick film of P(VDF-TrFE) copolymer were subjectively different from the 18 nm thick film. While bulk-like properties exhibited the nucleation and domain growth as the cause of polarization switching, 18 nm thick film exhibited purely intrinsic switching kinetics with a true threshold field. Vizdrik et al. [76] simulated the switching kinetics in P(VDF-TrFE) LB film with thickness of 30 monolayer. It was observed that the film experienced a pronounced slowing of polarization switching over six orders of magnitude in close proximity of coercive field which is distinct from the extrinsic switching that lacks true coercive field with increased field or temperature. The extrinsic switching is associated with the activation of nucleation and is a function of frequency. If the nucleation is non-existing, a very high coercive field is required to obtain the uniform polarization in ferroelectric crystal ideally, typically known as intrinsic switching and the associated threshold field is known as the intrinsic coercive field. Also, the intrinsic switching is not possible below the intrinsic coercive field as the constituent crystal dipoles are exceedingly harmonized and they tend to switch coherently or not at all. This type of switching is specifically observed in ultrathin P(VDF-TrFE) LB films. The reduced thickness of LB films apparently takes the edge off nucleation volume and therefore prohibits the occurrence of extrinsic switching. Notably, intrinsic switching process takes larger time (>1 s) as compared to extrinsic switching (works in microseconds) observed in thicker films and at lower field. Paramonova et al. [77] validated the intrinsic homogenous switching in PVDF/PVDF-TrFE Langmuir–Blodgett (LB) films using the molecular dynamic simulation method. Further, the intrinsic coercive field is independent of film thickness in PVDF-based LB film below ~15 nm, evincing the absence of finite size scaling below 15 nm [78, 79]. However, critical thickness for the intrinsic switching may vary in different polymer films because of diverse molecular structures. Theoretical modeling is a constructing way in guiding research for the dimensional effects in ferroelectricity. The nanoscale ferroelectrics constituted the switching kinetics contesting between extrinsic and intrinsic switching mechanism.

These mechanisms are associated with the film thickness, as the film thickness increases, domain mechanism carry the way, else the nucleation-independent switching mechanism is endured [80].

7. Summary and future outlook

Ferroelectrics with reduced dimension has exciting applications in modern electronics system, especially in medical engineering and material technologies [81]. The first challenge conveyed by nanoscale ferroelectrics for device application is the stability of ferroelectric properties at the desired ultralow-dimensional range. For the last few decades, tremendous effort, both theoretically and experimentally have been implied for finding stable ferroelectricity in nanoparticles at their maximum reduced dimensions. However, setting aside the academic cliché, the real scenario probably deals with the lacking of crucial steps toward the real-mass commercialization of nanoscale ferroelectrics. The science and technology of nano and ultra-nanoscale ferroelectrics is in infant stage. Numerous fundamental issues are still unsolved hampering the real-mass commercialization. It is expected that with the proper selection of material-system, minimizing intrinsic and extrinsic effects and the advancement in nanoscale characterization techniques, the possibility of scaling and size-effects could be minimized.

This chapter dealt with the ferroelectric phenomena emphasizing important functional parameters, such as phase transition temperature (T_c), polarization switching, coercive field (E_c), etc., taking the frame of reference of finite-size and scaling effect. The existence of critical dimensional range for ferroelectricity is limited by the experimental conditions, shape of the nanoparticles and the characterization techniques. Further, the theoretical analysis revealed that the rich set of complexities in the lower-dimensional scale of ferroelectrics were sensitively hung on structural, electrical and mechanical nature in their circumjacent. The pushing limit for perovskite ferroelectric crystal is as small as ~15 nm and the thinnest possible films were ~200 Å. Unlike nanoscaled ferroelectric ceramics system, the lower-dimensional polymer ferroelectric thin films are out of the way from the scaling effect. Langmuir–Blodgett deposition technique has produced high quality of ultrathin ferroelectric films of one monolayer thickness (~10 Å) of P(VDF-TrFE) ferroelectric polymer. Their long chain nature and the conformational variability countermanded the quantum confinement effect. This technique has opened a new frontier of finite-size effects on the atomic scale. Further, LB films also exhibited the two-dimensional properties of ferroelectrics by demonstrating that there is no supposed critical thickness in polymer ferroelectrics as films of only two monolayers (~1 nm) are ferroelectric with a transition temperature near that of the bulk material. However, the long-range cooperative ferroelectric interactions among dipoles are debilitated in otherwise customary ferroelectrics.

Acknowledgements

All authors gratefully acknowledge the financial support from the KIRAN Division, Ministry of Science and Technology, Department of Science and Technology (DST), Government of India through Project No. SR/WOS-A/PM-75/2018 (G) and Science and Engineering Research Board (SERB), Department of Science and Technology (DST), Government of India through Project No. EMR/2016/005281.

Conflict of interest

The authors declare that the research was conducted in the absence of any commercial or financial relationships that could be constructed as a potential conflict of interest.

Author details

Rolly Verma and Sanjeeb Kumar Rout*
Department of Physics, Birla Institute of Technology, Mesra, Ranchi, Jharkhand, India

*Address all correspondence to: skrout@bitmesra.ac.in

References

[1] Raj PM, Sharma H, Sitaraman S, Mishra D, Tummala R. System scaling with nanostructured power and RF components. Proceedings of the IEEE. 2017;**105**:2330-2346

[2] U. C. O. o. T. Assessment and U. S. C. O. o. T. Assessment. Miniaturization Technologies. Washington, DC: US Government Printing Office; 1991

[3] Ihlefeld JF, Harris DT, Keech R, Jones JL, Maria JP, Trolier-McKinstry S. Scaling effects in perovskite ferroelectrics: Fundamental limits and process-structure-property relations. Journal of the American Ceramic Society. 2016;**99**:2537-2557

[4] Palto S et al. Ferroelectric Langmuir–Blodgett films. Ferroelectrics Letters Section. 1995;**19**:65-68

[5] Bune A et al. Novel switching phenomena in ferroelectric Langmuir–Blodgett films. Applied Physics Letters. 1995;**67**:3975-3977

[6] Sorokin A, Palto S, Blinov L, Fridkin V, Yudin S. Ultrathin ferroelectric Langmuir-Blodgett films. Molecular Crystals and Liquid Crystals Science and Technology Section C-Molecular Materials. 1996;**6**:61-67

[7] Karasawa J, Sugiura M, Wada M, Hafid M, Fukami T. Ultra-thin lead titanate films prepared by tripole magnetron sputtering. Integrated Ferroelectrics. 1996;**12**:105-114

[8] Ishikawa K, Yoshikawa K, Okada N. Size effect on the ferroelectric phase transition in $PbTiO_3$ ultrafine particles. Physical Review B. 1988;**37**:5852

[9] Gao P et al. Possible absence of critical thickness and size effect in ultrathin perovskite ferroelectric films. Nature Communications. 2017;**8**:1-8

[10] Bune AV et al. Two-dimensional ferroelectric films. Nature. 1998;**391**: 874-877

[11] Waghmare UV. Theory of ferroelectricity and size effects in thin films. Thin Film Metal-Oxides: Springer. 2010:205-231

[12] Baettig P, Schelle CF, LeSar R, Waghmare UV, Spaldin NA. Theoretical prediction of new high-performance lead-free piezoelectrics. Chemistry of Materials. 2005;**17**:1376-1380

[13] Junquera J, Ghosez P. Critical thickness for ferroelectricity in perovskite ultrathin films. Nature. 2003; **422**:506-509

[14] A. Bratkovsky and A. Levanyuk, Abrupt appearance of the domain pattern and fatigue of thin ferroelectric films, in AIP Conference Proceedings, vol. 535(1): American Institute of Physics, 2000, pp. 218–228.

[15] Tadmor E, Waghmare U, Smith G, Kaxiras E. Polarization switching in PbTiO3: An ab initio finite element simulation. Acta Materialia. 2002;**50**: 2989-3002

[16] Nagarajan V et al. Size effects in ultrathin epitaxial ferroelectric heterostructures. Applied Physics Letters. 2004;**84**:5225-5227

[17] Buscaglia V, Randall CA. Size and scaling effects in barium titanate. An overview. Journal of the European Ceramic Society. 2020;**40**:3744-3758

[18] Shih WY, Shih W-H, Aksay IA. Size dependence of the ferroelectric

transition of small $BaTiO_3$ particles: Effect of depolarization. Physical Review B. 1994;**50**:15575

[19] Zhao Z et al. Grain-size effects on the ferroelectric behavior of dense nanocrystalline $BaTiO_3$ ceramics. Physical Review B. 2004;**70**:024107

[20] Naumov II, Bellaiche L, Fu H. Unusual phase transitions in ferroelectric nanodisks and nanorods. Nature. 2004; **432**:737-740

[21] Schilling A, Bowman R, Catalan G, Scott J, Gregg J. Morphological control of polar orientation in single-crystal ferroelectric nanowires. Nano Letters. 2007;**7**:3787-3791

[22] Naumov I, Bratkovsky AM. Unusual polarization patterns in flat epitaxial ferroelectric nanoparticles. Physical Review Letters. 2008;**101**:107601

[23] Szwarcman D, Vestler D, Markovich G. The size-dependent ferroelectric phase transition in $BaTiO_3$ nanocrystals probed by surface plasmons. ACS Nano. 2011;**5**:507-515

[24] Chattopadhyay S, Ayyub P, Palkar V, Multani M. Size-induced diffuse phase transition in the nanocrystalline ferroelectric $PbTiO_3$. Physical Review B. 1995;**52**:13177

[25] Almahmoud E, Kornev I, Bellaiche L. Critical behavior in ferroelectrics from first principles. Physical Review Letters. 2009;**102**:105701

[26] Geneste G, Bousquet E, Junquera J, Ghosez P. Finite-size effects in $BaTiO_3$ nanowires. Applied Physics Letters. 2006;**88**:112906

[27] Fong DD et al. Ferroelectricity in ultrathin perovskite films. Science. 2004; **304**:1650-1653

[28] Hao Y et al. Ferroelectric state and polarization switching behaviour of ultrafine $BaTiO_3$ nanoparticles with large-scale size uniformity. Journal of Materials Chemistry C. 2021;**9**:5267-5276

[29] Zhu D, Yan H, Tian S, Wang Z. Size, shape, and material effects in ferroelectric octahedral nanoparticles. Journal of Nanomaterials. 2021;**2021**

[30] Yan T, Shen Z-G, Zhang W-W, Chen J-F. Size dependence on the ferroelectric transition of nanosized $BaTiO_3$ particles. Materials Chemistry and Physics. 2006;**98**:450-455

[31] Nuraje N et al. Room temperature synthesis of ferroelectric barium titanate nanoparticles using peptide nanorings as templates. Advanced Materials. 2006;**18**: 807-811

[32] Xu R et al. Reducing coercive-field scaling in ferroelectric thin films via orientation control. ACS Nano. 2018;**12**: 4736-4743

[33] Mudinepalli VR, Feng L, Lin W-C, Murty B. Effect of grain size on dielectric and ferroelectric properties of nanostructured $Ba_{0.8}Sr_{0.2}TiO_3$ ceramics. Journal of Advanced Ceramics. 2015;**4**: 46-53

[34] Hong L, Soh A, Song Y, Lim L. Interface and surface effects on ferroelectric nano-thin films. Acta Materialia. 2008;**56**:2966-2974

[35] Koruza J, Groszewicz P, Breitzke H, Buntkowsky G, Rojac T, Malič B. Grain-size-induced ferroelectricity in $NaNbO_3$. Acta Materialia. 2017;**126**:77-85

[36] Branzi L, Back M, Cortelletti P, Pinna N, Benedetti A, Speghini A. Sodium niobate based hierarchical 3D perovskite nanoparticle clusters. Dalton Transactions. 2020;**49**:15195-15203

[37] Ayyub P, Chattopadhyay S, Pinto R, Multani M. Ferroelectric behavior in thin films of antiferroelectric materials. Physical Review B. 1998;**57**:R5559

[38] Zhao D et al. Depolarization of multidomain ferroelectric materials. Nature Communications. 2019;**10**:1-11

[39] Zhou Z, Wu D. Domain structures of ferroelectric films under different electrical boundary conditions. AIP Advances. 2015;**5**:107206

[40] Smith MB et al. Crystal structure and the paraelectric-to-ferroelectric phase transition of nanoscale $BaTiO_3$. Journal of the American Chemical Society. 2008;**130**:6955-6963

[41] Goswami AK. Theory on the effect of hydrostatic pressure on the permittivity and curie point of single crystal barium titanate. Journal of the Physical Society of Japan. 1966;**21**: 1037-1040

[42] Canedy C, Li H, Alpay S, Salamanca-Riba L, Roytburd A, Ramesh R. Dielectric properties in heteroepitaxial $Ba_{0.6}Sr_{0.4}TiO_3$ thin films: Effect of internal stresses and dislocation-type defects. Applied Physics Letters. 2000;**77**:1695-1697

[43] Polking MJ et al. Ferroelectric order in individual nanometre-scale crystals. Nature Materials. 2012;**11**:700-709

[44] Abolhasani MM, Azimi S, Fashandi H. Enhanced ferroelectric properties of electrospun poly (vinylidene fluoride) nanofibers by adjusting processing parameters. RSC Advances. 2015;**5**:61277-61283

[45] Wu T, Prasetya N, Li K. Recent advances in aluminium-based metal-organic frameworks (MOF) and its membrane applications. Journal of Membrane Science. 2020:118493

[46] Aktsipetrov O et al. Two-dimensional ferroelectricity and second harmonic generation in PVDF Langmuir–Blodgett films. Surface Science. 2000;**454**:1016-1020

[47] Pan M et al. Composite poly (vinylidene fluoride)/polystyrene latex particles for confined crystallization in 180 nm nanospheres via emulsifier-free batch seeded emulsion polymerization. Macromolecules. 2014;**47**:2632-2644

[48] Xiao Z et al. Synthesis and application of ferroelectric P(VDF-TrFE) nanoparticles in organic photovoltaic devices for high efficiency. Advanced Energy Materials. 2013;**3**: 1581-1588

[49] Chen X, Han X, Shen QD. PVDF-based ferroelectric polymers in modern flexible electronics. Advanced Electronic Materials. 2017;**3**:1600460

[50] Li M et al. Revisiting the δ-phase of poly(vinylidene fluoride) for solution-processed ferroelectric thin films. Nature Materials. 2013;**12**:433-438

[51] Chen S, Li X, Yao K, Tay FEH, Kumar A, Zeng K. Self-polarized ferroelectric PVDF homopolymer ultra-thin films derived from Langmuir–Blodgett deposition. Polymer. 2012;**53**: 1404-1408

[52] Yang L, Li X, Allahyarov E, Taylor PL, Zhang Q, Zhu L. Novel polymer ferroelectric behavior via crystal isomorphism and the nanoconfinement effect. Polymer. 2013;**54**:1709-1728

[53] Furukawa T. Ferroelectric properties of vinylidene fluoride copolymers. Phase Transitions: A Multinational Journal. 1989;**18**:143-211

[54] Fujisaki S, Ishiwara H, Fujisaki Y. Low-voltage operation of ferroelectric

poly(vinylidene fluoride-trifluoroethylene) copolymer capacitors and metal-ferroelectric-insulator-semiconductor diodes. Applied Physics Letters. 2007;**90**:162902

[55] Xu F et al. Facile preparation of highly oriented poly(vinylidene fluoride) uniform films and their ferro- and piezoelectric properties. RSC Advances. 2017;7:17038-17043

[56] Blinov L, Fridkin V, Palto S, Sorokin A, Yudin S. Ferroelectric polymer Langmuir films. Thin Solid Films. 1996;**284**:469-473

[57] Mohammadi B, Yousefi AA, Bellah SM. Effect of tensile strain rate and elongation on crystalline structure and piezoelectric properties of PVDF thin films. Polymer Testing. 2007;**26**:42-50

[58] Mandal D, Kim KJ, Lee JS. Simple synthesis of palladium nanoparticles, β-phase formation, and the control of chain and dipole orientations in palladium-doped poly(vinylidene fluoride) thin films. Langmuir. 2012;**28**:10310-10317

[59] Verma R, Rout S. Influence of annealing temperature on the existence of polar domain in uniaxially stretched polyvinylidene-co-hexafluoropropylene for energy harvesting applications. Journal of Applied Physics. 2020;**128**:234104

[60] Ducharme S, Reece TJ, Othon CM, Rannow RK. Ferroelectric polymer Langmuir–Blodgett films for nonvolatile memory applications. IEEE Transactions on Device and Materials Reliability. 2005;**5**:720-735

[61] Yin Z, Tian B, Zhu Q, Duan C. Characterization and application of PVDF and its copolymer films prepared by spin-coating and Langmuir–Blodgett method. Polymers. 2019;**11**:2033

[62] Kausar A. Survey on Langmuir–Blodgett films of polymer and polymeric composite. Polymer-Plastics Technology and Engineering. 2017;**56**:932-945

[63] Zhu H, Miyashita T, Mitsuishi M. Energy storage behaviors in ferroelectric capacitors fabricated with sub-50 nm poly(vinylidene fluoride) Langmuir–Blodgett nanofilms. Polymer Journal. 2019;**51**:795-801

[64] Ducharme S et al. Ultrathin ferroelectric polymer films. Ferroelectrics. 1997;**202**:29-37

[65] Scott J. Phase transitions in ferroelectric thin films. Phase Transitions. 1991;**30**:107-110

[66] Zhu H, Yamamoto S, Matsui J, Miyashita T, Mitsuishi M. Highly oriented poly (vinylidene fluoride-co-trifluoroethylene) ultrathin films with improved ferroelectricity. RSC Advances. 2016;**6**:32007-32012

[67] Chandra P, Dawber M, Littlewood P, Scott J. Scaling of the coercive field with thickness in thin-film ferroelectrics. Ferroelectrics. 2004;**313**:7-13

[68] Scott J. Models for the frequency dependence of coercive field and the size dependence of remanent polarization in ferroelectric thin films. Integrated Ferroelectrics. 1996;**12**:71-81

[69] Jung D, Dawber M, Scott J, Sinnamon L, Gregg J. Switching dynamics in ferroelectric thin films: An experimental survey. Integrated Ferroelectrics. 2002;**48**:59-68

[70] Fridkin V, Ievlev A, Verkhovskaya K, Vizdrik G, Yudin S, Ducharme S. Switching in one

monolayer of the ferroelectric polymer. Ferroelectrics. 2005;**314**:37-40

[71] Gaynutdinov R, Mitko S, Yudin S, Fridkin V, Ducharme S. Polarization switching at the nanoscale in ferroelectric copolymer thin films. Applied Physics Letters. 2011;**99**: 142904

[72] Mai M, Liu G, Zhu C, Ma X. Switching current characteristics of ferroelectric P(VDF-TrFE) copolymer films. Ferroelectrics. 2019;**550**:127-135

[73] Devonshire AF, XCVI. Theory of barium titanate: Part I. The London, Edinburgh, and Dublin Philosophical Magazine and Journal of Science. 1949; **40**:1040-1063

[74] Jiang B, Bursill L. Phenomenological theory of size effects in ultrafine ferroelectric particles of lead titanate. Physical Review B. 1999;**60**: 9978

[75] Tagantsev AK, Cross LE, Fousek J. Domains in Ferroic Crystals and Thin Films. New York, NY: Springer; 2010

[76] Vizdrik G, Ducharme S, Fridkin V, Yudin S. Kinetics of ferroelectric switching in ultrathin films. Physical Review B. 2003;**68**:094113

[77] Paramonova E et al. Polarization switching in ultrathin polyvinylidene fluoride homopolymer ferroelectric films. Ferroelectrics. 2017;**509**:143-157

[78] Ducharme S et al. Intrinsic ferroelectric coercive field. Physical Review Letters. 2000;**84**:175

[79] Li M, Katsouras I, Asadi K, Blom PW, De Leeuw DM. Low voltage extrinsic switching of ferroelectric δ-PVDF ultra-thin films. Applied Physics Letters. 2013;**103**:072903

[80] Wang J et al. Transition of the polarization switching from extrinsic to intrinsic in the ultrathin polyvinylidene fluoride homopolymer films. Applied Physics Letters. 2014;**104**:182907

[81] Garbovskiy Y, Zribi O, Glushchenko A. Emerging applications of ferroelectric nanoparticles in materials technologies, biology and medicine. Advances in Ferroelectrics. 2012:475-498

Chapter 16

Perovskites in Next Generation Memory Devices

Gregory Thien Soon How, Mohd Arif Mohd Sarjidan, Boon Tong Goh, Boon Kar Yap and Eyas Mahmoud

Abstract

Although perovskites are widely employed in other industries such as photovoltaics and light-emitting diodes (LEDs), digital technology is rapidly gaining pace in today's market and shows no signs of abating. As a result, the progress of system memory and memory storage has accelerated into new inventions. The invention of dynamic Random-Access Memory (RAM) in the 1960s laid the groundwork for today's multibillion-dollar memory technology sector. Resistive switching (RS) capabilities of perovskite-based materials such as perovskite oxides and metal halides have been extensively studied. Chemical stability, high endurance, quick writing speed, and strong electronic interaction correlation are some of the benefits of employing perovskites in RS devices. This chapter will investigate the progress of system memory and memory storage employing perovskites, the advantageous properties of perovskites utilized in memory devices, the various types of RS employing perovskites, as well as the research challenges that perovskite-based memory systems face in future commercial development.

Keywords: resistive switching, memory devices, memory storage, perovskites, non-volatile

1. Introduction

Perovskites, which have the same crystal structure as calcium titanium oxide ($CaTiO_3$), are semiconductor materials that have gained massive interest in various technology. When exposed to light, the structure and characteristics of these materials allow them to transfer electric change. These materials are very beneficial for system memory and memory storage in computer memory. Perovskites' resistive switching (RS) properties enable fast writing speed and long durability. Thus, RS is employed in the most recent computer memory technology, Resistive Random Access Memory (ReRAM), which is expected to replace flash memory [1, 2].

In memory devices, there are many forms of RS. Bipolar and unipolar switching, write-once-read-many (WORM) [3], and multilevel RS [4] are examples of these. This chapter will discuss the advancements in the use of perovskites for RS memory. Oxide perovskites, halide perovskites, and layered perovskites are some of the perovskite materials employed in this application. Lastly, some design challenges are discussed, and future work is proposed.

2. System memory and memory storage technology

Digital technology is being employed extensively in today's economy, and it shows no signs of abating. As a result, the progress of system memory and memory storage has accelerated, resulting in new improvements. The system memory is where the computer stores currently running applications and data. On the other hand, memory storage is generally for the goal of orderly retrieval and documentation. **Figure 1** depicts the chronology of the evolution of system memory and memory stage.

In 1932, Gustav Tauchek invented the drum memory technology which marked the beginning of system memory [5, 6]. The drum memory was cylindrical in form, with an outside covering comprised of recordable ferromagnetic elements, and it could store up to 500,000 bits, or 62.5 kilobytes of memory [7]. Eventually, in the mid-1940s, the delay line memory was found, which was a refreshable memory that used sequential access and was constructed of mercury. This unique mercury delay line was capable of transmitting data at a rate of around 5,000,000 binary digits per second. It was not until World War II that the United States Navy adopted the initial drum memory idea and refined it into the magnetic drum memory system. The magnetic core memory was then constructed using tiny toroidal ferrimagnetic ceramic ferrites. The memory was stored via an induced magnetic field, which could store one bit depending on the magnetization direction [8]. Twister memory, which used magnetic tape instead of rings to replace core memory, was introduced in 1968 at Bells Lab but received little attention [9]. The magnetic tape was intentionally chosen to enable magnetization only down the length of the tape. As a result, only one point of the twistor would have the proper field direction to ever get magnetized. On the other hand, bubble memory which is a sort of non-volatile memory employs a small layer

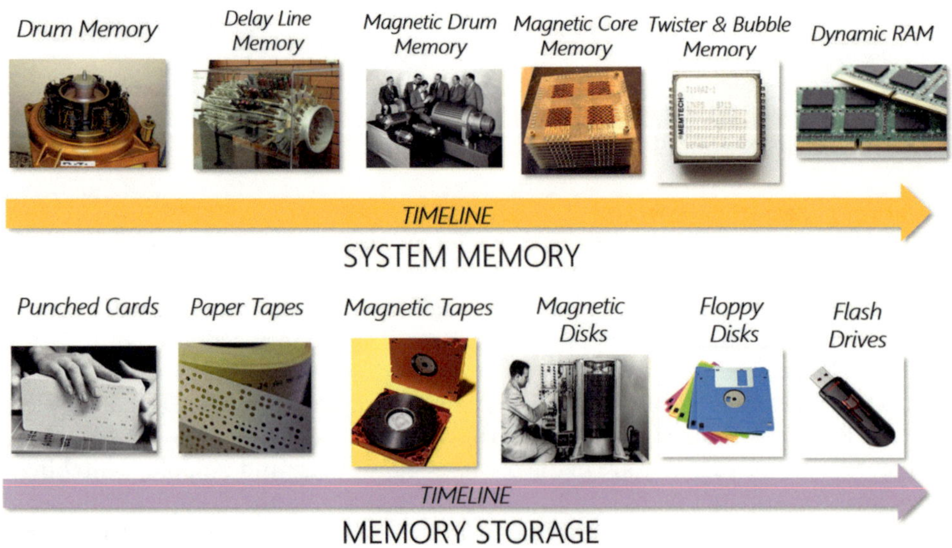

Figure 1.
A timeline depicting the progress of system memory and memory storage over time.

of magnetic material in its fabrication due to the influence of an external magnetic field. This contains little magnetized patches known as bubbles or domains, each of which may retain one bit of data [10, 11]. Similarly, bubble memory also suffered the same fate as twister memory since both were eclipsed by the development of dynamic RAM.

The invention of dynamic Random-Access Memory (RAM) in the 1960s laid the groundwork for today's multibillion-dollar memory technology sector. Every sort of memory technology described above is rendered obsolete by the discovery of RAM. The earliest architecture of dynamic RAM was a square array with a capacitor and a transistor for each data bit [12]. In today's technology, a broad range of RAM technologies have been researched for their commercialization potential. The advancement of memory storage technology began in 1976 with the usage of punched cards, with certain holes on them as a set of instructions for digital programs [13]. It was then refined further into punch tapes or paper tapes. Similarly, paper tapes were developed to replace punched cards, which were considerably more convenient since they provided a continuous set of data or instructions without the need to insert punched cards one at a time. The substance used to create the paper tapes was then altered and replaced with magnetic materials.

Magnetic tapes were significantly easier to use and could contain far more data than paper tapes. This has transformed the broadcasting industry by allowing live broadcasts to be recorded and replayed at any time [5]. It was not until 1969 that memory storage technology was substantially influenced by the invention of magnetic discs, which can store databases and vast volumes of data. As a result, floppy discs were inspired by magnetic discs, which were portable and generally available to the public. Flash drives, which are based on Erasable Programmable Read Only Memory (EPROM) and Electrically Erasable Programmable Read Only Memory (EEPROM) technologies, are no longer rare in today's technology [14, 15]. Future predictions for system memory and memory storage technologies have focused on a few advancements, including the ReRAM technology, which was projected to replace flash memory.

3. Resistive switching in memory devices

3.1 Bipolar and unipolar switching

The rapid switching speed, lower power consumption, and excellent scalability, ReRAM has emerged as the most encouraging choice for the future use of non-volatile devices [1, 2]. A dual terminal ReRAM device has an insulating layer wedged between two conducting electrodes. An external electric field can cause the memory cell to flip between two resistance states, known as the low resistance state (LRS, ON state) and high resistance (HRS, OFF state) state [16]. ReRAM is divided into two switching modes: unipolar and bipolar. The polarity of the switching voltage is inconsequential in unipolar switching, whereas in bipolar switching, the electrical polarity required to change from an HRS to an LRS is the inverse of that required to switch from an LRS to an HRS. Switching materials such as organics, binary oxides, and perovskite oxides, have been studied to achieve ReRAM with reliable switching, high ON/OFF current ratio, and long retention period [17–19].

Non-volatile memory (NVM) is a classification of computer memory that cannot be deleted or removed even when the power is turned off. In today's technology, NVMs are typically utilized for long-term data storage or secondary storage. Flash drives, magnetic storage devices, ferroelectric RAM, magnetoresistive RAM, ReRAM, and optical discs are examples of NVM devices. In today's storage technology, NAND (Boolean operator and logic gate) flash memory are widely used in most goods. However, the demand for faster writing speeds, higher density, and lower cost drives new research into developing technologies such as ReRAM. In the late 1990s and early 2000s, researchers investigated ReRAM technology, which allows RS between two resistance states to be exchanged via a thin film layer [20]. It works on the principle of applying a high voltage across a dielectric to transition from insulating to conductive qualities via a conduction filament pathway. Classification of ReRAM mechanism as memristors is controversial [21]. Theoretically, their RS curves differ from each other. **Figure 2** shows a typical current-voltage (*I-V*) characteristic obtained for a ReRAM device [2]. Such classifications of ReRAM and memristors are still a source of contention today.

The memristor is the fourth fundamental circuit element identified, after resistors, capacitors, and inductors. Chua et al. suggested the notion of memristor in 1971 [22], and it was validated in 2008. There are several materials having memristive characteristics which have been discovered since 1962. Memristors, like resistance in Ohms, are defined as:

$$M = \frac{d\varphi}{dq} \tag{1}$$

where φ is magnetic flux and *q* is electric charge. As observed in its theoretical memristor curves, memristor curves are typically non-linear devices. Real-world memristors, on the other hand, have a comparable *I-V* characteristic to ReRAM. Some researchers hypothesized that such behavior was caused by conducting filament [23], active memristor [24], and non-zero crossing [25]. ReRAM curves, on the other hand, can be classified as bipolar or unipolar switching. To generate the RS curve, unipolar switching uses the same polarity of the swept bias with variable magnitudes, but bipolar switching requires various polarities [26]. ReRAM devices, in general, consist of an insulator layer sandwiched between two electrodes. Conduction pathways were formed by the flow of charge carriers alternating between the cathode and anode,

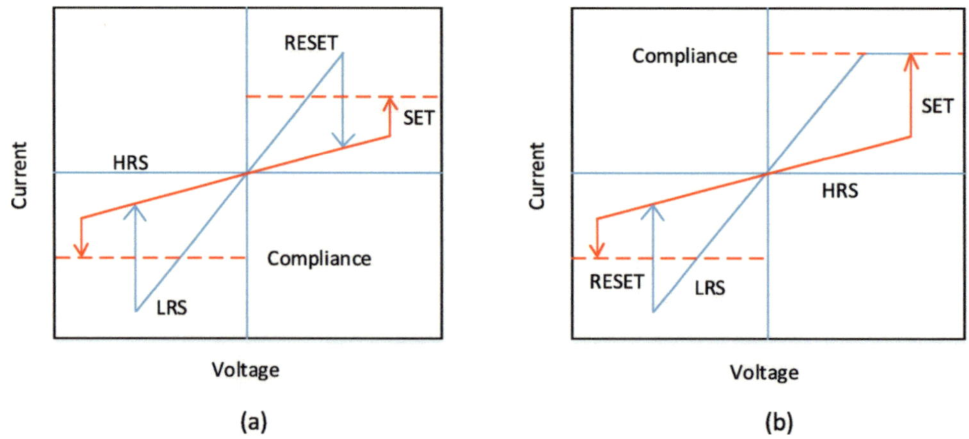

Figure 2.
*(a) The predicted and typical types of (a) unipolar switching and (b) bipolar switching ReRAM curves. Adapted from **Figure 4** [2].*

which was induced by several physical mechanisms that are still commonly utilized today. Furthermore, the conduction methods differ depending on the materials utilized and the device's overall architecture. Szot et al. were the first to detect conductive filament using an electron microscope, which they attributed to filament build-up and rupture [27].

3.2 Write-once-read-many (WORM)

WORM devices, in general, are devices in which data that has been written cannot be manipulated or removed. WORM memory is used in fields such as healthcare, security, taxation, and accounting where the data cannot be tampered with or updated to secure information. The most prevalent WORM devices in use today are the Compact Disc Recordable (CD-R) and Digital Versatile Disc Recordable (DVD-R). Furthermore, the "read-many" element implies that the device's data can be read an infinite number of times, with the only limitation being the device lifetime. The WORM memory pixels are read according to the rows, with an unwritten pixel labeled as logical "0" and a written pixel labeled as logical "1" [3].

Today's WORM memory technology is based on the electrically or laser programmed fuse WORM type. However, in current WORM research, the emphasis was shifted to organic materials or solution process techniques in order to achieve rapid switching rate, lower power consumption,, large storage density, simplicity, and being cost-effective, which has been dubbed WORM RS [28, 29]. Unlike ReRAM, the intrinsic features of WORM RS were sufficient to oppose the applied electric field, resulting in an irreversible shift. A typical WORM RS *I-V* curve obtained when a voltage is supplied is shown in **Figure 3** [30]. The change between resistance states (OFF to ON) is an irreversible process, indicating WORM features.

3.3 Multilevel resistive switching

The potential of multilayer RS effect has been documented in several inorganic materials due to its superior memory performance, but the difficult fabrication

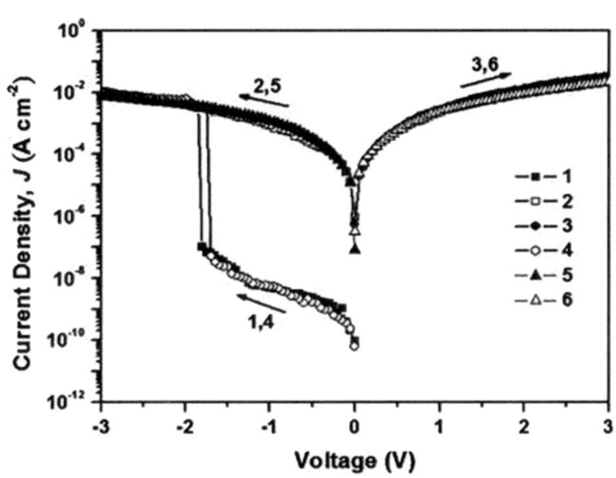

Figure 3.
A typical WORM curve transiting from OFF to ON process. Adapted from ***Figure 3*** *[30].*

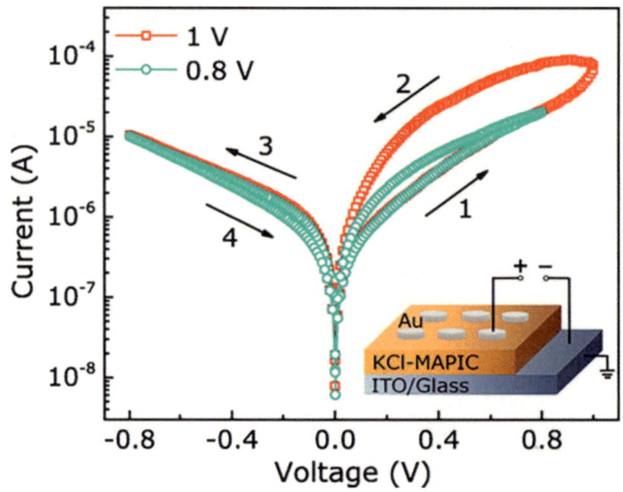

Figure 4.
I-V curves of Au/KCl-MAPIC/ITO/glass indicating two V_{SETs} of 0.8 V and 1.0 V. Adapted from ***figure 3*** *[36].*

method and stiffness restrict the development for ReRAM [31–33]. Organometal halide perovskites (OHPs) have recently sparked a lot of interest in the ReRAM community because to the high flexibility, variable band gaps, huge absorption coefficients, and long electron-hole diffusion length [34, 35]. Furthermore, OHPs highly feature defect-tolerant, simple, and cost-effective solution-processed procedures for fabricating the OHPs layers. As a result, they envision it being used in multilevel RS, which is advantageous for multilevel data storage. Typically, these devices feature numerous resistance states that can be changed within the device. **Figure 4** shows a typical multilevel perovskite memory consisting of a all-inorganic $CsPb_{1-x}Bi_xI_3$ perovskite film [36]. This reveal that multilevel RS was accomplished by altering the reset stop voltages.

4. Recent advancements of perovskites in memory devices

4.1 Oxide perovskites in memory devices

In recent oxide perovskites based works, a $Ag/BaTiO_3/Nb{:}SrTiO_3$ ferroelectric tunnel junction (FTJ) with the quickest switching speed of 600 *ps* has been constructed [37]. When the sub-nanosecond switching action is maintained at 112.85°C, the device exhibits great temperature resilience. In addition, the gadget established 32 states or 5 bits of states for each cell, which is regarded to be the highest in the class. The combination of a high carrier concentration in the $Nb{:}SrTiO_3$ electrode and the low work function of silver metal has resulted in significant increases in operation speed with a low current density of 4×10^3 A cm^{-2}. The device performance can be seen through the rapid resistance switching at normal temperature depicted in **Figure 5.**

The advancement of fabrication for $BaTiO_3$ has been established by growing this material on pre-deposited $SrTiO_3$ substrate using the epitaxy technique to form a free-standing film. Then, they are transferred onto a silicon substrate for integration into complementary metal-oxide-semiconductor (CMOS) devices [38]. Not only

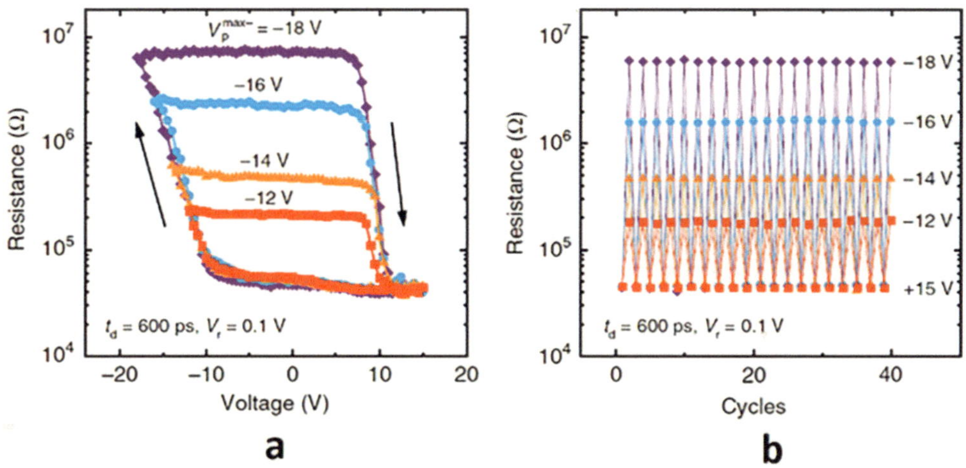

Figure 5.
(a) Graph of resistance vs. driven voltage (V_d), and (b) common resistance switching between ON state and various high-resistance states at pulse duration, t_d = 600 ps. Adapted from ***Figure 2*** *[37] .*

the performance of the device is comparable with previously reported work [39], but it also exhibits a large ON state current with good bias voltage measurement of 0.2 V. This enables a non-destructive readout during the operation process. With an optimal depiction of $SrRuO_3/BaTiO_3$ dual buffer layers, highly structured $BiFeO_3$ films were prepared oriented on flexible mica substrates [40]. The $BiFeO_3$ films possess a stable ferroelectric polarization with over 100 bending cycles at a 5 mm radius. Therefore, it results in the continuously controllable resistance memristor characteristics which suggests the feasibility for solid synaptic devices. Moreover, using interactive supervised learning, the handwritten digits reveals excellent recognition accuracy valued at 90% in artificial neural network simulations which displays the potential for flexible ferroelectric memristors in wearable devices (data storage and computation). Ferroionic tunnel junctions have been suggested to make a huge electroresistance in ReRAM based $BaTiO_3$ [41]. In low-resistance states, it works as a ferroelectric tunnel junction and as a Schottky junction which is due to changes within the interface caused by a field. This device significantly employs the ferroelectric barrier ($BaTiO_3$) and Nb-doped $SrTiO_3$ as the bottom electrode. The giant electroresistance result in ON/OFF ratios of 5.1×10^7 and 2.1×10^9 at room temperature and 10 K, respectively. The movement of oxygen vacancies from polarization reversal caused by the bias voltage could significantly alter the dimension of the interface barriers [42].

The ferroelectric manipulation of spin-filtering $BaTiO_3/CoFe_2O_4$ composite barriers has been demonstrated in multiferroic FTJ synapses [43]. By manipulating the polarization switching of $BaTiO_3$, it is possible to establish long-term memory and a constant conductance change achieving a 544,400% ON/OFF current ratio. On a crossbar neural network, supervised learning simulations applies the Spike-Timing Dependent Plasticity (STDP) outcomes as a database for weight training which achieved recognition accuracy rates above 97%. As a result, there is an approximately 10-fold shift in tunneling magnetoresistance ratio including a turnaround relying on the resistance state of the electrodes when the polarization is switched. A novel approach to multiferroic neuromorphic devices with energy-saving electrical exploitation is provided by these studies, notably the switchable spin polarization.

Additional FTJ device options include ferroelectric oxide-grown spinel ferrite barriers, which open a wider range of potential applications.

On the other hand, $SrTiO_3$ has been used to develop the Pt/CeO_2/Nb: $SrTiO_3$ heterostructure which demonstrates outstanding memory behavior with a highest RS ratio of 3×10^4 [44]. Under the irradiation of an ultraviolet 405 nm laser beam, an obvious photoresponse was detected, which also corresponds to substantial switching characteristics in a high resistance state. This device demonstrates light-manipulated RS and voltage dependent photoresponse, which are two different types of RS. The RS and photoresponse features were contributed from the Schottky barrier at the Pt/CeO_2 interface, as well as the electron trapping, and de-trapping caused by oxygen vacancies at the interface. An ultrathin (6.2 nm) ferroelectric $La_{0.1}Bi_{0.9}FeO_3$(LBFO) layer has been introduced on a 0.7 wt% Nb-doped $SrTiO_3$ (001) single-crystal substrate to form a Pt/$La_{0.1}Bi_{0.9}FeO_3$/Nb-doped $SrTiO_3$ heterostructure [45]. The ferroelectricity of the LBFO film was extremely high, but the coercive field was extremely low. By adjusting the thickness of the LBFO film, it was possible to produce a high resistance OFF/ON ratio of up to 2.8×10^5 for the Pt/LBFO (6.2 nm)/NSTO heterostructure. Moreover, the heterostructure exhibited multi-level storage and outstanding retention properties, as well as steady bipolar resistance switching behavior, which is suitable for application in ferroelectric memristors. On the $LaBiFeO_3$/Nb-doped $SrTiO_3$ interface, the resistance switching behavior has been demonstrated to be caused by a modulating impact of ferroelectric polarization turnaround on both the breadth of the depletion area and the potential barrier's heigh.

Using pulsed-laser deposition (PLD) technology, epitaxial $BiFeO_3$ (BFO) thin films were fabricated to produce the robust in-plane domain dynamic process created when applied under the influence of external electric fields [46]. It has also been noticed that the retention and repeatability are good, particularly at high temperatures. Besides, by forming a heterostructure consisting of indium tin oxide (ITO), BFO, and strontium ruthenium oxide (SRO), an optically triggered non-volatile memory has been demonstrated [47]. In comparison to traditional devices, in which optical excitations often increase conductivity, the constructed structure demonstrates a significant drop in conductivity (1×10^{-4}) following laser illumination at wavelengths 405, 532, and 1064 nm, indicating that the device is poorly conductive. Additionally, optical stimuli may be used to reset the negative optoelectronic memory, and an electrical pulse could be used to establish the memory. It was discovered that this property could be inhibited by annealing in an oxygen-rich environment, but that it could be restored by annealing in an oxygen-depleted atmosphere. Based on investigations of the transport and dielectric characteristics, it has been determined that the optical/electrical RS behavior observed at the ITO/BFO interface is caused by the potential profile's modulation at the ITO/BFO interface caused by optical and electrical excitations.

Fundamental research into the reversible topotactic phase change between the insulating brownmillerite (BM) phase and the conducting perovskite structure is critical for the creation of RS memories. Using $SrFeO_x$ as a model, the system demonstrated that in the ON state, $SrFeO_3$ nanofilaments are produced and stretch essentially through the BM $SrFeO_{2.5}$ matrix, and that in the OFF state, they are ruptured, indicating indisputably the presence of a filamentary RS process [48]. The nanofilaments are roughly 10 nm in diameter, permitting for the first time the downscaling of Au/$SrFeO_x$/$SrRuO_3$ RS devices to the 100 nm range. They have exceptional performance, with an ON/OFF ratio of up to 10^4, a retention time of more than 105 s, and an endurance of up to 10^7 cycles, among other things.

On the other hand, an investigation into the process of irreversible RS conversion from bipolar to unipolar process is conducted in a capacitor model composed of $SrZrO_3$, TiO_x/Pt, which has been produced on a substrate composed of Pt, Ti, SiO_2, and SiO_2 [49]. The *I-V* properties of bipolar RS memory were seen in the RS operating voltage range spanning from +2.5 V to −1.9 V, and the memory exhibits outstanding durability and retention characteristics. An additional forming step happens when the voltage bias is raised above +4 V, irrevocably changing the RS mode from bipolar to unipolar. This process occurs as the voltage bias is raised above +4 V. In this study, two materials are combined with two different switching processes to create an RS memory with acceptable properties in various current regions. Furthermore, a bipolar RS characteristic was investigated in a Pt/$LaNiO_3$/Nb: $SrZrO_3$/Cu structure, where the Cu and LNO layers function as capping and buffering layers, respectively [50]. It was feasible to achieve a high endurance performance for the bistable bipolar switching characteristic at room temperature, which was measured up to 1.2103 times. The Cu layer was utilized as a reservoir layer to change the distribution of oxygen vacancies and traps inside the films, resulting in a steady RS response, low operation voltage, and extended retention duration.

A nanocomposite consisting of $La_{0.7}Sr_{0.3}MnO_3$ (LSMO) and reduced graphene oxide (rGO) has been identified as a viable option for non-volatile memory applications in oxide electronic devices [51]. Individual component phases were identified because of the structural characterization process. A minimal switching speed of 1.1 μs and a relatively steady switch mechanism over 1000 switching cycles were obtained from the device. The switching behavior is also shown to be resilient against variable voltage sweeping rates. It is highlighted that the transportation of the oxygen ions at the SET and RESET voltages result in alterations of the resistance states due to the conduction filaments forming or rupturing. LSMO is also combined with BTO to form a heterostructure of non-volatile and reversible RS [52]. It was revealed that altering the electric field orientation caused variations in the LSMO layer's resistivity and metal-insulator transition temperature (TMI). When the BTO layer is subjected to a negative electric field, the resistivity for the accumulation state of hole carriers drops while the TMI for the accumulation state of hole carriers rises. When a positive electric field is applied to the BTO layer, the resistivity rises while the TMI falls for the hole carrier depletion condition.

4.2 Halide perovskites in memory devices

An organo-metal halide source has been shown to produce perovskite layers through one-step spin-coating technique for the construction of unipolar RS devices in a cross-bar array design utilizing a simple one-step spin-coating procedure [53]. With gold as the electrodes, these unipolar perovskite RS devices attain a high ON/OFF ratio of up to 10^8 while operating at a small operating voltage, with high stability, over 1000 writing cycles, and retention over 10^4 s. The memory devices were successfully incorporated into an 8 8 crossbar array design with up to 94 percent yield. Furthermore, as shown in **Figure 6**, the 1D-1R system of selective activation of memory cells was shown by eliminating crosstalk interference between nearby cells linked by external diodes.

An active layer for resistive memory has been developed using $CsPbBr_3$ single-crystal film (SCF) [54]. With an extremely high switching ratio of above 10^9 and a rapid switching speed of 1.8 s, the Ag/$CsPbBr_3$/Ag memory cells demonstrate repeatable RS. A large interface contact is produced when the metal/$CsPbBr_3$ SCF contact

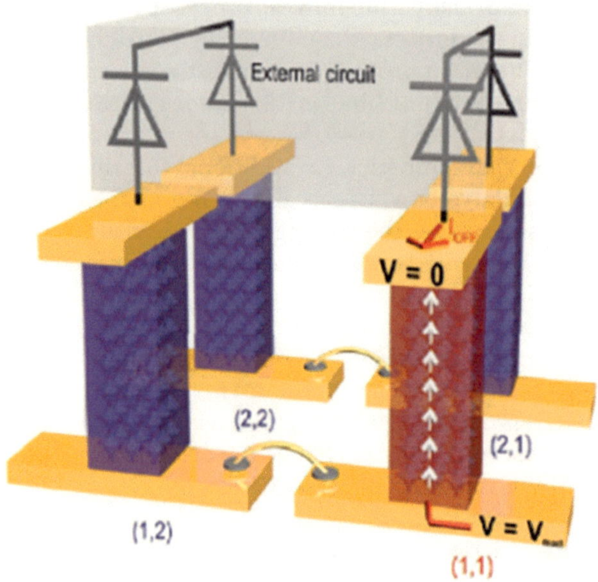

Figure 6.
A diagram of the identical reading procedure in a 2 × 2 array in a 1D-1R architecture with memory cells connected by external diodes. Adapted from Figure 4 [53].

has an interface S value of 0.50, indicating that a large interface contact has been established. Because of the high RS ratio at the interface, the high interface contact leads to the stable high resistance state (HRS), and the steady HRS leads to an ultra-high RS ratio. Besides, it has been demonstrated that the use of vacancy defects in lead halide perovskite structures may create excellent performance nano floating gate memory (NFGMs) [55]. A CdS nanoribbon (NR) surface was evenly covered with $CH_3NH_3PbBr_3$ nanocrystals (NCs) using a simple dip-coating procedure, resulting in a core-shell structure composed of CdS NR/$CH_3NH_3PbBr_3$ NC cores and shells. It is noteworthy that the device exhibited a very large memory window of up to 77.4 V and a long retention time of 12,000 s, as well as a high current ON/OFF ratio of 7×10^7 and long-term air stability for 50 days, all of which were attributed to the presence of sufficient carrier trapping states in $CH_3NH_3PbBr_3$ NCs.

On the other hand, a $CH_3NH_3PbI_{3-x}Cl_x$/FTO RS device structure has been proven to retain information in dual levels of resistance states generated by electrical probe stimulation [56]. The device with the silver probe demonstrates bipolar RS behavior after formation, with a 10^6 ON/OFF resistance ratio, showing that it is bipolar RS. The constructed probe-based memory cell has a minimum endurance of 10^4 cycles and a minimal retention length of 2×10^3 s, which are both good performance features. Thus, organic-inorganic lead halide perovskite (OILHP) materials are highly proposed as a feasible candidate for usage as a storage layer for probe-based storage memories. Thin polyethyleneimine (PEI) interfacial layers have been introduced between the layers to avoid direct contact between the perovskite layer and the top and bottom electrodes, resulting in a device structure consisting of ITO/PEI/$CH_3NH_3PbI_3$/PEI/metal [57]. This device can reach over 4000 durability cycles while maintaining a low operating voltage of around 0.25 V. Aside from that, the repeatability of memory switching behavior was proven across 180 devices manufactured using eight different device batching settings.

Recently, 2D/3D perovskite heterostructure films consisting of 2D perovskite (phenethylammonium lead iodide, PEA_2PbI_4)/3D perovskite ($MAPbI_3$) have been produced using a low-temperature all-solution technique [58]. The integration of 2D and 3D perovskite RS memories displayed remarkable efficiency, with an overall durability of 2700 cycles, an ON/OFF ratio of more than 10^6, and a reaction time of 640 μs. Both the expected activation energy for thermally aided ion hopping and the time-of-flight secondary ion mass spectrometry data suggested that the 2D perovskite layer effectively blocked Ag ion migration through into the 3D perovskite film. By putting n-butylammonium iodide above $CH_3NH_3PbI_{3-x}Cl_x$ ($MAPbI_{3-x}Cl_x$), another 2D/3D memory device was demonstrated [59]. The perovskite film is made in a single step by heating molten salt methylammonium acetate to room temperature and spinning it in the air. When compared to their 3D counterparts, RS memory devices with a 2D/3D perovskite heterostructure provide a significantly better switching window with an ON/OFF ratio of more than 10^3 while needing a lower operating voltage. The 2D/3D perovskite heterostructure is advantageous for manufacturing uniform-crystalline-grain, highly compact structures, and it can passivate defect states for the $MAPbI_{3-x}Cl_x$ film and interface, resulting in improved memory properties for both the film and the interface.

In contrast, the cube of $CsPbX_3$ was used in an Al/$CsPbCl_xBr_x$ (x = 3, 1.5, 0)/ITO/PET memory device that demonstrated a bipolar RS pattern at a low working voltage [60]. When compared to all other memory devices developed, the $CsPbBr_3$-based system has the most obvious RS qualities, such as the lack of an initial forming procedure, reproducibility, uniform switching, and a long retention period with a high ON/OFF ratio. The multilayer data storage potential of flexible memory devices may also be evaluated by making minor changes to current compliance and stopping voltage.

Recently, flexible wearable electronic materials and fiber-shaped resistive random access memory based on $MAPbI_3$, an organic-inorganic halide perovskite semiconductor, have been created utilizing a simple and cost-effective cheap deep coating process [61]. After refining the manufacturing settings, a well-arranged pinhole-free layer was coated on the aluminum fiber. The device has a bipolar RS feature with a roughly 10^6 ON/OFF ratio, a low working voltage, and a retention duration of more than 10^4 s. More crucially, the switching mechanism has remained nearly unchanged with up to a 45° bending angle.

In addition, our previous study investigated increasing the molar ratio of Pb to Ti by 5% in the $MAPbI_3$-TiO_2 layer, which increased V_{SET} and V_{RESET} values to 3.6 V and 1.1 V, respectively [59]. This study is regarded as a forerunner in understanding the use of a single layer $MAPbI_3$-TiO_2 in several applications. In the meanwhile, $Cs_3Sb_2I_9$ inorganic halide perovskite has been developed as a lead-free source of high ReRAM and artificial synaptic devices [62]. A vapor-assisted solution technique was used to create this 2D perovskite (VASP). The memristive devices not only feature reproducible bipolar RS with a massive ON/OFF ratio of 10_4 at a low working voltage of 0.4 V, but they also have superb retention over 10_4 s and extraordinary resistance to environmental degradation. The ReRAM devices show promise to produce phototunable memories and artificial synaptic devices with the capacity to perform concurrent processing and learning due to high light-matter interaction in the perovskite and an intrinsic electronic–ionic connection.

4.3 Layered perovskites in memory devices

Dion-Jacobson organic-inorganic halide perovskite (OIHP) has been proven as a resistive switching memory (RSM), with grain size varying to improve grain boundaries [63]. By adjusting the ratio of *N, N*-dimethylformamide to dimethyl sulfoxide in the reaction mixture, the grain structure of the OIHP may be easily controlled. The controlled grain sizes in RSM can alter the paths for halide ion migration, enabling for a shift in the ON/OFF ratio by modifying the grain size. Large memory applications also need the use of the cross-point array structure. However, because sneak-current paths may generate undesirable current flow across unselected memory cells in a cross-point array structure, it is critical to minimize leakage current from surrounding cells by integrating selector devices in the design. We demonstrate the usage of selector devices in combination with the devices to avoid sneak current paths in OIHP-based RSMs. These findings suggest that OIHP might be utilized in high-density memory applications.

An Al_2O_3/2D Ruddlesden-Popper perovskite (2D PVK) heterostructure dielectric architecture, on the other hand, has been utilized to create ambipolar SnO transistor-based non-volatile memories with multibit memory behavior and ultralong retention time of $>10^5$ s [64]. The unique storage features are attributed to the decreased gate leakage generated by the Al_2O_3 layer, as well as the hopping-like ionic transport in 2D PVK with changeable activation energy under various light intensities. Because of the photoinduced field-effect process, it is feasible to operate a top-gated transistor in the presence of light, which would not be possible in the absence of light. As a result, it has exceptional photoresponsive properties, such as an extremely high specific detector detectivity of 2.7×10^{15} Jones and a large bandwidth spectrum differentiating capacity (375–1064 nm). The shape of metal halide perovskite layers, rather than grain size, is critical for high-performance memory systems. By including the organic semiconductor 2,7-dioctyl[1]benzothieno[3,2-b]benzothiophene (C8-BTBT) into the perovskite formulation, the microstructure of solution-processed layered Ruddlesden–Popper-phase perovskite films based on phenethylammonium lead bromide ($(PEA)_2PbBr_4$) can be manipulated [65]. The hole is transported in the CB-BTBT, while the charge is stored in the $(PEA)_2PbBr_4$. With the combination of the $(PEA)_2PbBr_4$/C8-BTBT channels, the transistor-based memory device exhibited a huge record memory window over 180 V, large erase/write channel current ratio 10^4, excellent data retention, and hood durability over 10^4 cycles. Moreover, a Ruddlesden-Popper-phase strontium titanate, $SrO(SrTiO_3)_n$ (n = 1) and conventional perovskite $SrTiO_3$ have been combined to form a heterojunction thin-film on an FTO substrate using the sol-gel method [66]. The Au/Sr_2TiO_4/$SrTiO_3$/FTO/glass memory device performed a stable switching ratio over 10^2 under a high operating voltage of 8 V.

Generally, ion migration in the *I-V* hysteresis is known as the drawback of halide perovskite optoelectronic. This unwanted issue has been solved by employing layered Ruddlesden-Popper perovskites (RPPs) [67]. As a result, the memory devices of RPP with indices n = 5 show the largest ON/OFF ratio of 10^4, operated in low V_{SET}, in comparison to n = 1 and the 3D indices composition of perovskite, as presented in **Figure 7** In addition, the device can last for 500 cycles in an inert environment with data retention of 250 h. The data retention can be extended when the device is operated under 60% relative humidity. These results are due to the chemical interaction between moving ions and the external contacts, which results in a modification of the charge transfer barrier at the interface, which subsequently modifies the device's resistive states.

On the other hand, neuromorphic computing requires extremely minimal operating energy to provide huge parallel data processing that mimics the human brain. It is necessary to attain this aim by using resistive memory that is based on materials that have good ionic transport and operate at very low currents. Extremely low operating

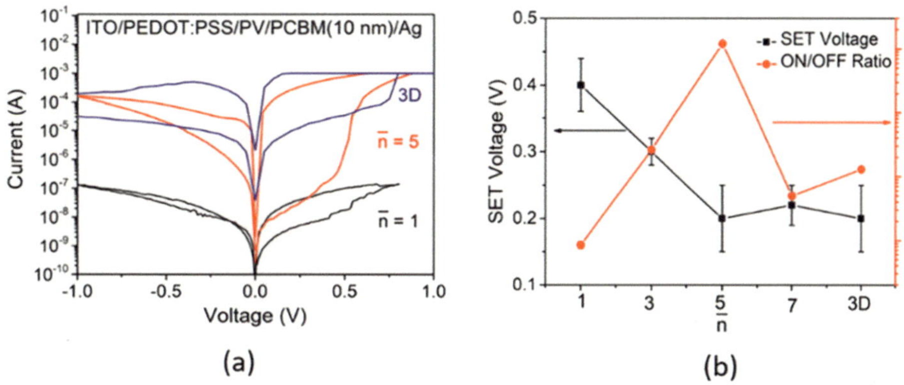

Figure 7.
Electrical characteristics of various ReRAM devices: (a) I-V curves for devices incorporating perovskites with varying n indices (n = 1, 5/3D) in an ITO/PEDOT:PSS/perovskite/PCBM/Ag setup (b) overview of SET voltages and ON/OFF ratios for various devices. Adapted from ***Figure 3*** *[67].*

current facilitates low-power operation by minimizing program, erasing, and read currents. The mixed electronic and ionic transport, as well as the ease with which they may be produced, make the 2D Ruddlesden-Popper phase hybrid lead bromide perovskite single crystals appealing materials for low operating current nanodevice applications [68]. The migration of bromide ions across the exfoliated 2D perovskite layer demonstrates ionic transport in the layer. Resistance memories with the lowest program currents down to 10 pA with a 100 ON/OFF ratio. Furthermore, the resistive memory demonstrated 400 fJ/spike synaptic functioning, which is comparable to the energy consumption required to convey information in the normal nervous system.

The use of lead-free perovskite through Aurivillius phase thin films have been suggested to improve their ferroelectric characteristics by modifying the growth, texture, and orientation [69]. In particular, liquid injection chemical vapor deposition (LI-CVD) was used to grow c-plane oriented $Bi_6Ti_3Fe_2O_{18}$ (B6TFO) functional oxide Aurivillius phase thin films on *c*-plane sapphire substrates, which were then annealed at 850°C to generate highly crystalline, well-textured single-phase Aurivillius plate-like shapes with 110 nm average film thickness and 24 nm roughness. Piezoresponse force microscopy (PFM) shows in-plane polarization enhanced by adjusting the deposition of *a*-axis oriented grains along the plane of the B6TFO films. Interestingly, the device shows a large and stable ferroelectric polarization switching under high operating temperatures of up to 200°C even after 20 h of PFM scanning. These investigations show the promise of B6TFO thin films for high-temperature piezoelectric applications and non-volatile ferroelectric memory applications.

It is also important to highlight that multiferroic materials with associated ferroelectric and ferromagnetic order characteristics might be used to store data by writing bits electrically and reading them magnetically. For example, Aurivillius phase $Bi_6Ti_{2.8}Fe_{1.52}Mn_{0.68}O_{18}$ (B6TFMO) produced by chemical solution deposition (CSD) shows magnetoelectric coupling at ambient temperature [70]. The in-plane ferromagnetic signature can be enhanced by manipulating the deposition method using the liquid injection chemical vapor deposition technique, which is related to the formation of Aurivillius phase [71]. Under magnetoelectric coupling, the ferroelectric switching volume increased by up to 14 percent as compared to CDS-grown films, and irreversible and reversible magnetoelectric domain switching was observed. This demonstrates that B6TFMO thin films are a viable choice for in-plane RAM applications as well as future high data storage multistate memory devices.

5. Design challenges of perovskites based memory devices

Understanding the RS properties is critical when building a RS device. According to published research, RS is often associated with oxygen or halide vacancies, metallic defects, and dislocations in perovskite thin films [72]. The kind of perovskites, film thickness, inclusion of dopants, and selection of bottom/top electrodes are all basic elements to consider in memory design construction. To fabricate a stable and efficient perovskite film device, preparation processes of perovskite thin films and understanding of the interaction between each layer are critical.

Wearable gadgets are gaining popularity among researchers in the age of wearable technology. Conventional electrodes, such as metals or transparent conducting oxides, are inflexible and readily shattered under stress. On the other hand, perovskites may be prepared via low temperature solution processing methods, Flexible devices could be considered while memory performance is maintained as published for study involving polyethylene naphthalate (PEN) or polyethylene terephthalate (PET) substrates [73, 74].

For further development of halide perovskites with high performance RS memory, the stability issue must be addressed due to the sensitivity to heat and moisture. This obstacle must be overcome to compete with metal oxide-based memory, which are more stable and easier to manufacture. To slow down the deterioration process while boosting stability, thick encapsulation might be applied to the devices. This layer may consist of a thin metal oxide or polymer-based layer [75]. Similar, to perovskite solar cells, the usage of lead compound is a key drawback in halide perovskite-based RS devices. Since lead is a poisonous chemical, the environmental impact of lead leaking may be disastrous. As a result, researchers are investigating lead alternatives such as lead-free perovskites. One example that researchers are looking at is tin-based perovskites, which have similar RS capability. Ji et al. discovered a lead-free all-inorganic cesium tin iodide perovskite ($CsSnI_3$) [76]. The (Ag or Au)/PMMA/$CsSnI_3$/Pt/SiO_2/Si bipolar RS could represent an ecologically acceptable option.

6. Conclusions

RS memory is one of the most sophisticated techniques for next-generation storage class memory, with lower power consumption, high density, and better performance. RS devices are regarded as one of the viable technologies for next-generation non-volatile memory. In addition to the well-studied usage of perovskite in perovskite solar cells, the use of perovskites in memory devices might be a fascinating subject to examine. Perovskites of various classes, such as perovskite oxides, perovskite halides, and layered perovskites, can be used in various RS devices. There are still numerous options to investigate for RS devices. The interactions of these perovskites with other elements are currently understudied and need to be investigated further. However, they still face significant challenges in entering the commercial sector. Although memory performance is advancing at a rapid pace, fundamental challenges in stability, reproducibility, and real-world applications are being addressed. As a result, perovskite may continue to play a significant role in dominating the memory storage sector soon.

Acknowledgements

This work was financially supported by the AUA-UAEU Joint Research Grant Project (IF016-2021 and G00003485), UNITEN BOLD grant J5150050002/20021170, Fundamental Research Grant Scheme (Project No.: FP113-2019A), Geran Putra-Inisiatif Putra Muda (GP-IPM/2018/9667000) and SATU Joint Research Scheme (ST002-2021).

Author details

Gregory Thien Soon How[1,2], Mohd Arif Mohd Sarjidan[1], Boon Tong Goh[1*], Boon Kar Yap[3,4] and Eyas Mahmoud[5]

1 Low Dimensional Materials Research Center, Faculty of Science, Department of Physics, Universiti Malaya, Kuala Lumpur, Malaysia

2 Centre for Advanced Devices and Systems, Faculty of Engineering, Multimedia University, Cyberjaya, Selangor, Malaysia

3 Electronic and Communications Department, College of Engineering, Universiti Tenaga Nasional, Kajang, Selangor, Malaysia

4 Institute of Sustainable Energy, Universiti Tenaga Nasional, Kajang, Selangor, Malaysia

5 Department of Chemical and Petroleum Engineering, United Arab Emirates University, Al Ain, United Arab Emirates

*Address all correspondence to: gohbt@um.edu.my

References

[1] Kamiya K, Yang MY, Magyari-Köpe B, Nishi Y, Shiraishi K. Modeling of resistive random access memory (RRAM) switching mechanisms and memory structures. In: Advances in Nonvolatile Memory and Storage Technology. Amsterdam: Elsevier; 2014. pp. 262-284e. DOI: 10.1533/9780857098092.2.262

[2] Zahoor F, Azni Zulkifli TZ, Khanday FA. Resistive random access memory (RRAM): An overview of materials, switching mechanism, performance, multilevel cell (mlc) storage, modeling, and applications. Nanoscale Research Letters. 2020;**15**:90. DOI: 10.1186/s11671-020-03299-9

[3] Möller S, Perlov C, Jackson W, Taussig C, Forrest SR. A polymer/semiconductor write-once read-many-times memory. Nature. 2003;**426**:166-169. DOI: doi.org/10.1038/nature02070

[4] Ge S, Wang Y, Xiang Z, Cui Y. Reset voltage-dependent multilevel resistive switching behavior in $CsPb_{1-x}Bi_xI_3$ perovskite-based memory device. ACS Applied Materials & Interfaces. 2018;**10**:24620-24626. DOI: 10.1021/acsami.8b07079

[5] Klein D. The history of semiconductor memory: From magnetic tape to NAND flash memory. IEEE Solid-State Circuits Magazine. 2016;**8**:16-22. DOI: 10.1109/MSSC.2016.2548422

[6] Huskey H. Chronology of computing devices. IEEE Transactions on Computers. 1976;**C-25**:1190-1199. DOI: 10.1109/TC.1976.1674587

[7] Auerbach IL, Eckert JP, Shaw RF, Sheppard CB. Mercury delay line memory using a pulse rate of several megacycles. Proceedings of the IRE. 1949;**37**:855-861. DOI: 10.1109/JRPROC.1949.229683

[8] North B, Nash O. Magnetic core memory reborn. Access. 2011:1-13

[9] Ellerbruch D. A new memory device-the twister. IRE Transactions on Component Parts. 1959;**6**:42-44. DOI: 10.1109/TCP.1959.1136273

[10] Suzuki R. Recent development in magnetic-bubble memory. Proceedings of the IEEE. 1986;**74**:1582-1590. DOI: 10.1109/PROC.1986.13670

[11] Juliussen JE. Bubble memory as small mass storage. Microelectronics and Reliability. 1977;**16**:427-430. DOI: 10.1016/0026-2714(77)90441-3

[12] Siddiqi M. Dynamic RAM: Technology advancements. 2017. DOI: 10.1201/b13005

[13] Kaur R. A Journey of digital storage from punch cards to cloud. IOSR Journal of Engineering. 2014;**4**:36-41. DOI: 10.9790/3021-04343641

[14] Coughlin T. A timeline for flash memory history [the art of storage]. IEEE Consumer Electronics Magazine. 2017;**6**:126-133. DOI: 10.1109/MCE.2016.2614739

[15] Schenk T, Pešić M, Slesazeck S, Schroeder U, Mikolajick T. Memory technology—A primer for material scientists. Reports on Progress in Physics. 2020;**83**:086501. DOI: 10.1088/1361-6633/ab8f86

[16] Fadeev AV, Rudenko KV. To the issue of the memristor's HRS and LRS states degradation and data retention time. Russian MicroElectronics.

2021;**50**:311-325. DOI: 10.1134/S1063739721050024

[17] Kumar D, Aluguri R, Chand U, Tseng TY. Metal oxide resistive switching memory: Materials, properties and switching mechanisms. Ceramics International. 2017;**43**:S547-S556. DOI: 10.1016/j.ceramint.2017.05.289

[18] Hwang B, Gu C, Lee D, Lee J-S. Effect of halide-mixing on the switching behaviors of organic-inorganic hybrid perovskite memory. Scientific Reports. 2017;7:43794. DOI: 10.1038/srep43794

[19] Wang Y, Lv Z, Zhou L, Chen X, Chen J, Zhou Y, et al. Emerging perovskite materials for high density data storage and artificial synapses. Journal of Materials Chemistry C. 2018;**6**:1600-1617. DOI: 10.1039/C7TC05326F

[20] Akinaga H, Shima H. Resistive random access memory (ReRAM) based on metal oxides. Proceedings of the IEEE. 2010;**98**:2237-2251. DOI: 10.1109/JPROC.2010.2070830

[21] Gale E. TiO_2-based memristors and ReRAM: Materials, mechanisms and models (a review). Semiconductor Science and Technology. 2014;**29**:1-29. DOI: 10.1088/0268-1242/29/10/104004

[22] Strukov DB, Snider GS, Stewart DR, Williams RS. The missing memristor found. Nature. 2008;**453**:80-83. DOI: 10.1038/nature06932

[23] Gale E, Costello B de L, Adamatzky A. Filamentary extension of the mem-con theory of memristance and its application to titanium dioxide sol-gel memristors. In: 2012 IEEE Int. Conf. Electron. Des. Syst. Appl., 5-6 November 2012, Malaysia. New York: IEEE; 2012. pp. 86-91. DOI: 10.1109/ICEDSA.2012.6507822

[24] Itoh M, Chua LO. Memristor oscillators. International Journal of Bifurcation and Chaos. 2008;**18**:3183-3206. DOI: 10.1142/S0218127408022354

[25] Valov I, Linn E, Tappertzhofen S, Schmelzer S, van den Hurk J, Lentz F, et al. Nanobatteries in redox-based resistive switches require extension of memristor theory. Nature Communications. 2013;**4**:1771. DOI: 10.1038/ncomms2784

[26] Molina-Reyes J, Hernandez-Martinez L. Understanding the resistive switching phenomena of stacked $Al/Al_2O_3/Al$ thin films from the dynamics of conductive filaments. Complexity. 2017;**2017**:8263904. DOI: 10.1155/2017/8263904

[27] Szot K, Speier W, Bihlmayer G, Waser R. Switching the electrical resistance of individual dislocations in single-crystalline $SrTiO_3$. Nature Materials. 2006;**5**:312-320. DOI: 10.1038/nmat1614

[28] Song Y, Chen Y, Jiang X, Ge Y, Wang Y, You K, et al. Nonlinear few-layer MXene-assisted all-optical wavelength conversion at telecommunication band. Advanced Optical Materials. 2019;7:1-9. DOI: 10.1002/adom.201801777

[29] Hsu C, Tsao C, Lin Y. Write-once-read-many-times characteristic of InZnO oxide semiconductor. IEEE Transactions on Electron Devices. 2018;**65**:978-985. DOI: 10.1109/TED.2018.2798710

[30] Wang K-L, Liu Y-L, Lee J-W, Neoh K-G, Kang E-T. Nonvolatile electrical switching and write-once read-many-times memory effects in functional polyimides containing triphenylamine and 1,3,4-oxadiazole moieties. Macromolecules. 2010;**43**:7159-7164. DOI: 10.1021/ma1006446

[31] Rabbani P, Dehghani R, Shahpari N. A multilevel memristor–CMOS memory cell as a ReRAM. Microelectronics Journal. 2015;**46**:1283-1290. DOI: 10.1016/j.mejo.2015.10.006

[32] Ge S, Guan X, Wang Y, Lin C, Cui Y, Huang Y, et al. Low-dimensional lead-free inorganic perovskites for resistive switching with ultralow bias. Advanced Functional Materials. 2020;**30**:2002110. DOI: 10.1002/adfm.202002110

[33] Wang W, Li Y, Yue W, Gao S, Zhang C, Chen Z, et al. Study on multilevel resistive switching behavior with tunable ON/OFF ratio capability in forming-free ZnO QDs-based RRAM. IEEE Transactions on Electron Devices. 2020;**67**:4884-4890. DOI: 10.1109/TED.2020.3022005

[34] Dittrich T, Lang F, Shargaieva O, Rappich J, Nickel NH, Unger E, et al. Diffusion length of photo-generated charge carriers in layers and powders of CH_3NH_3 PbI_3 perovskite. Applied Physics Letters. 2016;**109**:073901. DOI: 10.1063/1.4960641

[35] Yoo EJ, Lyu M, Yun J-H, Kang CJ, Choi YJ, Wang L. Resistive switching behavior in organic-inorganic hybrid $CH_3NH_3PbI_{3-x}Cl_x$ perovskite for resistive random access memory devices. Advanced Materials. 2015;**27**:6170-6175. DOI: 10.1002/adma.201502889

[36] Lv F, Ling K, Zhong T, Liu F, Liang X, Zhu C, et al. Multilevel resistive switching memory based on a $CH_3NH_3PbI_{3-x}Cl_x$ film with potassium chloride additives. Nanoscale Research Letters. 2020;**15**:126. DOI: 10.1186/s11671-020-03356-3

[37] Ma C, Luo Z, Huang W, Zhao L, Chen Q, Lin Y, et al. Sub-nanosecond memristor based on ferroelectric tunnel junction. Nature Communications. 2020;**11**:1439. DOI: 10.1038/s41467-020-15249-1

[38] Lu D, Crossley S, Xu R, Hikita Y, Hwang HY. Freestanding oxide ferroelectric tunnel junction memories transferred onto silicon. Nano Letters. 2019;**19**:3999-4003. DOI: 10.1021/acs.nanolett.9b01327

[39] Guo R, Wang Z, Zeng S, Han K, Huang L, Schlom DG, et al. Functional ferroelectric tunnel junctions on silicon. Scientific Reports. 2015;**5**:12576. DOI: 10.1038/srep12576

[40] Sun H, Luo Z, Zhao L, Liu C, Ma C, Lin Y, et al. $BiFeO_3$-based flexible ferroelectric memristors for neuromorphic pattern recognition. ACS Applied Electronic Materials. 2020;**2**:1081-1089. DOI: 10.1021/acsaelm.0c00094

[41] Li J, Li N, Ge C, Huang H, Sun Y, Gao P, et al. Giant electroresistance in ferroionic tunnel junctions. IScience. 2019;**16**:368-377. DOI: 10.1016/j.isci.2019.05.043

[42] Jin HW, Wang Z, Yu W, Wu T. Optically controlled electroresistance and electrically controlled photovoltage in ferroelectric tunnel junctions. Nature Communications. 2016;**7**:10808. DOI: 10.1038/ncomms10808

[43] Yang Y, Xi Z, Dong Y, Zheng C, Hu H, Li X, et al. Spin-filtering ferroelectric tunnel junctions as multiferroic synapses for neuromorphic computing. ACS Applied Materials & Interfaces. 2020;**12**:56300-56309. DOI: 10.1021/acsami.0c16385

[44] Xie S, Pei L, Li M, Zhu Y, Cheng X, Ding H, et al. Light-controlled resistive switching and voltage-controlled photoresponse characteristics in the $Pt/CeO_2/Nb:SrTiO_3$ heterostructure.

Journal of Alloys and Compounds. 2019;**778**:141-147. DOI: 10.1016/j.jallcom.2018.11.161

[45] Dai W, Li Y, Jia C, Kang C, Li M, Zhang W. High-performance ferroelectric non-volatile memory based on La-doped $BiFeO_3$ thin films. RSC Advances. 2020;**10**:18039-18043. DOI: 10.1039/D0RA02780D

[46] Qiao X, Geng W, Sun Y, Zheng D, Yang Y, Meng J, et al. Robust in-plane polarization switching in epitaxial $BiFeO_3$ films. Journal of Alloys and Compounds. 2021;**852**:156988. DOI: 10.1016/j.jallcom.2020.156988

[47] Yang N, Hu C-Z, Ren Z-Q, Bao S-Y, Tian B-B, Yue F-Y, et al. Nonvolatile negative optoelectronic memory based on ferroelectric thin films. ACS Applied Electronic Materials. 2020;**2**:1035-1040. DOI: 10.1021/acsaelm.0c00066

[48] Tian W, Lu H, Li L. Nanoscale ultraviolet photodetectors based on onedimensional metal oxide nanostructures. Nano Research. 2015;**8**:382-405. DOI: 10.1007/s12274-014-0661-2

[49] Ju H, Yang MK. Duality characteristics of bipolar and unipolar resistive switching in a Pt/$SrZrO_3$/TiO_x/Pt stack. AIP Advances. 2020;**10**:065221. DOI: 10.1063/5.0010045

[50] Shao F, Lv ZL, Ren ZY, Zhang LP, Zhao GL, Teng J, et al. High endurance of bipolar resistive switching in a Pt/$LaNiO_3$/Nb:$SrZrO_3$/Cu stack: The role of Cu modulating layer. Chemical Physics Letters. 2020;**739**:137040. DOI: 10.1016/j.cplett.2019.137040

[51] Kumari K, Kumar A, Kotnees DK, Balakrishnan J, Thakur AD, Ray SJ. Structural and resistive switching behaviour in lanthanum strontium manganite - reduced graphene oxide nanocomposite system. Journal of Alloys and Compounds. 2020;**815**:152213. DOI: 10.1016/j.jallcom.2019.152213

[52] Li TX, Li R, Ma D, Li B, Li K, Hu Z. Resistive switching behaviors in the $BaTiO_3$/$La_{0.7}Sr_{0.3}MnO_3$ layered heterostructure driven by external electric field. Journal of Magnetism and Magnetic Materials. 2020;**497**:165879. DOI: 10.1016/j.jmmm.2019.165879

[53] Kang K, Ahn H, Song Y, Lee W, Kim J, Kim Y, et al. High-performance solution-processed organo-metal halide perovskite unipolar resistive memory devices in a cross-bar array structure. Advanced Materials. 2019;**31**:1804841. DOI: 10.1002/adma.201804841

[54] Li L, Chen Y, Cai C, Ma P, Ji H, Zou G. Single crystal halide perovskite film for nonlinear resistive memory with ultrahigh switching ratio. Small. 2022;**18**:2103881. DOI: 10.1002/smll.202103881

[55] Jiang T, Shao Z, Fang H, Wang W, Zhang Q, Wu D, et al. High-performance nanofloating gate memory based on lead halide perovskite nanocrystals. ACS Applied Materials & Interfaces. 2019;**11**:24367-24376. DOI: 10.1021/acsami.9b03474

[56] Shaban A, Joodaki M, Mehregan S, Rangelow IW. Probe-induced resistive switching memory based on organic-inorganic lead halide perovskite materials. Organic Electronics. 2019;**69**:106-113. DOI: 10.1016/j.orgel.2019.03.019

[57] Wu X, Yu H, Cao J. Unraveling the origin of resistive switching behavior in organolead halide perovskite based memory devices. AIP Advances. 2020;**10**:085202. DOI: 10.1063/1.5130914

[58] Lee S, Kim H, Kim DH, Bin KW, Lee JM, Choi J, et al. Tailored 2D/3D halide perovskite heterointerface for substantially enhanced endurance in conducting bridge resistive switching memory. ACS Applied Materials & Interfaces. 2020;**12**:17039-17045. DOI: 10.1021/acsami.9b22918

[59] Xia F, Xu Y, Li B, Hui W, Zhang S, Zhu L, et al. Improved performance of CH_3 $NH_3PbI_{3-x}Cl_x$ resistive switching memory by assembling 2d/3d perovskite heterostructures. ACS Applied Materials & Interfaces. 2020;**12**:15439-15445. DOI: 10.1021/acsami.9b22732

[60] Paul T, Sarkar PK, Maiti S, Chattopadhyay KK. Multilevel programming and light-assisted resistive switching in a halide-tunable all-inorganic perovskite cube for flexible memory devices. ACS Applied Electronic Materials. 2020;**2**:3667-3677. DOI: 10.1021/acsaelm.0c00719

[61] Shu P, Cao X, Du Y, Zhou J, Zhou J, Xu S, et al. Resistive switching performance of fibrous crosspoint memories based on an organic–inorganic halide perovskite. Journal of Materials Chemistry C. 2020;**8**:12865-12875. DOI: 10.1039/D0TC02579H

[62] Paramanik S, Maiti A, Chatterjee S, Pal AJ. Large resistive switching and artificial synaptic behaviors in layered $Cs_3Sb_2I_9$ lead-free perovskite memory devices. Advanced Electronic Materials. 2022;**8**:2100237. DOI: 10.1002/aelm.202100237

[63] Park Y, Lee J-S. Controlling the grain size of Dion–Jacobson-phase two-dimensional layered perovskite for memory application. ACS Applied Materials & Interfaces. 2022;**14**:4371-4377. DOI: 10.1021/acsami.1c20272

[64] Tian Q, Hong R, Liu C, Hong X, Zhang S, Wang L, et al. Flexible SnO optoelectronic memory based on light-dependent ionic migration in Ruddlesden–Popper perovskite. Nano Letters. 2022;**22**:494-500. DOI: 10.1021/acs.nanolett.1c04402

[65] Gedda M, Yengel E, Faber H, Paulus F, Kreß JA, Tang M-C, et al. Ruddlesden–Popper-phase hybrid halide perovskite/small-molecule organic blend memory transistors. Advanced Materials. 2021;**33**:2003137. DOI: 10.1002/adma.202003137

[66] Li J, Tang X-G, Liu Q-X, Jiang Y-P, Li W-H, Tang Z-X. Interfacial resistive switching properties of $Sr_2TiO_4/SrTiO_3$ heterojunction thin films prepared via sol-gel process. Ceramics International. 2021;**47**:18808-18813. DOI: 10.1016/j.ceramint.2021.03.216

[67] Solanki A, Guerrero A, Zhang Q, Bisquert J, Sum TC. Interfacial mechanism for efficient resistive switching in Ruddlesden–Popper perovskites for non-volatile memories. Journal of Physical Chemistry Letters. 2020;**11**:463-470. DOI: 10.1021/acs.jpclett.9b03181

[68] Tian H, Zhao L, Wang X, Yeh Y-W, Yao N, Rand BP, et al. Extremely low operating current resistive memory based on exfoliated 2D perovskite single crystals for neuromorphic computing. ACS Nano. 2017;**11**:12247-12256. DOI: 10.1021/acsnano.7b05726

[69] Faraz A, Deepak N, Schmidt M, Pemble ME, Keeney L. A study of the temperature dependence of the local ferroelectric properties of c-axis oriented $Bi_6Ti_3Fe_2O_{18}$ Aurivillius phase thin films: Illustrating the potential of a novel lead-free perovskite material for high density memory applications. AIP Advances. 2015;5:087123. DOI: 10.1063/1.4928495

[70] Keeney L, Maity T, Schmidt M, Amann A, Deepak N, Petkov N, et al. Magnetic field-induced ferroelectric switching in multiferroic Aurivillius phase thin films at room temperature. Journal of the American Ceramic Society. 2013;**96**:2339-2357. DOI: 10.1111/jace.12467

[71] Faraz A, Maity T, Schmidt M, Deepak N, Roy S, Pemble ME, et al. Direct visualization of magnetic-field-induced magnetoelectric switching in multiferroic aurivillius phase thin films. Journal of the American Ceramic Society. 2017;**100**:975-987. DOI: 10.1111/jace.14597

[72] Panda D, Tseng TY. Perovskite oxides as resistive switching memories: A review. Ferroelectrics. 2014;**471**:23-64. DOI: 10.1080/00150193.2014.922389

[73] Ye H, Sun B, Wang Z, Liu Z, Zhang X, Tan X, et al. High performance flexible memristors based on a lead free $AgBiI_4$ perovskite with an ultralow operating voltage. Journal of Materials Chemistry C. 2020;**8**:14155-14163. DOI: 10.1039/D0TC03287E

[74] Gu C, Lee JS. Flexible hybrid organic-inorganic perovskite memory. ACS Nano. 2016;**10**:5413-5418. DOI: 10.1021/acsnano.6b01643

[75] Hwang B, Lee J-S. Hybrid organic-inorganic perovskite memory with long-term stability in air. Scientific Reports. 2017;**7**:673. DOI: 10.1038/s41598-017-00778-5

[76] Han JS, Van LQ, Choi J, Kim H, Kim SG, Hong K, et al. Lead-free all-inorganic cesium tin iodide perovskite for filamentary and interface-type resistive switching toward environment-friendly and temperature-tolerant nonvolatile memories. ACS Applied Materials & Interfaces. 2019;**11**:8155-8163. DOI: 10.1021/acsami.8b15769